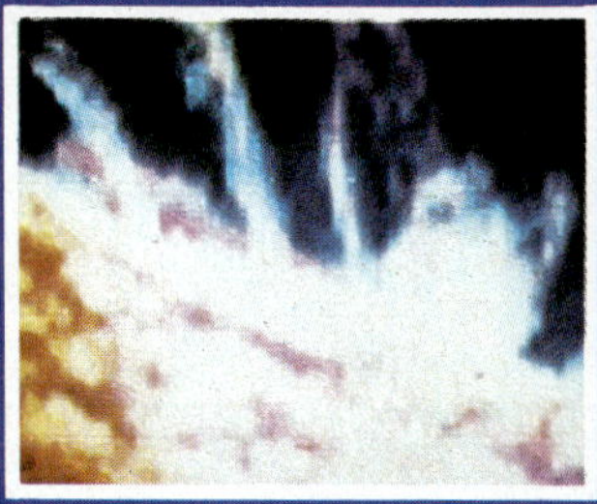

ALLYN AND BACON

GENERAL SCIENCE

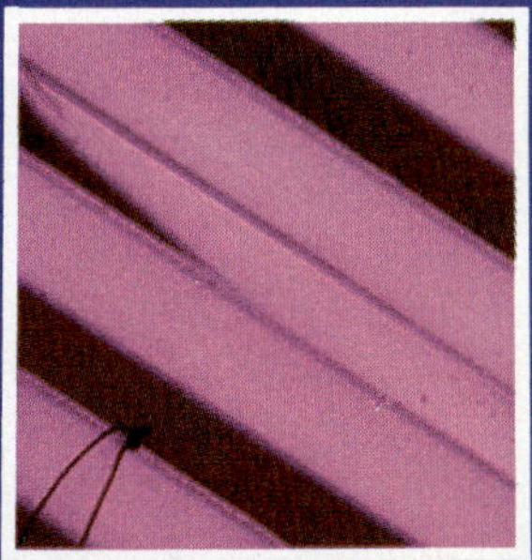

The quotation on the cover is attributed to Isaac Newton, who formulated the laws of motion discussed in Chapter 16.

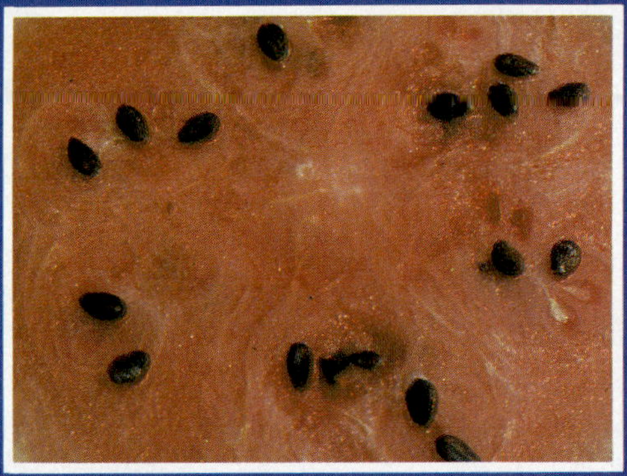

400
300
APPROXIMATE VOLUME

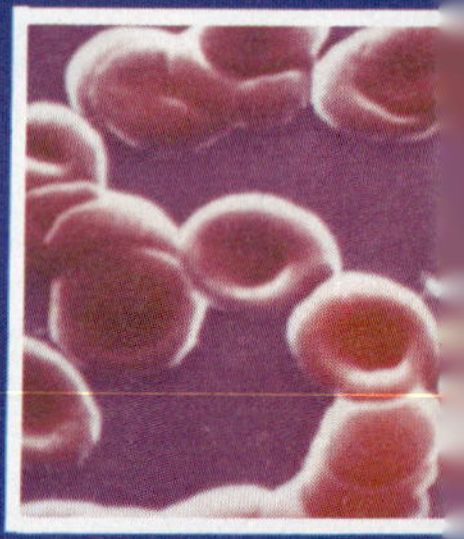

ALLYN AND BACON

GENERAL SCIENCE

CAROLYN SHEETS BROCKWAY
Science Teacher
George Washington High School
Cedar Rapids, Iowa

ROBERT GARDNER
Chair, Science Department
Salisbury School
Salisbury, Connecticut

SAMUEL F. HOWE
Educational Consultant and Writer
formerly Science Teacher
Friends' Central School
Philadelphia, Pennsylvania

ALLYN AND BACON, INC., Newton, Massachusetts

Rockleigh, NJ   Atlanta   Warrensburg, MO   Dallas   Rancho Cordova, CA
London   New South Wales, Australia   Markham, Ontario

Staff Contributors

LOIS B. ARNOLD
Editorial Development

CHRISTOPHER VALENTE
Art and Design Direction

MARTHA E. BALLENTINE
Preparation Services Manager

ANNIE PUCILOSKI
Book Manufacturing

***Freelance Staff*:**

CAROL H. ROSE
Design and Production Services

SUSAN VAN ETTEN
Photo Research

***Cover Design*:**

HANNUS DESIGN ASSOCIATES AND CHRISTOPHER VALENTE

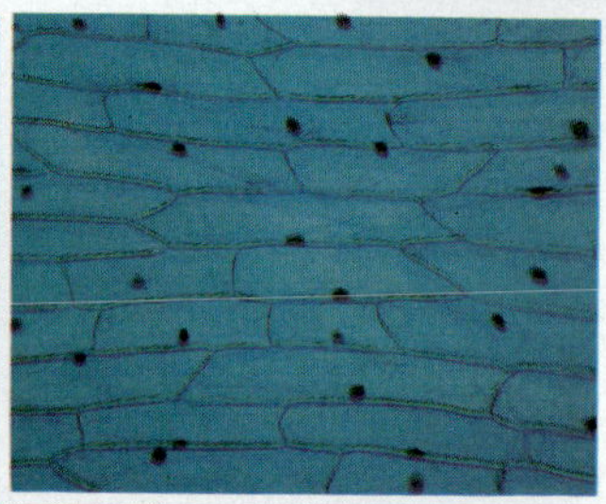

ISBN 0-205-08130-4

Printed in the United States of America

1 2 3 4 5 6 7 8 9 93 92 91 90 89 88 87 86 85 84

Acknowledgements

Teacher Reviewers

JOHN J. SHERIDAN
Biology Teacher
Boca Ciega Senior High School
Gulfport, Florida

ANN S. BURDEN
Chair, Science Department
North Pitt High School
Bethel, North Carolina

LINDA B. KNIGHT
General Science/Earth Science Teacher
Bloomington High School North
Bloomington, Indiana

LARRY A. SYRON
Chair, Science Department
Gresham Union High School
Gresham, Oregon

Consultants

SCIENCE CONTENT

RICHARD L. ROTUNDO
Department of Embryology
Carnegie Institution of Washington
Baltimore, Maryland

W.T. LIPPINCOTT
Department of Chemistry
The University of Arizona
Tucson, Arizona

LOU WILLIAMS PAGE
Geologist and Author
Houston, Texas

CLIFFORD E. SWARTZ
Department of Physics
State University of New York
Stony Brook, New York

SHARON C. THOMAS
John G. Diefenbaker High School
Calgary, Alberta

READING

PATRICIA TERRELL
School of Education
East Carolina University
Greenville, North Carolina

MAINSTREAMING/HANDICAPPED ADAPTATIONS

DENNIS W. SUNAL
Division of Education
West Virginia University
Morgantown, West Virginia

ACTIVITIES SAFETY

RONALD J. KENDALL
Environmental Toxicology Services, Inc.
Bellingham, Washington

Unit One • Life Science 12–143

Chapters

1

2

3

Unit Two • Chemistry 144–267

6

7

8

Contents

Unit Three • Earth and Space Science 268–395

Unit Four • Physics 396–529

Prologue

Introduction to Science

These life scientists are studying disease in young elm trees.

P-1

Approaches to Problems

Figure P-1 *Chemists, earth scientists, and physicists work in many different settings.*

Have you ever wondered how scientists are able to launch the space shuttle at just the right speed so it will orbit the Earth and return safely? Whether there is life on other planets? Why human organ transplants have become more successful than ever before? Or what causes cancer? If you have, then you have begun to think as a scientist does.

Scientists are curious. They are always asking questions and seeking answers. Scientists try to solve problems and learn new things about the world and universe. Often they seek solutions to problems that affect the lives of us all.

Scientists work in four major areas of study. Life scientists, or biologists, study living things. Chemists investigate the make-up, structure, properties, and reactions of matter. Earth and space scientists study the Earth and the objects in space. Physicists explore the interactions between matter and energy. Each area of study includes several specialties. For example, life science includes ecology, which is the study of the interactions between living things and their environment.

The process of science is complex, but it involves thinking, observing, experimenting, discovering and demonstrating. Often a scientist acquires a new insight into the workings of nature. He or she then finds a way to test the validity of that insight. Scientific discoveries have been made in many ways. Isaac Newton is said to have discovered the theory of gravity by thinking about why an apple fell from a tree and hit him. Alexander Fleming discovered penicillin by accident. But then, of course, he tested his discovery experimentally.

scientific method

However, science usually involves a systematic approach to problem solving. The approach is sometimes called the **scientific method.** Actually, it is a planned, logical way to develop and test ideas. The ideas to be tested are tentative explanations for what scientists have observed. Carefully controlled testing is what makes knowledge gained in science different from knowledge acquired in other fields. But even after testing, results must be verified by other scientists. All explanations can be changed or replaced by better explanations in the future.

P-2

Dutch Elm Disease:
An Unsolved Problem

A good way to understand the scientific method is to look at a real problem that scientists are trying to solve and to follow what they do. One such problem, which scientists have been working on for many decades, is Dutch elm disease. It is a serious plant disease that has killed many elm trees all over North America and Europe.

At one time, thousands upon thousands of stately elms shaded streets and yards across North America and Europe. Then about 60 years ago, people began to observe that the elms were dying in unusually large numbers. Figure P-2 shows some victims of Dutch elm disease.

Scientists found that a fungus, carried from tree to tree by bark beetles, causes the disease. (See Figure P-3.) *Fungi* (plural of fungus) are living things, or *organisms*, that do not move about. They obtain their food from other organisms

Figure P-2 *The symptoms of Dutch elm disease begin with the wilting of leaves at the tips of branches. Eventually the leaves die and drop off, leaving the trees bare.*

Figure P-3 *Bark beetles feeding in the crotch of a pair of elm twigs. During feeding, the beetles infect healthy elm trees with the disease-causing fungus.*

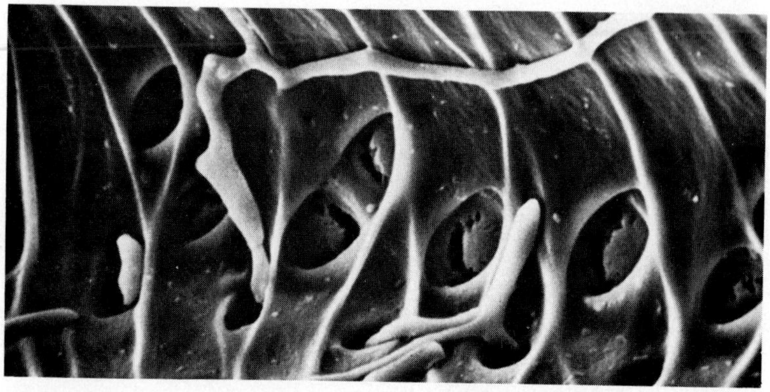

Figure P-4 *Threads of the Dutch elm disease fungus growing along the inner surface of an elm tree's water-carrying tube. Branches of the threads have grown into pits in the tube wall. Magnification is about 1400 diameters.*

that are either living or dead. Figure P-4 shows the Dutch elm disease fungus growing in elm wood. It kills the trees by preventing water from flowing within the trees' water-carrying tubes.

Two biologists, Gary Strobel and Don Myers, have been investigating the problem of Dutch elm disease. To them, the problem was clear. It was to find an agent that would stop the growth of the fungus, thus preventing it from killing the trees. Identifying or stating the problem clearly is often a first step in scientific investigation. If a problem is complex, it is broken into several parts and each part is investigated separately.

The next step is to collect information related to the problem. One way to do that is to read about the work done by other scientists. Or, a scientist can communicate with others who have investigated the problem.

Strobel and Myers had studied plants for many years. They knew from experience that certain bacteria produce antibiotics that can kill fungi. *Bacteria* (singular, bacterium) are microscopic organisms. However, Strobel and Myers needed to know which particular bacterium is effective against the elm disease fungus. They read even more in technical articles and talked to other scientists. They also did exploratory work in the laboratory to learn more about bacteria that produce substances that can kill fungi.

They were then ready to take an important step in the problem-solving process. They had enough information to make an educated guess as to a possible solution. Such a guess, or tentative explanation, is called a **hypothesis**. A hypothesis is based on facts related to the problem. It was Strobel's and Myers' hypothesis that an approach to treating Dutch elm disease would be to use a bacterium that kills the fungus causing the disease. They set out to find one.

In order for a hypothesis to be of use, it must be testable. Testing a hypothesis is done by means of experiments. An **experiment** is a way of testing that makes use of measurements or observations. Experiments may be done in a few simple steps or they may take years of complicated work.

A certain type of thinking is required to test a hypothesis. A scientist predicts a result of an experimental test. If the result occurs, the hypothesis is supported. Strobel and Myers predicted that if a bacterium were found that kills the fungus, placing that bacterium in healthy trees would prevent the disease in those trees.

They searched for bacteria that might be effective against the Dutch elm disease fungus. One bacterium, shown in Figure P-5, was found to be particularly effective in killing the elm disease fungus in the laboratory. The bacterium's effect on the fungus is shown in Figure P-6.

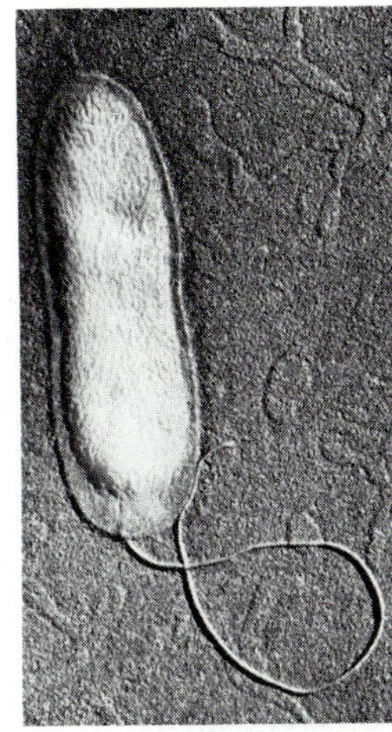

Figure P-5 *The bacterium discovered to be effective against the elm disease fungus. Magnification is about 15,000 diameters.*

hypothesis
(hy PAHTH uh sis)

experiment

Figure P-6 *How the bacterium tested affects the Dutch elm disease fungus. The lighter-colored layer on top is the fungus. The smaller circles are groups of the bacterium. The bacterium produced a substance that prevented growth of the fungus in the circular areas around the groups.*

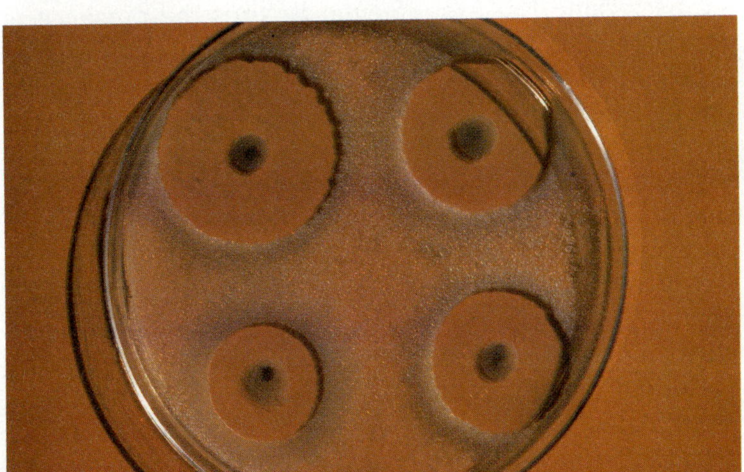

The scientists then went to a greenhouse in which small, healthy elm trees were growing. They injected the effective bacterium into the healthy trees. They observed that the bacterium established itself in the trees, and that there was little or no harm to either the bacterium or the trees. So far so good.

controlled experiment

Next, they tried to determine if the bacterium would prevent the fungus from causing Dutch elm disease. To find out, Myers and Strobel chose two groups of small elm trees on which to perform a controlled experiment. A **controlled experiment** is made up of two test groups. Everything about the two groups is the same, except for the one factor that scientists are testing.

experimental factor

experimental group

control group

The factor being tested is called the **experimental factor**. The group that is subjected to the experimental factor is called the **experimental group**. The group that is not subjected to the experimental factor is called the **control group**. The use of a control group helps assure that the changes observed during the experiment are actually caused by the experimental factor. The control group provides a standard of comparison for the results.

To conduct their controlled experiment, Strobel and Myers used this reasoning: If the bacterium is injected into the healthy trees of the experimental group and the trees are then exposed to the fungus, the trees should remain healthy. However, when the control group trees are exposed to the fungus, those trees should contract Dutch elm disease. That is because the control group is not protected by the bacterium. The experimental factor is the presence of the bacterium.

observations

data

When performing an experiment, it is important to keep accurate records of **observations**: measurements, changes observed, or other bits of information. These observations are called **data**. The purpose of keeping records is to help organize data so that patterns, if any exist, can be seen. Patterns in data support a hypothesis when they are consistent with it. Or, patterns can prove a hypothesis false when they are inconsistent with it. If Strobel and Myers had found in the controlled greenhouse experiment that the bacterium did not prevent the disease, their hypothesis would have been incorrect. A useless hypothesis must be discarded or revised. In such a case, it may be necessary to make a new hypothesis and to test it.

P-3

Dutch Elm Disease:
Experimental Results

What Strobel and Myers found from their greenhouse experiment was very encouraging. After eight weeks, none of the small trees treated with the bacterium showed symptoms of Dutch elm disease. All the untreated control trees, however, did show symptoms of the disease. (See Figure P-7.)

At this point, the hypothesis became a theory. A **theory** is a hypothesis that has been supported by the results of experiments. A theory is an explanation of observations that sets the stage for an even wider range of experimental testing. Strobel's and Myers' theory was that the bacterium they had tested is effective in preventing Dutch elm disease. **theory**

When scientists complete an investigation, they report their findings. This allows other scientists to repeat their experiments. If the same results are obtained, the theory

Figure P-7 *The antifungal effect of the bacterium in living elm saplings. The area around the injection point in the treated tree (left) is clear of fungus. The wood of the control tree (right) is discolored by the fungus.*

Table P-1: Scientific Method

Steps	*Dutch elm disease example*
Carefully identify the problem under study.	The problem was to find an agent that would stop the growth of the fungus, thus preventing it from killing the elm trees.
Collect all information that may help in solving the problem.	Strobel and Myers researched the facts and communicated with other scientists.
Form a hypothesis or possible solution to the problem.	Dutch elm disease could be prevented by using a bacterium that kills the fungus.
Plan and conduct a controlled experiment or series of observations.	Scientists injected one group of healthy trees with the bacterium, but not a second, control group. Both groups of trees were exposed to the fungus.
Record the data.	Members of the first group did not contract the disease; members of the second group did.
Formulate a theory based on the facts discovered.	The bacterium tested is effective in preventing Dutch elm disease.
Other, independent scientists test the theory again.	Scheffer successfully repeated the experiment with European elms in the Netherlands.

being tested receives additional support. If the results differ, the theory usually must be changed, or even discarded. This is one way the body of scientific knowledge changes and grows. Table P-1 summarizes the steps that can lead to the formulation and testing of a theory, using the example of Dutch elm disease.

Strobel's and Myers' theory received important support from an experiment conducted in the Netherlands. In that experiment, biologist Ruud Scheffer treated field-grown

European elm trees with the bacterium. Then he attempted to infect them with the disease-causing fungus. After two growing seasons, Scheffer observed that 90% of the treated trees did not show any evidence of Dutch elm disease. Figure P-8 shows dramatically how treatment with the bacterium protects against the disease. The untreated control trees all died.

Figure P-8 *(a) Elm tree treated with the bacterium, then given the fungus. (b) Control tree that received only the fungus. The time after receiving the fungus is the same in both photos.*

Reporting experimental findings allows other scientists to use them. Work in science often has practical applications. Strobel and Myers and Scheffer reported their findings at meetings, in scientific papers, and in news releases. A company that makes garden products heard about their work. It then investigated the possibility of providing the effective bacterium in a form useful to the public. More research of this kind could eventually help save our remaining large elm trees.

The process of scientific investigation is never finished. Every ending is a new beginning. Every time a scientist answers a question about nature, many new questions arise. These questions lead to new hypotheses that are then tested experimentally.

Strobel and Myers found a new approach to the problem of Dutch elm disease. But their work raises new questions. For instance, will the bacterium work in large trees or in trees in which the disease is already established? Can similar procedures be used to treat other plant diseases? If so, further benefits will result from their research.

The problem of Dutch elm disease has been only partially solved. Strobel's and Myers' theory that the bacterium tested is effective in saving elms from Dutch elm disease is still being tested and is being further developed through experimentation.

Some theories, however, become so well established by experimental evidence in many different situations over many years that they become scientific laws. A **scientific law** is a precise statement of a relationship in nature—so precise that it can be expressed mathematically. For example, in climates where the weather varies greatly between summer and winter, variation in growth causes visible rings to form within the trunks of trees. It is a scientific law that each ring in a tree's trunk represents one year of its growth. Using this law, scientists reported one California bristle-cone pine tree as being 4900 years old.

scientific law

Understanding The Prologue

Main Ideas

1. Scientists are always asking questions and trying to solve problems. They develop and test ideas in a planned, logical way called the scientific method.
2. Scientists often start to investigate a problem by identifying or stating the problem clearly, collecting information relating to the problem, and forming a hypothesis.
3. A hypothesis is a tentative explanation, or solution to a problem, that can be tested by means of an experiment. A hypothesis shown by an experiment to be incorrect must be discarded or revised.
4. An experiment is a way of testing that makes measurements or observations in the natural world.
5. A controlled experiment uses two test groups to test the effects of one experimental factor. The experimental group is subjected to the experimental factor; the control group is not.
6. During an experiment, scientists keep accurate records of data in order to see whether the data support the hypothesis.
7. A theory is a hypothesis backed by experimental results.
8. Scientists report their experimental results so they can be checked by other scientists.
9. A scientific law is a precise statement of a relationship in nature that can be expressed in mathematical terms.

Vocabulary Review

From the following list, choose the term that best completes each of the statements. Write your answers on a separate sheet of paper.

bacterium	experimental group
biologists	fungus
control group	hypothesis
controlled experiment	observations
data	organisms
experiment	scientific law
experimental factor	scientific method
	theory

1. People who study living things are called _?_.
2. The observations recorded in an experiment are called _?_.
3. A(n) _?_ is a tentative explanation for a problem found in nature.
4. The part of an experiment that provides a standard against which to compare results is the _?_.
5. A(n) _?_ is a living thing that does not move about and that obtains its food from other living things.
6. A precise statement of a relationship in nature that can be expressed mathematically is a(n) _?_.
7. A test of a hypothesis in which observations and measurements are made is a(n) _?_.
8. A(n) _?_ is a hypothesis supported by experimental results.
9. The part of a controlled experiment that varies is the _?_.
10. A planned, logical way to develop and test ideas is called _?_.

Prologue Review

Write your answers on a separate sheet of paper.

Know The Facts

1. Dutch elm disease is caused by a __?__. (bacterium, fungus, beetle)
2. __?__ carry the disease-causing agent from elm tree to elm tree. (bark beetles, fungi, bacteria)
3. Strobel and Myers discovered a __?__ that is effective against the cause of Dutch elm disease. (tree, fungus, bacterium)
4. Strobel's and Myers' __?__ was supported by the experimental work of Scheffer in the Netherlands. (scientific law, theory, control group)
5. A hypothesis that fails to get experimental support must be __?__. (reported to other scientists, discarded or revised, solved)
6. In Strobel's and Myers' greenhouse experiment, Dutch elm disease in the control trees __?__. (was checked, was changed, spread unchecked)
7. In order for experimental results to be accepted, other scientists must be able to __?__. (solve them, repeat them successfully, control them)
8. In Scheffer's field experiment, Dutch elm disease in the control trees __?__. (was checked, was changed, spread unchecked)

Understand The Concepts

9. What process did Strobel and Myers follow to identify an agent that is effective against Dutch elm disease?
10. Describe the experiments Strobel and Myers conducted to show that the agent they selected is effective against Dutch elm disease.
11. What was the experimental factor in Strobel's and Myers' controlled experiment? In what way was the experimental factor related to the control group?
12. Why is controlled experimentation important in scientific investigation?
13. In what ways might a test of a hypothesis lead to additional experimental work?
14. What is the relationship between hypothesis and theory?
15. Explain the meaning of the term scientific law.

Challenge Your Understanding

16. Describe the evidence that convinced Strobel and Myers that they had found an agent effective against Dutch elm disease.
17. Often there is more than one way to solve a problem. What other approach(es) to the problem of Dutch elm disease might Strobel and Myers have tried?

Projects

1. Several control measures are in use against Dutch elm disease. Write a report covering all of these measures.
2. The use of a living organism to control another organism is called biological control. Report to your class on two biological controls that have worked as planned and two that have failed.

Unit 1 · Life Science

This life scientist is at work observing the habits of a tropical gannet.

Chapter

1

Introduction to Life

An underwater scene in the Red Sea.

1-1

Characteristics of Life

Goal

To identify the characteristics that define life.

Most people would say that the fish shown at the left are living and the rock is not. But what about the coral? Like a rock, the coral stays in one place most of its life and its hard, outer covering makes it appear lifeless. But if you watched the coral for a while you would notice long, slender *tentacles* reaching out of the rocklike covering and moving about in search of food. You would also discover that the coral reacted in different ways to the surroundings. For example, the coral would probably pull back its tentacles when touched by an unfamiliar object. You would also notice that the coral grew and produced new corals over time. Corals display all the characteristics that define life. They are one example of a living thing, or **organism.**

organism

Scientists agree that there is a set of characteristics that distinguishes living things from nonliving things. These are:

1. Movement (of the organism, its parts, or materials inside of it)
2. Getting and using food
3. Growth and repair of body parts
4. Reproduction
5. Response to changes in the surroundings

Nonliving things can have one or two of the characteristics of life. A car moves, a computer responds, and an icicle grows. But only organisms carry out *all* life activities.

Checkpoint

1. What is an organism?
2. Name the life activities carried out by organisms.
3. How do scientists know that corals are alive?

TRY THIS Set up a freshwater aquarium at home or in your classroom. Learn how to feed and care for the organisms. Observe how the various organisms carry out their life activities. Keep a record of your observations.

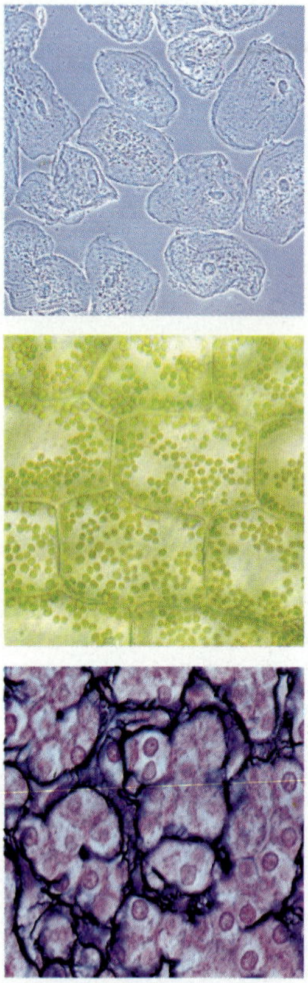

Figure 1-1 *Three kinds of cells. Human cheek cells, stained (top).* Elodea *leaf cells (middle). Monkey liver cells (bottom).*

1-2

The Cell

Goals

1. To describe a cell and its parts.
2. To compare plant and animal cells.

The invention of the microscope in the 1600's by Anton van Leeuwenhoek (LAY vun hook) allowed people to observe what was once invisible to the unaided eye. An English scientist, Robert Hooke, examined cork bark through a microscope. He noticed that the bark was divided into tiny, boxlike compartments, which he called *cells*.

cell

Many years after Hooke's observations, scientists concluded that all living things are made up of cells and that cells are produced only by other cells. A **cell** is the smallest unit able to carry out all the life activities of an organism. Having cells is a characteristic of life. In fact, it is the characteristic that makes all the other life activities possible. You yourself are made up of many different cells.

Some cells, such as the yolk of an ostrich's egg, are quite large. Most cells, however, can be seen only with the aid of a microscope. This is because they are smaller than the point of a pin. Cells may differ in size, shape, and function, but all cells have certain features in common. Figure 1-1 shows cells from different organisms.

There are three main parts, or *structures*, in every cell: (1) the *cell membrane*, (2) the *nucleus*, and (3) the *cytoplasm*. Each structure carries out certain *functions*, or performs certain "jobs." Study the cells shown in Figure 1-2, as you read about the different cell structures.

cell membrane

The **cell membrane** is a thin layer that surrounds the parts of a cell and keeps them together. The cell membrane protects the contents of a cell and gives the cell its shape. Certain materials move into and out of the cell through spaces in the cell membrane. Food, water, gases, and other materials that are small enough, move through the cell membrane by diffusion. **Diffusion** is the movement of a substance from an area where there is more of it to an area where there is less. The cell membrane also acts like a "guard," preventing some substances from entering and others from leaving the cell.

diffusion

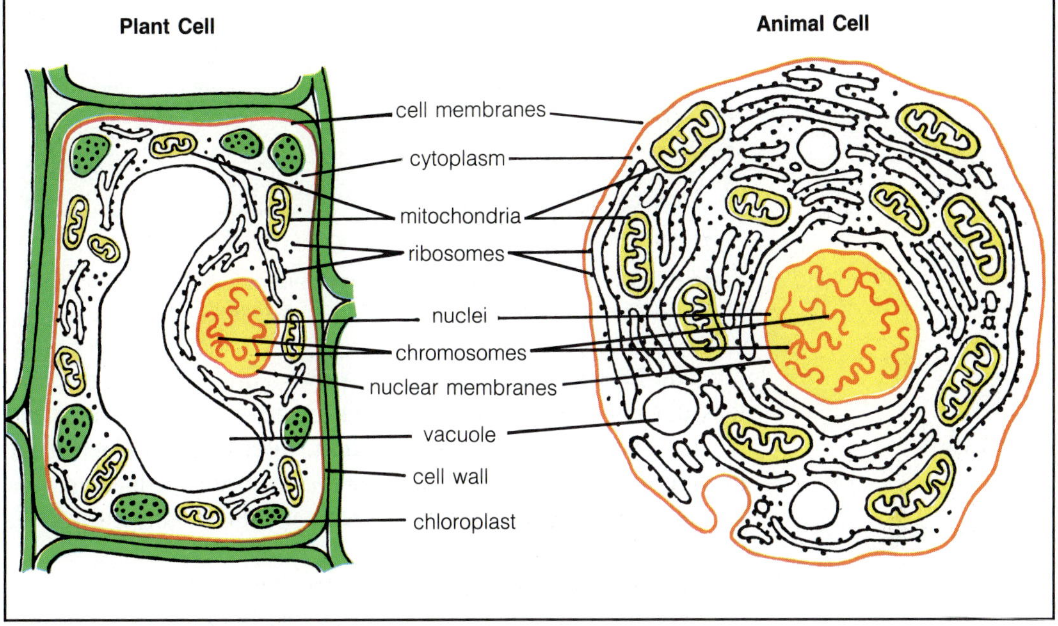

Figure 1-2 *A generalized plant cell and a generalized animal cell contain many of the same structures. Notice the large, liquid-filled vacuole in the plant cell.*

The **nucleus** is the "control center" of the cell. It is surrounded by a thin layer called the *nuclear membrane*. The nuclear membrane separates the nucleus from the other parts of the cell. Only certain substances pass through the nuclear membrane. Inside the nucleus are chromosomes. **Chromosomes** are long, thin fibers containing instructions that control the activities of the cell.

nucleus
(NOO klee us)

chromosomes
(KROH muh sohmz)

Cytoplasm makes up most of the "body" of the cell. It contains many small structures, which carry out cell activities.

cytoplasm

Ribosomes (RY buh sohmz) are structures in the cytoplasm that manufacture complex substances called *proteins* (PROH teenz). The "building materials" of cells and the substances that control many cell activities are made of protein.

Storage areas in the cytoplasm are called *vacuoles*. They may contain water, food, or waste materials. Animal cells usually have a few small vacuoles. Plant cells may contain many large, liquid-filled vacuoles or a single, large vacuole that fills most of the cell. Water-filled vacuoles help support a plant cell. When the vacuoles lose water, the plant wilts.

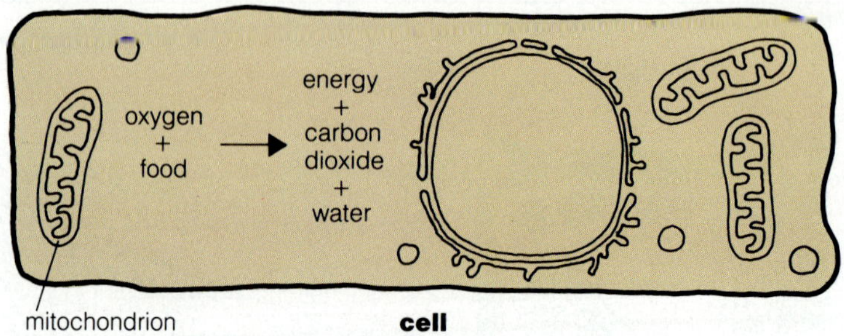

Figure 1-3 *Respiration takes place in the mitochondria. This process provides the energy for an organism's life activities.*

respiration

The energy for movement, growth, and other life activities comes from food. Energy is made available from food by a process known as **respiration.** Respiration takes place in cell structures called *mitochondria* (my tuh KAHN dree uh). The shape and number of mitochondria in different cells varies.

Oxygen and carbon dioxide are gases found in air and water. During respiration, oxygen combines with food in the mitochondria, releasing energy, carbon dioxide, and water. (See Figure 1-3.) Respiration goes on continually in all cells.

chlorophyll (KLOR uh fil)

chloroplasts (KLOR uh plastz)

Most plant cells contain a green substance called **chlorophyll** that enables them to produce food. Chlorophyll is found in structures called **chloroplasts.** Animal cells do not contain chlorophyll and, therefore, cannot produce their own food.

cell wall

Another structure found only in plant cells is the cell wall. The **cell wall** provides protection and support for the plant cell. Cell walls contain a tough, fibrous material called *cellulose* (SEL yuh lohs), which remains long after the cells have died. The cork "cells" that Robert Hooke studied were really the empty cell walls left behind after the cells of a tree died.

Checkpoint

1. What is a cell?
2. Describe the functions of the following parts of a cell: (a) the cell membrane, (b) the cytoplasm, (c) the nucleus, (d) mitochondria, (e) vacuoles.
3. How are plant cells different from animal cells?
4. What is respiration? Where does it occur?

TRY THIS

Make a three-dimensional model of a cell and its parts using materials you can find at home.

1-3

One-Celled Organisms

Goal

To describe the characteristics of some one-celled organisms.

Some organisms consist of only one cell. Special structures enable one-celled organisms to carry out all the life activities performed by multi-celled organisms.

The *paramecium* (pa ruh MEE see um) is a slipper-shaped, one-celled organism that lives in ponds and streams. (See Figure 1-4.) A stiff layer around the cell membrane gives the paramecium a definite shape. Paramecia (plural of paramecium) are propelled by rows of tiny, beating hairlike structures, called **cilia.** The movement of a paramecium is like a boat being propelled by many oars.

Paramecia respond to conditions in their surroundings. When a paramecium bumps into an object in its path, it turns and swims off in a different direction. Paramecia tend to move towards areas that contain food and away from areas of extreme heat or extreme cold.

Figure 1-5 *Imagine a paramecium's cilia propelling it like the oars of these Roman galleys.*

cilia
(SIL ee uh)

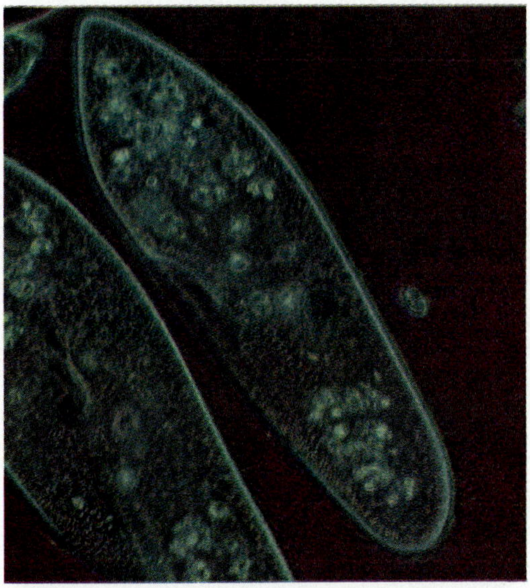

Figure 1-4a *Use Figure 1-4b to help you identify the structures of the paramecia shown here.*

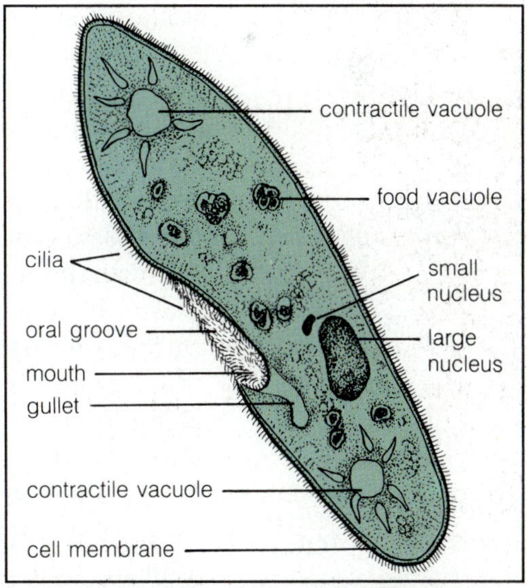

Figure 1-4b *Notice the contractile vacuoles, the oral groove, the gullet, and the many cilia on the paramecium.*

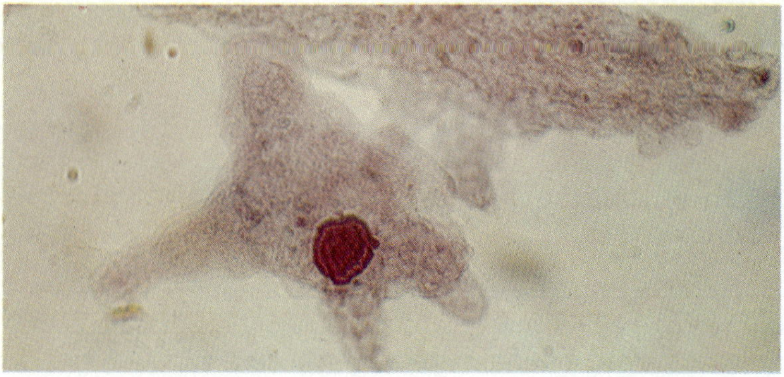

Figure 1-6 *An amoeba's shape is constantly changing as the flowing cytoplasm forms fingerlike projections called* pseudopods.

Smaller one-celled organisms and other tiny bits of food are taken in by a paramecium through an *oral groove*. Beating cilia lining the groove pull in food and sweep it toward the *mouth* and *gullet*. At the end of the gullet, the food is enclosed in a food vacuole, where it is digested. The food vacuole floats through the cytoplasm, releasing food in different areas of the cell. Undigested food particles diffuse out through the cell membrane.

Water and waste products collect in two *contractile* (kun TRAK tul) *vacuoles*, located at opposite ends of the cell. These vacuoles stretch and swell to hold water and wastes. When full, the vacuoles contract, "pumping" their contents into the water outside the cell.

A paramecium reproduces by dividing into two identical cells that are copies of the parent.

Another one-celled organism that lives in pond water is the *amoeba* (uh MEE buh). (See Figure 1-6.) Unlike a paramecium, an amoeba does not have a definite shape. The cytoplasm of an amoeba is constantly moving. The moving cytoplasm pushes against the cell membrane, causing it to bulge out in fingerlike projections. The rest of the cell flows along behind the projections. An amoeba is constantly changing shape as it moves.

Figure 1-7 *An amoeba surrounds its food, trapping it in a food vacuole. Digestive juices change the food into a form that can be used by the amoeba for energy.*

An amoeba takes in food by surrounding it. (See Figure 1-7.) Food is trapped in a food vacuole, which breaks away from the cell membrane and floats through the cytoplasm. The digested food is absorbed by the cytoplasm and is used for energy, growth, and repair. Wastes pass out through the cell membrane.

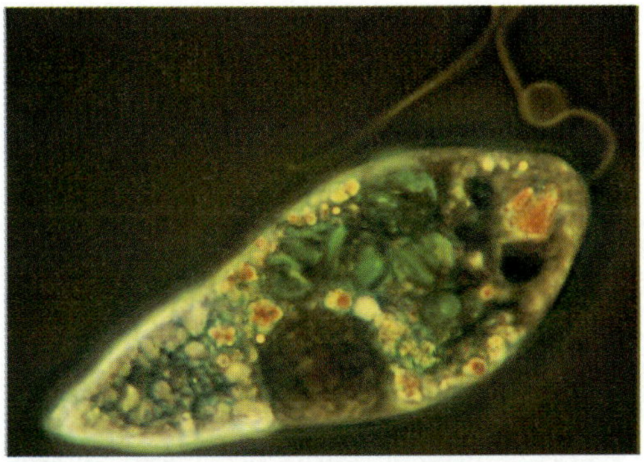

Figure 1-8 *Euglenas make their own food. They move by means of a flagellum.*

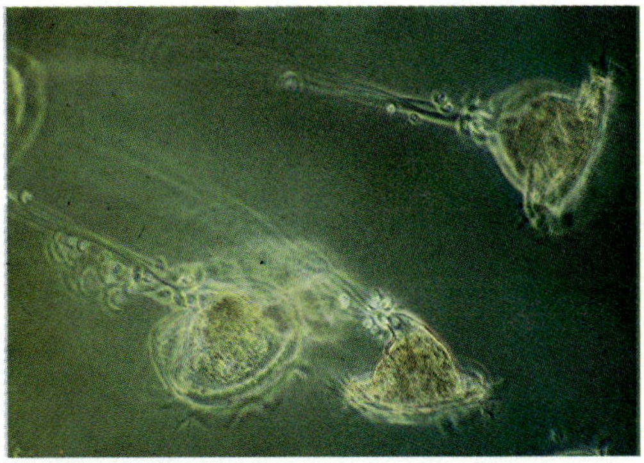

Figure 1-9 *A ring of beating cilia creates a whirlpool that pulls in food from the water.*

Amoebas respond to their surroundings. When conditions are unfavorable for life, such as when a pond dries up, an amoeba is able to form a protective covering called a *cyst* (sist). The amoeba becomes inactive inside the cyst and may remain this way for years. When conditions are favorable, the cyst dissolves and the amoeba carries on.

Like a paramecium, an amoeba reproduces by dividing. One *parent* cell divides, forming two identical cells in less than an hour.

There are thousands of different kinds of one-celled organisms. Each kind of organism has unique ways of carrying out many of its life activities. For example, *euglenas* (yoo GLEE nuhz), the one-celled organisms shown in Figure 1-8, have chlorophyll and can make their own food in sunlight. But during periods of darkness, euglenas take in food from their surroundings. They move by means of a whiplike tail, called a *flagellum* (fluh JEL um). Other one-celled organisms attach themselves to objects in the water. *Vorticella* (vor tuh SEL uh) "spring" up and down to capture food, as shown in Figure 1-9.

TRY THIS

Collect water from a nearby pond or stream. If you can, carefully take some water from the bottom. Make slides of the pond water and look at them under a microscope. Use books to identify the variety of one-celled organisms that you find.

Checkpoint

1. How does a paramecium move? How does it get food?
2. Describe how an amoeba takes in food.
3. How does an amoeba reproduce?
4. How are euglenas different from paramecia?

1-4

Levels of Organization

Goal

To describe the levels of organization in a multi-celled organism.

Some organisms, such as blue whales or giant redwood trees, contain billions of cells. Multi-celled organisms are made up of many different kinds of cells. Each cell carries out the same basic life activities. But each *kind* of cell is *specialized*, or performs a special function, as well. Different kinds of cells have different structures, which allow the cells to specialize. Observe the similarities and differences among the cells shown in Figure 1-10.

tissue

In multi-celled organisms, cells function in groups. A **tissue** is a group of cells that are alike in appearance and function. Multi-celled organisms contain many kinds of tis-

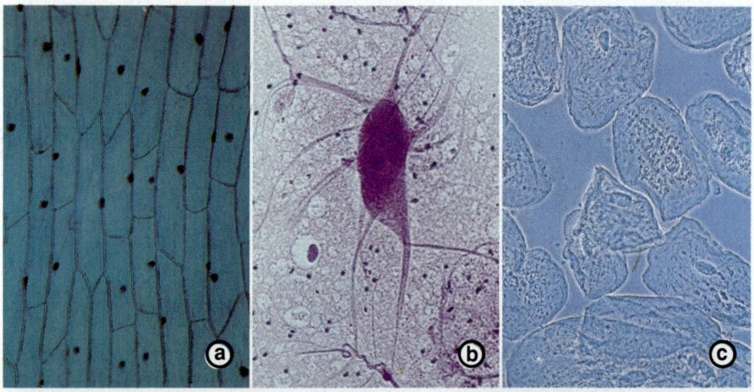

Figure 1-10 *Observe the similarities and differences among: (a) onion skin cells, (b) ox nerve cell, (c) human cheek cells.*

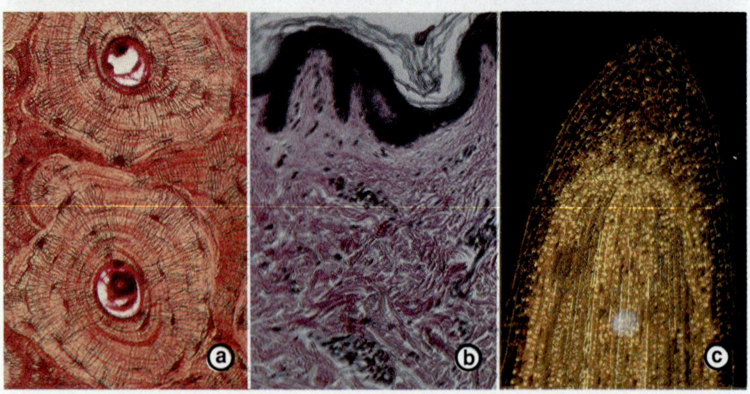

Figure 1-11 *Different kinds of tissue: (a) bone tissue, (b) skin tissue, (c) onion seedling root-tip.*

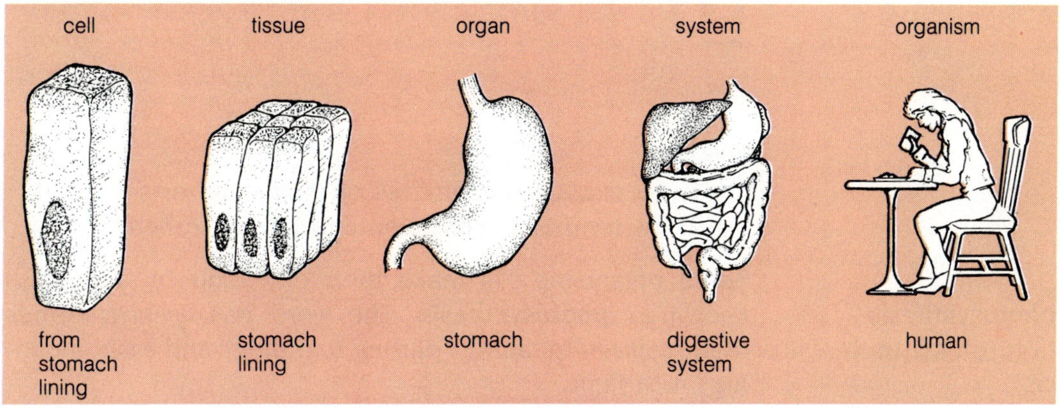

Figure 1-12 *Levels of organization of a human.*

sue. Blood, muscle, skin, and bone are examples of human tissue. Each kind of tissue has a specific function. For example, muscle tissue moves body parts and skin tissue protects the body from harmful substances. You can think of these specializations as a "division of labor."

Plants have different kinds of tissue, too. For example, leaves contain a skinlike protective tissue and tissues that transport water and minerals. Figure 1-11 shows some different kinds of tissue.

organ

Two or more different types of tissue functioning together make up an **organ.** Each organ has a specific function. The heart is an organ that pumps blood through the body. It contains muscle tissue, skin tissue, nerve tissue, connective tissue, and blood tissue. Plants have organs, too. Roots, leaves, flowers, and stems are examples of organs in plants.

system

Organs function together, making a **system.** The *circulatory system* is made up of the heart, blood vessels, and blood. It functions to move materials through the body. The *digestive system* contains organs that change food into a form in which it can be used by the body. In an organism, all the systems together enable the organism to survive. Figure 1-12 shows the levels of organization of a human.

Checkpoint

1. What is a tissue? Name two kinds of tissue.
2. What is meant by the term *specialized cell*?
3. Name the levels of organization in a multi-celled organism, beginning with a cell.

1-5

Photosynthesis

Goals

1. To explain the process of photosynthesis.
2. To compare respiration and photosynthesis.

photosynthesis (foh toh SIN thuh sis)

Some organisms can make their own food by a process known as **photosynthesis.** The word *photosynthesis* comes from *synthesis,* meaning "putting together," and *photo,* meaning "with light."

Green plants and some one-celled organisms contain structures that trap energy from sunlight. The "solar collectors" are *chloroplasts,* which contain the substance *chlorophyll.* The trapped energy is used in converting carbon dioxide and water into a type of sugar called *glucose* (GLOO kohs). Plants use glucose as food. Figure 1-13 shows the steps involved in photosynthesis. Notice that oxygen is given off in the process.

When plants make more food during photosynthesis than they use, the extra food is stored in the form of *starch.* Plants use the stored starch when they run out of sugar or when photosynthesis cannot take place. Plants store energy in molecules of protein, fat, and oil also. Plant-eating animals use the substances stored in plants as food.

Compare the diagram of photosynthesis with the diagram of respiration, Figure 1-3. Note that the two processes seem to be "opposites." The materials produced by one process are the starting materials for the other.

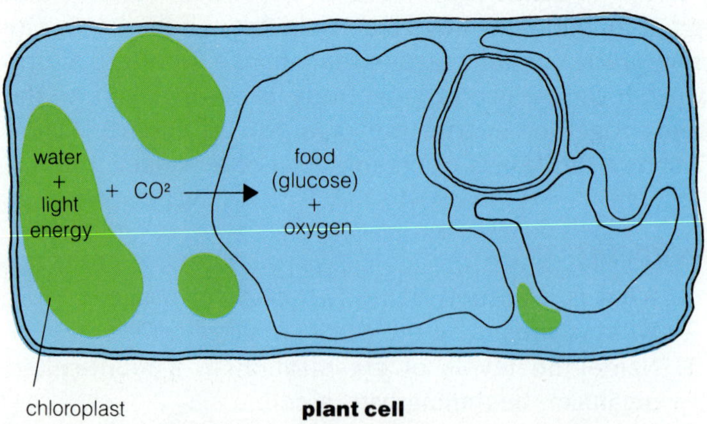

***Figure 1-13** Photosynthesis takes place in the leaves and stems of most plants.*

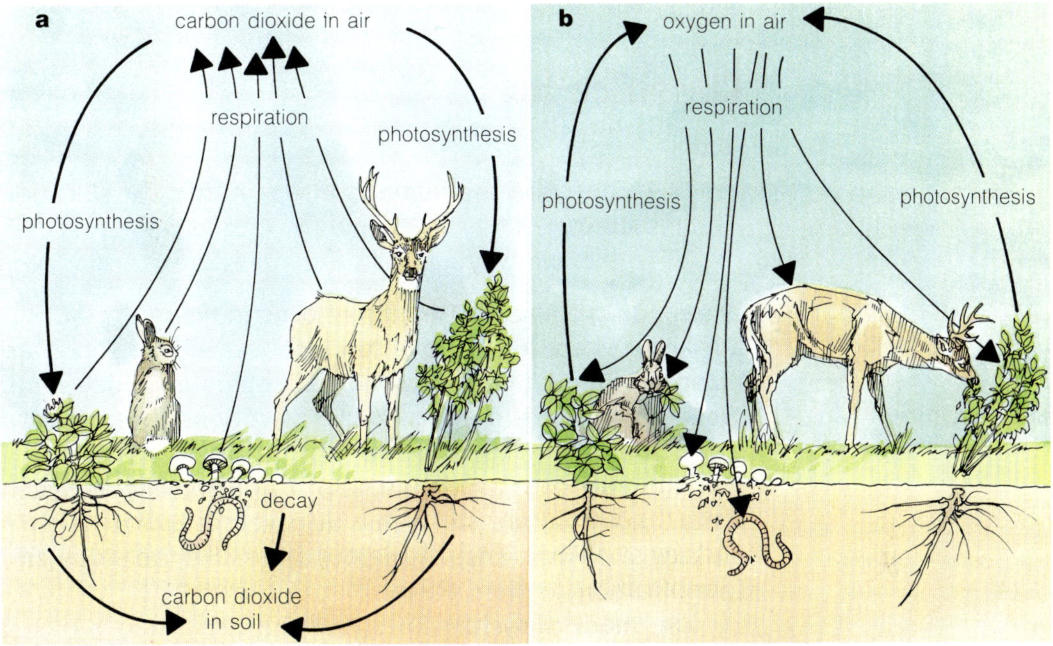

Figure 1-14a *The carbon cycle.*

Figure 1-14b *The oxygen cycle.*

There is a limited amount of oxygen and carbon on Earth. These materials are constantly being reused, or *recycled*. Figure 1-14 shows how carbon and oxygen are recycled by organisms. Notice that all organisms use oxygen given off in photosynthesis during respiration. Plants use the carbon dioxide produced during respiration for photosynthesis.

The exchange of oxygen and carbon dioxide among organisms is of great interest to space scientists, also. They are considering using plants in space colonies. A large number of plants might be able to provide astronauts with enough oxygen and fresh food to survive in space for a long time.

Checkpoint

1. What is photosynthesis? Which organisms are capable of photosynthesis?
2. Which substances are necessary for photosynthesis to take place? What are the products of photosynthesis?
3. Compare the processes of respiration and photosynthesis.

1-6

Classification

Goals

1. To describe the features of a scientific classification system.
2. To explain why a classification system is used.

classification

When you go into a supermarket or variety store, you usually know where to look for a product because the goods are organized into groups or categories. Separating objects into groups by similarities is called **classification.** Classifying is a way of organizing information.

There are over a million known kinds of living things on Earth, and more are still being discovered. Scientists have *classified* all living things into groups to make the study and identification of them easier.

The modern system for classifying organisms is based on a system developed by an 18th-century Swedish scientist, Carolus Linnaeus (lih NEE us). Linnaeus studied the characteristics of thousands of organisms. He set up a classification system based on the similarities and differences he observed.

Linnaeus divided all the organisms into two large groups, which he called *kingdoms*. According to Linnaeus' system, all organisms belonged to the *Plant Kingdom* or to the *Animal Kingdom*. As more precise tools and methods for studying organisms became available, scientists realized that all organisms did not fit neatly into the two-kingdom system. Scientists agreed that additional categories had to be added. Most modern classification systems have more than two kingdoms. The system presented in the next section of this chapter has five kingdoms.

Table 1-1: Levels of Classification

kingdom
phylum
class
order
family
genus
species

All scientific classification systems for organisms are divided into a series of levels. The first level is called a *kingdom*. A kingdom is a large group containing many organisms. The members of a kingdom have some broad, general characteristics in common. Each kingdom is divided into smaller groups called *phyla* (FY luh) (singular, *phylum*, FY lum). These are divided into classes; classes are divided into orders. Family, genus, and species follow. Table 1-1 shows the seven levels of classification.

Members of a *species* (SPEE sheez) share many similarities. They are so much alike that they are able to mate and produce young, which in turn can produce young. Moving from the kingdom level to the species level, the groups become smaller and the organisms in each group more alike.

Scientists classify organisms based on similarities in structure. Note the similarities between the domestic housecat and the mountain lion, shown in Figure 1-15. They are both classified as members of the *Animal Kingdom* because they are multi-celled and get their food by eating other organisms. The cat and mountain lion are in the phylum *Chordata* (kor DAH tuh) and the subphylum *Vertebrata* (vur tuh BRAH tuh) because they have a series of bones, or *vertebrae* (VUR tuh bree), along their backs.

Mammalia (ma MAY lee uh) is the class of animals that includes the cat and mountain lion. Mammals have a body covered with hair or fur, they breathe with lungs, and give birth to live young who are fed with milk. Cats and mountain lions are in the order *Carnivora*. Carnivores have strong jaws and sharp teeth, which help them to catch and eat other animals.

Notice that the mountain lion and cat have low-slung bodies, small round heads, and claws. Scientists classify them in the family *Felidae* (FEE lih day), with other animals that have catlike features. But mountain lions and housecats do not belong to the same genus.

Figure 1-15a *A mountain lion.*

Figure 1-15b *A housecat.*

The cat and mountain lion are similar enough to be classified in the same genus, *Felis* (FEE lis). Animals in this genus are night-hunters. They have claws that can be extended and pulled back inside a cover, large feet, and well-developed eyes and ears. The cat and mountain lion are different enough from each other, however, to be categorized in separate species. The cat is in the species *domesticus* (duh MES ti kus) and the mountain lion in the species *concolor*.

In some parts of North America, a mountain lion is called a puma (PYOO muh) or a cougar. A crayfish (which is not even a fish) is called a crawfish or crawdad, depending on where you live. To avoid confusion, and to make world-wide communication possible, scientists have given a scientific name to every known organism.

The scientific name of an organism is its genus name (beginning with a capital letter) followed by its species name (beginning with a small letter). When many of the scientific names were invented, Latin was understood and used by scientists all over the world. Latin names were given to many plants and animals at that time and are still being used today.

Scientific names sometimes include words that describe the organisms, such as *Sequoia gigantea* (si KWOY uh jy gan tee uh) for *giant* redwood, *Alligator mississippiensis* for American alligator, and *Megabombus pennsylvanicus* for bumblebee. Occasionally, an organism is named after the person who studied or discovered it. Milkweed plants are in the genus *Fernaldia* (fur nal DEE uh), after the botanist Merritt Lyndon Fernald. The system of naming organisms is called the *binomial* (by NOH mee ul) *system*. Binomial means "two names."

Checkpoint

1. Define classification. What is the function of the scientific classification system for organisms?
2. Why do most modern classification systems have more than two kingdoms?
3. List the seven levels of the scientific classification system for organisms.
4. *Canis lupus* (KAY nis LOO pus) is the scientific name for a wolf. Which is the genus name? Which is the species name?

TRY THIS

Find out the name of the flower, tree, or animal representing your state or province. Use reference books to find out its scientific name.

1-7

Variety of Life

Goals

1. To identify the general characteristics of the five kingdoms.
2. To give examples of organisms in each kingdom.

Moneran Kingdom (moh NAIR an)

The **Moneran Kingdom** is made up of the smallest one-celled organisms, the blue-green algae (AL jee) and the bacteria. (See Figure 1-16, left.) Monerans have no nuclear membranes, mitochondria, or chloroplasts. Monerans do have nuclear material.

The largest group of monerans is the *bacteria*. Bacteria can live almost anywhere. They have been found in deep soil, in hot springs, and in the bodies of animals. For example, *E. coli* (KOH ly) bacteria live in human intestines. They appear to be necessary for proper digestion and for elimination of waste products. Some bacteria are harmful. Different varieties of bacteria cause diseases such as pneumonia, strep throat, and food poisoning.

Protist Kingdom (PROH tist)

The **Protist Kingdom** contains one-celled organisms that have nuclear membranes, such as the paramecium and amoeba. (See Figure 1-16, right.) Most protists live in water

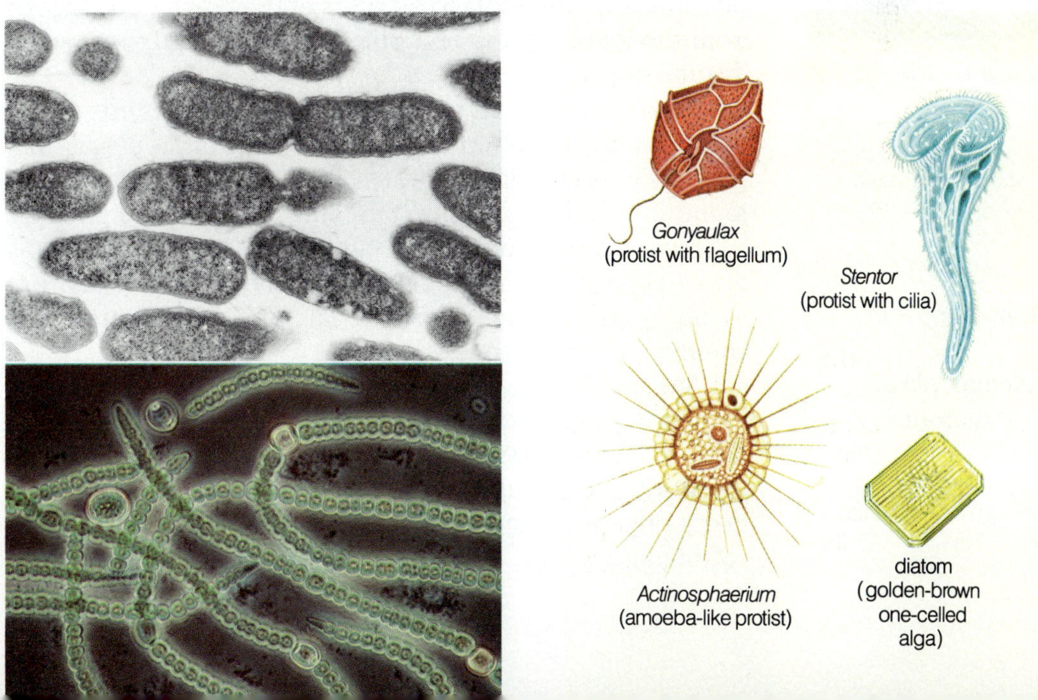

Figure 1-16 *Left, photographs of monerans:* E. coli *(top), a bacterium;* Anabaena *(bottom), a blue-green alga. Right, drawings of protists.*

or in moist places. One-celled algae, a kind of protist, are an important food source for many water organisms. They give off large amounts of oxygen, through photosynthesis.

Fungi Kingdom
(FUHN jy)

Mushrooms and *molds* are probably the best-known members of the **Fungi Kingdom.** Fungi (plural of fungus) were once classified as plants because they have cells with cell walls and they do not move. But fungi do not have chlorophyll and must obtain their food from other organisms. Figure 1-17 shows some examples of different fungi. Yeasts are fungi used by people in making bread and wine.

Figure 1-17 *Fungi. (top) Mushrooms growing on a rotting log. (bottom) Bracket, or shelf, fungi often grow and feed on old tree trunks.*

Plant Kingdom

The **Plant Kingdom** is made up of multi-celled organisms with chloroplasts. Examples are shown in Figure 1-18. Plants make their own food by photosynthesis and have cells with cell walls. The Plant Kingdom can be divided into two large groups—plants with transport tubes and plants without them.

Many-celled algae are plants without transport tubes. They are found in ponds, oceans, streams, and in damp places. Some algae may be made up of only a few cells, while others may grow to be 10 m in length. Mosses and liverworts are small plants that also do not have transport tubes. They grow in moist, shady places.

vascular plants
(VAS kyuh lur)

Plants that do have a system of transport tubes are called **vascular plants.** Vascular plants have two types of transport tubes. Water and minerals move through one set of tubes from the roots to all parts of the plant. Food moves from the leaves to the rest of the plant through another set of tubes.

Ferns, gymnosperms (JIM nuh spurmz), and angiosperms (AN jee uh spurmz) are three groups of vascular plants. Ferns grow in moist, shaded areas and are commonly found in woods and along the banks of rivers and streams.

Most gymnosperms have needlelike leaves, which stay green all year long. Gymnosperms produce seeds that develop in cones. Pines, spruces, larches, and firs are some common gymnosperms.

Angiosperms are flowering plants. Grasses, fruit trees, garden flowers, and vegetable plants are just a few examples of angiosperms. Angiosperms make up a large part of the world's food supply. They provide food and shelter for many organisms and are used for products such as paper, cotton, and linen.

Figure 1-18 *The Plant Kingdom. (a)* Spirogyra, *a green algae commonly found in ponds. (b)* Fucus, *a brown algae often found growing on northern rocky coasts. (c) Moss. (d) Liverworts. (e) Ferns. (f) A sunflower. (g) A lodgepole pine tree with cones. (h) A crabapple tree in bloom.*

Animal Kingdom

Members of the **Animal Kingdom** are multi-celled organisms that get food by eating other organisms. Animal cells do not contain chlorophyll or cell walls. The Animal Kingdom is divided into many phyla.

Sponges belong to the phylum of animals known as *Porifera* (puh RIF ur uh). Their bodies are made up of groups of cells attached to a skeletonlike framework. Sponges take in bits of food from the water flowing through their *pores*, or openings. (See Figure 1-19.) The phylum name *Porifera* means "having pores."

Animals with a saclike body and tentacles surrounding their mouth opening are in the phylum *Coelenterata* (sih LEN tuh rah tuh). Hydras, jellyfish, coral, and sea anemones (uh NEM uh neez) have the stinging cells on their tentacles that are characteristic of organisms in this phylum.

Figure 1-19 *Sponges: (a) yellow tube sponges. Coelenterates: (b) sea anemones and (c) a jellyfish. Worms: (d) a planarian flatworm. Molluscs: (e) a land snail, (f) a sea scallop, and (g) a squid.*

Figure 1-20 Echinoderms: (a) starfish and (b) sea urchins. Arthropods: (c) a cucumber beetle. (d) a centipede— often found around rotting logs and under rocks. (e) a giant red millipede. (f) an orchard spider. (g) a green crab.

There are three phyla of worms: *flatworms, roundworms,* and *segmented worms*. Some worms, such as tapeworms or leeches, live in or on another organism, obtaining energy from it. They are called *parasites*.

Most animals classified in the phylum *Mollusca* (mah LUHS kah) have a hard shell surrounding their soft body parts. A few, like the squid, octopus, and sea slug do not have a shell. Molluscs live in water or in damp places. Figure 1-19 shows a few kinds of molluscs.

Animals with a star-shaped body and spine-covered skin are called *echinoderms* (ih KY nuh durmz). *Echinoderm* means "spiny-skin." (See Figure 1-20.)

About 75% of all known animals belong to the phylum *Arthropoda*. Arthropods have a hard, external skeleton, a segmented body, and jointed legs. Members of this phylum are very important to people. For example, people raise bees for the large quantities of honey they produce. Crabs, lobsters, and shrimp are food for many people. Figure 1-20 shows representatives of the five classes of arthropods.

Figure 1-21 *Vertebrates include fish, amphibians, reptiles, birds, and mammals.*

Chordata is the phylum humans belong to. *Vertebrata*, a subphylum of the chordates, is divided into the seven classes shown in Figure 1-21. All vertebrates have an internal skeleton that includes a vertebral column (backbone) and limbs.

Checkpoint

1. Which kingdom contains one-celled organisms without a nuclear membrane?
2. Describe the general characteristics of protists.
3. What characteristics do members of the Fungi Kingdom have in common? Name three fungi.
4. Describe the general characteristics of the Plant Kingdom. What are the three groups of vascular plants?
5. What features do all animals have in common?

1-8

Viruses

ENRICHMENT

Goals

1. To describe viruses.
2. To explain immunization.

There is one group of tiny life forms that has puzzled scientists for many years. These life forms seem to have some of the properties of living things, but they do not have cells and they cannot carry out life activities on their own. These tiny "things," called *viruses*, are classified in a group by themselves.

Viruses are so small that they cannot be seen with a light microscope. For many years scientists could only guess about the existence of viruses, based on their observations of the damage the viruses caused. With the invention of powerful electron microscopes, scientists could finally study the structure of these mysterious life forms. Figure 1-22 shows several kinds of viruses.

Some scientists consider viruses to be alive, while others do not. Viruses seem to use neither food nor oxygen. They lack most of the parts of a cell and cannot reproduce on their own. Viruses need a host cell in order to become active. They do not grow, reproduce, or carry on respiration outside of the host cell.

Figure 1-22 *Three kinds of viruses: (a) bacteriophage attacks bacterium, (b) tobacco mosaic viruses infect tobacco leaves, (c) the viruses that cause measles.*

When a virus enters a cell, it seems to "overpower" the cell's control system. The infected cell begins making copies of the virus, instead of following the normal instructions on the cell chromosomes. The cell becomes a "virus factory," producing hundreds of new viruses. Eventually the host cell bursts, releasing the viruses, which then attack other cells.

Viruses generally invade specific kinds of cells. For example, the polio virus attacks one kind of cell in the brain and spinal cord. Viruses cause diseases in a number of animals and plants. They are responsible for the common cold, measles, polio, chicken pox, mumps, influenza, and smallpox. Scientists now believe that some viruses may remain inactive, or *dormant*, in a cell for many years, until a change in the cell causes the virus to become active. It is thought that some types of cancer may be caused by viruses such as these.

Doctors use a process, called *immunization*, to prevent certain diseases caused by viruses. When you are immunized, you swallow or receive an injection of a small amount of weakened or dead viruses, called a *vaccine* (vak SEEN). The vaccine causes the body to prepare its own defense system for fighting that particular type of virus. Within a few days, the body's natural defense system has produced substances called *antibodies* that can recognize and destroy a specific type of virus if it enters the blood. There are vaccines for polio, mumps, measles, and smallpox. Through the systematic use of vaccines in North America, smallpox, polio, and measles have been nearly wiped out.

Checkpoint

1. What characteristics make viruses seem alive? How are viruses different from living things? From nonliving things?
2. Describe what happens when viruses enter living cells. What kinds of cells do different viruses invade?
3. Name two diseases caused by viruses.
4. What is a vaccine?
5. How does immunization prevent disease? What diseases have been prevented through immunization?

Zeroing In On Life's Secrets

The first microscope used to magnify objects was a *simple* microscope. It consisted of a single lens, or magnifying glass, which magnified an object ten to twenty times its normal size. In the 1600's the *compound* microscope was invented, using two lenses mounted at each end of a tube. It magnified objects hundreds of times more than the simple microscope.

Most modern microscopes are compound microscopes. They use a source of light to illuminate the objects being viewed. Because of the properties of light, a light microscope cannot go beyond a magnification of 400–1000 times normal size without losing the clear details of the image.

The invention of the *electron* microscope made it possible to observe a specimen at magnifications 200,000 times its normal size. With the electron microscope, scientists could see details never before visible. Compare the pictures taken through a light microscope (Figure 1) and an electron microscope (Figure 2).

There are two kinds of electron microscope. Both use a beam of negatively charged particles, called *electrons*, instead of light.

The transmission electron microscope aims a beam of electrons at thin slices of the specimen being studied. The specimen is coated with chemicals and is inserted into a part of the microscope that has had all the air removed from it. A limitation of the transmission electron microscope is that things that are alive cannot be observed.

The scanning electron microscope makes possible three-dimensional images, such as the one shown in Figure 2. Its use is not limited to the viewing of thin slices of material. An electron beam is passed over the surface of a specimen, producing images on a screen. Using the scanning electron microscope, scientists have been able to work with some live organisms.

Figure 1

Figure 2

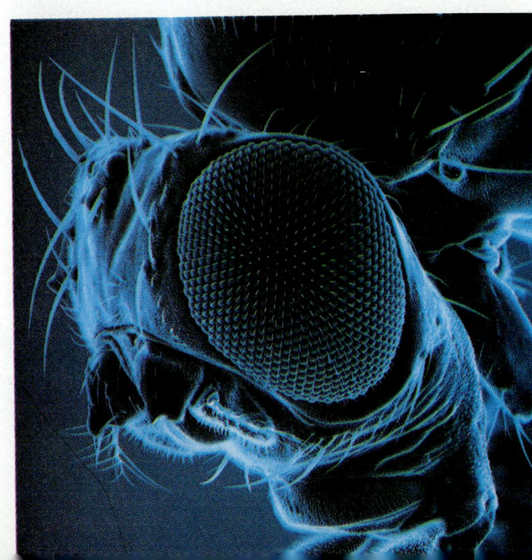

Understanding The Chapter 1

Main Ideas

1. The characteristics that define life include: movement, growth and repair, getting and using food, reproduction, responding to changes in the surroundings, and having cells.
2. A cell is the smallest unit able to carry out all the life activities of an organism.
3. Cells contain a cell membrane, cytoplasm, and a nucleus. In addition, plant cells have cell walls and chloroplasts.
4. Respiration is the process of obtaining energy from food.
5. In multi-celled organisms, cells make up tissues, tissues form organs, and organs function together in systems.
6. Photosynthesis is the process by which some organisms make food, using water, carbon dioxide, chlorophyll, and energy absorbed from the sun.
7. Scientists have organized all known organisms into a seven-level classification system.
8. The scientific name of an organism is its genus name followed by its species name.
9. Organisms are classified in one of the five kingdoms: Moneran, Protist, Fungi, Plant, and Animal.
10. Viruses show some characteristics of life. They need a host in order to become active.

Vocabulary Review

From the following list, choose the term that best completes each of the statements. Write your answers on a separate piece of paper.

Animal Kingdom	Moneran Kingdom
cell	nucleus
cell membrane	organ
cell wall	organism
chlorophyll	photosynthesis
chloroplast	Plant Kingdom
chromosomes	Protist Kingdom
cilia	respiration
classification	ribosomes
cytoplasm	system
diffusion	tissue
Fungi Kingdom	vacuoles
mitochondria	vascular

1. _?_ is the process by which energy is released from food.
2. _?_ is the substance in a plant cell that absorbs energy from the sun during photosynthesis.
3. _?_ makes up the body of a cell.
4. Paramecia move as a result of the beating action of _?_.
5. Another name for a living thing is _?_.
6. A _?_ is the smallest unit able to carry on the life activities of an organism.
7. A _?_ is a group of specialized cells, which together carry out a specific function.
8. _?_ is the separation of objects into groups by similarities.
9. _?_ are storage areas in cells.

10. Respiration takes place in structures called __?__ in cells.
11. The __?__ surrounds an animal cell and protects it.
12. The __?__ is the tough, outer-covering of a plant cell.
13. The __?__ are fibers containing instructions that control cell activities.
14. Some substances move through a cell membrane by __?__.
15. The __?__ is made up of one-celled organisms with cell membranes.

Chapter Review

Write your answers on a separate piece of paper.

Know The Facts

1. Respiration is a process that results in the production of __?__. (food, oxygen, carbon dioxide)
2. During photosynthesis, energy is absorbed by the substance called __?__. (chromosomes, chlorophyll, carbon dioxide)
3. __?__ is given off during photosynthesis. (carbon dioxide, oxygen, water)
4. Mushrooms are part of the __?__ Kingdom. (Moneran, Protist, Fungi)
5. __?__ are found in plant cells, but not animal cells. (vacuoles, mitochondria, chloroplasts)
6. Flowering plants are called __?__. (vascular plants, gymnosperms, angiosperms)
7. Amoebas get food by __?__. (surrounding it, beating their cilia, springing up and down)
8. A group of bone cells functioning together forms bone __?__. (tissue, organs, systems)
9. A __?__ is the largest division in a classification system. (species, kingdom, genus)
10. The scientific name of an organism is its __?__ and species name. (genus, order, kingdom)
11. __?__ do not have a membrane around their nucleus. (amoebas, bacteria, green algae)
12. A paramecium is an example of a __?__. (protist, fungus, moneran)
13. When a virus enters a cell, it causes the cell to __?__. (grow, produce more viruses, become inactive)
14. __?__ are given to people to protect them from disease-causing organisms. (antibodies, vaccines, vacuoles)
15. The __?__ phylum is made up of animals with hard shells and soft bodies. (moneran, protist, mollusc)
16. Animals have many cells and __?__. (eat other organisms for food, produce food by photosynthesis, produce oxygen by respiration)

Understand The Concepts

17. What are the characteristics a scientist uses to define life?
18. Draw a diagram of an animal cell. Label the parts. Explain the function of each part.
19. Draw a plant cell. Label the parts. Explain the function of each part.
20. Describe three structures and specialized areas found in the cytoplasm of a cell. What are their functions?

21. Compare the ways a paramecium, an amoeba, and a euglena move and get food.
22. List the levels of organization of a multi-celled organism, in order, beginning with a cell.
23. Describe the differences between photosynthesis and respiration.
24. Define classification. Give three examples of things that are classified.
25. Name the five kingdoms of organisms. Give an example of one in each kingdom.
26. Describe the characteristics of viruses. How do vaccines prevent viral diseases?

Challenge Your Understanding

27. If an unknown "thing" was discovered on a space expedition, how would scientists decide whether or not it was a form of life? If it was alive how would they go about classifying it?
28. All housecats do not look alike. Why are they in the same species?
29. What are the advantages of using a five-kingdom classification system instead of a two-kingdom system? Some biologists have used a three-kingdom system. What organisms do you think are in each of those kingdoms?
30. Which cells in a plant do not carry on photosynthesis? How do they get food?

Projects

1. Find out the scientific names of some organisms, such as a dog, mosquito, corn plant, and rhinoceros beetle. If possible, find out what the names mean and why the names were chosen.
2. Collect hydras, planarias, and other small organisms from a pond or stream. Find out how to feed and care for them. Observe the organisms through a microscope. Keep a record of your observations.
3. Do a report on fungi that are helpful and fungi that are harmful to people. Make a bulletin-board display.
4. Invite a representative from the local public health service to speak to your class about bacterial diseases, their cure, and how to prevent them.
5. Find out how penicillin was discovered and by whom.
6. Collect a variety of seashells. Sort them into groups. Use reference books or field guides to identify them. Label each with its common name and genus and species name. Display the shells in class.
7. Collect electron microscope photographs from magazines and newspapers. Prepare a bulletin-board display.
8. Some research scientists study life processes by growing and observing cells in a laboratory. Find out how cell cultures are maintained and what scientists can learn by studying them.

Chapter

2

Heredity

Offspring resemble their parents because many of their traits are inherited.

2-1

From Generation to Generation

Goal

To identify traits that are passed from generation to generation.

Look at the photograph of the plants in Figure 2-1. Pay particular attention to the shapes of the leaves, and the colors and shapes of the flowers. What similar features do you see? What differences do you see?

traits

Specific features of an organism, such as the shape and texture of a leaf or the structure of a flower, are called **traits** or characteristics. Traits that are passed from parents to their offspring are called *inherited traits*. Offspring resemble their parents because many of their traits are inherited.

Agricultural scientists often try to improve the traits of a species of farm animals or crops. They try to produce crops that have a better taste or that are more convenient. For example, the navel orange was bred to be seedless and the tangelo was developed for its sweet flavor and thin, easy-to-peel skin.

Information about every inherited trait is contained in the chromosomes. Remember that chromosomes are structures in the nucleus of each cell. The chromosomes function

Figure 2-1 *Observe the similarities and differences among the flowers. What differences do you see?*

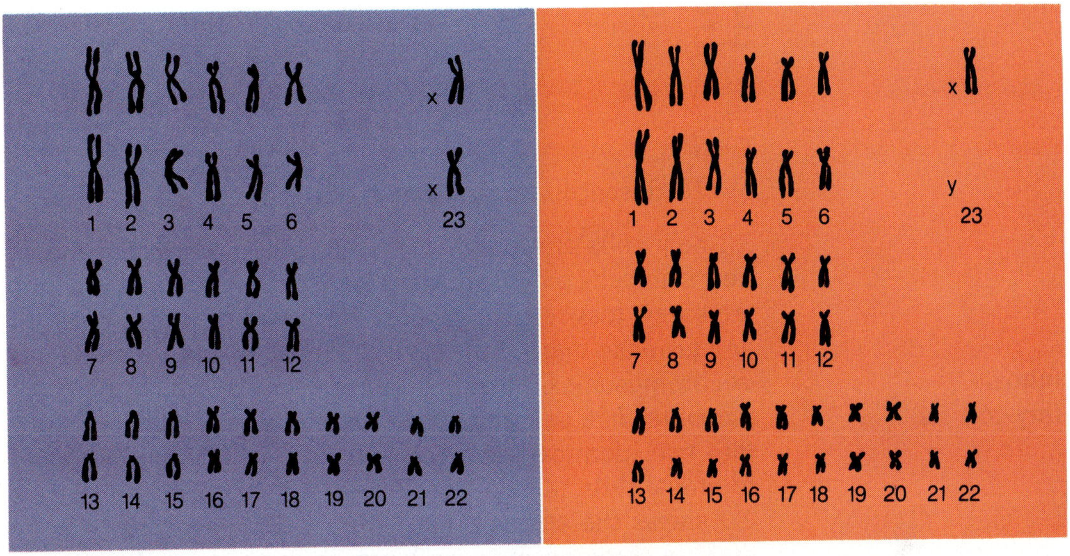

Figure 2-2 *The chromosomes of a human female (left) and a human male (right). The 46 chromosomes have been arranged in 23 pairs.*

like an architect's plan of a building. The plan tells where floors, walls, plumbing, and electric wires are to be located. It also specifies the building's shape and appearance. Like a building's plan, the chromosomes contain information that specifies the shape and structure of an organism. Your chromosomes contain information that controls inherited traits, such as eye color, height, shape of the mouth, and hair color.

Each species of organism has a certain number of chromosomes. Table 2-1 lists the number of chromosomes for some different kinds of organisms. Figure 2-2 shows the 46 human chromosomes arranged in 23 pairs.

Table 2-1: Chromosome Numbers

Organism	*Number*
mosquito	6
housefly	12
corn	20
hydra	32
human	46
potato	48
dog	78

Checkpoint

1. Where in each cell is information about inherited traits contained?
2. Name three inherited traits in people.
3. How many chromosomes are in human cells?

TRY THIS

One inherited trait is tongue-rolling, or the ability to roll the tongue into a U-shape. Find out how many members of your class have the tongue-rolling trait. Are there more rollers or nonrollers? Which appears to be the more common trait?

2-2

Mitosis

Goal

To describe the process of mitosis.

mitosis
(my TOH sis)

replicate

All new cells arise from old ones. When you cut yourself, new skin cells grow, replacing the damaged ones. Oak trees grow new leaves each spring. A fish begins life as one cell and grows to many. Each new cell must have a complete set of chromosomes. **Mitosis** is the process by which the chromosomes in a cell duplicate themselves and then separate into two identical sets. The cell then divides, forming two identical cells. Mitosis occurs in recognizable stages. Figure 2-3 shows the stages of mitosis.

The chromosomes **replicate**, or copy themselves in the period just *before* mitosis begins. Each chromosome attaches to its copy, forming a two-part chromosome. (See Figure 2-3a and Figure 2-3b.)

In the first stage of mitosis the nuclear membrane surrounding the chromosomes disappears. (See Figure 2-3c.) A series of thin strands, called *spindle fibers*, begins to form. The chromosomes take on a thicker appearance.

Figure 2-3 *Mitosis is a continuous process. It does not stop and start between stages. Mitosis can occur in minutes or several hours.*

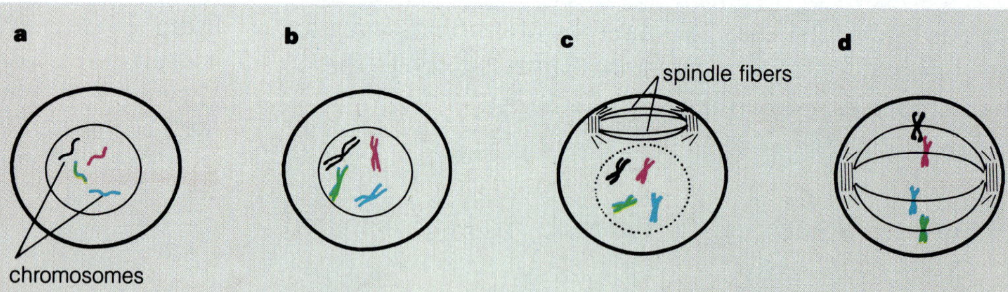

a: animal cell before mitosis

b: the chromosomes replicate

c: the nuclear membrane disappears and spindle fibers begin to form

d: the spindle fibers attach to the chromosomes lined up in the middle of the cell

The second stage of mitosis begins with the chromosomes lining up in the middle, or *equator*, of the cell. The spindle fibers attach to the chromosomes, as shown in Figure 2-3d.

During the third stage, the two-part chromosomes separate and are pulled by the spindle fibers to opposite ends of the cell. At the end of this stage, there is an identical set of chromosomes at opposite ends of the cell. This stage is shown in Figure 2-3e.

The spindle fibers disappear in the final stage of mitosis. A new nuclear membrane forms around each group of chromosomes. Mitosis is usually followed by *cell division*. The cell splits in half along the equator, producing two identical cells. (See Figure 2-3f and Figure 2-3g.)

Mitosis and cell division are continuous processes. There are no stops and starts between different stages. The processes ensure that cell characteristics will be maintained from one cell generation to the next.

Checkpoint

1. What is mitosis?
2. When do the chromosomes in a cell replicate?
3. Draw and label the stages of mitosis, in order.
4. What happens during the last stage of mitosis?

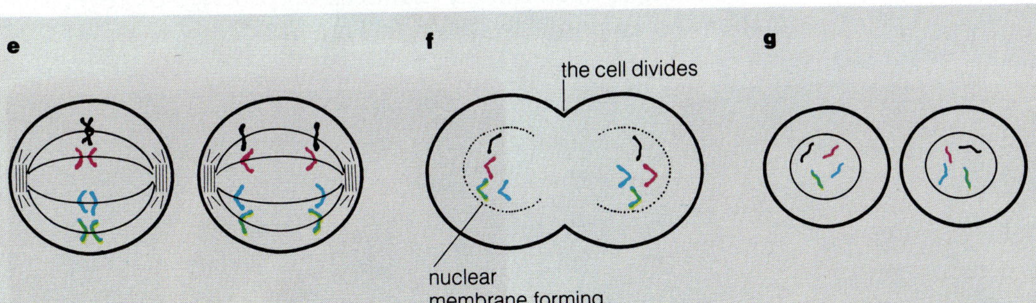

the two-part chromosomes separate

the chromosomes are at opposite ends of the cell

the final stage of mitosis

both cells have identical sets of chromosomes

2-3

Asexual Reproduction

Goal

To describe several methods of asexual reproduction.

reproduction

asexual reproduction

Inherited traits are passed from parents to offspring as a result of reproduction. **Reproduction** is the process by which organisms produce new organisms. One type of reproduction is asexual reproduction. Through **asexual reproduction**, new organisms are produced by a single parent. The offspring is an exact copy of the parent.

Most one-celled organisms reproduce by a kind of asexual reproduction called fission. During *fission*, the parent cell undergoes mitosis and cell division, resulting in two identical *daughter cells*. (See Figure 2-4.)

Hydras, and other relatives of the jellyfish, reproduce by budding. During *budding*, cells in the side of an adult hydra form a young hydra, called a *bud*. (See Figure 2-5.) The bud is a copy of the parent. When it is large enough, the bud separates from the parent. Buds may go on to become parent cells themselves.

Some organisms can reproduce by a process known as regeneration. *Regeneration* is the ability to replace missing parts, or a whole new organism from a small body part. Planaria, starfish, and many plants can reproduce by regeneration. An example is shown in Figure 2-6.

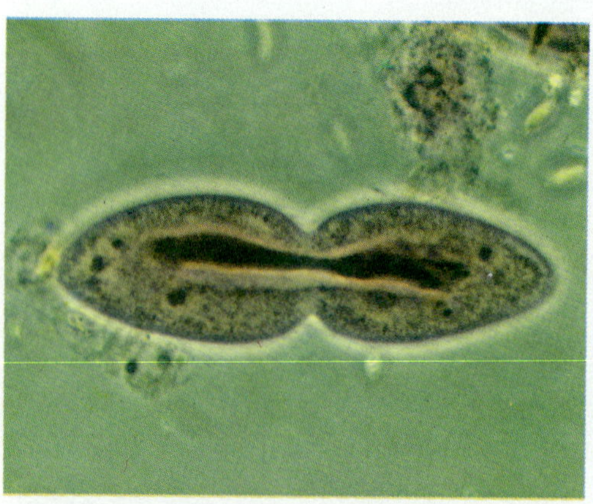

Figure 2-4 *A paramecium undergoing fission. Fission results in two identical cells.*

Figure 2-5 *Hydras often reproduce by budding. The young hydra is a copy of the parent.*

Figure 2-6 *This starfish is regenerating an arm. Some starfish can regenerate a whole new body from a small piece of one arm.*

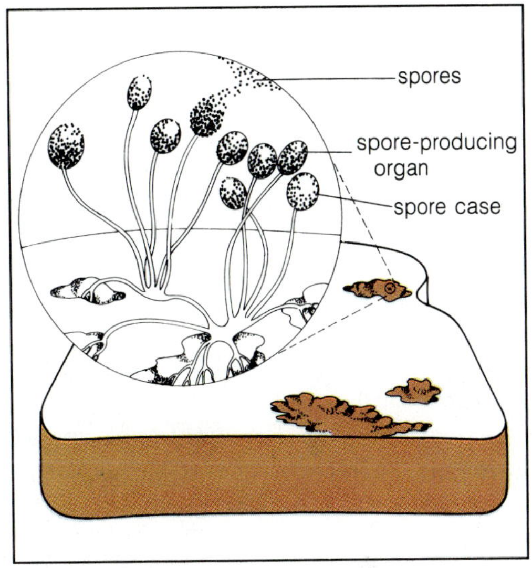

Figure 2-7 *When a spore case breaks open, millions of spores are released. Spores grow only when they are in the right surroundings.*

spores

Bread molds and mushrooms form reproductive cells called **spores**. Millions of spores are produced and stored in *spore cases*. (See Figure 2-7.) When the spore cases break open, the spores are released. Each spore is protected by a tough covering. If a spore lands in an area with favorable conditions, the covering dissolves and the spore develops into a new organism. Spores can remain *dormant*, or inactive, for a long time until "activated" by the right conditions.

Checkpoint

1. Define asexual reproduction.
2. Describe the differences between fission and budding. Give examples of each.
3. Describe how a bread mold reproduces.

TRY THIS

Carefully rub a piece of fresh bread on the floor, window, or desktop. Moisten the bread and place it inside a plastic bag. Seal the bag and put it in a dark place for several days. Moisten it from time to time if it dries out. Watch for the appearance of bread mold. **CAUTION:** Be careful not to touch the molds.

2-4

Sexual Reproduction

Goals

1. To describe how sexual reproduction differs from asexual reproduction.
2. To explain the function of meiosis.
3. To describe reproduction in flowering plants.

sexual reproduction

gametes (GAM eetz)

meiosis (my OH sis)

Sexual reproduction is the creation of a new organism by joining one male sex cell with one female sex cell. Male sex cells, or **gametes**, are called *sperm* and female gametes are called *eggs*. Sexual reproduction occurs in many plants and animals, including birds, mammals, fish, and flowering plants. Through sexual reproduction parents contribute chromosomes (in the sex cells) to offspring. Thus, offspring show characteristics of their parents.

Meiosis is a kind of cell division that results in cells with half the usual number of chromosomes. Gametes are produced through meiosis. All gametes have half the number of chromosomes as other cells in an organism. You may remember that humans have 46 chromosomes in all their

Figure 2-8 *Meiosis occurs only in egg and sperm parent cells. Each sperm or egg cell has half the number of chromosomes as the parent. This sequence shows meiosis occurring in a sperm parent cell.*

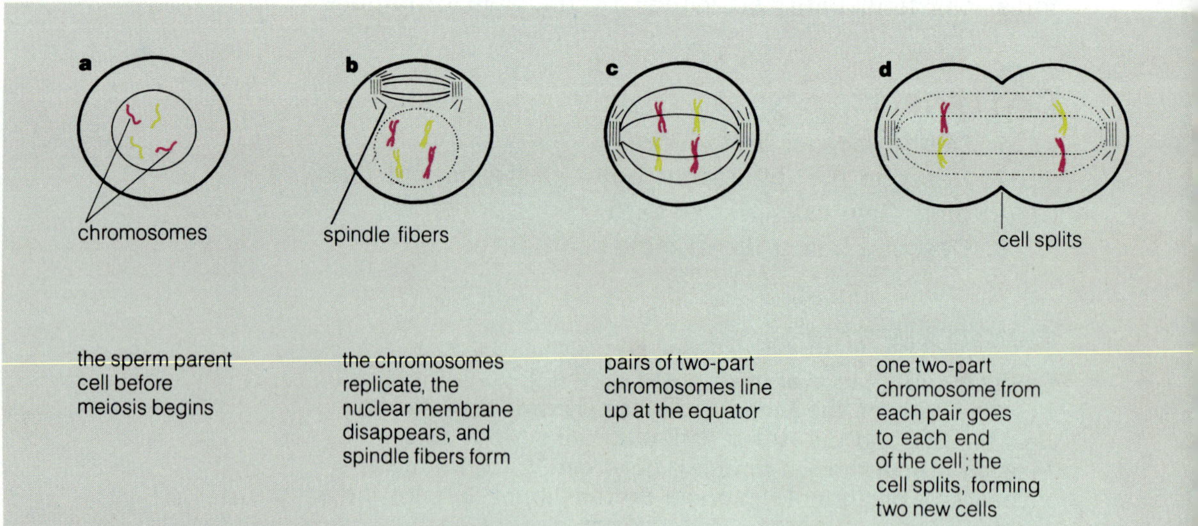

body cells. The sex cells have only 23 chromosomes. Because of meiosis, when two sex cells join, the offspring have the same number of chromosomes as their parents.

Meiosis occurs only in cells that produce the sex cells. In males, sperm parent cells undergo meiosis and form sperm cells. (See Figure 2-8.) During meiosis, each sperm parent cell divides twice, producing four sperm cells. Each sperm cell has half the number of chromosomes as the parent cell.

In females, the egg parent cell undergoes meiosis. The parent cell divides twice, creating four cells. Only one cell becomes an egg. The other cells are discarded by the body. The egg cell contains half the number of chromosomes as the parent cell.

fertilization

zygote (ZY goht)

During sexual reproduction, a sperm cell joins an egg cell by a process known as **fertilization**. Fertilization is complete when the egg nucleus and sperm nucleus fuse. The fertilized egg is called a **zygote**. Following the "blueprint" in its chromosomes, the zygote undergoes mitosis many times, forming a multi-celled organism. After numerous cell divisions, the zygote is called an *embryo* (EM bree oh). Cells in the embryo grow and develop into specialized organs.

You can study an example of sexual reproduction by observing the process in flowering plants. Flowers contain the reproductive organs of flowering plants. Locate each part on the flower shown in Figure 2-9.

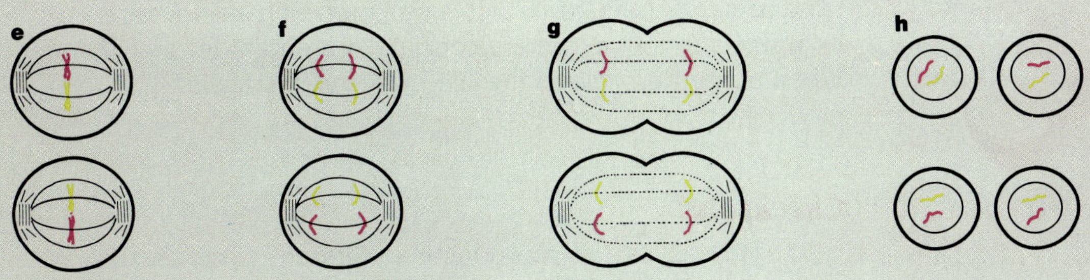

e: the chromosomes in the offspring cells line up and become attached to the spindle fibers

f: the two-part chromosomes separate and move to opposite ends of the cell

g: the two cells divide, forming four cells

h: at the end of meiosis there are four sperm cells, each containing half as many chromosomes as the sperm parent cell

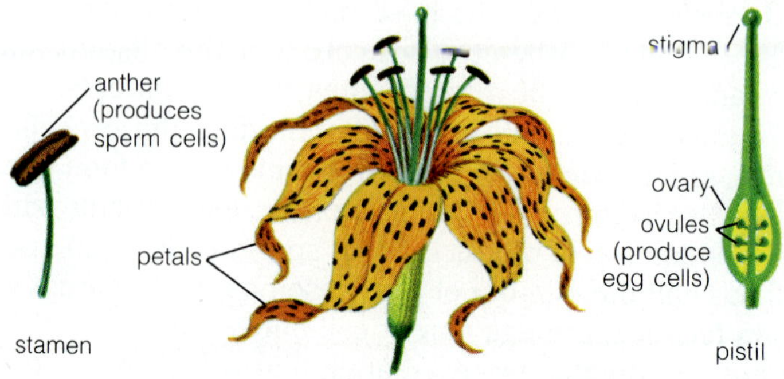

Figure 2-9 *The reproductive structures of a tiger lily.*

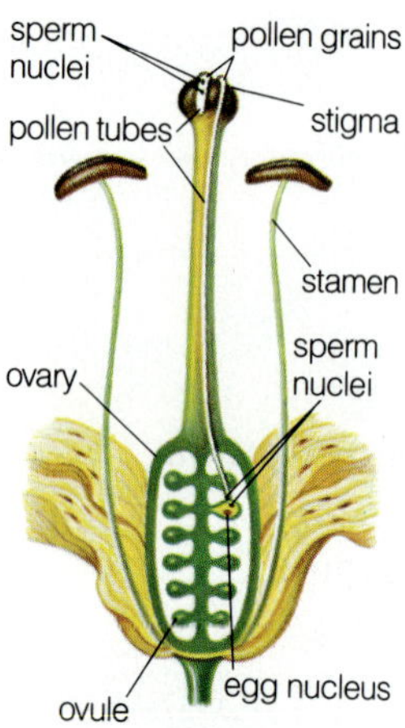

Figure 2-10 *Pollination makes fertilization of the egg possible.*

The male reproductive structure is called the *stamen* (STAY mun). *Pollen grains* are produced by meiosis in the top of the stamen, called the *anther*. The sperm cells are contained in pollen grains. The female reproductive structure is called the *pistil*. Egg cells are formed by meiosis in the parts of the pistil called the *ovules* (OH vyoolz).

pollination

In order for fertilization to occur, a pollen grain (containing sperm) must be transported from the stamen to the pistil. This process is known as **pollination**. Wind or insects carry pollen to the sticky top of the pistil, called the *stigma*. (See Figure 2-10.) Each grain grows a *pollen tube* from the top of the stigma to the *ovary*. The sperm nuclei in the pollen grain travel down the tube and into an ovule. Fertilization occurs when a sperm nucleus and egg nucleus fuse.

Tissues in the ovule form a food-storage layer and a protective covering, or *seed*, around the zygote. The ovary develops into a fruit around the seed. (See Figure 2-11.) A seed remains dormant until it is transported by wind, water, or animals to a place where conditions are right for growth. When a seed *germinates*, the life cycle begins again.

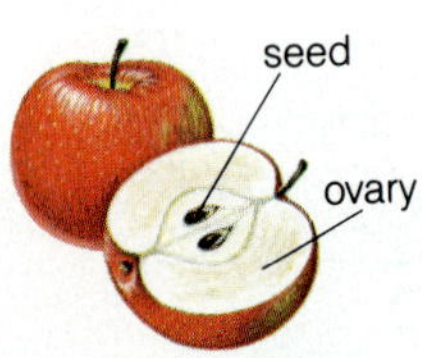

Figure 2-11 *A fruit is the ripened ovary of a plant. It may contain one or more seeds.*

Checkpoint

1. How is sexual reproduction different from asexual reproduction?
2. What are gametes? What are male gametes called? What is the name given to female gametes?
3. What is the function of meiosis? How is meiosis different from mitosis?
4. Explain the process of pollination and the steps leading to seed-formation in a flowering plant.

2-5

Dominant and Recessive Traits

Goal

To describe the inheritance of dominant and recessive traits.

In the mid-1800's an Austrian monk named Gregor Mendel examined inherited traits in pea plants. Mendel discovered that certain traits occurred in two forms. He studied the way certain trait pairs, such as yellow versus green pods and round versus wrinkled peas, were inherited. Figure 2-12 shows some of the traits Mendel observed. Mendel was able to understand and predict the pattern of these inherited traits.

In his experiments, Mendel *crossbred* pea plants by placing pollen from plants with green pods on the stigmas of plants with yellow pods. He crossbred plants with other traits as well. Mendel grew the seeds that resulted from each cross and observed the pattern of inherited traits. He discovered that when a pea plant with green pods was

Figure 2-12 *Gregor Mendel studied these inherited traits in pea plants. Shown here are five of the seven pairs of traits that Mendel studied.*

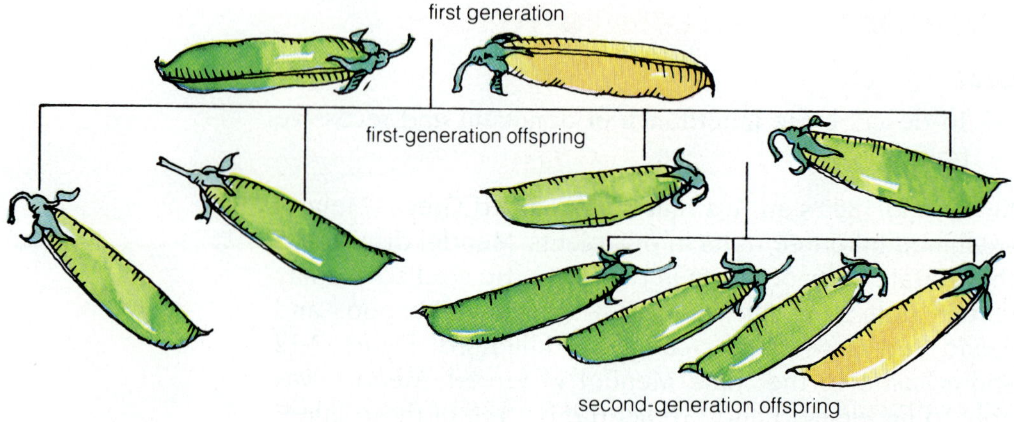

Figure 2-13 *A summary of the results of Mendel's studies. The "lost" trait reappeared in second-generation offspring.*

crossbred with a pea plant that had yellow pods, the first-generation offspring all had green pods. The yellow-pod trait seemed to "disappear." Mendel called the green-pod trait *dominant* because the trait "dominated" or overpowered the yellow-pod trait. He called the yellow-pod trait *recessive* because the trait receded or disappeared.

When first-generation offspring (all green-pod plants) were crossed with each other, the second-generation offspring included green-pod plants and yellow-pod plants. The "lost" yellow-pod trait had reappeared. Figure 2-13 summarizes the results of Mendel's experiments.

From his observations, Mendel concluded that something in pea plants controlled each trait. He called the controllers "factors." Mendel also concluded that some traits were controlled by a *pair* of factors.

Mendel was one of the first scientists to suggest the existence of factors that are responsible for the inheritance of traits. Though he described certain actions of the factors accurately, he never discovered that the factors were contained on the chromosomes.

Checkpoint

1. Give examples of dominant and recessive traits.
2. What happened to the recessive traits in the first-generation offspring Mendel experimented with?
3. Where in a cell are Mendel's "factors" found?

2-6

Genes and Chromosomes

Goals

1. To describe the function of genes.
2. To explain the difference between a pure trait and a hybrid trait.
3. To identify the sex chromosomes and explain the inheritance of sex-linked traits.

Scientists now know that the "factors" Mendel described are genes. A **gene** is a segment of a chromosome. There are hundreds of genes on each chromosome. All inherited traits are controlled by one or more genes. These traits include those that you yourself inherited from your parents and grandparents. A human gamete carries thousands of different genes. In sexually reproducing organisms, such as humans, half of the genes in a zygote come from each parent through the gametes.

gene
(jeen)

In body cells, almost all genes occur in pairs. Letters are used to represent the different forms of a gene in a gene pair. A **dominant gene** is a gene that is always expressed when it is present in an organism. A **recessive gene** is expressed only when the dominant form of the gene is absent. A dominant gene is represented by a capital letter. A recessive gene is represented by the small letter of the dominant gene. The ability to roll your tongue in a U-shape is a dominant trait, symbolized by the letter *T*. Small *t* is used to represent the non-tongue-rolling trait.

dominant gene
recessive gene

If both genes of a pair are identical, the trait that is expressed is called a **pure trait**. People with two genes for tongue-rolling, *TT*, are tongue-rollers. People with the two recessive genes, *tt*, are non-tongue-rollers. If the two genes are different, the trait is called a **hybrid trait**. People with the hybrid gene pair, *Tt*, are tongue-rollers because tongue-rolling is a dominant trait.

pure trait

hybrid trait

Mendel realized that some pea plants exhibited pure traits. The traits shown in Figure 2-12 were pure traits in the first generation Mendel studied. The plants with green pods had the gene pair *GG*. Yellow-pod plants had the gene pair *gg*. When Mendel crossed a pure green-pod plant with a pure yellow-pod plant, all the offspring had green pods.

a

	G	G
g	Gg	Gg
g	Gg	Gg

b

	G	g
G	GG	Gg
g	Gg	gg

Figure 2-14 *(a) All first-generation offspring are hybrid. (b) Yellow-pod color appears in plants with two recessive genes.*

Figure 2-14a shows the results of Mendel's first-generation crosses in a table called a *Punnett square*. The symbols for the genes of one parent are listed along the side of the square. The symbols for the genes from the other parent are written across the top of the square. Each possible combination of genes that could be inherited by the offspring is listed inside the Punnett square. You can count the number of times a combination appears in the Punnett square to predict the likelihood of that trait occurring in the offspring.

All the offspring of Mendel's first-generation crosses were *hybrid* for the pod-color trait. Since the green-color gene (*G*) is dominant, all the first-generation offspring had green pods.

Figure 2-14b summarizes the results of the crosses between first-generation offspring. Plants with yellow pods "reappeared" in second-generation offspring. Recessive traits, such as yellow-pod color, are expressed only when *both* genes in a pair are recessive.

sex chromosomes

In many plants and animals one pair of chromosomes contains genes that determine the sex of the organism. These chromosomes are called the **sex chromosomes**. Look back at the human chromosomes pictured in Figure 2-2. Pair 23 is the sex-chromosome pair.

In females, both sex chromosomes appear identical. Females have an XX combination. In males, the sex-chromosome pair contains two chromosomes that appear different. The larger one is an X chromosome and the smaller, hook-shaped one is a Y chromosome.

The sex of a new organism depends on the sex chromosome in the sperm cell at fertilization. Since the sex chromosomes of a female are identical, all eggs produced by a female contain an X chromosome. In males, meiosis results in the production of sperm cells containing either a single X chromosome or a single Y chromosome. The sex of the offspring is determined by whether an X- or Y-carrying sperm fertilizes the egg. (See Figure 2-15.) A sperm carrying an X chromosome will result in a female (XX) offspring when joined with an egg cell. A sperm carrying a Y chromosome will result in a male (XY) offspring when joined with an egg cell.

Some traits are controlled by genes that are found on the sex chromosomes. In humans, color vision is one exam-

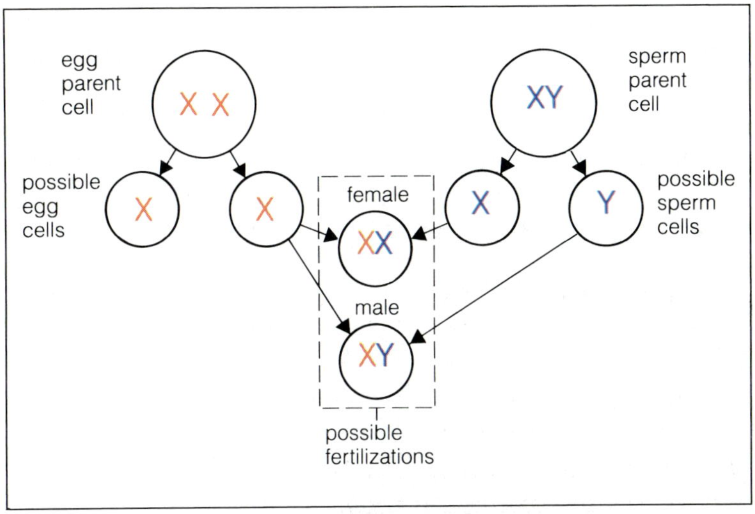

Figure 2-15 *The sex of a new organism depends on whether an X- or Y-carrying sperm fertilizes the egg.*

ple of a **sex-linked trait**. The gene for color vision is carried on the X chromosome only. Y chromosomes have no gene for color vision. Color vision in males is inherited from the mother.

sex-linked trait

Since the gene for colorblindness is recessive, a female with only one normal gene for color vision would still have normal color vision. The woman would be a *carrier* of colorblindness — she would not show the trait, but she could pass it on to her offspring. (See Figure 2-16a.) A male with the gene for colorblindness would be colorblind. (See Figure 2-16b.) Colorblind females are rare. About 20 times as many males as females are colorblind.

Hemophilia is a blood disease that occurs mostly in males, as a result of a recessive, sex-linked gene. People with hemophilia lack a necessary substance that helps blood to clot. A person with hemophilia bleeds heavily, even from a small cut or scratch.

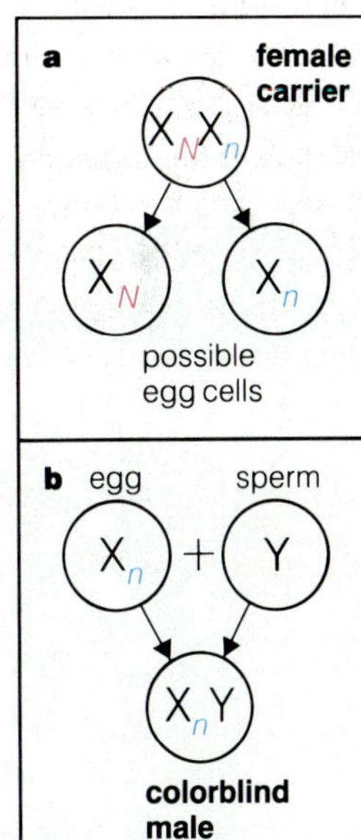

Figure 2-16 *(a) A carrier has normal color vision. (b) A colorblind male inherits the trait from his mother.*

Checkpoint

1. What is a gene?
2. What is the difference between a pure and a hybrid trait?
3. Which chromosome pair is different in males and females? How are the pairs different?
4. Give one example of a sex-linked trait.

2-7

DNA

Goals

1. To describe the structure of DNA.
2. To identify the role of DNA in inheritance.

DNA

Chromosomes contain a substance called **DNA**, which stands for *deoxyribonucleic acid*. DNA contains the information that controls all cell activity. Scientists came up with a model of the structure of DNA that helps explain how DNA functions. According to the model, a DNA molecule is made up of small repeating units joined together in a long, twisted, ladderlike structure. (See Figure 2-17.)

The two sides of the "ladder" making up a DNA molecule contain alternating sugar (deoxyribose) and phosphate units. Pairs of chemicals called *bases* are connected to the sides of the ladder, forming "rungs." There are four types of bases: adenine (AD n een), thymine (THY meen), guanine (GWAH neen), and cytosine (SY toh seen). Adenine pairs with thymine and guanine with cytosine. (See Figure 2-18.)

Figure 2-17 *This model shows the twisted, ladderlike structure of a DNA molecule.*

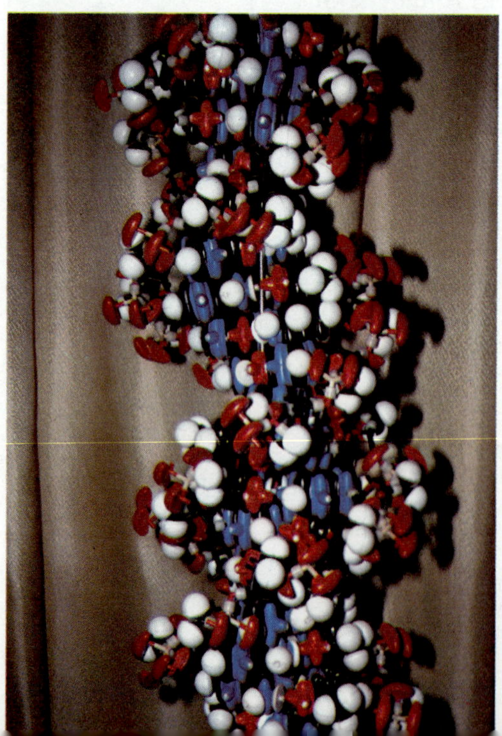

Figure 2-18 *Notice the repeating sugar-phosphate units and the paired bases.*

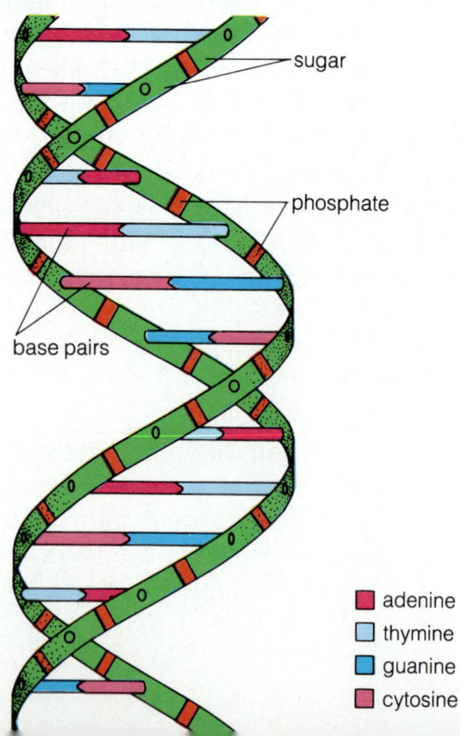

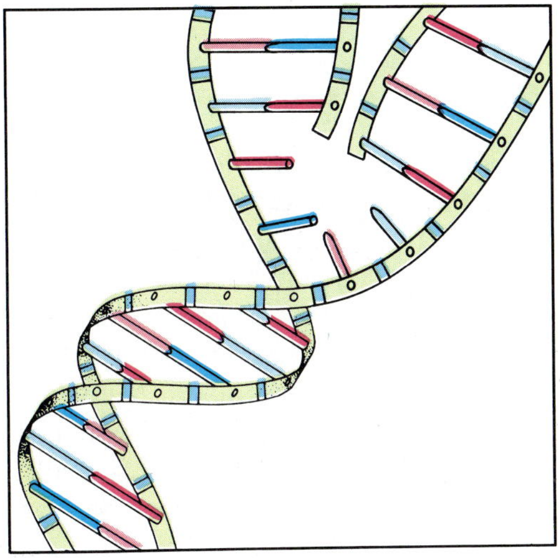

Figure 2-19 *A DNA molecule unwinds and a copy of each "missing" half is made.*

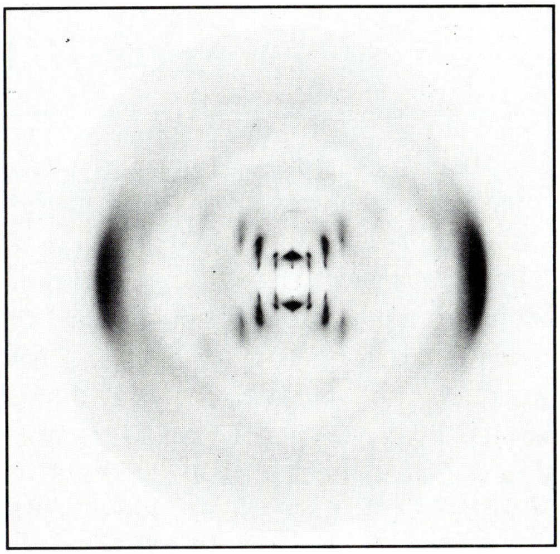

Figure 2-20 *X-ray studies of DNA helped scientists figure out its structure.*

A molecule of DNA contains millions of base pairs. The order in which the base pairs occur is responsible for the traits of an organism. Each gene is a portion of the long series of base-pair rungs in the DNA ladder. Genes control the production of proteins, which are responsible for the growth, development, and functioning of an organism. Every organism has a different arrangement of genes (base pairs). This makes each organism different from any other.

During mitosis, the strands of DNA reproduce, or replicate. In this way, each new cell contains all the necessary genetic information. In the replication process, the DNA molecule unwinds and separates into two separate parts, as shown in Figure 2-19. Each half of the DNA molecule acts as a *template*, or pattern, for the missing half. (See Figure 2-19.) Two copies of the original molecule are formed.

James Watson and Francis Crick first proposed the model of DNA in 1953. They described the *double helix* (HEE liks) after examining x-ray pictures of DNA like the one in Figure 2-20.

TRY THIS

Build a model of a DNA molecule out of objects you can find at home, such as colored toothpicks and styrofoam balls. Try to follow the illustration in Figure 2-18.

Checkpoint

1. Draw a diagram of a model of a DNA molecule.
2. How do DNA molecules determine inherited traits?
3. What happens to DNA during mitosis?

Genetic Engineering

Scientists have learned how to slice, splice, and rearrange DNA from different organisms. The DNA produced is called *recombinant* DNA because different parts have been "recombined," or rejoined, to form a new kind of DNA molecule. Recombinant DNA technology has made it possible to isolate specific human genes and attach them to bacterial DNA. Researchers have been able to get the bacterial cells to produce important substances.

One success that researchers have had is with the production of *insulin*. Insulin is normally produced in humans by the *pancreas*. People who cannot produce enough insulin on their own must receive injections of insulin every day. Using recombinant DNA techniques, researchers have been able to remove the insulin-producing gene from a sample of human DNA and join it to the DNA inside E. *coli* bacteria. During cell division, the insulin-producing gene is replicated, along with the bacterial DNA, and is passed on to future generations of bacteria. "Fooled" by the new DNA, the E. *coli* cells follow the "instructions" in the genes and produce human insulin. Until now, the only source of insulin was from an animal pancreas.

The photographs show a piece of "foreign" DNA being inserted into the DNA of a common type of bacteria. Bacteria are being used for most recombinant DNA research because they are easy to grow and they reproduce rapidly. Colonies of bacteria can be made to function like tiny "chemical factories," programmed to produce any one of a variety of important substances.

Researchers envision many ways in which recombinant DNA techniques can be used to solve medical problems. Bacteria have been developed that can produce human *interferon* (in tur FEER ahn), a substance that protects cells from attack by viruses. Another strain of bacteria has been "programmed" to break up oil and other petroleum products. These bacteria may someday be used to clean up large oil spills.

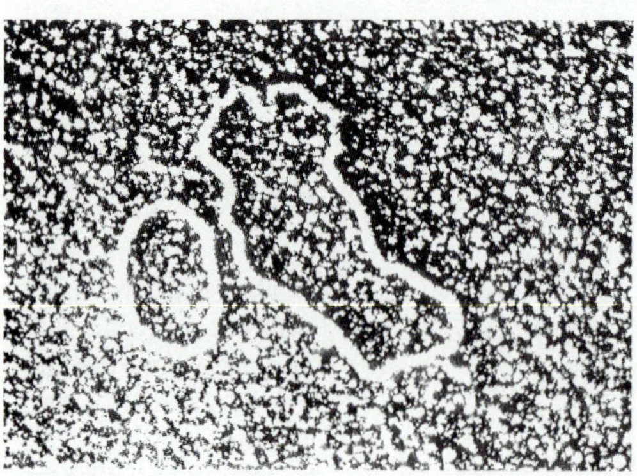

Figure 1a

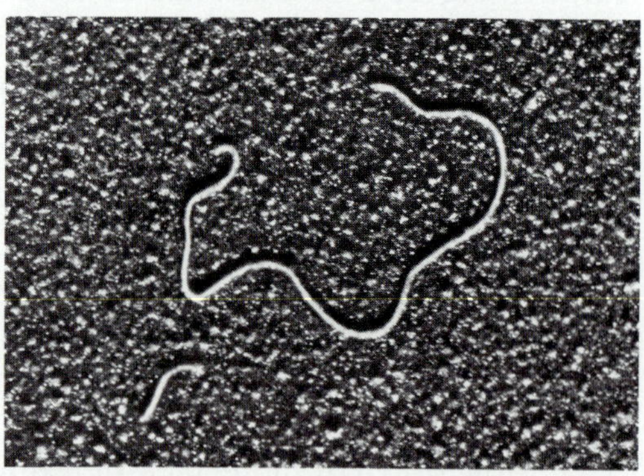

Figure 1b

2-8

Mutations

Goal

To identify what mutations are and how they occur.

mutation (myoo TAY shuhn)

A **mutation** is a change in a gene or chromosome. Mutations usually happen during the replication of DNA. A *gene mutation* usually involves a change in one base in a DNA molecule. *Albinism* is caused by a gene mutation. Albinos are missing the substance that gives color to the hair, skin, and eyes. (See Figure 2-21.)

The expression of a gene with a mutation may result in a trait that affects the survival of an organism. For example, a plant with a mutation may be more susceptible to diseases.

Occasionally, the expression of a mutation results in a trait that increases the survival ability of an organism. For example, mutations in some insects have made them less susceptible to chemical pesticides. Other mutations have resulted in bacteria that are resistant to certain *antibiotics*, or medicines. If an organism with a gene mutation survives and reproduces, it will pass the mutation on to its offspring.

Figure 2-21 *This albino deer is missing the substance that gives color to the hair, skin, and eyes.*

Chromosome mutations usually involve many genes. During mitosis or meiosis, parts of chromosomes may be broken off and "lost" or they may get reattached in the wrong place. Chromosome mutations are often fatal.

Some mutations occur spontaneously. In most organisms the mutation rate is very low. Certain substances have been found to increase the rate of mutation in cells. Radioactive materials, x rays, ultraviolet light, tars (in cigarette smoke), and some chemicals are known causes of mutations in laboratory animals.

Checkpoint

1. What is a mutation? What causes mutations to occur in organisms?
2. What is the difference between a gene mutation and a chromosome mutation?
3. How can a mutation affect the survival ability of an organism?

2-9

ENRICHMENT

Plant and Animal Breeding

Goal

To describe how people have developed breeds of plants and animals with desirable traits.

Agricultural scientists and farmers often try to improve the traits of crops or farm animals by a process called *breeding*. In the breeding process, a scientist crosses, or mates, two organisms in order to produce offspring with the desirable traits of both parents.

Selective breeding involves choosing, or selecting, individuals with the most desirable traits in a species and breeding them over several generations. By selectively breeding organisms over a period of years, the desirable traits of many organisms are combined in a few individuals. The white Leghorn chicken, a variety that lays many eggs, was produced by selective breeding. This method of breeding has a drawback. The continued breeding of offspring for many generations may result in the passing along (and maintenance) of some undesirable traits along with the favored ones.

Figure 2-22 *The beefalo is a cross between a buffalo and a cow. Someday people may be eating beefalo burgers.*

Hybridization, or *crossbreeding*, is the crossing of two different, but related, kinds of organisms to produce a new kind (hybrid) of organism. The desired traits of both parents are combined in the hybrid. For example, scientists have crossed cattle and buffalo to produce steer with the desirable traits of both. The resulting *beefalo*, seen in Figure 2-22, can live on ordinary grass, like the buffalo. Beefalo fatten faster and yield more meat than standard breeds of cattle.

Agricultural breeding is not a new technique. Figure 2-23 shows the improvements that have occurred in corn plants as a result of many years of breeding. Scientists believe that the modern corn plant was once similar to the teosinte (tee uh SIN tee) plant that grows wild in Southern Mexico. It is believed that Native Americans bred teosinte in fields with another type of corn, one with softer kernels and more kernels per ear. The two types of corn probably crossbred naturally, producing a better-tasting hybrid. As a result of continued breeding, the sweet corn eaten today contains approximately 1000 kernels per ear.

Figure 2-23 *The teosinte plant (left) is believed to be an ancestor of modern corn (right).*

Recent advances in plant breeding have resulted in the production of crops with greater nutritional value. Fruits and vegetables have been developed that can withstand mechanical harvesting with little damage.

The breeding process can often be quite slow. For example, farmers attempting to grow plants that produce more soybeans must wait until the crop is fully grown to see if the desired improvement has occurred.

Researchers in Canada are looking for ways to speed up the breeding process. Left on its own, a cone-bearing tree may take 20 to 30 years to produce cones. Using a method called *cone induction*, the scientists "fool" pine, fir, spruce, and hemlock trees into sprouting cones in three to five years. This method may someday be used to repopulate forests that have been cut down or lost to forest fires.

Checkpoint

1. What is meant by selective breeding?
2. Give an example of a plant or animal variety that has been improved as a result of breeding.
3. How was the beefalo developed?

Understanding The Chapter 2

Main Ideas

1. Inherited traits are passed on to offspring through the genes.
2. Mitosis results in the production of two identical cells that are copies of the parent cell.
3. Reproduction is the process by which organisms produce offspring.
4. Asexual reproduction, through fission, budding, regeneration, or spore formation, is carried out by a single parent.
5. Sexual reproduction is the creation of a new organism by joining a male sex cell with a female sex cell.
6. Meiosis is the cell division process that produces gametes with half the number of chromosomes as body cells.
7. A dominant gene is always expressed when it is present in an organism. A recessive gene is expressed only when the dominant gene is absent.
8. Sex chromosomes determine the sex of an offspring.
9. Chromosomes contain DNA.
10. DNA is a long molecule with a twisted, ladderlike structure. The order of base pairs in DNA determines an organism's traits.
11. A mutation is a change in a gene or chromosome.
12. The traits of crops or farm animals can be improved by selective breeding and hybridization.

Vocabulary Review

From the following list, choose the term that best completes each of the statements. Write your answers on a separate piece of paper.

asexual reproduction	pollination
budding	pure trait
DNA	recessive gene
dominant gene	regeneration
fertilization	replicate
fission	reproduction
gametes	sex chromosomes
gene	sex-linked trait
hybrid trait	sexual reproduction
meiosis	spores
mitosis	traits
mutation	zygote

1. A __?__ is created when a sperm nucleus and an egg nucleus unite.
2. A change in a gene or chromosome is called a __?__.
3. A segment of a chromosome that determines a specific trait is called a __?__.
4. __?__ is the process in which chromosomes are duplicated and then divided between two cells.
5. A pair of identical genes is expressed as a __?__.
6. __?__ is the general process by which a single parent produces offspring.
7. Another name for reproductive cells is __?__.

8. __?__ is the transfer of pollen from a stamen to a pistil.
9. __?__ results in sex cells with half the number of chromosomes as in body cells.
10. An amoeba reproduces by a type of asexual reproduction known as __?__.
11. A __?__ is always expressed when it is present in an organism.
12. The type of reproduction in which organisms produce offspring by joining two gametes is called __?__.
13. A trait that is expressed when the two genes in a pair are different is known as a __?__.

Chapter Review

Write your answers on a separate piece of paper.

Know The Facts

1. In a hybrid gene pair, such as W*w*, the W stands for the __?__ trait. (recessive, pure, dominant)
2. Budding and fission are examples of __?__ reproduction. (asexual, sexual, both sexual and asexual)
3. __?__ produces cells with half the number of chromosomes found in most body cells. (mitosis, fission, meiosis)
4. Information about all inherited traits is contained on __?__. (spindle fibers, genes, mutations)
5. The joining of an egg cell nucleus and a sperm cell nucleus is known as __?__. (fertilization, regeneration, mitosis)
6. The beefalo is an example of an animal developed by __?__. (selective breeding, crossbreeding, cross-pollination)
7. In flowering plants, sperm cells are contained in the __?__. (stigma, pollen, ovary)
8. The symbol *ss* stands for a __?__ trait. (pure dominant, hybrid recessive, pure recessive)
9. In humans the sex chromosomes are X and Y. The __?__ combination produces a female. (XX, YY, XY)
10. Sex-linked genes are responsible for __?__. (colorblindness, hemophilia, both)
11. The arrangement of __?__ on DNA molecules determines an organism's traits. (bases, sugars, phosphates)
12. After meiosis, human gametes normally contain __?__ chromosomes. (43, 23, 46)

Understand The Concepts

13. Draw and describe the stages of mitosis.
14. Identify each of the following: genes, chromosomes, DNA.
15. Draw a diagram of the DNA model.
16. Explain the differences between sexual and asexual reproduction. Give an example of an organism that reproduces each way.
17. How is the process of meiosis different from mitosis?
18. In the gene pair S*s*, which letter represents the dominant gene? Which letter stands for the recessive gene? Is this a hybrid or a pure-gene pair?

19. Under what conditions will a recessive trait be expressed?
20. Describe two methods agricultural scientists use for developing new varieties of plants and animals.
21. Draw and label the reproductive structures in a flower. Explain the function of each part.
22. Describe the process of pollination and fertilization.

Challenge Your Understanding

23. Use a Punnett square to show the possible outcomes of a cross between two flowers, each with the same gene pair for color, V*v*. How many pure combinations result? How many hybrid pairs are there? Suppose V represents the gene for the violet-red form of the flower and *v* represents the gene for the white flower form. How many combinations result in plants with violet-red flowers? Which are they? How many combinations result in white flowers?
24. How can a knowledge of reproduction and gene-inheritance help scientists to improve the traits of crops and farm animals?
25. Explain why individuals with the same parents have some traits that are alike and some traits that are different.
26. A sex-linked gene controls the clotting ability of blood. Explain how a couple with normal clotting factors can have a son with hemophilia.
27. Genetic counselors do research and counsel couples about their chances of having a child with a genetic disorder. What kind of background and training would genetic counselors need to have?

Projects

1. Do a report on the ways genetic engineering is being used to produce substances that are important to people.
2. Prepare a report on the new varieties of crops and livestock produced by selective breeding and crossbreeding. Discuss the advantages of some of the new varieties.
3. Invite a local public health official to speak to your class about inherited diseases. Find out about tests to detect carriers of genetic diseases.
4. Collect water from a pond or stream. Make slides of the water and observe them under a microscope. Look for examples of organisms undergoing fission, budding, or regeneration.
5. Obtain red- and white-eyed varieties of fruit flies (*Drosophila melanogaster*) from a biological supply house. Mate them over several generations. Keep a record of the number of offspring with each trait. Which trait do you think is dominant? Which is recessive? Explain, based on your observations.
6. Hold a plant exchange day in class. Learn how to produce new plants from parts of old plants.
7. Use magazines and newspapers to find out about substances that cause mutations. Prepare a bulletin-board display.

Chapter

3

The Human Body

This gymnast must exercise control over the muscles in her body.

3-1

Skeleton and Muscles

Goal

To describe what bones and muscles are like and how they work in the body.

As you turn a page in this book, the bones and muscles in your hand and arm work together. Your hand and arm bones are part of your body's framework, or *skeleton*. (See Figure 3-1.) A tall building contains a similar framework that supports the floors and walls. Your skeleton supports the muscles and organs of your body. Without a skeleton, humans would have a very different form.

The skeleton protects certain organs of the body. The skeleton serves, also, as a point of attachment for the muscles, making movement possible.

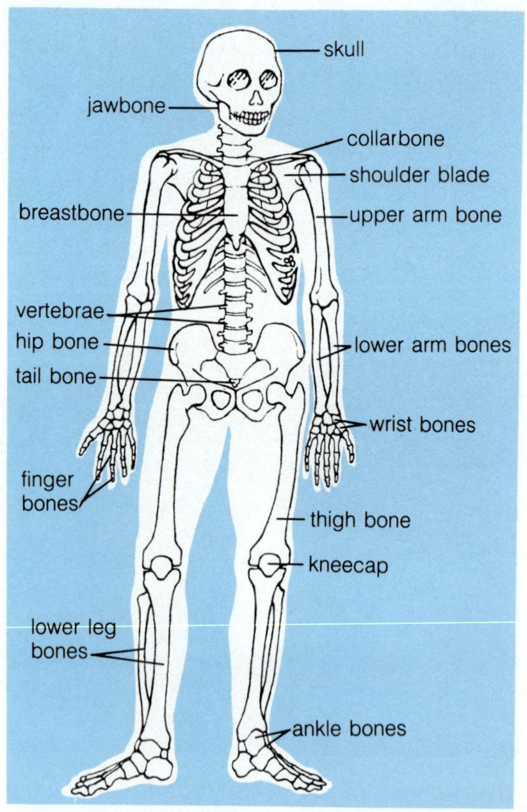

Figure 3-1 *Some bones protect soft body parts. For example, the lungs are protected by the ribs and the brain is protected by the skull. Other bones, such as the thigh bones, support the body and aid movement.*

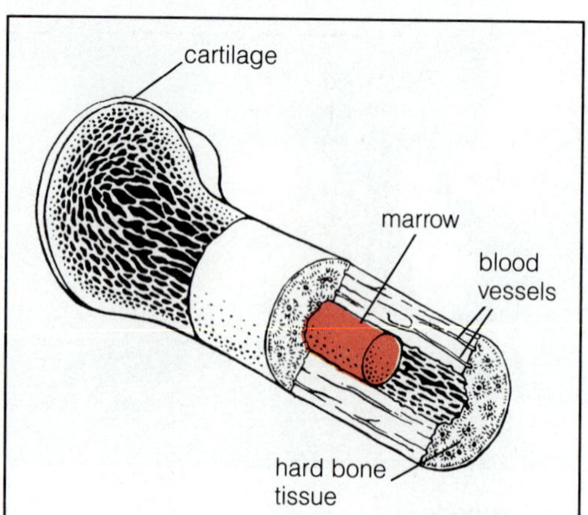

Figure 3-2 *Bones are light in weight in comparison to their strength because they are porous, not solid.*

Your skeleton contains 206 bones. Bone tissue is made up of living cells surrounded by a hard, calcium-rich covering. Blood vessels provide nourishment to the bone cells and remove wastes from them. (See Figure 3-2.) **Cartilage** makes up the stiff, yet flexible, part of the skeleton. It is found in the tip of the nose, in the outer ear, and at the ends of many bones. Cartilage cushions the ends of bones, and allows bones to move smoothly against one another.

cartilage

joint

ligaments

The point at which two bones meet is called a **joint**. There are many types of joints in the skeleton. Some are shown in Figure 3-3. Bones that have grown together form *immovable joints*. Figure 3-3 shows the fused bones that make up the skull. *Hinge joints*, such as the elbow and knee, operate like the hinge on a door. They permit back-and-forth movement only. *Ball-and-socket joints*, located in the shoulder and hip, permit movement in many directions. *Gliding joints* are found between the individual bones of the back, allowing the back to bend and twist.

Most bones are held together at movable joints by bands of tough, stretchy tissue called **ligaments**. (See Figure 3-4.) Ligaments enable bones to move, yet keep bones from separating from one another at the joint. The bone-ends in a movable joint are lubricated by a fluid that keeps them from grinding against each other.

A sprain occurs when ligaments are stretched or torn. A muscle strain usually involves soreness, but no damage to the tissue.

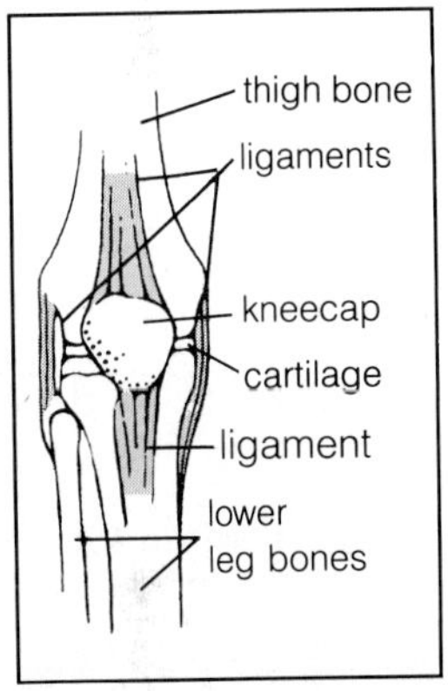

Figure 3-4 *Ligaments enable bones to move.*

Figure 3-3 *The different joints make a variety of movements possible. Joints in the skull begin to seal at age forty.*

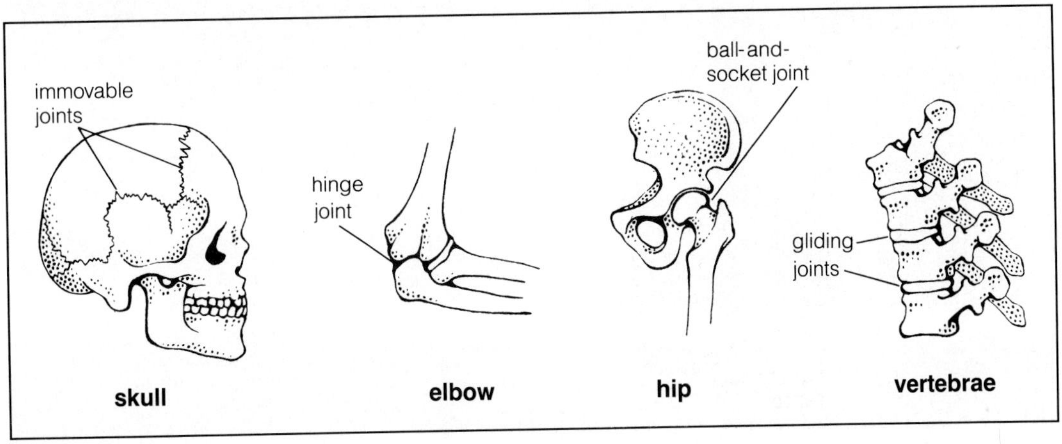

Your body contains over 600 muscles. All your body movements are made possible by the action of muscles. Some muscles move the skeleton. Others help to move blood and food through the body. All muscles function by *contracting*, becoming shorter and thicker.

voluntary muscles

involuntary muscles

Muscles are classified as voluntary or involuntary. **Voluntary muscles** function when you want them to. The muscles of the arm, leg, and tongue are voluntary muscles. **Involuntary muscles** are not under your conscious control. Involuntary muscles are found in the stomach, blood vessels, and pupil of the eye. Some muscle fibers in the body are always contracted, giving the body muscle *tone*. One group of fibers contracts, while another relaxes.

There are three types of muscle cells in the body: (1) striated (STRY ay tid), (2) smooth, and (3) cardiac. (See Figure 3-5.)

Voluntary muscles are made up of cells that are striped, or striated, in appearance. *Striated muscle* moves the skeleton and body parts, such as the lips, tongue, and face. Striated muscles are capable of two kinds of actions. They can move in relatively short and rapid contractions or they contract more slowly, producing a strong force.

Muscles that move bones work by pulling. When a muscle contracts it pulls a bone in one direction. Pairs of muscles make movement of bones in two directions possible. For example, the *biceps* and *triceps* are a pair of muscles located in the upper arm. (See Figure 3-6.) Each muscle is

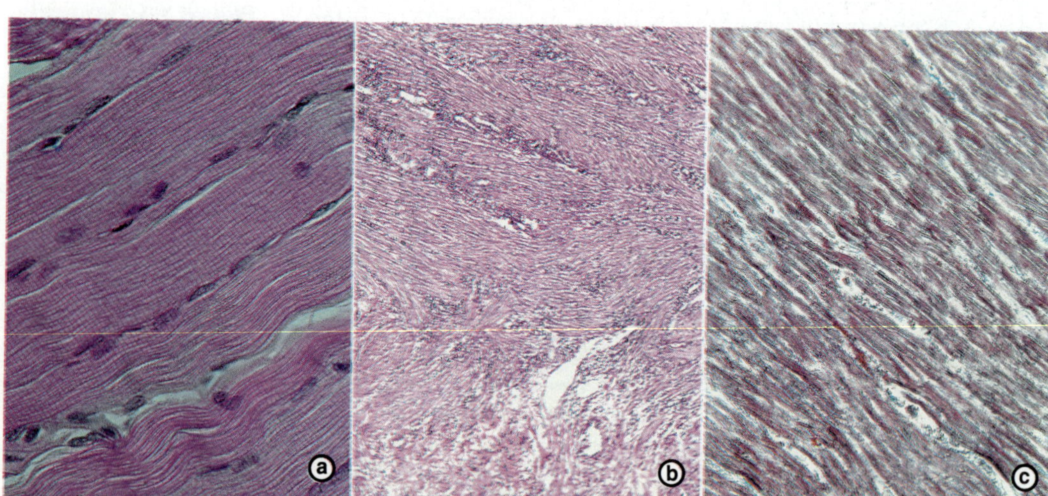

Figure 3-5 *Muscles: (a) Striated. (b) Smooth. (c) Cardiac.*

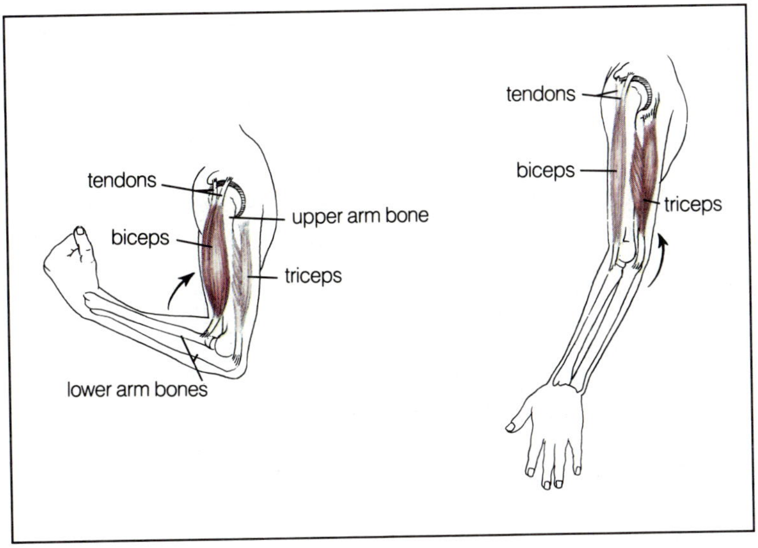

Figure 3-6 *Muscles that move bones work in pairs. When one muscle contracts, the other stretches. The muscles move the bone in opposite directions.*

attached to a bone by bands of tissue called **tendons**. When the biceps contracts, the forearm is pulled up and the arm bends. When the triceps contracts, the arm straightens.

tendons

Most involuntary muscles are made up of smooth muscle cells. *Smooth muscle* does not have bands or stripes. Smooth muscle is capable of slow contractions over a long period of time. For example, smooth muscle in the digestive system may contract continually for hours, churning and grinding the food inside.

A third type of muscle cell occurs in the heart only and is called *cardiac muscle*. Cardiac muscle has the strength and appearance of striated muscle and the endurance of smooth muscle. It contracts about 70 times a minute and 100,000 times each day. It slows down when you sleep but it does not stop as long as you are alive.

TRY THIS

Analyze your body as you walk slowly across the room. Which muscles are contracting and which joints are moving as you walk?

Checkpoint

1. Describe three functions of the human skeleton.
2. Name two bones that protect soft body parts.
3. What two types of tissue is the skeleton made of?
4. Name two kinds of joints. Tell where they occur in the skeleton.
5. Explain the differences between voluntary and involuntary muscles.
6. Describe how voluntary muscles work in pairs.

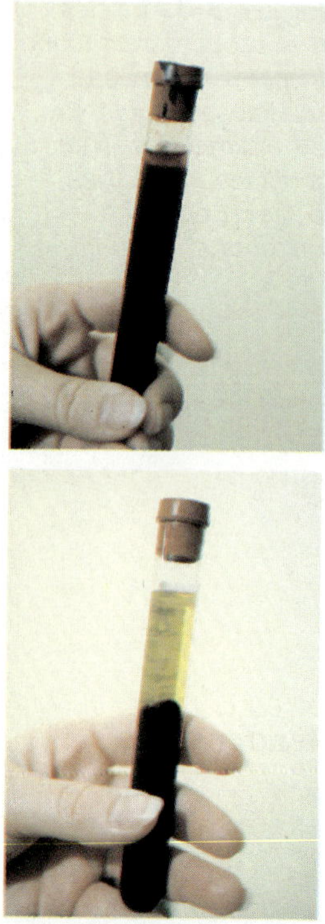

Figure 3-7 (a) Fresh, whole blood (top). (b) Separated blood (bottom). The pale-yellow liquid is plasma.

3-2

Circulatory System

Goal

To identify parts of the circulatory system and their functions.

blood

plasma
(PLAZ muh)

Blood is a liquid that carries various substances to and from body cells. For example, it supplies cells with food and oxygen and carries away wastes. The tubes through which blood flows are the *blood vessels*. The muscular pump that pushes, or circulates, blood throughout the body is the *heart*. The heart, blood, and blood vessels are the major parts of the *circulatory system*.

The circulatory system of an average adult contains five liters of blood. About half of the blood is a clear liquid called **plasma**, made up mostly of water. (See Figure 3-7.) Plasma carries food, wastes, and other materials.

In addition to plasma, blood contains three kinds of cells: (1) red blood cells, (2) white blood cells, and (3) platelets. The 25 trillion *red blood cells* give blood its characteristic red color. These cells carry oxygen from the lungs to all cells in the body. Red blood cells also carry some carbon dioxide from the cells to the lungs. The rest of the carbon dioxide is dissolved in the plasma. Red blood cells are made in *bone marrow*—the soft tissue in the center of many bones. A red blood cell lives for about 120 days. New red blood cells are being made all the time, replacing the ones that are "worn out." A highly magnified view of red blood cells can be seen in Figure 3-8.

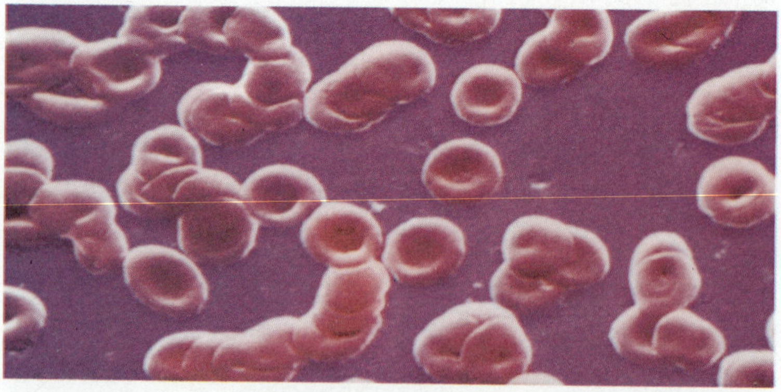

Figure 3-8 Red blood cells are doughnut-shaped cells without nuclei.

Your body contains approximately 35 billion *white blood cells*, one for every 600 to 800 red blood cells. There are several kinds of white blood cells. One kind uses pseudopods to surround and destroy foreign bacteria in the body. (See Figure 3-9.) Another kind produces *antibodies*, which combine with foreign bacteria, viruses, and harmful substances. White blood cells and antibodies enable the body to "fight" infection and disease.

Platelets are tiny pieces of cells produced in the bone marrow. They play an important role in the clotting of blood. When tissues are damaged, platelets release a substance that starts a chain of chemical reactions in the damaged area. A network of fibers forms that traps platelets and other blood cells, creating a clot. The clot stops the loss of blood from the damaged area. (See Figure 3-10.)

Your heart is a muscular pump about the size of your fist. It is located in the center of your chest with the bottom end tilted toward the left side of your body. Your heart pumps blood through a network of more than 100,000 km of blood vessels. This is more than twice the distance around Earth at the equator.

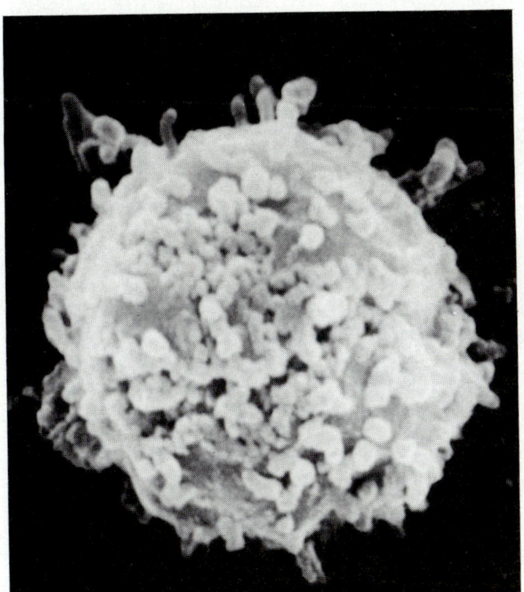

Figure 3-9 *Like the amoeba, a white blood cell has pseudopods to surround objects, such as foreign bacteria.*

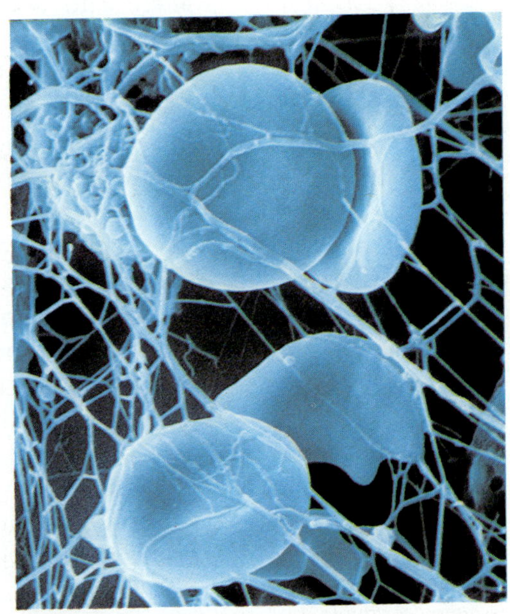

Figure 3-10 *Red blood cells are trapped in a fiber network. The clot that is created closes off the cut and stops it from bleeding.*

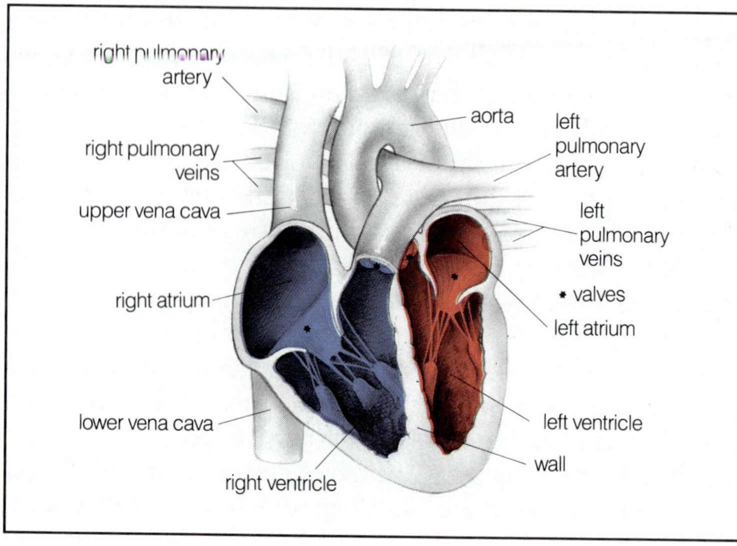

Figure 3-11 *Vena cavas bring oxygen-poor blood to the heart. Pulmonary arteries carry blood to the lungs. Pulmonary veins bring oxygen-rich blood from the lungs to the heart.*

Look at the diagram of a human heart in Figure 3-11. As you read about each part, try to locate it in the diagram. The heart has a thick wall separating the right side from the left. Each side of the heart is divided into two areas or chambers. Each upper chamber is called an *atrium* (A tree um, plural, *atria*). The two lower chambers are called *ventricles*. Each chamber acts as a pump, pushing the blood through the body. Valves between each atrium and ventricle allow blood to flow through the heart in one direction and keep blood from flowing backward in the heart. There are also valves between the ventricles and the blood vessels leading away from the heart. The closing of these two sets of valves creates a lub-dub sound, which can be heard with a stethoscope.

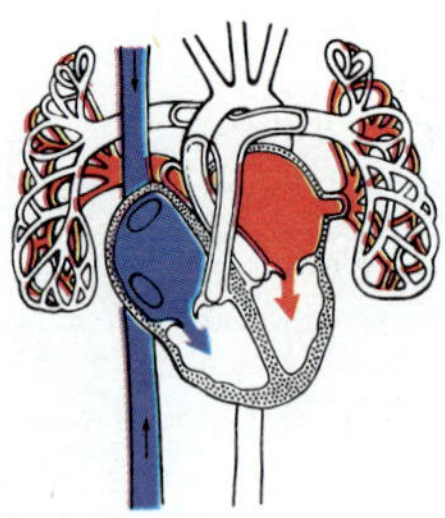

a

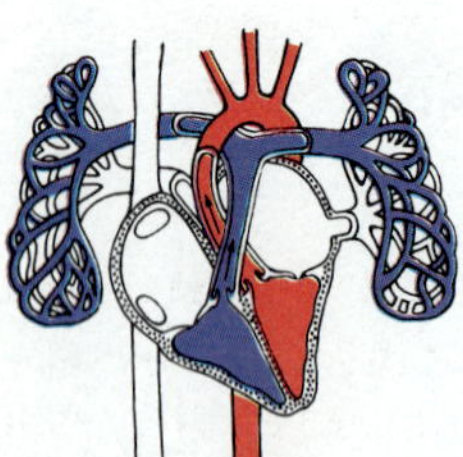

b

Figure 3-12 *The heart is called a double pump because it sends blood through two circulatory "loops."*

During a normal heartbeat, the right and left atria contract at the same time, pumping blood into the relaxed right and left ventricles. (See Figure 3-12a.) While the atria relax, the right and left ventricles contract. The right ventricle forces blood to the lungs. The left ventricle pumps blood to the rest of the body. (See Figure 3-12b.)

You can trace a drop of blood as it flows through the body. Refer to Figure 3-11 as you follow the path of blood. Begin your journey at the right atrium. Here, carbon dioxide-rich blood from the body enters the heart. A contraction pumps it into the right ventricle. The atrium relaxes as the ventricle contracts, pumping the blood to the lungs. Oxygen and carbon dioxide are exchanged between the blood and the air in the lungs. The oxygen-rich blood leaves the lungs and enters the left atrium. From here, a contraction pumps the blood into the left ventricle. Finally, a strong contraction pumps the blood out through the *aorta* (a OR tuh), a large, branching blood vessel. Blood vessels branching from the aorta carry blood to all parts of the body.

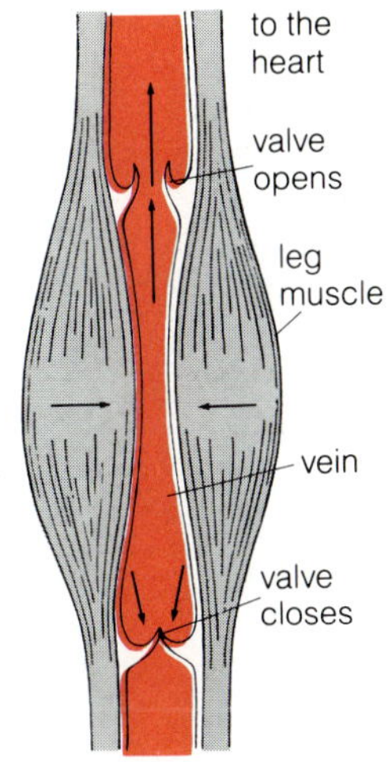

Figure 3-13 *When leg muscles contract, the vein is squeezed, pushing the blood along. The valves slam shut if the blood begins to flow backwards.*

Blood flows throughout the body in a series of vessels that are like one-way streets in a city. There are three types of blood vessels in the body: (1) arteries, (2) capillaries, and (3) veins (vaynz). Except for the pulmonary arteries, **arteries** carry oxygen-rich blood away from the heart toward the cells. Arteries have thick, muscular, elastic walls, which must withstand the pressure, or force, of the blood being pumped from the heart. The thick walls of an artery stretch and contract as the blood is pushed by the rhythmic contractions of the heart. The contraction of the elastic artery walls pushes the blood along in spurts.

arteries

Millions of narrow, thin-walled **capillaries** connect the arteries and veins. This network reaches every cell. Food, oxygen, and wastes are exchanged across the thin walls of the capillaries and the body cells.

capillaries

Except for the pulmonary veins, **veins** carry blood containing little oxygen and a lot of carbon dioxide back toward the heart. The walls of veins are thinner and less muscular than the walls of the arteries. By the time blood reaches the veins, the pressure is very low. Movement of blood in the veins is helped along by the contractions of nearby muscles. (See Figure 3-13.) Veins contain tiny valves that prevent blood from flowing backwards. These valves allow blood to move only toward the heart.

veins

Checkpoint

1. Name the major parts of the circulatory system.
2. Name the components of human blood. Describe the function of each.
3. Trace a drop of blood through the body beginning at the right side of the heart.
4. What is the function of the capillaries?
5. How do arteries and veins differ?

TRY THIS

Find your pulse at your wrist using your first two fingers but not your thumb. Count the number of times the artery wall stretches in one minute to find your heart rate.

3-3

Breathing

Goals

1. To describe the actions involved in breathing.
2. To explain the function of breathing.

You breathe twelve to sixteen times each minute, yet most of the time you are unaware of it. Though you can command yourself to breathe or to hold your breath, most of the time your brain controls your breathing automatically.

The function of breathing is to put oxygen into the blood and remove carbon dioxide from it. Oxygen is used by the cells to obtain energy from food through cellular respiration. Carbon dioxide and water are given off as a result.

trachea
(TRAY kee uh)

During breathing, air is warmed as it enters the nose or mouth. (See Figure 3-14.) Dust and other particles are trapped by tiny hairs and moist surfaces as the air passes through the nose and throat and enters the **trachea**. The trachea, or windpipe, is an air tube that is held open by rings of stiff cartilage. It divides into two *bronchi* (BRAHNG ky, singular, *bronchus*), each of which enters one of the two lungs. The bronchi branch into smaller and smaller tubes called *bronchioles*.

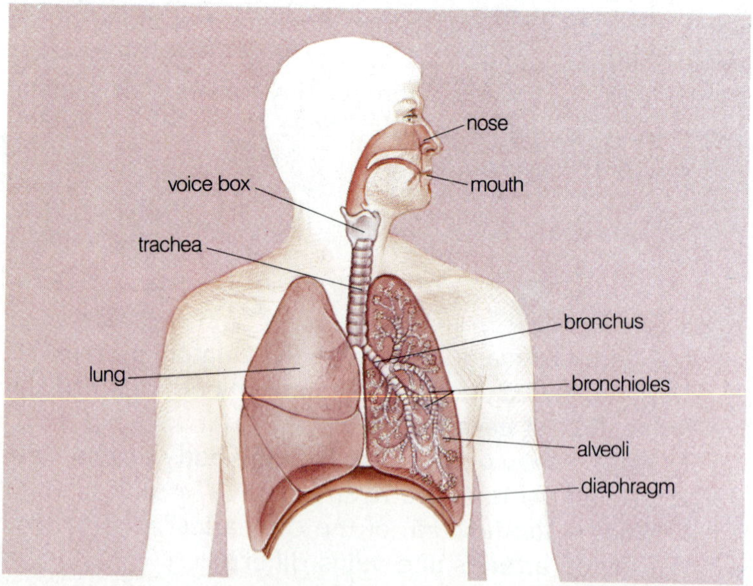

Figure 3-14 *This view shows a cut-away picture of one of the lungs. The lungs are part of the respiratory system.*

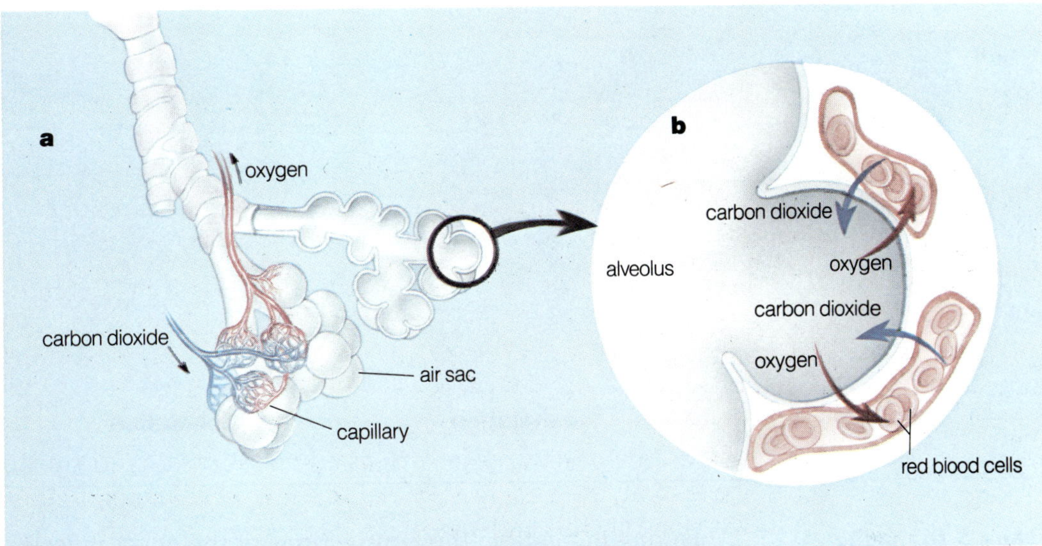

Figure 3-15 *There are over 300 million alveoli in each lung. Alveoli provide a large surface area for the exchange of gases. Oxygen and carbon dioxide are exchanged between the air in an alveolus and the blood in the surrounding capillaries.*

A cluster of moist, thin-walled air sacs called **alveoli** (singular, *alveolus*) lies at the end of each bronchiole. There are hundreds of millions of alveoli in each lung. The thin-walled alveoli are surrounded by tiny capillaries. (See Figure 3-15a.) The exchange of carbon dioxide and oxygen occurs between the thin walls of the alveoli and the capillaries. (See Figure 3-15b.)

alveoli
(al VEE uh ly)

In the exchange process, carbon dioxide brought from all parts of the body diffuses from the blood in the capillaries into the air in the alveoli. At the same time, oxygen diffuses from the air in the alveoli to the blood. The oxygen attaches to red blood cells in the capillaries. The red blood cells transport the oxygen to all the body cells.

Fresh air must be brought in to the alveoli constantly. Breathing makes this possible. The process of breathing consists of two actions, inhalation and exhalation. *Inhalation* is movement of air into the lungs. *Exhalation* is the movement of air out of the lungs. The lungs have no muscles and are unable to move air in and out on their own. The **diaphragm**, a strong flat muscle located below the lungs at the base of the ribs, and the chest muscles control inhalation and exhalation.

diaphragm
(DI uh fram)

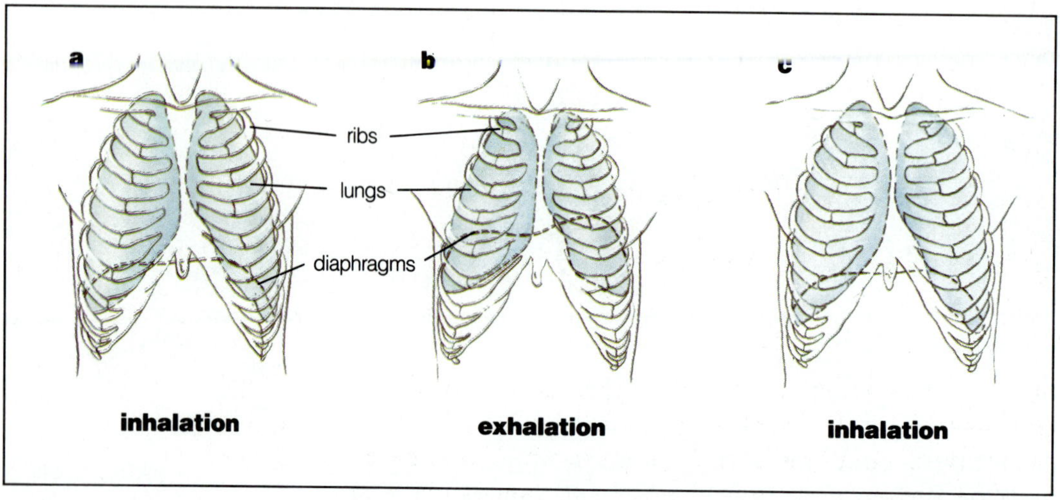

Figure 3-16 *(a) The ribs move up and the diaphragm moves down. (b) The diaphragm and chest muscles relax. Air is forced out of the lungs. (c) Air rushes in, filling the lungs.*

During inhalation, the contractions of the chest muscles pull the ribs up. At the same time, the diaphragm contracts and moves downward. (See Figure 3-16a.) These two movements increase the volume of the chest cavity and reduce air pressure in the lungs. Air pressure inside the lungs becomes lower than air pressure outside the body. Air from outside rushes in, causing the lungs to expand.

Exhalation occurs when the chest muscles relax and the ribs move down. The diaphragm relaxes at the same time and moves upward. These movements make the chest cavity smaller. Air is forced out and the lungs return to their original size. (See Figure 3-16b.)

Exhaled air is not pure carbon dioxide. It contains a mixture of gases, including oxygen, and can be used by the body if inhaled a second time. This is why mouth-to-mouth resuscitation can save the life of someone who has stopped breathing. During resuscitation, the rescuer exhales air into the mouth of the victim.

Checkpoint

1. What is the function of breathing?
2. Trace the pathway of oxygen as it enters your body from the air until it reaches your cells.
3. Describe the role of the diaphragm and chest muscles during inhalation and exhalation.
4. What is oxygen used for in the cells?

3-4

Digestion and Waste Removal

Goals

1. To explain how food is digested and absorbed by the body.
2. To describe the structure and function of the excretory system.

digestion

All living things need food. Food supplies materials for cell growth and repair, and energy for all life activities. Food cannot be used by cells in the form in which it is eaten. It must be broken down into simpler substances first.

The breaking down of food into smaller and simpler substances is called **digestion**. There are two types of digestion, mechanical and chemical. *Mechanical digestion* breaks food into smaller pieces. The teeth digest food mechanically, by tearing and grinding it up. *Chemical digestion* changes the small pieces of food into simpler substances. Chemical digestion occurs in the stomach (and other organs) when digestive juices break food down from large molecules to smaller ones. The digestive juices are aided by chemicals called *enzymes* that speed the breakdown of food.

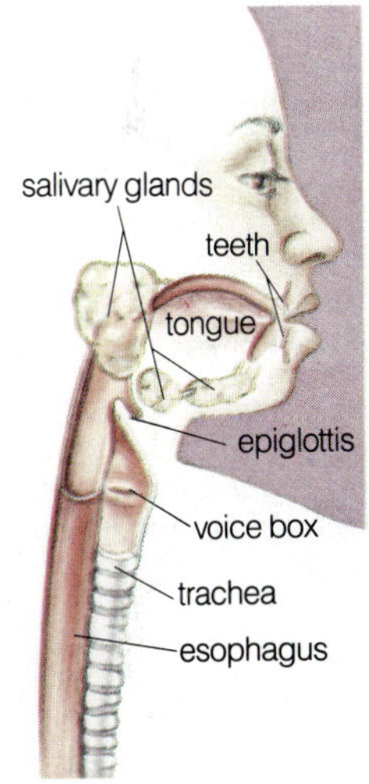

Figure 3-17 *Digestion begins in the mouth.*

Digestion and absorption of food take place in the *digestive system*. Food enters the digestive system through the mouth. (See Figure 3-17.) There, teeth start the digestive process by grinding food into small pieces. The tongue mixes the food with *saliva* (suh LY vuh), a watery liquid released into the mouth by the *salivary* (SAL uh vehr ee) *glands*. Saliva moistens and lubricates food, making it easier to swallow. An enzyme in saliva begins the digestion of starches.

esophagus (ih SAHF uh gus)

When food is swallowed, it goes down the throat and into the **esophagus**, a long tube that connects the throat and the stomach. A flap called the *epiglottis* (ep ih GLAHT is) drops down like a trap door, covering the trachea, to keep food from getting lodged there. (Within the trachea is the voice box.) The esophagus is about 25 cm long. Its walls are made of smooth muscle and can stretch to accommodate large pieces of food. The walls of the esophagus contract rhythmically, moving food to the stomach. (See Figure 3-18.)

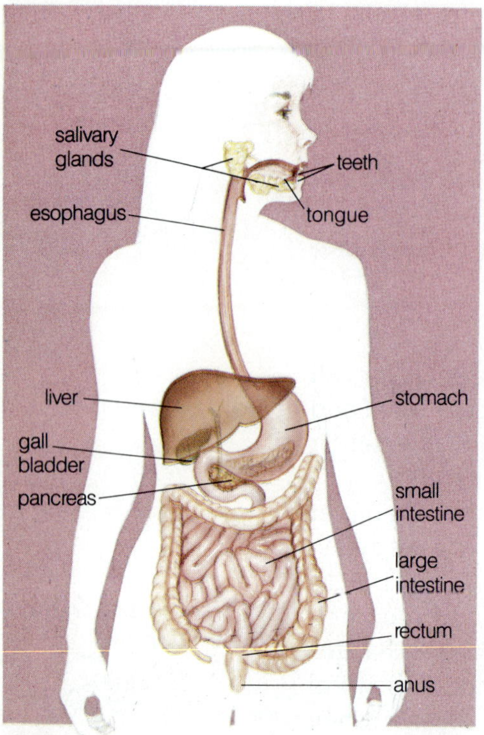

Figure 3-18 *The digestive system is made up of organs that form a long tube. The food tube would be about 9 m long if it were stretched out.*

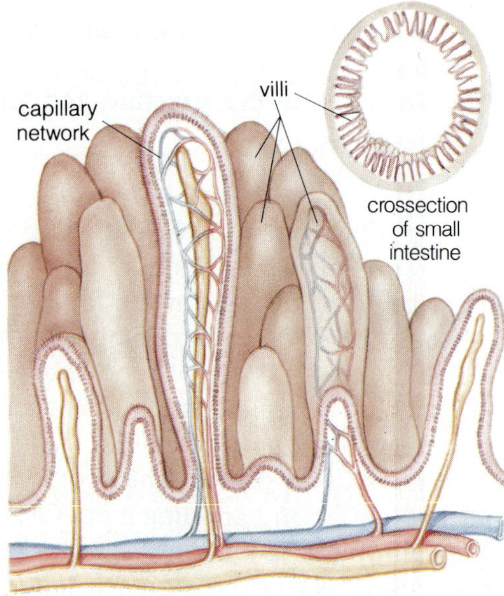

Figure 3-19 *The inner surface of the small intestine is folded into villi. Many more cells are exposed to digested food than if it were a smooth tube.*

The walls of the stomach contain smooth muscle. Muscle contractions churn and grind up the food, mixing it with *gastric juice,* given off by cells in the stomach lining. Gastric juice contains hydrochloric acid and an enzyme that digests protein. As the digestive juices mix with the food, it begins to look like a thick milkshake. This partially digested food is squeezed out of the stomach in spurts by muscular contractions and moved into the small intestine.

A thick mucous lining in the stomach protects it from being digested. If the stomach releases too much hydrochloric acid, ulcers can result. Ulcers are areas where the stomach lining is "eaten away" and sore.

Most digestion takes place in the small intestine. Juices from the pancreas (PANG kree us) and liver flow into the small intestine through tiny tubes. One of the juices from the liver is stored in a baglike organ called the *gall bladder.*

TRY THIS

Enzymes in your saliva digest large starch molecules into smaller sugar molecules. Place a piece of unsalted cracker in your mouth and leave it there for three minutes. How does the taste change?

The gall bladder releases *bile* which helps break up fats in the small intestine. All the juices combine with enzymes given off in the small intestine and complete the digestion of food.

The walls of the small intestine are lined with fingerlike projections called **villi**. (See Figure 3-19.) The villi contain networks of capillaries. Digested food molecules diffuse into the capillaries at the villi. The food is carried by the blood to all body cells.

villi
(VIL eye)

Not all the food you eat can be digested. Undigested food enters the two-meter long large intestine as a watery mass. Most of the water is absorbed by the large intestine and reused by the body cells. The undigested food collects at the end of the large intestine before being passed out through the rectum and the anus (AY nus).

Respiration of food produces waste products. If these wastes build up, they could poison the body. Cellular waste products are removed from the body by the *excretory system*.

Urea (yu REE uh) is a waste product formed in the liver. Blood carries urea from the liver to the **kidneys**, two bean-shaped organs located on each side of the spine at the lower back. (See Figure 3-20.)

kidneys

The kidneys contain millions of tiny tubes surrounded by capillaries. Urea and water pass out of the blood and into these tubes, forming **urine**. Urine passes through the *ureters* (yu REE turz) and into the *urinary bladder*. The bladder, which has walls of smooth muscle, stores urine. When it is full, it contracts and urine is excreted through the *urethra*.

urine
(YUR in)

Checkpoint

1. Define digestion. Explain the differences between mechanical and chemical digestion.
2. What are enzymes? Name two places in the body where they are released.
3. Trace a piece of food through the digestive system, beginning at the mouth.
4. Where does diffusion of food molecules into the blood take place?
5. What is the function of the excretory system?
6. List the parts of the excretory system and their function.

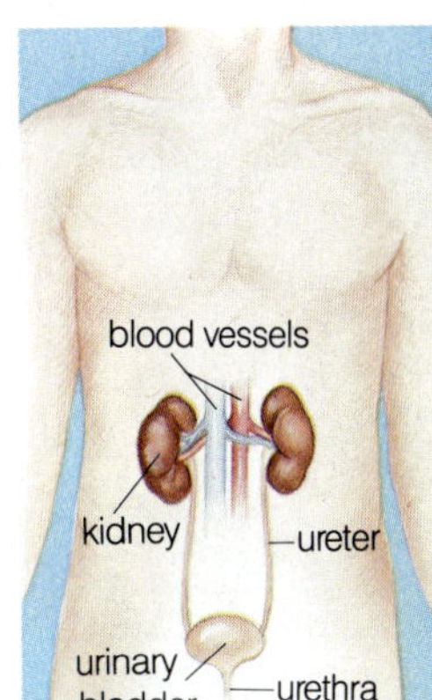

Figure 3-20 *The excretory system.*

3-5

Nervous System

Goals

1. To identify the parts of the nervous system.
2. To explain the function of the nervous system.

The *nervous system* controls and coordinates all the activities of the body. The brain and spinal cord make up the *central* nervous system. Nerves that branch from the central nervous system to the rest of the body make up the *peripheral* (puh RIF ur ul) nervous system. (See Figure 3-21.) The nervous system is similar to a telephone system. Messages are received and then responded to. Any "message" that produces a response is called a **stimulus** (plural, *stimuli*).

stimulus (STIM yuh lus)

neurons (NUR ahnz)

impulse

Neurons, or nerve cells, respond to stimuli from inside or outside the body. There are different kinds of neurons in the body, but all have some features in common. (See Figure 3-22.) A message, or **impulse**, enters a neuron through its branchlike parts called *dendrites* (DEN drytz). From the dendrites, the impulse passes through the *cell body* and along the *axon* (AK sahn). Impulses move in only one direc-

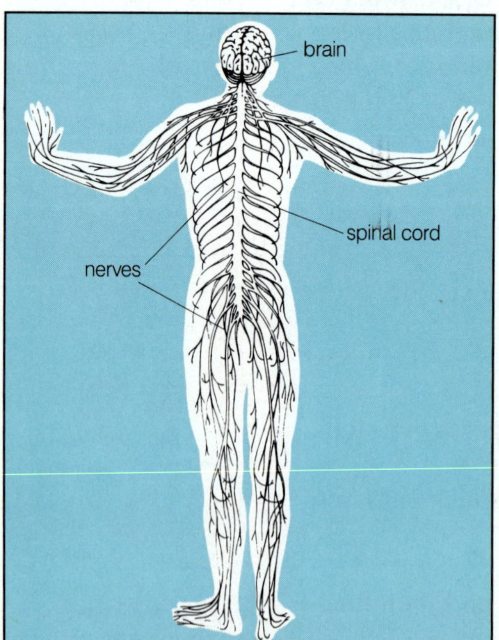

Figure 3-21 *Nerves go from the spinal cord to almost every part of the body and back again.*

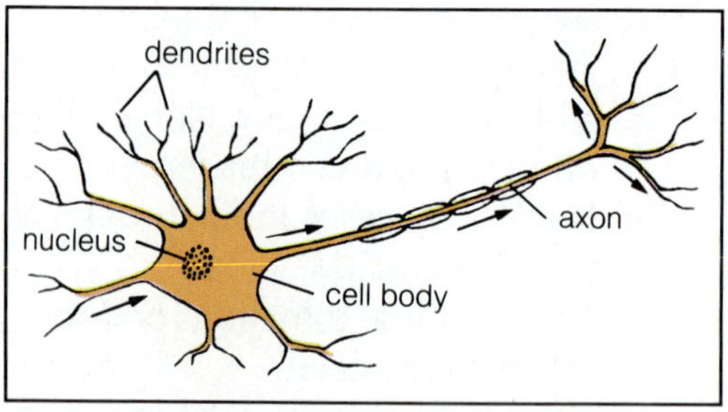

Figure 3-22 *Impulses move along a neuron from the dendrites, through the cell body, and out along the axon.*

tion through a neuron. Some impulses move at speeds of 130 meters per second. That is like traveling across the length of a football field in one second!

Neurons are not attached to one another. There is a gap where the ends of the axon of one neuron meet the ends of the dendrites of another neuron. This gap is called a **synapse** (See Figure 3-23.) Impulses are transferred from one neuron to the next across the synapse by chemicals. When the chemicals produced in the axon ends of one neuron reach the dendrites of the next neuron, the transfer is complete. The chemicals carrying the impulse across the synapse break apart quickly, making way for chemicals carrying next impulse. Impulses move in a fraction of a second.

synapse
(SIN aps)

The nervous system is made up of three types of neurons: (1) sensory, (2) association, and (3) motor. *Sensory neurons* carry impulses from *sense organs* to the spinal cord and brain. Sense organs include the eyes, ears, nose, tastebuds on the tongue, and the skin. Sensory neurons carry impulses from internal organs, too. *Association neurons* are "connectors." They receive impulses from sensory neurons or other association neurons and send the impulses to motor neurons. Most *motor neurons* transmit impulses to muscles from the brain and spinal cord.

Most responses are controlled by the brain. Three regions of the brain are shown in Figure 3-24. The **cerebrum** is

cerebrum
(suh REE brum)

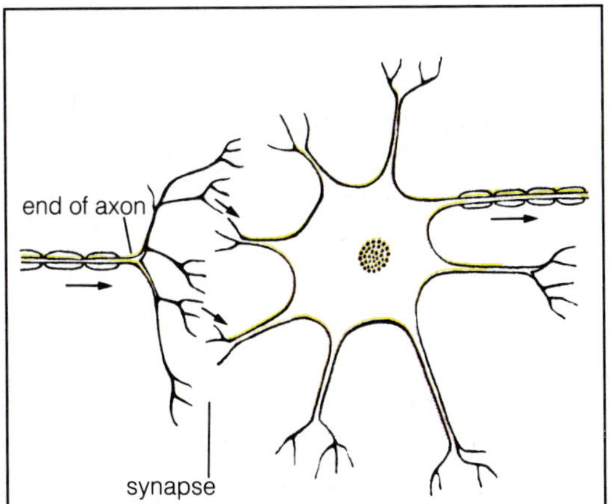

Figure 3-23 *A small electric charge can be detected during impulse transmission.*

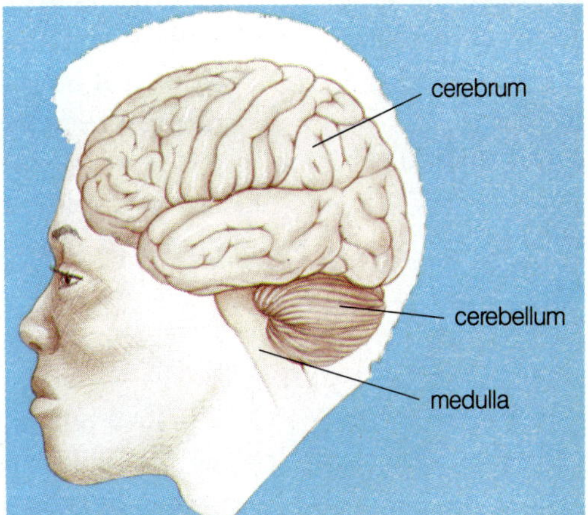

Figure 3-24 *Three main parts of the brain. Each part controls certain functions.*

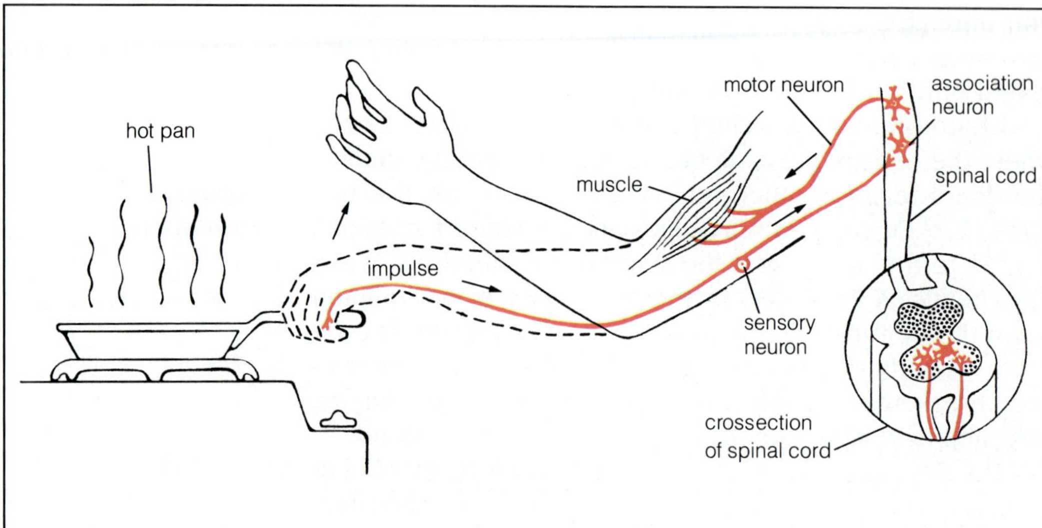

Figure 3-25 *Many reflexes, such as this one, protect people from harm. Time is saved by reacting automatically rather than thinking about what to do.*

the largest part of the brain. It controls the senses, memory, learning, thought, speech, and voluntary movement. The **cerebellum** is just below the cerebrum. It coordinates nerve impulses to and from the cerebrum. Muscle coordination and balance are controlled by the cerebellum, too. The **medulla**, located at the base of the brain, controls involuntary activities such as heartbeat, breathing, and the contraction of muscles in the digestive system.

cerebellum (sehr uh BEL um)

medulla (muh DUHL uh)

reflexes

All responses do not originate in the brain. Some responses, called **reflexes**, are controlled by the spinal cord. A reflex is an automatic response to a stimulus. Figure 3-25 shows what happens during a reflex response. The person does not have to stop and think about what to do. A sensory neuron receives the impulse. It transmits the impulse to an association neuron in the spinal cord. The impulse is transferred rapidly to the motor neuron controlling the arm muscles and a response occurs.

Checkpoint

1. What is the function of the nervous system?
2. What are the 3 types of neurons that make up the nervous system? Describe the function of each.
3. Describe how an impulse is transmitted from one neuron to another.
4. What is a reflex?

Windows on the Brain

The human brain has always been a difficult part of the body to study. It is an organ that cannot be touched or looked at without dangerous surgery. But in recent years, scientists have developed several safer, less painful ways to "look inside" the skull.

One new method is a form of x-ray photography called a *computerized axial tomography scan*, or CAT scan. During a CAT-scan examination, a fan-shaped beam of x rays is passed through the organ being studied. The x rays are detected by a scanner as they emerge at different angles from the body. A computer analyzes the x-ray pattern and "translates" it to produce a detailed cross-section map of the organ. Figure 1 shows a CAT-scan picture of a human brain.

Another tool that doctors use, the *electroencephalogram*, or EEG, takes advantage of the fact that neurons in the brain give off electrical energy in certain patterns. Doctors paste electrodes on the patient's scalp that detect the tiny amount of electrical activity given off. The measurements are recorded on strips of paper like the ones shown in Figure 2. A trained physician can recognize patterns such as those caused by epilepsy.

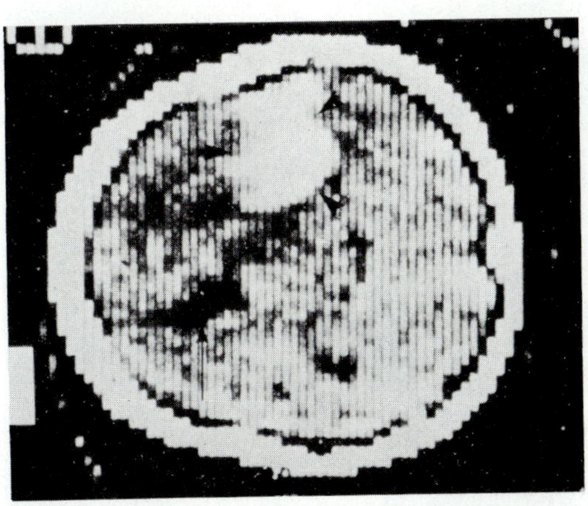

Figure 1

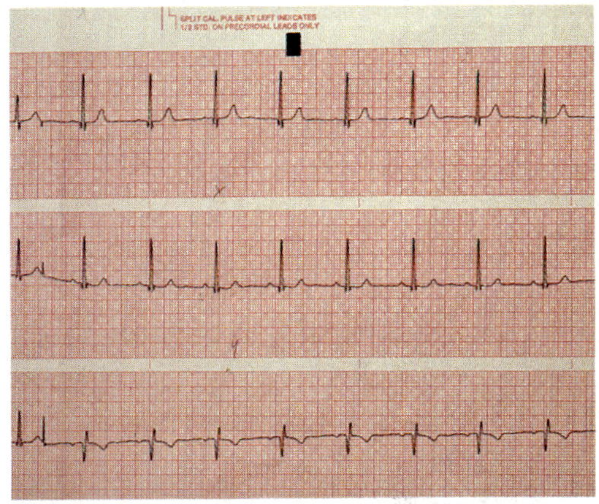

Figure 2

The newest test for problems inside the brain is called the *positron-emission tomography scan*, or PET scan. Like the CAT scan, it maps certain areas of the brain. But like the EEG, it can measure the level of activity occurring within the brain.

For the PET-scan test, a patient either inhales or is injected with a radioactive substance used by the brain. As the brain cells use the substances, high-energy gamma rays are given off, which are detected by scanners. A computer "translates" the information from the scanners into a picture of brain-cell activity.

3-6

Endocrine System

Goal

To describe the function of different hormones.

glands

hormones

Your body has two control and coordination systems. One is the nervous system. The other is the *endocrine* (EN duh krin) *system*. The endocrine system is made up of **glands**, structures that produce chemicals called **hormones**. (See Figure 3-26.) Hormones are released directly into the blood by the glands and travel to different body parts where they have specific effects. Hormones act as chemical "messengers." Each hormone causes changes in specific areas of the body. Table 3-1 describes some of the body's hormones, where they are made, and their function. For example, *thyroxin* is made by the thyroid gland. It controls metabolism. *Metabolism* is the process by which energy is produced and used in the body.

Figure 3-26 *The hypothalamus is part of the brain. It releases substances that turn endocrine glands "on" and "off."*

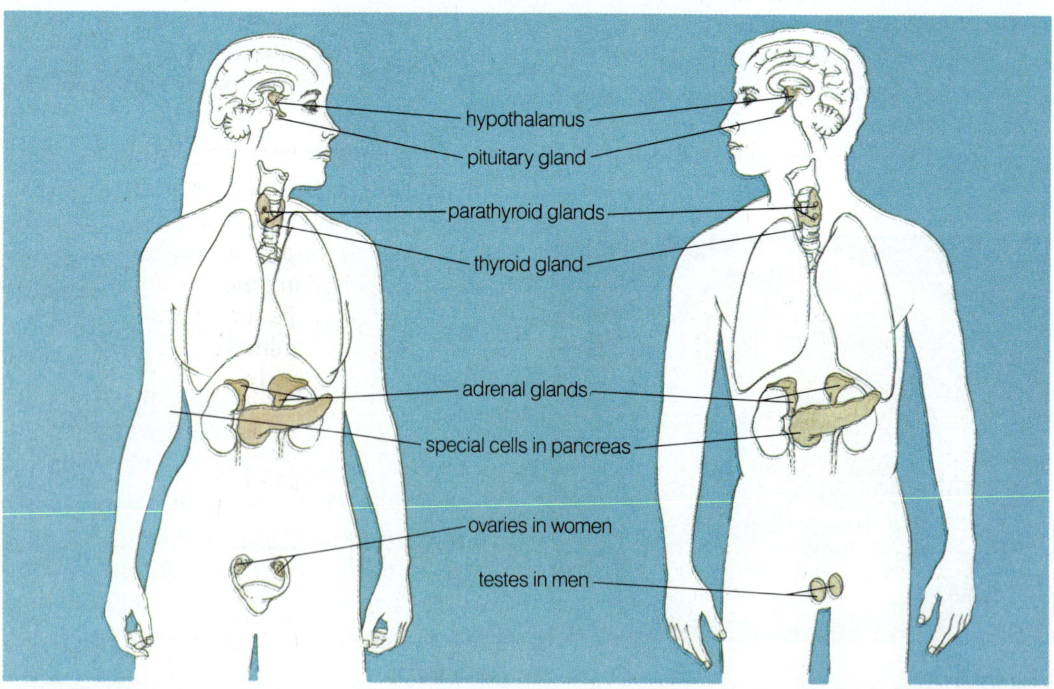

Table 3-1: Some Glands and Their Hormones

Glands	Hormones	Functions
pituitary	growth hormone	controls growth of bones too little: dwarfism too much: giantism
	other pituitary hormones	control release of hormones from other glands; regulate water balance in the body; help regulate blood pressure
thyroid	thyroxin	controls release of energy from food too little: sluggishness, overweight too much: nervousness, overactivity
parathyroid	parathyroid hormone	controls the balance of calcium in the body
pancreas	insulin, glucagon	regulates blood-sugar level and use of sugar by cells too little insulin: diabetes (sugar cannot be used properly by cells)
adrenals	adrenalin	maintains "fight or flight" response by increasing blood pressure and heartbeat rate and making more food available to cells for energy
	corticoids	regulate minerals and water level in body
ovaries (OH vuh reez)	estrogen (ES truh jun)	controls development of female sex characteristics
testes (TES teez)	testosterone (tes TAHS tuh rohn)	controls development of male sex characteristics

The endocrine system does not act as swiftly as the nervous system. However, hormones can be released continuously, in tiny amounts, causing long-term effects on the body. The production of too little or too much of a hormone can cause serious problems. For example, the production of too little growth hormone in children results in a condition called *dwarfism*. The production of too much growth hormone results in *giantism*.

The endocrine and nervous systems work together in coordinating body activities. The pituitary (pih TOO uh tehr ee) gland, located in the brain, is one link between the nervous and endocrine systems. It is affected by the release of substances from the part of the brain called the *hypothalamus* (hy poh THAL uh mus). The pituitary gland controls the release of hormones from other endocrine glands and is often called the "master gland."

If an angry dog chased after you, your nervous system would respond by sending impulses to your heart and blood vessels. Your heartbeat would speed up and blood flow to the striated muscles would increase. At the same time, nerves would send impulses to the adrenal glands to release *adrenalin*. Adrenalin speeds up the heartbeat and increases the flow of blood to the brain and muscles also. The changes caused by the nerves and the adrenalin would enable you to run faster. This reaction by the body is called the "fight or flight" response. It is another example of how the endocrine and nervous systems work together. The rapidly acting nervous system responds first. The slower-acting endocrine system strengthens the response of the nervous system and maintains it for as long as necessary for you to escape the dog.

Checkpoint

1. Why is the pituitary gland called the "master gland"?
2. What is a hormone?
3. Which hormone regulates growth? Where is it produced?
4. Give one example of how the nervous system and endocrine system work together to coordinate body activities.

3-7

Alcohol, Tobacco, and Other Drugs

ENRICHMENT

Goal

To identify the effects of alcohol, tobacco, and other drugs on the body.

Sometimes people use substances to alter their mood in some way. These substances are called *drugs*. Some people use drugs to relax, or to become more alert. Others use drugs to increase their self-confidence or block out unpleasant feelings. Substances that can be useful as medicines are often called drugs, also. They are known as *medicinal* drugs. Aspirin, penicillin, and antihistamines are some common medicinal drugs.

Drugs may be legal or illegal. Legal drugs have been researched by scientists to determine safe dosages and side effects. Legal drugs are obtained by a physician's prescription or by purchasing them "over the counter." No drug should be used without careful thought. People should find out what the side effects are and whether the drug might affect their ability to drive, operate machinery, or concentrate.

Drugs can become habit-forming. If a person becomes physically dependent on a drug, that is called an *addiction*. He or she will experience discomfort if the drug is removed. Sometimes a person with an addiction develops a tolerance for a drug. The person has to increase the amount of the drug each time to get the same effect.

One of the most commonly used drugs is *alcohol*. It is found in beer, wine, and "hard liquor." Even in small quantities alcohol depresses, or slows down, the central nervous system, decreasing the number and intensity of nerve impulses. Alcohol slows reaction time and decreases muscular coordination.

Drinking too much alcohol can damage the liver and may lead to a disease called *alcoholism*. Alcoholism is an addiction to alcoholic beverages.

Nicotine is another widely used drug. It is most commonly found in tobacco. Nicotine is habit-forming and has harmful effects on the cells in the trachea. It increases the blood pressure as well. Tars and certain gases in tobacco smoke

Table 3-2: Some Commonly Used Drugs

Drugs	Effects on the Body	Harmful Side Effects
nicotine	increases heartbeat rate, blood pressure	increases risk of heart disease, mouth and lung cancer, and emphysema
alcohol	relaxant; slows reaction time; may cause drowsiness, loss of coordination, and slurred speech	excessive use can cause alcoholism, liver and kidney damage, nutritional deficiencies, and emotional problems
caffeine	increases heartbeat rate and blood pressure; reduces feeling of fatigue	loss of sleep; upset stomach; increases blood pressure
amphetamines ("uppers")	reduce appetite, stimulate nervous system	nervousness, lack of sleep, severe weight loss
barbiturates ("downers")	relieve tension; cause sleep	cause drowsiness, poor muscular coordination, and decreased reaction time; excessive use causes liver or kidney damage; mixing with certain other drugs or accidental overdose may cause death
morphine codeine opium heroin	cause drowsiness; relief from pain; cause sleep	addictive; produce loss of consciousness; overdose can cause death
marijuana	relaxes; causes drowsiness; decreases muscular coordination	interferes with concentration and muscular coordination; smoke may cause lung cancer
cocaine	increases heartbeat rate; raises blood pressure	causes feelings of depression and fatigue and desire for more shortly after use; sleeplessness; loss of appetite; behavior changes

damage the sensitive alveoli of the lungs. (See Figure 3-27.) The tars also contain a variety of harmful chemicals. Some of them have been linked to cancer, a disease that involves the abnormal reproduction of certain body cells. Cancer cells grow quickly, often destroying and replacing healthy tissue. Smokers have a higher risk of getting lung cancer than nonsmokers.

Marijuana is a commonly used illegal drug. Since it is usually smoked, marijuana can cause damage to the lungs similar to that caused by tobacco. Marijuana can interfere with short-term memory and learning. Substances in marijuana affect coordination and concentration. It can be unsafe to drive a car or operate machinery after smoking marijuana. Marijuana contains over 300 different chemicals. Some of them have been linked to cancer. Scientists have only begun to understand the effect of marijuana on the body.

Certain risks are involved in using drugs illegally. First, there is the risk of overdose. When physicians prescribe medications they take into account the patient's age, weight, and general health. The dosage is adjusted according to these factors. Drug abusers often do not know safe dosages, nor can they be certain of the purity of a drug bought on the street. Street drugs might contain other harmful substances. Certain combinations of drugs, such as barbiturates and alcohol, or an overdose could be deadly. Table 3-2 summarizes the effects of commonly used drugs.

Research continues on the short- and long-term effects of many drugs. Because of the dangerous nature of many drugs, much of the research is done with mice and rats. People who abuse drugs are making experimental subjects of themselves.

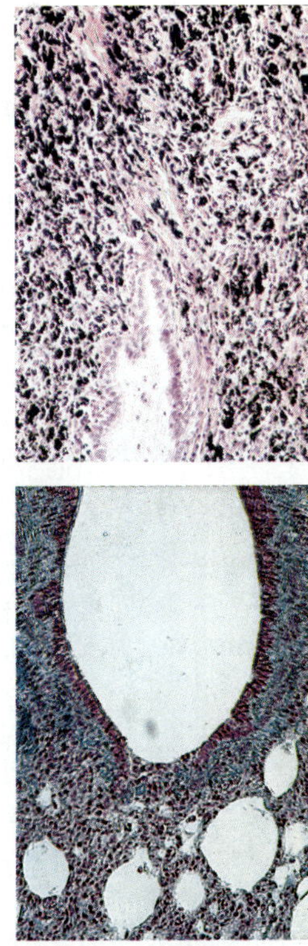

Figure 3-27 *Compare lung tissue from a nonsmoker (top) with lung tissue from a smoker (bottom).*

Checkpoint

1. What is an addiction? Name one drug that can be addictive.
2. Which substances in tobacco smoke are harmful? What effects do they have on the body?
3. What factors does a physician consider when prescribing a medication? Why can drugs taken without a doctor's advice be dangerous?

Understanding the Chapter 3

Main Ideas

1. The skeleton is a framework of bones and cartilage that provides protection and support, and enables us to move.
2. Striated, smooth, and cardiac are the three types of muscle cells.
3. Muscles move bones and other body parts by contracting.
4. Blood moves food, oxygen, and waste products through the body.
5. The heart pumps blood through arteries, capillaries, and veins.
6. Blood consists of plasma, red blood cells, white blood cells, and platelets.
7. Oxygen and carbon dioxide are exchanged between the capillaries and the alveoli in the lungs.
8. Digestion is the breakdown of food into simpler substances that can be used by cells.
9. The kidneys remove waste and water from the blood and form urine. Urine is excreted through the urethra.
10. The brain and spinal cord control all the activities of the body.
11. Impulses are transmitted by the nervous system along neurons.
12. Hormones are released into the blood stream by the endocrine glands. Hormones control growth, development, and other activities.
13. There are serious health risks involved in using alcohol, tobacco, and other drugs.

Vocabulary Review

From the following list, choose the term that best completes each of the statements. Write your answers on a separate piece of paper.

alveoli
arteries
blood
capillaries
cartilage
cerebellum
cerebrum
diaphragm
digestion
esophagus
glands
hormones
impulse
involuntary muscles
joint
kidneys
ligaments
medulla
motor neurons
neurons
plasma
reflexes
sensory neurons
stimulus
synapse
tendons
trachea
urine
veins
villi
voluntary muscles

1. Your _?_ filter wastes and water from the blood.
2. _?_ are air sacs in the lungs.
3. _?_ attach muscles to bones.
4. The gap between two neurons is called the _?_.
5. Tough, stretchy, bands of tissue called _?_ connect bone to bone.
6. _?_ are single nerve cells.
7. The _?_ is the part of the brain that controls the senses.
8. _?_ is the clear liquid in blood.
9. The _?_ is the muscle at the base of the ribs that aids in breathing.
10. _?_ usually carry oxygen-rich blood away from the heart.

11. __?__ are chemicals released directly into the blood, which cause changes in specific areas of the body.
12. __?__ is the stiff, flexible material found in the tip of the nose, outer ears, and ends of bones.
13. __?__ produce hormones.
14. The __?__ is a tube that carries food to the stomach.

Chapter Review

Write your answers on a separate piece of paper.

Know The Facts

1. Food is moved to the __?__ from the stomach. (esophagus, small intestine, liver)
2. __?__ muscle moves bones in the skeleton. (striated, smooth, cardiac)
3. __?__ are the blood vessels where the exchange of food, oxygen, and carbon dioxide takes place at the cells. (arteries, veins, capillaries)
4. __?__ are the part of blood that carries oxygen. (platelets, white blood cells, red blood cells)
5. Information carried by all nerve cells is called __?__. (reflexes, synapses, impulses)
6. The __?__ system removes waste products from the body. (excretory, endocrine, nervous)
7. Muscles move body parts by __?__. (ligaments, contracting, stretching)
8. Inhalation and exhalation are controlled by the contraction and relaxation of the __?__. (lungs, alveoli, diaphragm)
9. Digested food diffuses into the blood at the __?__. (pancreas, villi, large intestine)
10. A __?__ is anything that causes a reaction by the body. (stimulus, neuron, reflex)
11. The __?__ gland controls the release of hormones from other endocrine glands. (salivary, pituitary, adrenal)
12. __?__ slows reaction time and affects muscular coordination. (alcohol, nicotine, adrenalin)
13. __?__ are waste products that are excreted by the body. (oxygen and carbon dioxide, urea and platelets, carbon dioxide and urea)
14. __?__ is a commonly used drug that reduces fatigue and increases heartbeat rate and blood pressure. (alcohol, caffeine, bile)

Understand The Concepts

15. Explain how striated muscles work in pairs.
16. Describe the differences between voluntary and involuntary muscles.
17. Compare the functions of red blood cells, white blood cells, and platelets.
18. Describe the actions involved in breathing.
19. Name three places in the digestive system where food is acted upon by enzymes.
20. Draw and label the parts of the digestive system.
21. Beginning at the right atrium, trace a drop of blood on its journey through the heart, to a cell in the big toe, and back to the heart.
22. Explain the functions of the bladder and urethra.
23. Draw a diagram of two neurons and a synapse. Label the axons, dendrites, cell bodies, and synapse.

Use arrows to show the direction in which an impulse would move along both neurons.

24. Name three regions of the brain. Describe some of the functions carried out by each region.

25. Give an example of how the nervous system and endocrine system work together.

26. What are some of the dangers of using drugs illegally?

Challenge Your Understanding

27. Pretend that you heard someone call your name. How would you respond to that stimulus? Trace the path of the impulses that lead to your response, beginning with sensory neurons in one or more sense organs.

28. Anemia is a condition in which the blood does not contain enough red blood cells. What effects would anemia have on the body?

29. The lungs, kidneys, and large intestine rid the body of certain waste products. Tell (a) what waste products each gets rid of and (b) what system each of these organs is part of.

30. Food goes through a mechanical and chemical digestive process. How does mechanical digestion aid chemical digestion?

31. Imagine that your small intestine was smooth instead of lined with villi. How would absorption of food be different?

32. Some diseases are caused by too little of a certain hormone. How would you test for such a condition?

Projects

1. Artificial parts are being used to replace many worn-out or damaged organs. Find out about replacement parts that are being used safely now in humans.
2. Arrange for a representative of your local Red Cross to teach your class how to perform CPR and the Heimlich Maneuver.
3. Find out about the effects of diet and exercise in preventing heart disease.
4. Using materials you can find at home, construct models of hinge and ball-and-socket joints. Include ligaments in your models.
5. Research and write a report about blood diseases. Find out how they are contracted, how each affects the body, the treatment or cure, and how each can be prevented.
6. Driving a car under the influence of alcohol is very dangerous. Find out how your state or province treats drunk drivers.
7. Write to the American Hearing Association or Canadian Council on Deafness for information on sign language. Demonstrate signing to your class.
8. Sleep is one of the ways the body has for "recharging" itself. Scientists have been studying people in sleep labs to learn more about its effects. Look for recent articles in magazines and newspapers about sleep. Prepare a report on sleep for your class.
9. Interview a police officer about drug and alcohol laws in your state or province.

Chapter 4

Ecology

A biologist examines insects found in an alfalfa-field ecosystem.

4-1

Ecosystems

Goal

To identify the parts of an ecosystem.

A family of beavers depends upon a pond and its surroundings for survival. (See Figure 4-1.) An organism's surroundings, consisting of both living and nonliving things, are called an **environment**. All organisms interact with their environment.

environment

Air, water, and soil are three nonliving parts of the beavers' environment. They provide the things beavers need to stay alive. For example, the beavers take in oxygen from the air, drink water from the pond, and feed on plants that grow in the soil. Living parts of the environment, such as the small trees and bushes, provide food and building material for the beavers.

community

Many of the organisms in an area, such as a beaver pond, interact with one another. These interacting organisms make up a **community**. Trees, reeds, fish, beavers, and one-celled organisms are part of the pond community.

ecosystem
(EK oh sis tum)

The pond community and the nonliving parts of the pond environment can be thought of as one interacting unit. The unit is called an *ecosystem*. An **ecosystem** is made up of a community and the nonliving parts of the environment.

An ecosystem is a model invented by people as a way of talking about a particular area. Materials such as food, water, and oxygen are exchanged, and energy is transferred throughout an ecosystem. A tiny ecosystem can exist in a

Figure 4-1 *Many different organisms live in and around a pond. Some live at the pond's edge, others live on the surface of the water, and some live in the muddy pond bottom.*

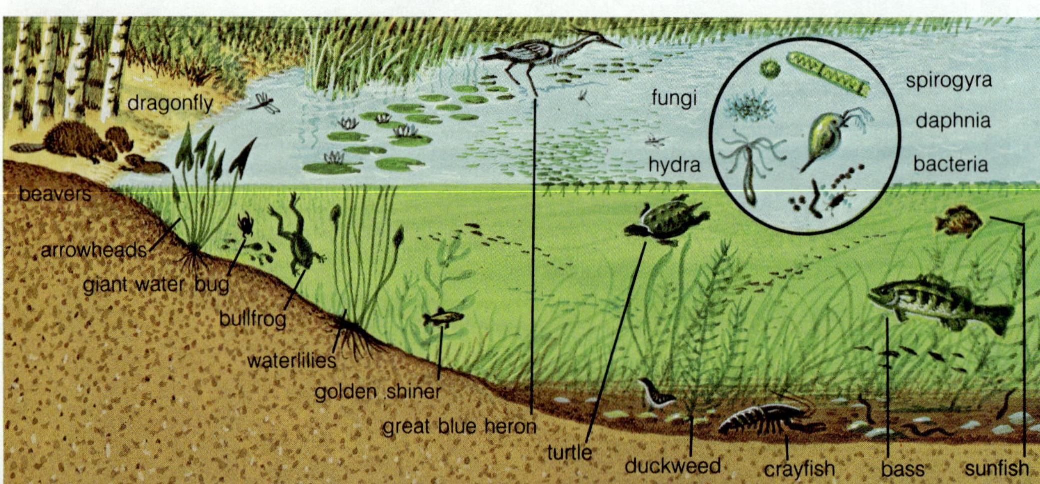

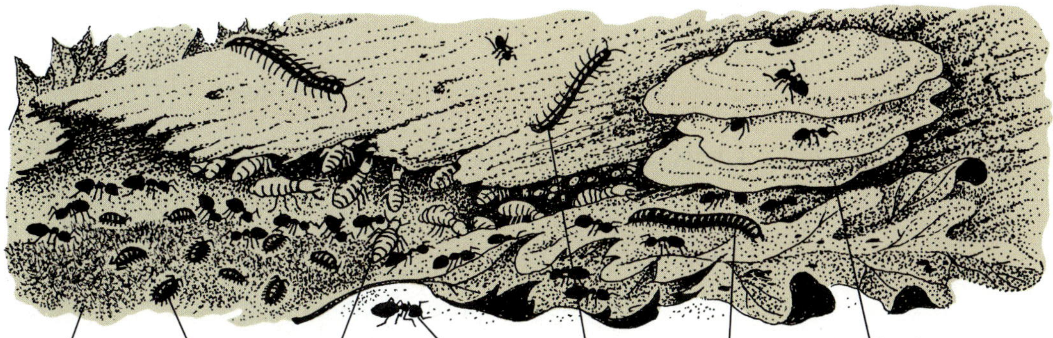

Figure 4-2 *Many different organisms live in and around a rotting log. The ecosystem contains several plants and animals.*

puddle in your backyard. Or an ecosystem can be very large, such as a forest. Figure 4-2 shows an ecosystem you are likely to find in and around a rotting log.

The place where an organism lives in an ecosystem is called a **habitat**. Each ecosystem contains many different habitats. The habitat of a crayfish in a pond ecosystem is along the bottom of the pond. The habitat of a pine tree may be among other trees in a forest ecosystem.

habitat

Different kinds of organisms can live in the same habitat. For example, tadpoles, minnows, and duckweed may share the same habitat—the edge of a pond.

Within a habitat, each species has its own way of living. Each different way of living is called a **niche**. A niche includes the kind of food an organism uses, the places it goes to find food, and the way it protects itself from enemies. No two species have the same niche, but niches can overlap. For example, in a pond ecosystem, a fish may eat insects that move across the surface of the water and insects that move below the surface. A frog may eat insects that land on small plants along the pond's edge and insects that move on the surface of the water. The feeding locations of the frog and the fish overlap, but each has other sources of food as well. Each has a different niche.

niche
(nich)

Checkpoint

1. What is an ecosystem? Give two examples of ecosystems.
2. Name three nonliving parts of an ecosystem.
3. Define community.
4. Explain the difference between a habitat and a niche.

TRY THIS

Describe the ecosystem in which you live. Name as many living and nonliving things as you can think of that are part of your ecosystem.

4-2

Food Chains and Food Webs

Goals

1. To define and give examples of food chains and food webs.
2. To identify the roles of producers, consumers, and decomposers in food chains.

food chain

producers

consumers

Energy is trapped in ecosystems by plants and some protists during photosynthesis. The energy is transferred when one organism eats another. In a pond ecosystem algae capture and store energy from the sun. The stored energy is passed on to minnows that eat the algae. A heron feeding on the minnows obtains the energy stored in the small fish. A kind of chain is set up.

A set of eating interactions within an ecosystem is called a **food chain**. A food chain shows how the energy stored in food is passed from one organism to another. Figure 4-3 shows some examples of pond food chains.

Every food chain begins with the sun. Energy from the sun is converted to food during photosynthesis. Organisms that manufacture food are called **producers**. Plants and some protists are producers. Organisms that do not produce their own food are called **consumers**. Consumers eat other organisms.

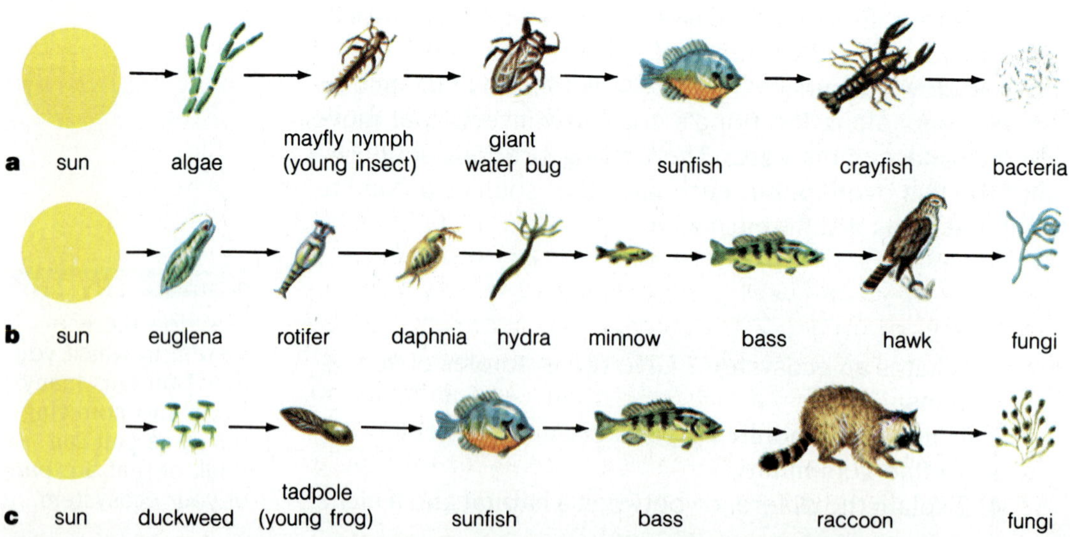

Figure 4-3 *All food chains begin with the sun. Energy from sunlight is trapped during photosynthesis. The arrows show how energy is transferred when one organism eats another. Food chains usually end with bacteria or fungi.*

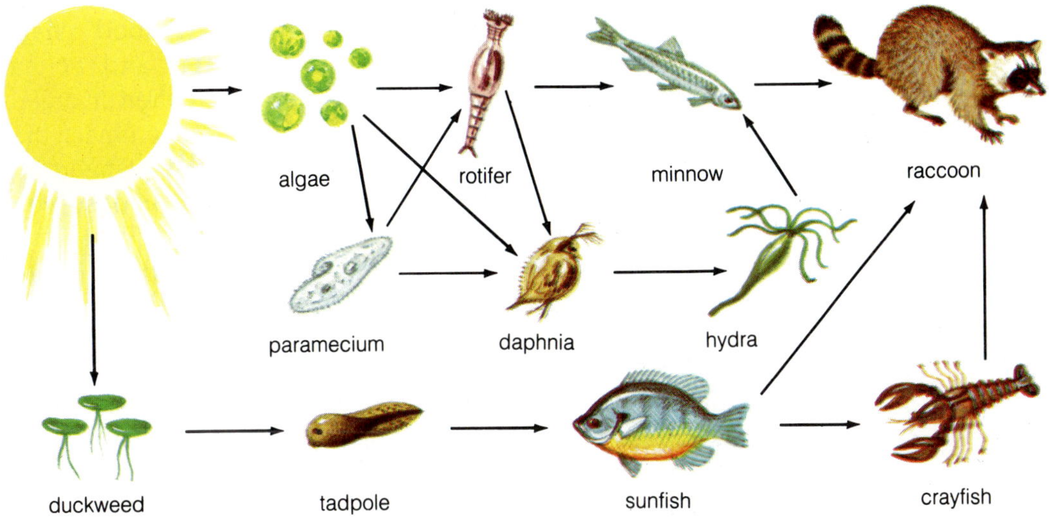

Figure 4-4 *Most organisms eat more than one type of food, as shown in this simplified food web. The complete food web would contain many more organisms.*

First-order consumers eat producers. The young insect, rotifer, and tadpole, shown in Figure 4-3, are first-order consumers. *Second-order consumers* feed on first-order consumers. The giant water bug, daphnia, and sunfish are second-order consumers. *Third-order consumers,* such as the sunfish, hydra, and bass, eat second-order consumers. Most organisms eat more than one kind of food. An organism, such as a sunfish, may be a third-order consumer in one food chain and a second-order consumer in another.

food web

Look at the food chains shown in Figure 4-4. Note that some species of organisms are in more than one food chain. Food chains that overlap make up a **food web**. A food web shows the complex interactions among producers and consumers.

Some organisms feed on plants only. Plant-eaters are known as *herbivores* (HUR buh vorz). Caterpillars, rabbits, sheep, and mice are common herbivores. All herbivores are first-order consumers.

Carnivores are organisms that eat protists or animals. Wolves, frogs, spiders, and snakes are carnivores. All carnivores are second- or higher-order consumers. Organisms that feed on both plants and animals are known as *omnivores* (AHM nuh vorz). Bears, chimpanzees, and raccoons are omnivores. Human beings eat both plants and animals too. They are omnivores also. An omnivore can be a first-order consumer in one food chain and a second- or third-order consumer in another.

decomposers

Decomposers are the final link in a food chain. **Decomposers** break down dead organisms and animal wastes, using them as food. A food chain ends when an organism dies and is decomposed. *Scavengers* are animals that feed on dead organisms also. Some common scavengers are vultures, coyotes, crayfish, and crows. The action of scavengers is fast, while the action of decomposers is slow.

Decomposers do not use all materials in dead organisms or wastes as food. Through the action of decomposers, important materials are returned to the soil, water, and air. Nitrogen is a substance needed by all organisms for making proteins. It is released into the air and soil through the process called *decomposition*. Most organisms cannot use nitrogen gas from air directly. *Nitrogen-fixing* bacteria in the soil and in the roots of some plants change nitrogen gas into a substance that can be used by plants.

Plants take up the nitrogen substance that has been added to the soil by bacteria. The nitrogen is used in building proteins. Animals obtain the plant proteins when they eat the plants. Some of this nitrogen is returned to the air and soil through animal wastes. The remainder is stored in the animal and will be returned to the soil by decomposers when the animal dies. The reuse, or *recycling*, of nitrogen is called the *nitrogen cycle*. (See Figure 4-6.)

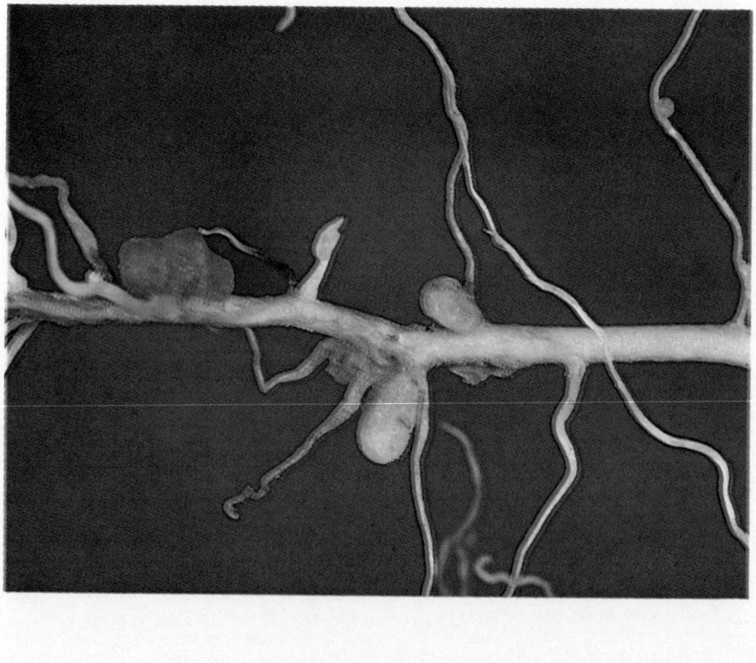

Figure 4-5 *Nitrogen-fixing bacteria live in the nodules, or lumps, on the roots of this clover plant. Beans, peas, peanuts, and soybeans also contain nodules. Farmers use these plants to replace nitrogen removed from soil by other plants.*

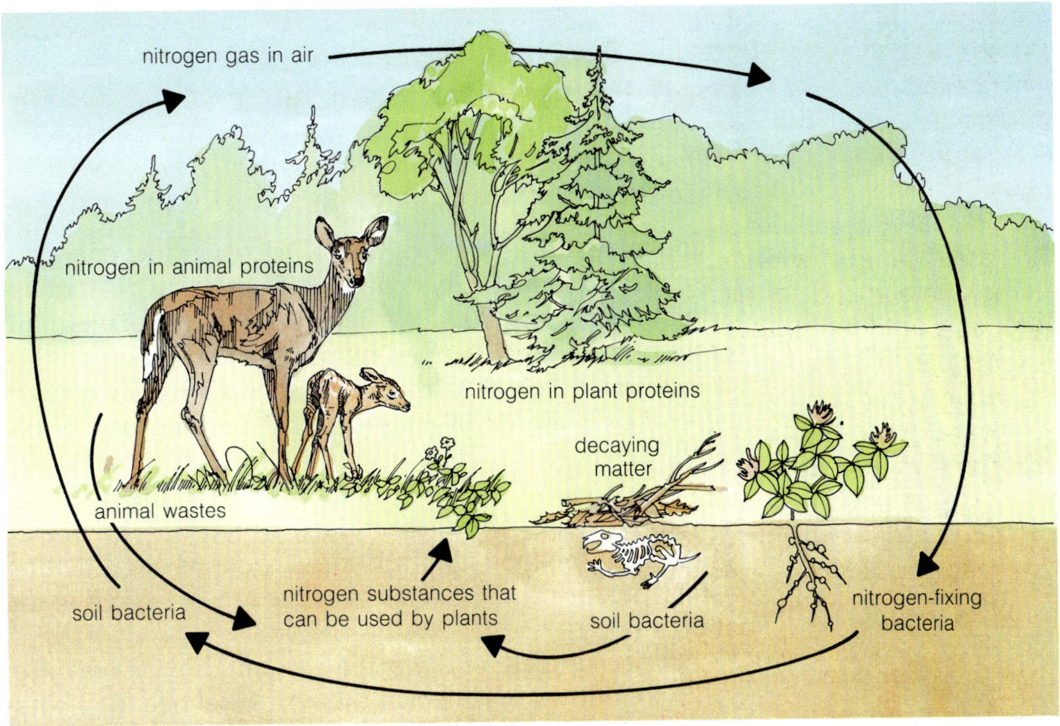

Figure 4-6 *The nitrogen cycle on land. Plants take in nitrogen substances from the soil. Animals get nitrogen from the food they eat.*

Checkpoint

1. What is a food chain?
2. Draw a complete food chain. Label the producer, first-order consumer, second-order consumer, third-order consumer, and decomposer.
3. Explain the difference between a food chain and a food web.
4. What is the role of a producer in a food chain? a consumer? a decomposer?
5. Describe how nitrogen is recycled in an ecosystem.

TRY THIS

Discover which organisms live in your backyard, schoolyard, or local park. Draw a quick sketch or write a description of each one. Use reference books to identify the organisms. Then draw a food web that is likely to occur in the area you studied.

4-3

Food Pyramids

Goal

To trace the transfer of energy in a food pyramid.

food pyramid

A **food pyramid** is another way of showing how energy is passed along in an ecosystem. (See Figure 4-7a.) The transfer of energy is not 100% efficient. On the average, nine tenths of the energy trapped by plants during photosynthesis is used for their life activities. The remaining one tenth of the energy is stored in materials that become part of the plants' cells.

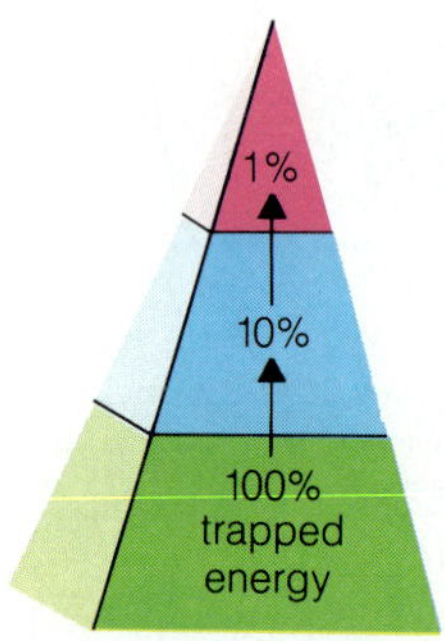

Figure 4-7a *Each level of a food pyramid contains one tenth of the energy of the level below it.*

When a rabbit eats the plants, it takes in the materials and stored energy. The rabbit uses up most of the energy gained from the food it eats in carrying out its life activities. A small amount of the energy from the food is stored in the rabbit's body.

About one tenth of the energy acquired by an organism is available to an organism at the next level of a food pyramid. Scientists refer to this "loss" of energy as you move up the pyramid as the *ten percent law.*

A food pyramid can also be used to represent the *number* of organisms it takes to meet the energy needs at each level of a particular food chain. (See Figure 4-7b.) For example, tens of thousands of plants are required to keep alive several hundred rabbits during a single year. Several hundred rabbits keep alive a single fox. Notice that the organisms become fewer in number and larger in size at each higher level in a food pyramid. Since each organism in any food chain takes in only a small amount of energy from every organism it eats, a consumer must eat a large number of producers or lower-order consumers to survive.

Figure 4-7b *The number of consumers a food pyramid can support depends on the number of producers at the base.*

Checkpoint

1. What is a food pyramid?
2. The transfer of energy in a food chain is not 100% efficient. Explain.
3. How much energy is available to each higher level of a food pyramid? Explain.

4-4

Populations

Goals

1. To describe four processes that cause a population to change in size.
2. To identify limiting factors in an ecosystem.

A meadow community may contain hundreds of field mice. These mice make up a population. A **population** is the number of organisms of the same kind, or species, living within a certain area at a particular time. Some other examples of populations in a meadow are the population of meadow grass, the population of grasshoppers, and the population of garter snakes.

population

Most populations change in size over time. There are four processes that can cause a population to change: (1) immigration, (2) emigration, (3) birth, and (4) death. *Immigration* means movement of organisms into a particular area. *Emigration* means movement out of an area. Immigration increases the size of a population. Emigration decreases it.

The population of fur seals on the Pribilof (PRIB uh lawf) Islands off the North American coast increases each spring due to immigration. (See Figure 4-8.) Adult females usually give birth shortly after arriving on the islands, increasing the population even more. The seals leave the island (emigrate) in late September, moving south for the winter. By December the population of seals on the islands is nearly zero. The seals return to the breeding grounds again the following spring.

Figure 4-8 *Adult male seals, called bulls, arrive on the island first and establish their territories. Female seals, called cows, arrive later. The young seals, called pups, are born on the island.*

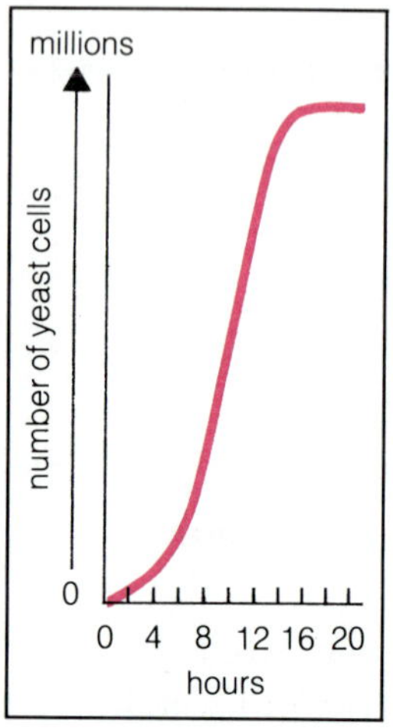

Figure 4-9 *A graph of a yeast-cell population. The curve rises when the population is increasing and flattens out when the population levels off.*

The amount of food available places a limit on the number of organisms an ecosystem can support. If food is plentiful, many individuals are likely to live long enough to produce offspring. The offspring are likely to survive to an age at which they, too, can produce offspring. New individuals might move into the area to take advantage of the food supply. So the population will increase.

If the population increases faster than the food supply, some organisms will not have enough to eat. They may weaken, become susceptible to disease, and die. Some individuals may move out of the area in search of food. Others may die of starvation.

limiting factor

Any factor that keeps a population from growing indefinitely is called a **limiting factor**. Food, space, climate, the presence of enemies, and disease are some limiting factors in an ecosystem. In most ecosystems, the supply of resources needed for survival is limited. Organisms may compete for a share of soil, sun, water, food, oxygen, or shelter.

Most populations do not grow indefinitely. Eventually one or more limiting factors slow the growth rate. Figure 4-9 shows a graph of a yeast-cell population in a closed container. As wastes built up and food ran out, the population size leveled off. The number of yeast cells being produced became equal to the number of yeast cells that were dying.

Figure 4-10 *Changes in population size usually occur around a set point (dotted line).*

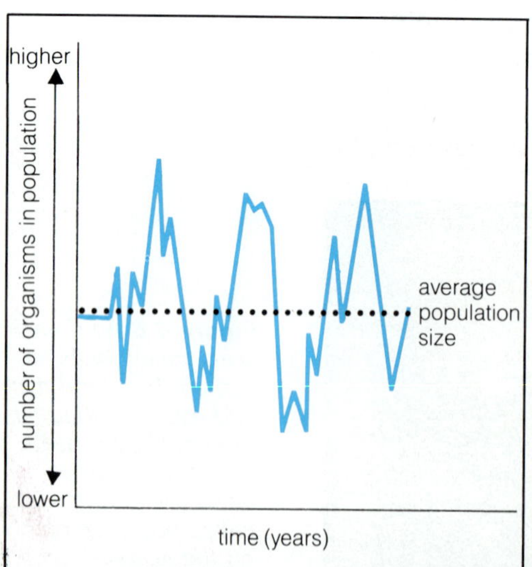

Figure 4-11 *Changes in rabbit and lynx populations usually occur in a ten-year cycle.*

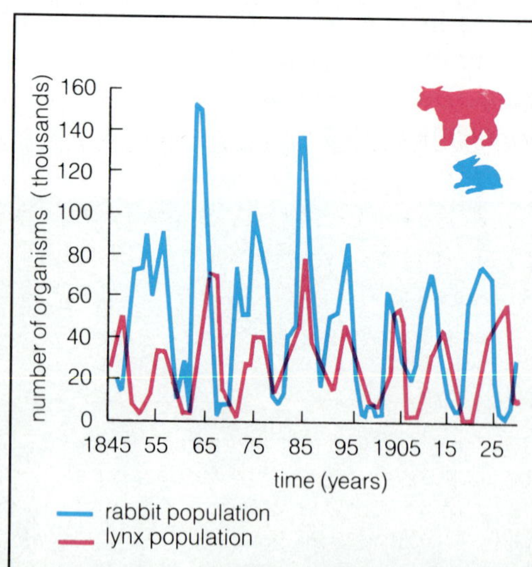

Most populations rise and fall over time but the *average* population size stays the same from year to year. Figure 4-10 shows how a graph of most populations would look if they were measured over time.

Populations in a community affect each other. If one population changes greatly in size, other populations in the community may change also. For example, if the number of field mice in a meadow increased because of a plentiful food supply, more snakes or owls might come to the area to feed on the mice. The snake or owl population in the meadow would increase. Organisms, such as snakes and owls, that catch and eat other organisms are called **predators**. The organism that is caught and eaten is called the **prey**. As predators increase, prey decreases.

predators
(PRED uh turz)

prey
(pray)

In communities that have only a few populations, a change in the size of one population will have a dramatic effect on the size of the other populations. Figure 4-11 shows what happens in a community where there are just a few populations. The lynx is a predator of rabbits. As the rabbit population grows, the lynxes have more to eat. More lynxes survive and produce young, increasing the lynx population. As more and more rabbits are eaten, the rabbit population drops off. Fewer rabbits live long enough to produce offspring. With a smaller rabbit population, the lynx population has less food. Some lynxes die of starvation. As the lynx population decreases, the rabbit population starts to increase, beginning the cycle again.

Checkpoint

1. What are the four processes that cause a population to change in size?
2. Name two limiting factors that might keep a population from growing.
3. What is a predator? What is a prey?

TRY THIS

Obtain two identical pots with soil. In the first pot, plant 2 radish seeds. In the second pot, plant 12 radish seeds. Place the pots in a sunny window. Water the plants when the soil feels dry. Compare the plants in both pots after several weeks. Explain the results in terms of limiting factors.

4-5

Succession

Goal

To describe the process of succession in an ecosystem.

An ecosystem may change over time. New populations can move into an area, replacing old ones. The new community that is formed has an effect on the nonliving parts of the ecosystem. For example, if thick grasses began to grow on a dry, abandoned field, they would form a protective cover over the surface of the soil. The grasses would keep the soil from being blown away by wind and would increase the soil's ability to hold water. As some grass plants died and decomposed, they would enrich the soil with valuable materials. Eventually the soil would become a suitable habitat for a variety of plant species. The dry field might become a moist, rich meadow or forest over time. The grass plants cause changes in the ecosystem that eventually lead to their own replacement. The process by which an ecosystem

Figure 4-12 *(a) Pioneer organisms establish the first community. (b) Floating pondweeds and rooted plants appear. (c) As the pond bottom builds up, thick growths of plants take hold. A wide variety of organisms live in and around the pond. (d) After many years, land organisms replace the pond organisms.*

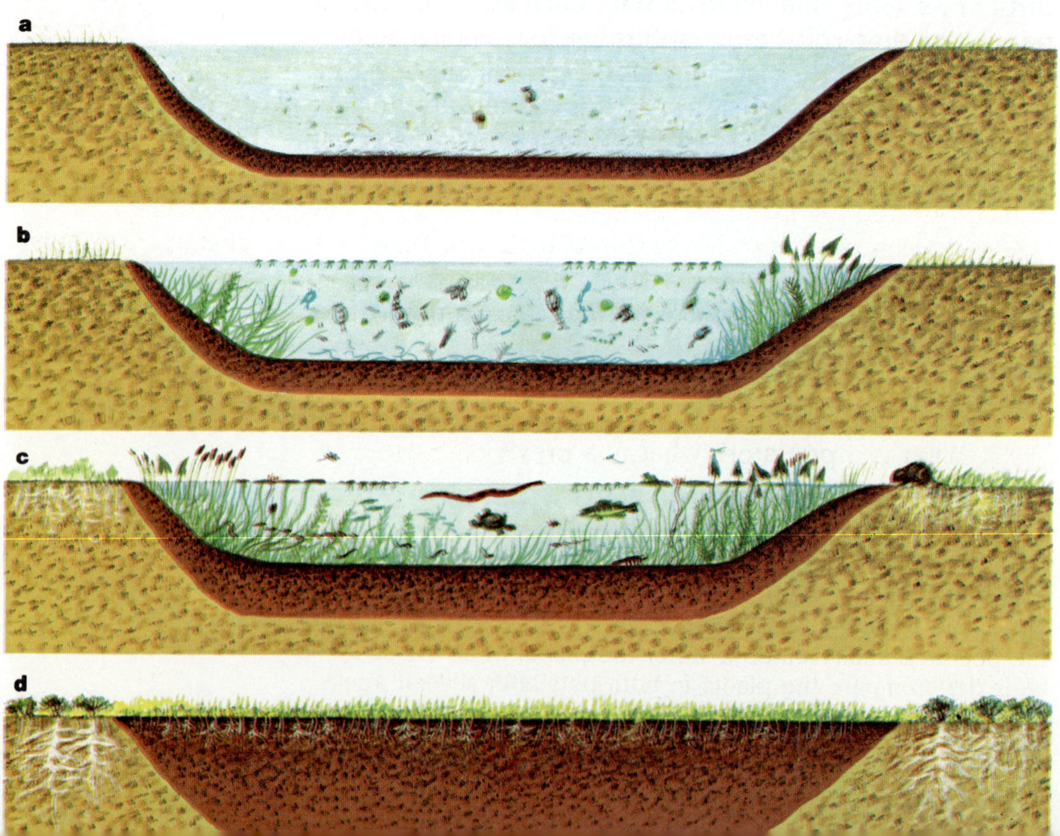

changes over time is called **succession**. Succession takes place gradually over many years. Figure 4-12 shows succession in a pond ecosystem.

succession

The first community to develop in a bare area is called the **pioneer community**. Spores of fungi and bacteria, one-celled organisms, and plant seeds carried by wind probably established the pioneer community in the pond.

pioneer community

As the community multiplied and grew, it had an effect on the pond. Organisms died and were decomposed, adding mineral-rich matter to the bottom of the pond. The accumulated decayed matter acted as fertilizer, permitting the growth of algae and other vegetation. Animals immigrated to the pond to feed on the plants. Animal wastes and decaying material began to fill in the pond. It grew more shallow. Reeds, cattails, and lilies grew at the edge of the shallow water. Spores and eggs of other organisms were transported on the fur and feathers of animals. The community eventually contained a variety of plants, one-celled organisms, insects, fish, amphibians, reptiles, birds, and mammals.

After many years, the pond became filled in with sand, mud, and materials that had decayed into soil. Grasses and small shrubs began to grow where the pond once was. As land organisms gradually replaced pond organisms, the pond changed into a meadow. The meadow may change into a forest through continued succession.

Succession goes on in stages. Each stage may last for a short time or for many years. Eventually a *climax community* is formed. A **climax community** is a community that does not change any more. Individuals are replaced, but the same kinds of organisms remain in a climax community for many, many years. Only natural disasters, sudden large climate changes, or human intervention change a climax community. A climax community usually contains many habitats and many different species making up large, complex food webs. Oak-hickory and beech-maple forests are two examples of climax communities.

climax community

Checkpoint

1. What is a pioneer community?
2. Describe the succession of a pond to meadow.
3. What is the final stage in succession?

4-6

Habitat Destruction

Goal

To identify the factors that can harm a habitat.

When one habitat in an ecosystem is damaged or destroyed, it can affect all the organisms in the ecosystem. The "balance" of an ecosystem may be threatened by two things: natural disasters and the actions of people.

A volcano can damage or destroy an ecosystem. Ash, lava, and poisonous gases from a volcano may kill plant and animal populations. It may take months for organisms to slowly reappear in an area affected by a volcano. (See Figure 4-13.)

In 1883 a volcanic eruption left the island of Krakatoa, in the Pacific Ocean, covered with lava and ash. No life remained on the island. Twenty-five years later, 263 species of animals were part of the island ecosystem. Forest fires, earthquakes, tidal waves, and floods are other naturally occurring events that can damage or destroy an ecosystem. Some ecosystems, such as forests, may take hundreds of years to become re-established.

Figure 4-13a *This is the area around a volcano in Washington state, Mount St. Helens, immediately after it erupted in 1980.*

Figure 4-13b *Once a pioneer community is established, succession will lead to repopulation of the area.*

Human activities, such as the building and operating of factories and machinery, mining, agriculture, and the search for fuel can result in the production of harmful substances in the environment. The accumulation of these harmful substances in the environment is known as **pollution**. Pollution is generally categorized by the type of resource that is affected. Air pollution, water pollution, and land pollution are the three major kinds of pollution.

pollution

Air pollution may occur as gases, solid particles, or liquids. A major cause of air pollution is the burning of *fossil fuels*, such as coal, gas, and oil, for energy. Air pollution damages the tissues of organisms that are exposed to it. (See Figure 4-14.)

Figure 4-14 *Air pollution can have a harmful effect on plants and trees.*

Sulfur dioxide is a *pollutant* produced during the burning of coal and high-sulfur oil. When sulfur dioxide reacts with water in the atmosphere, it forms dilute sulfuric acid which falls to Earth as *acid rain*. In many parts of the world acid rain has built up in lakes and streams, changing the chemical balance of the water, and killing fish and other organisms that live there. Salamanders are disappearing in some areas because their eggs and skin are harmed by acid rain. The pollutants that cause acid rain can be carried by strong winds far away from where they were formed.

Automobile exhaust produces many harmful gases. Some of these gases react in the presence of sunlight and form *smog*. Smog contains a dangerous mixture of gases. Carbon monoxide, one gas in smog, prevents oxygen from getting into the blood stream. Other substances in smog are believed to cause cancer. In some cities, people with breathing problems cannot go outside when the smog is particularly bad.

Water pollution is caused by the dumping of industrial wastes, human wastes, garbage, and other substances into ponds, streams, rivers, and oceans. Some of the wastes are *biodegradable*. That means that they can be decomposed by bacteria and fungi and become part of the environment. Wastes that are *nonbiodegradable* do not decay. Often they harm the habitats of the ecosystem where they are dumped.

TRY THIS Cover several slides with petroleum jelly and leave them outside in different areas for 24 hours. Examine the slides and account for the differences.

Some of the harmful pollutants in water are taken in by the plants and other producers that live there. When the plants are eaten by consumers, the harmful substances are passed along to them. Lead and mercury are two waste materials that have been found in the tissues of organisms living in polluted waters. People have become seriously ill and some have died as a result of mercury poisoning from eating contaminated fish. Figure 4-15 shows how a pollutant can get passed along in a food chain.

Cooperation among government agencies, industry, and communities has resulted in the cleanup of several rivers that were once thought to be hopelessly polluted. For example, today people swim and fish in the Willamette River in Oregon—a river once filled with sludge and slime.

The transportation of oil and the drilling for oil at sea have caused a serious type of water pollution known as an *oil spill.* Since oil and water do not mix, the oil forms a slick on the surface of the water. The feathers of birds that land on the water often get covered with the oil, preventing the birds from flying and causing many to drown. Fish die when their gills become covered with the thick, black oil. When the oil is finally broken up by the action of ocean waves, it sinks to the bottom as black tar balls. The tar balls may cover and kill clams, shrimps, lobsters, and the eggs of many

Figure 4-15 *Notice how mercury is passed along a food chain. The mercury collects in the tissues of organisms and is transferred when one organism eats another.*

sea animals. Oil spills are very expensive and difficult to clean up.

People have changed ecosystems in other ways. *Strip mining* is a process used to obtain coal that is buried in the ground. In this process, topsoil and plants are removed from the surface of the Earth. Very few organisms can survive in an area that has been stripped. Some old strip mines have been restored with improved soil, grass, and other plants. (See Figure 4-16.)

Figure 4-16 *A reclaimed coal strip mine in West Virginia.*

Laws have been passed in an effort to control the kinds and amount of pollutants that are put into an ecosystem. Factories and power plants must now have antipollution devices to reduce the amount of pollutants that go into the air. Automobile engines have been redesigned to reduce pollutants in exhausts and to burn fuel more efficiently. Chemical dumping is prohibited in many areas and some companies are now required to pay the cost of cleaning up waterways they have polluted.

Efforts are being made to use less fossil fuel and other limited resources. *Conservation* means that resources will last longer and less damage will be done to ecosystems. For example, people are using nonpolluting sources of energy, such as the sun and wind, instead of relying on gas and oil only. Many companies are recycling glass, paper, and metals in an effort to conserve limited resources and to reduce the amount of waste that gets put into the environment.

Checkpoint

1. Describe one natural event that can destroy an ecosystem.
2. Define pollution. Name one source of air pollution. Describe what steps are being taken to prevent this type of pollution.
3. What is acid rain? How does it affect ecosystems?
4. What steps have been taken to protect ecosystems?

TRY THIS Place a clean bucket outside during a rainstorm. Measure the acidity of the rainfall with *p*H paper. If it measures below 7.0, the rain is acidic. If it is below 5.6, it may be having a harmful effect on the environment. Try to find out the source of the acid rain.

4-7

ENRICHMENT

Endangered Species

Goal

To identify some of the factors that lead a species to become endangered or extinct.

extinct

endangered species

The growing need for land and resources has placed people in competition with many other species on Earth. In some cases, this has resulted in the *extinction* of a species. When a species no longer exists it is said to be **extinct**. Those species that are close to extinction are called **endangered species**. Figure 4-17 shows some endangered species.

A species may become extinct because of natural changes in the environment. But in the last few hundred years human activities have brought about the extinction of many species. The clearing of forest land for farms and houses, and the filling in of swamps and marshes, have resulted in the loss of breeding sites, hiding places, and food

Figure 4-17 *Human activities, such as the clearing of land and unrestricted hunting, have led to the near extinction of thousands of species such as these.*

supplies for many organisms. The destruction of habitats has contributed to the extinction of the eastern cougar and the passenger pigeon.

Many plant species are threatened by extinction also. In some countries, whole forest ecosystems were destroyed when people cut down trees for use in building houses and ships. The Nile lotus and the Egyptian papyrus are two plants that became endangered as a result of changes in the irrigation system of the Nile River.

Various kinds of pollution can also threaten the survival of a species. DDT is a chemical that used to be sprayed on crops and around marshes and swamps in order to kill insects. People did not realize that DDT would harm other organisms as well. DDT stays in the tissues of organisms that ingest it. Figure 4-18 shows how DDT was passed along in one food pyramid.

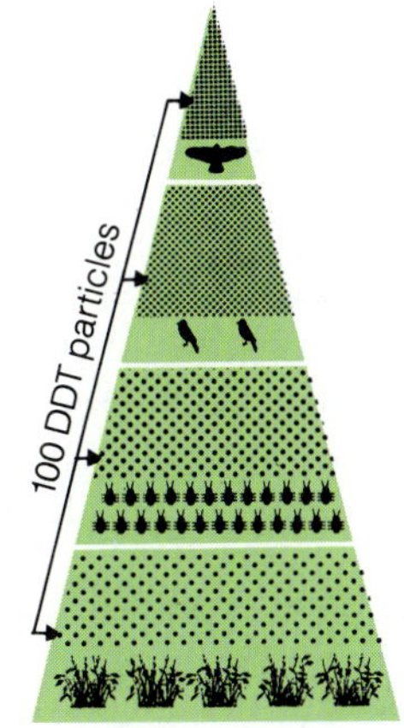

Figure 4-18 *DDT becomes more concentrated moving up the food pyramid. At each higher level, the same amount of DDT particles is distributed among fewer organisms. So each organism receives a larger amount.*

Consumers that ate large numbers of DDT-contaminated organisms accumulated dangerous amounts of DDT in their tissues. Scientists discovered that DDT in the tissues of peregrine falcons and bald eagles caused the females to produce eggs with weak, thin shells. Their eggs broke easily, killing the chicks inside. Through the efforts of conservationists, the birds were saved from extinction. DDT is now outlawed in many countries.

Many steps have been taken to protect endangered species. Laws have been enacted making it illegal to kill, trap, or collect most endangered species. The breeding grounds of many species have been protected, also. Other laws restrict the import and production of products made from parts of endangered species, such as their fur, hide, and tusks. Many wildlife reserves and zoo-breeding programs have been established to protect and increase the number of individuals in some endangered species.

Checkpoint

1. What factors can lead to the extinction or endangering of a species?
2. Name one species that has become extinct. Name an endangered species.
3. What steps have been taken to protect endangered species?

TRY THIS

Find out if there is an endangered plant or animal species in your state or province. What is being done to help the species survive?

Whooping Cranes

The secretive whooping crane is an endangered species that nearly became extinct because of unrestricted hunting and the loss of feeding and breeding sites.

A bird refuge was created in 1937 in Texas to provide a place where the cranes could live unharmed. The cranes' winter migration route was protected by legislation making it illegal to hunt the cranes.

Despite the protective measures, the population of cranes did not increase. Scientists began to study the cranes' life cycle, hoping to find out how the cranes could be saved from extinction. They discovered that the cranes formed tight family structures known as *pair bonds*. Most males and females mated for life. Females laid two eggs each season, but usually only one survived.

Each winter the crane pairs and their offspring migrated to a warm climate and established winter feeding territories. However, many young birds did not survive the long, slow migration. Competition for nesting sites and the difficult migration kept the crane population from growing.

Recently, scientists have been experimenting with ways to increase the population of cranes. (See Figure 1.) United States and Canadian wildlife officials have gathered extra eggs from whooping-crane nests in Canada and have placed them in incubators at a wildlife center in Maryland. The chicks that have hatched and survived make up a small flock of captive cranes. Scientists hope to raise, and eventually breed, the captive cranes.

Researchers have also used sandhill cranes, a close relative of the whooper, as "foster parents" for whooping cranes. Whooping-crane eggs, placed in the nests of sandhill cranes, have been successfully hatched. Several whoopers have been raised by the sandhills. (See Figure 2.)

Figure 1

Figure 2

Understanding The Chapter

4

Main Ideas

1. An ecosystem is made up of a community and the nonliving parts of the environment.
2. The place where an organism lives is its habitat. Its way of living there is its niche.
3. A food chain shows how the energy stored in food is passed from one organism to another. Overlapping food chains form food webs.
4. Decomposers break down dead organisms and animal wastes. Through the action of decomposers, materials are returned to an environment.
5. Only one tenth of the energy acquired by an organism is available to an organism at the next level of a food pyramid.
6. A population is the number of organisms of the same species living in a certain area at a particular time.
7. The four processes that cause a population to change in size are immigration, emigration, birth, and death.
8. The process by which an ecosystem changes over time is called succession. The climax community is the final stage in succession.
9. Pollution is the accumulation of harmful substances in the environment.
10. The changes people have made in ecosystems have led to the extinction and endangering of some plant and animal species.

Vocabulary Review

From the following list, choose the term that best completes each of the statements. Write your answers on a separate piece of paper.

climax community	habitat
community	limiting factor
consumer	niche
decomposer	pioneer community
ecosystem	pollution
endangered	population
environment	predator
extinct	prey
food chain	producers
food web	succession
food pyramid	

1. A(n) __?__ breaks down dead organisms and animal wastes.
2. __?__ make food by photosynthesis.
3. In an ecosystem, each organism's way of living is called its __?__.
4. A(n) __?__ is made up of all the living things in an area.
5. A(n) __?__ is a group of overlapping food chains in an ecosystem.
6. The place where an organism lives in an ecosystem is called a(n) __?__.
7. __?__ is the presence of harmful substances in the environment.
8. A species that no longer exists is said to be __?__.
9. The process by which an ecosystem changes over time is called __?__.
10. The organism that a predator eats is called its __?__.

11. An organism that eats producers is called a(n) ___?___.
12. A(n) ___?___ is made up of a community and the nonliving parts of the environment.

Chapter Review

Write your answers on a separate piece of paper.

Know The Facts

1. All the living things in an area make up a ___?___. (population, niche, community)
2. In an ecosystem, different kinds of organisms cannot have the same ___?___. (habitat, niche, community)
3. Another name for a plant-eater is ___?___. (producer, second-order consumer, first-order consumer)
4. All food chains begin with ___?___. (a plant, the sun, a producer)
5. Animals that are both plant-eaters and animal-eaters are called ___?___. (omnivores, carnivores, prey)
6. About ___?___ of the energy acquired by an organism is available to one at the next level of a food pyramid. (one fourth, one half, one tenth)
7. A food ___?___ shows how energy is "lost" when it is passed from organism to organism. (chain, pyramid, web)
8. Two processes that can *increase* the size of a population in an area are ___?___. (immigration and birth, immigration and emigration, increased food supply and more predators)
9. Anything that keeps a population from growing indefinitely is called ___?___. (a limiting factor, succession, a niche)
10. After a fire turns a forest into a bare area, the first community to develop is known as the ___?___ community. (predator, pioneer, climax)
11. A pond may change into a meadow through ___?___. (extinction, a climax community, succession)
12. Wastes that can be decomposed are said to be ___?___. (nonbiodegradable, biodegradable, pollutants)
13. A species that no longer exists is ___?___. (endangered, a climax community, extinct)
14. All the interacting populations in an area and their environment form ___?___. (a community, an ecosystem, a habitat)
15. Nitrogen in soil is changed into a substance that can be used by plants through the action of ___?___. (fungi, scavengers, bacteria)

Understand The Concepts

16. Define population. Name two populations that may be found in a pond ecosystem.
17. Draw a pond food chain. Label the producers, the consumers, and the decomposers.
18. What is the role of a producer in a food chain?
19. What is the role of a decomposer in a food chain?
20. What is a food chain?
21. What is a food pyramid designed to show?
22. What are the four processes that cause a population to change in size?

23. Explain the difference between an organism's habitat and its niche.
24. Name three limiting factors in a forest ecosystem. Explain how each factor can keep a population from growing.
25. Define succession. Explain what is meant by the term *pioneer community*. Name two organisms that might belong to a pioneer community.
26. What is a climax community? What would cause a climax community to change?
27. What is acid rain? What causes it? How does it affect wildlife?
28. Define pollution. Name one kind of air pollution. Tell how it is caused and describe its effect on the environment.
29. Name three factors that have led to the extinction or endangering of a species.

Challenge Your Understanding

30. All herbivores are first-order consumers, but not all first-order consumers are herbivores. Explain this statement.
31. The number of species living in an ecosystem is limited by the number of available niches. Explain.
32. Which group probably has a greater variety of niches—insects or whales? Explain.
33. Which food chain can feed the most people? Explain. A. sun⟶soybeans⟶people. B. sun⟶soybeans⟶cattle⟶people. (Hint: Draw two food pyramids. Put the same amount of soybeans in each food pyramid.)
34. Explain why the changes brought about by organisms in an ecosystem during succession often lead to their own replacement.

Projects

1. Obtain an old aquarium or a large, wide-mouthed jar. Collect the organisms needed to set up a small ecosystem. Observe the interactions among the organisms. Keep a record of your observations. Note the changes that take place over time. Return the organisms to where you found them when you have completed the project.
2. Investigate what is being done to protect endangered species of whales. Prepare a report for your class. Find out what individuals can do to help save whales.
3. Find out if there is a recycling program in your town or city. Find out how your school can participate in it. Arrange to have bottles and cans from the school cafeteria brought to the recycling center each week.
4. Do a report on the following kinds of interactions that take place in ecosystems: mutualism, commensalism, and parasitism. Look for examples of these interactions in your community.
5. Find out how the human population has changed in your community, state or province, and country in the last 50 years. What projections are being made about the size of each population in the future? Are the populations expected to increase, decrease, or stay the same?

Chapter 5

Distribution of Life

Life is distributed worldwide. This is an African savanna.

5-1

Adaptations

Goal

To identify some plant and animal adaptations.

Any trait that enables an organism to live in a particular environment is called an **adaptation.** Certain body parts, or structures, are adaptations that help organisms get food. Teeth and mouth parts are examples of *food-getting adaptations* in animals. Herbivores, such as horses and cows, have large, flat, grinding teeth. Carnivores, such as dogs and cats, have sharp, pointed teeth that help them tear meat into small pieces. (See Figure 5-1.) Omnivores, such as humans, have sharp, cutting teeth in the front of the mouth and flat, grinding teeth in the back. Look at the pictures of birds' beaks in Figure 5-2. The shape of each bird's beak determines the type of food it eats and how it gets the food.

adaptation

Figure 5-1 *Teeth are an adaptation that help an animal get and use a particular food.*

herbivore

carnivore

omnivore

Figure 5-2 *Three different kinds of feeders: a water-feeder, a seed-eater, and a meat-eater.*

water-feeder

seed-eater

meat-eater

Figure 5-3 *The Venus' flytrap captures insects with its leaves. Digestion and absorption take place slowly.*

Plants also have adaptations that enable them to get food. Chloroplasts, in plant cells, make photosynthesis possible. Some plants, such as the Venus' flytrap, produce a sweet-smelling, sticky substance that attracts insects. When an insect lands on the sticky surface, the leaves close up, trapping the insect. (See Figure 5-3.) The insect is then digested, providing the plant with nitrogen and other nutrients.

Some adaptations protect organisms. For example, quills are an adaptation that protect a porcupine from enemies. Contrary to popular belief, a porcupine does not shoot its quills, but it can imbed them deeply in an enemy with the slap of its tail. Claws, shells, and spines are other *protective adaptations* in animals. So is the foul-smelling spray of a skunk.

Some animals have colors, shapes, or textures that make it possible for them to blend in with their surroundings. Examples of *camouflage* are shown in Figure 5-4.

instincts

Behavior is the way an organism responds to its internal and external environments. Behaviors can be adaptations. **Instincts** are behaviors an organism is born with. Instinctive behavior does not have to be learned. For example, a baby albatross pecks at its parent's bill when it feels hungry. The parent responds by spitting up food. (See Figure 5-5.) Both the pecking of the baby and the providing of food by the parent are examples of instinctive behavior. The instincts help baby albatrosses survive.

Figure 5-4 *The animals in these photographs are hard to distinguish from their surroundings because of camouflage.*

Figure 5-5 A baby albatross pecks at its parent's bill when it feels hungry.

Figure 5-6 The viceroy butterfly (on the left) looks like the monarch butterfly (on the right).

In areas that have cold, snowy winters, many animals travel, or *migrate*, to warmer places. Migration is an instinctive behavior that helps organisms survive harsh conditions.

The ability to learn is an adaptation. Sometimes learned behavior becomes automatic by being repeated a number of times. For example, you learned to write your name by practicing. Now you do it without effort. Learning how to do something saves time. Learning can also protect an organism from harm. For example, a bird gets sick and vomits when it eats a monarch butterfly. It learns to recognize and avoid the monarch. Birds also avoid the nonpoisonous viceroy butterfly because it looks like the monarch. (See Figure 5-6.)

Checkpoint

1. What is an adaptation?
2. Describe two food-getting adaptations in animals.
3. Give one example of an adaptation in plants.
4. Describe one behavioral adaptation in animals. Explain how it helps the animal survive.

TRY THIS

Visit a zoo or wildlife preserve. Select one animal for observation. Make a list of as many adaptations as you can. Pay special attention to structural adaptations for movement, protection, and feeding. What behavioral adaptations do you observe?

5-2

Biomes

Goals

1. To explain how climate affects biomes.
2. To describe the features of tundra and tropical rain forest biomes.

Picture a "zone of life" that covers the entire Earth. It extends approximately 100 m above the land and several meters underground. In areas that are covered by water, the zone of life begins at the surface of the water and goes to a depth of 200 m. Conditions in the zone of life vary.

climate

The **climate** of an area is the average weather over a long period of time. Climate is one factor that determines the kinds of organisms that can live in a given area. Scientists have divided Earth into three major climate zones: *polar, temperate,* and *tropic* zones.

Regions in the polar zones have long, cold winters and short, cool summers. The coldest winter temperature on Earth, −88°C, was reported in the polar zone. The tropic zone is warm or hot most of the time. Seasons in the tropic zone are marked by changes in the amount of rainfall more than by changes in temperature. The average temperature may change by only a degree or two between summer and winter. The temperate zone has four seasons during which the weather gradually shifts from colder to warmer and then back again.

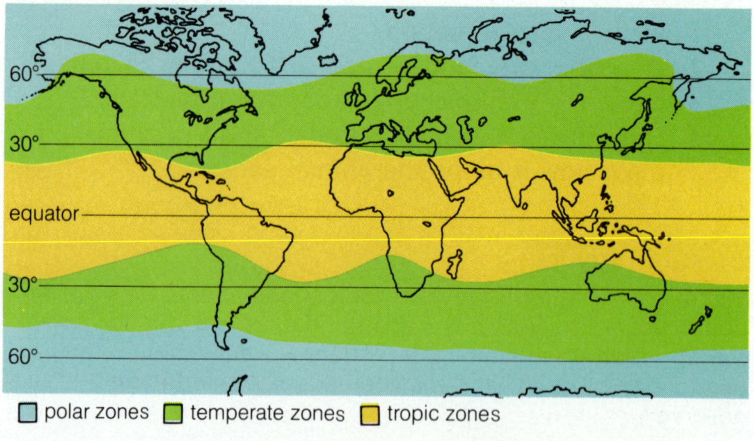

Figure 5-7 *Earth's major climate zones. Find the climate zone you live in.*

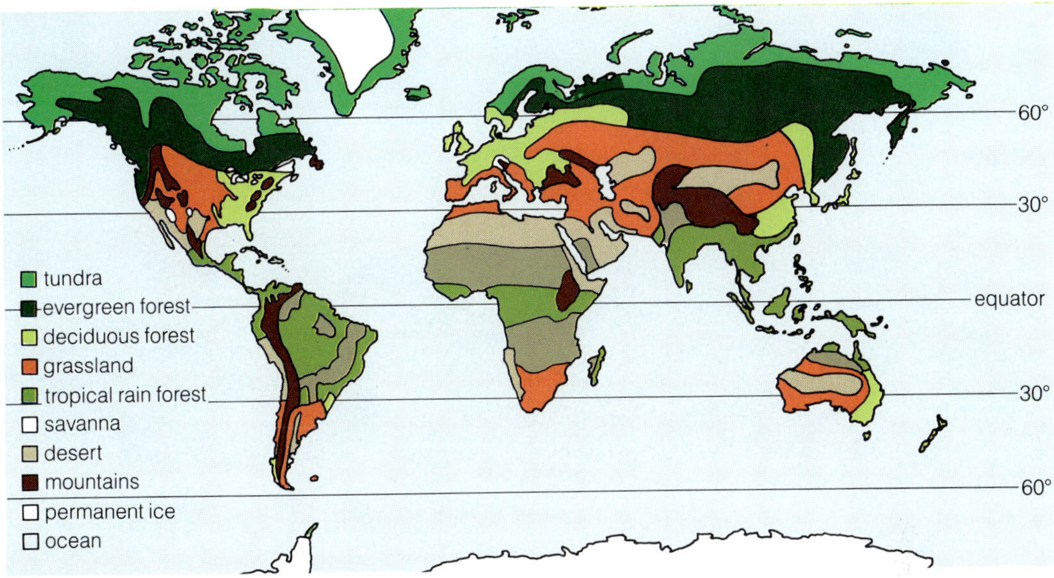

Figure 5-8 *The world's biomes. Boundaries between biomes are drawn on this map as a way of organizing them for study. In nature, biomes usually overlap.*

Look at the map in Figure 5-7. Notice that the climate zones occur in belts circling Earth. Within each climate zone there are large areas of land scattered around the globe that contain similar communities. These areas are called **biomes.** Each biome contains plants and animals that have adaptations that help them survive in that particular area. Figure 5-8 shows the distribution of the world's biomes.

Compare the maps in Figures 5-7 and 5-8. Notice how climate zones and biomes are related. The climate determines the types of plants growing in an area. The plants provide food for certain animals. The climate, then, determines the characteristics of a biome. Wherever the climates are the same, the biomes are similar. A biome may contain one or more ecosystems.

biomes

tundra

permafrost

Find the location of the tundra biome on the map in Figure 5-8. The **tundra** is an area in the polar climate zone that receives little energy from the sun. The tundra has low temperatures and little rainfall. During the winter months, much of the tundra receives *no* sunlight. Tundra soil is frozen solid in winter. During the summer, only the top few centimeters of soil thaw, permitting plants with shallow root systems to take hold. The frozen soil below the surface is called **permafrost.** The tundra has a short growing season for plants. Most tundra plants grow close to the ground—they never become very tall. Large plants would not survive the

TRY THIS

Look at the maps in Figures 5-7 and 5-8. Locate your state or province. In which climate zone and biome do you live? Find out the average yearly temperature and amount of rainfall for your area.

Figure 5-9 *The plants and animals shown in this portion of the tundra biome are typical of those organisms found in the tundra in the summer.*

cold, strong winds of the tundra. Figure 5-9 shows some of the organisms that live in the tundra biome.

The tropical rain forest biome is located in the tropic zone. Direct sunlight, heavy rainfall, and constant high temperatures provide a climate that is well suited for a great variety of plant species. (See Figure 5-10.) Tall trees form a thick canopy, shading the ground below. Shade-tolerant trees and long, climbing vines make up a layer of life beneath this canopy. Rootless plants, which absorb water from the moisture-rich air, live on the bark of trees. Such plants are called *epiphytes* (EP uh fyts).

There is a tremendous variety of niches in the tropical rain forest. Many animals, including tree-dwelling varieties of reptiles, amphibians, and mammals, live in "layers" among the thick vegetation. Huge numbers of insects, fungi, and bacteria live on the remains of plants and animals on the forest floor.

Figure 5-10 *The tropical rain forest biome supports a great number and variety of organisms, including the organisms that are shown in this illustration.*

Checkpoint

1. Name the three climate zones on Earth.
2. Explain how the climate of an area affects a biome.
3. Describe the characteristics of the tundra biome.
4. Describe the climate of the tropical rain forest biome.
5. Describe one animal adaptation and one plant adaptation to life in the tropical rain forest biome.

5-3

Forest Biomes

Goals

1. To describe the characteristics of deciduous and evergreen forest biomes.
2. To identify some adaptations of forest organisms.

Forest biomes in the temperate zone are characterized by ample rainfall, seasonal temperature changes, and day length that varies with the season. There are two types of forest biomes in North America: *deciduous forest* and *evergreen forest* biomes.

deciduous (dih SIJ oo us)

Trees that lose their leaves in response to shortening periods of daylight are called **deciduous** trees. The deciduous forest biome contains many trees, such as maple, oak, and hickory, that lose their leaves each autumn. The fallen leaves form a thick layer of "forest litter" on the ground, which is slowly broken down by decomposers.

Trees use large amounts of water during photosynthesis. Some water escapes through openings in the leaves. During the winter, when the ground is frozen and cannot absorb water, the leafless trees use and lose very little moisture. Losing leaves is an adaptation that helps deciduous trees stay alive through the winter.

A variety of wildflowers and shrubs grow in the deciduous forest. These plants grow and bloom early each spring, before the tree leaves have grown back. The canopy of trees shades much of the sunlight from the forest floor in late spring and summer.

In the deciduous forest, each "layer" of plant life has different adaptations. The adaptations enable plants to survive the given amounts of sunlight and moisture in each layer of the forest. For example, mosses and ferns have structures that allow them to grow successfully on the damp, shady, forest floor. (See Figure 5-11.)

The large number of producers in the deciduous forest provide food for a large number of consumers. Deer, mice, pheasants, and quail feed on the leaves, berries, and seeds of plants on the forest floor. Second-order consumers, such as woodpeckers, feed on insects living in the bark of trees.

Figure 5-11 *Plants and animals typically found in a deciduous forest biome are shown in this view of a deciduous forest in the early fall. Try to identify some of the plants and animals.*

Hawks, owls, and foxes hunt at night for small animals such as mice, shrews, and rabbits.

In the fall, mammals, such as mice and squirrels, gather and store food. Woodchucks and skunks develop thick layers of fat. These adaptations, and others, help many animals survive the cold winter months when food is scarce. Some birds and insects migrate from the forest to warmer climates where food is more plentiful. Small animals, such as snakes and chipmunks, spend the winter in burrows in a sleeplike state called **hibernation.** During hibernation an animal's body temperature is lower and its heartbeat and breathing rates decrease. Hibernation allows an animal to survive the winter on very little energy. In the spring, the animal "wakes up."

hibernation

The evergreen forest biome occurs in the northern parts of the North American continent, Europe, and Asia. The **evergreen forest** is made up mostly of trees that stay green all year long, such as pine, fir, and spruce. Evergreen trees have thin needles, instead of broad leaves. This adaptation prevents the loss of water during cold winter months. In many parts of the forest, evergreens grow close together, blocking out sunlight from the forest floor year round. Few ground plants grow in the very dense shade of an evergreen forest.

evergreen forest

Elk, moose, squirrels, and finches are common herbivores in the evergreen forest. Caribou, migrating from the northern tundra, make their home in the evergreen forest in winter. The snowshoe hare, another evergreen forest dweller, has a brown coat in spring. The coat gradually turns white in winter, helping the hare blend in with its snowy surroundings. The hare's main predator is the lynx. Hawks, wolves, and owls are some other predators in the evergreen forest. (See Figure 5-12.)

Checkpoint

1. Compare and contrast the deciduous forest biome and the evergreen forest biome.
2. Describe two adaptations that help animals survive the winter in the deciduous forest biome.
3. Describe some of the adaptations that help deciduous and evergreen trees survive in winter.

Figure 5-12 *Plants and animals typically found in an evergreen forest biome are shown in this view of an evergreen forest in the spring. This biome covers large areas of Canada.*

5-4

The Desert Biome

Goals

1. To describe the features of the desert biome.
2. To identify some adaptations of desert plants and animals.

The most obvious feature of the desert biome is its lack of moisture. In some desert areas of North America, rainfall may be only 12 cm a year. (Compare this to the average 100 cm of rainfall in a forest biome.) When it does rain, the water evaporates very quickly. The desert biome generally contains sandy soil that does not absorb much water. The soil is often blown about by strong winds because it has little, or no, vegetation holding it down.

Temperatures in the desert tend to vary dramatically from very hot in the afternoon to very cold at night. There might be as much as a 35°C temperature difference in a twelve-hour period.

To avoid the heat of the blazing sun, many desert animals sleep in the shadows of cacti and rocks or in cool, underground burrows. They become active at night when the desert cools off.

Figure 5-13 shows some organisms of the desert biome. All desert organisms have adaptations for living in an area with very little moisture. For example, some desert plants have a wide network of shallow roots that quickly absorbs large quantities of water when it rains. Many desert plants have a thick, waxy, outer layer that reduces water loss.

Leaves are absent in many desert plants, reducing the possibility of water loss to the atmosphere. Photosynthesis occurs in the rounded and flattened stems of many desert plants. Plants such as cacti are covered with sharp spines that prevent animals from robbing moisture that has been stored inside. The harsh desert conditions cause plants to grow slowly.

Desert animals also have adaptations that help them obtain and conserve water. Some animals, such as mice, have kidneys that produce highly concentrated urine, containing little water. Beetles and mice store food underground, where it often absorbs small amounts of moisture

from the soil. Some desert animals survive without drinking water. Their cells use the water produced during cellular respiration.

Figure 5-13 *Plants and animals that are well adapted to living in a desert biome must be able to survive on a limited supply of water. Some plants and animals found in a desert biome are shown here.*

Checkpoint

1. Name two features of the desert biome.
2. What is the function of the waxy, outer layer of a cactus? Name another adaptation in desert plants.

5-5

The Grassland Biome

Goal

To identify the features of the grassland biome.

The grassland biome is found throughout the temperate zone. Grasslands receive seasonal rainfall averaging 25–75 cm each year—less than a forest, but more than a desert.

Few trees grow in the grassland biome. Grasses, such as wheat, corn, clover, and barley, are the dominant vegetation. Most of the world's grain crops are grown on what used to be wild grasslands. Parts of the grassland biome are known as the "breadbaskets" of the world.

Large numbers of herbivores make their home in the grassland biome. Harvest mice, prairie dogs, and ground squirrels eat the seeds and leaves of grassland plants. They live in underground burrows, protected from the coyotes, badgers, snakes, and owls that hunt them. (See Figure 5-14.)

Antelopes and wild buffalo were once the dominant grassland mammals. They have been replaced by domesticated sheep and cattle.

Figure 5-14 *Some of the plants and animals that are typically found in a grassland biome are shown here. Grasses provide food for a number of animals. Other animals feed on the herbivores.*

Where tropical dry seasons are long and severe, trees grow far apart. **Savannas** are grasslands that contain scattered trees. They are found in South America and Africa. Plants in the savanna have adaptations that help them survive the long, dry winter. Zebras, gazelles, and antelopes are large hoofed mammals that feed on grasses in the savanna. Figure 5-15 shows some other large herbivores of the savanna. Many of them are able to run at high speeds, helping them escape the leopards, cheetahs, and other predators that hunt there. Hyenas and vultures survive as scavengers in the savanna.

savannas
(suh VAN uhz)

Checkpoint

1. Where is the grassland biome found?
2. What is the dominant vegetation type in the grassland biome?
3. Name two herbivores that live in the grassland biome. Name two predators in the grassland biome.
4. Why are parts of the grassland biome known as the "breadbaskets" of the world?
5. Name two animals that live in the savanna. Describe some of the adaptations that help these animals survive there.

Figure 5-15 *Some of the plants and animals that are found in a savanna are shown in this illustration. A savanna is a grassland that contains scattered trees.*

5-6

ENRICHMENT

Mountain Biomes

Goal

To identify the features of mountain biomes.

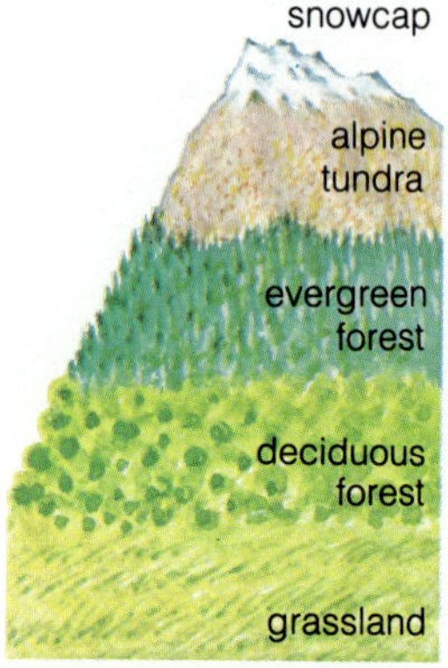

Figure 5-16 *The distribution of biomes on a mountain. The climate of each biome depends on the altitude and the amount of rainfall it receives.*

A high mountain contains many biomes. (See Figure 5-16.) Hiking up a mountain from the base to the top is like moving across the face of Earth from the equator to one of the poles. Every 100 m of elevation is like moving one degree in latitude toward a pole.

The biome at the base of a mountain could be grassland. As you moved up the mountain, you might pass through a deciduous forest biome and, higher up, an evergreen forest biome. Above the evergreen forest, you might find an area of stunted, or short, trees. A biome known as *alpine* tundra might occur near the top. **Alpine tundra** is a mountain biome that resembles Arctic tundra except that there is no permafrost or long period of darkness. (See Figure 5-17.) Many species of plants and animals are the same in both alpine and Arctic tundra biomes. The top of the mountain may be covered with a permanent snowcap.

alpine tundra

The major factor that causes the climatic differences among mountain biomes is the difference in *altitude*, or height, at each level on the mountain. In general, for each 500 m in altitude, the temperature decreases about 2.7°C. The amounts of rainfall and wind vary at different altitudes in mountain biomes.

No mountain biome is very far from any other mountain biome. In the fall, animals migrate from biomes higher up on the mountain to forest biomes below. They spend the winter where food is more plentiful. The animals return to the evergreen forest and alpine tundra in the spring.

Figure 5-17 *Alpine tundra is typically ablaze with bright flowers in the spring.*

Checkpoint

1. How is an alpine tundra different from Arctic tundra?
2. Name two climatic conditions that vary with altitude among mountain biomes.
3. Name three different biomes that may be found on a mountain.

5-7

Life Zones in the Ocean

Goals

1. To describe life zones in the ocean.
2. To identify adaptations that help organisms survive in the ocean.

Biome is a term used to describe *land* areas with a particular climate and communities. Since oceans are water environments with no definable climate, they are not usually called biomes. Oceans do, however, contain a large variety of life. Each species has adaptations that enable it to live in one of the life zones in the ocean.

The *coastal* zone is an area of shallow water containing an abundance of different organisms. (See Figure 5-18.) Some varieties of seaweed in the coastal zone contain air-filled bladders that allow them to float near the surface of the water where light is plentiful. Burrowing animals, such as crabs and clams, and bottom-dwelling fish are other examples of coastal-zone organisms.

Every day the area around thc shore is covered and uncovered by tides. At low tide, the ocean floor near the shore is exposed. At high tide, the area is covered with water. The tides rise and fall twice a day. The zone between high and low tides is called the *intertidal* zone.

Figure 5-18 *Coastal zones contain a large number of different organisms. Some are shown in this illustration.*

Organisms that live in the intertidal zone have adaptations that help them survive the daily changes in their environment. Part of the day, the organisms are exposed to the hot sun, air, and wind. Intertidal-zone organisms, such as barnacles and mussels, have tough, outer shells that remain closed when exposed, keeping in moisture. Many organisms live in *tide pools*—small areas near the shore that are left filled with water when the tide uncovers the rocks. (See Figure 5-19.) At times, intertidal-zone organisms must withstand the pounding surf. Some organisms, such as seaweed, barnacles, and snails, have structures that help them cling tightly to rocks. When covered with water, barnacles and mussels open their shells in order to feed.

Millions of tiny producers carry on photosynthesis in the upper layer of the *open-ocean* zone. Algae form the beginning of long ocean food chains. The abundance of producers in the open ocean supports large, complex food webs, which include animals as small as the microscopic copepod and as large as the gray whale.

Figure 5-19 *Tide pools are small water-filled areas that are often present in the intertidal zones. A number of different organisms live in and around tide pools. Some of these organisms are shown in the illustration.*

Figure 5-20 *A number of unusual adaptations are found among the organisms that live in the deep-water zone of the ocean.*

The *deep-water* zone of the ocean contains organisms with adaptations for surviving in the cold and dark. Because sunlight does not reach the deep water, producers cannot survive there. Many organisms in the deep water live on the remains of organisms that drift down from the upper ocean.

Some deep-water organisms produce and give off a type of light through chemical reactions in their bodies. This process of giving off light is known as *bioluminescence*. Some bioluminescent fish use light to attract food or to frighten off predators. Figure 5-20 shows some deep-water organisms.

Checkpoint

1. Why is the ocean not called a biome?
2. Name four life zones found in an ocean. Describe each one.
3. Pick one ocean animal and one ocean plant. Describe the adaptations that help these organisms to survive in their environment.

TRY THIS

Visit a beach or an aquarium or pet store that has saltwater fish tanks. Observe the different kinds of organisms. Describe the adaptations different organisms have for food-getting, movement, and protection.

FEATURE ARTICLE

Life in the Deep

A deep-ocean community has been discovered where scientists once believed no life forms could exist. These organisms were found around volcanic vents in the Pacific Ocean. The vents, which resemble giant chimneys, are part of the oceanic ridge system. This is an area where the sea floor is spreading apart. Materials that will form new crust pour out of the vents.

This environment would seem to be hostile to most organisms. It is located more than 2500 m beneath the surface—too deep for light to reach and too dark for photosynthesis to take place. Water that has been heated to temperatures up to 350°C by volcanic rocks spews forth from the vents. The water pressure around the vents is 200 times the pressure of Earth's atmosphere at sea level. Scientists were astounded to find rich, thriving communities in this part of the ocean floor.

Figure 1

Figure 2

Scientists discovered that the producers in this community are bacteria. They use a sulfur-containing substance in the vent water, instead of light, as an energy source for making food. The bacteria form the beginning of all the vent-community food chains. Giant clams, fish, crabs, corals, and tube worms are some of the other organisms scientists discovered in this hot-water, deep-ocean ecosystem. (See Figures 1 and 2.) All the organisms have adaptations that make them particularly well suited to the unusual environment. Some of the species found deep in the ocean had never been seen by people before.

Understanding The Chapter 5

Main Ideas

1. Adaptations are traits that enable an organism to live in a particular environment.
2. Biomes are large land areas in different parts of the world that have similar climates and communities.
3. The climate of an area determines the features of a biome.
4. The tundra biome has long cold winters, short cool summers, and a very short growing season.
5. Heavy rainfall, direct sunlight, and constant high temperatures are characteristics of the tropical rain forest biome.
6. The deciduous forest biome contains trees that lose their leaves each autumn in response to shortening periods of daylight.
7. The evergreen forest biome contains trees with needles that stay green all year long.
8. Desert-biome organisms have adaptations for living in an area with little moisture and daily large temperature changes.
9. Grassland biomes support large numbers of herbivores. People use grasslands for growing crops and raising cattle.
10. A mountain contains many biomes from the base to the top.
11. The coastal zone, intertidal zone, open ocean, and deep ocean are four life zones in the ocean.

Vocabulary Review

From the following list, choose the term that best completes each of the statements. Write your answer on a separate piece of paper.

adaptation	hibernation
alpine tundra	instincts
biomes	permafrost
climate	savannas
deciduous	tide pools
evergreen forest	tundra

1. Trees that lose their leaves each autumn are called __?__ trees.
2. Behaviors that an animal is born with are called __?__.
3. __?__ is a sleeplike state that enables animals to conserve energy in the winter.
4. __?__ is soil that is frozen all year long.
5. The flat, grinding teeth of an herbivore are an example of a structural __?__.
6. Large land areas with similar climates and communities are called __?__.
7. The average weather of an area over a long period of time is called the __?__.
8. The biome that contains trees with needles that stay green all year long is the __?__.
9. A biome that is found only on the top of a mountain is called the __?__.

10. The biome that has a short growing season and soil that is frozen year round is the __?__.

Chapter Review

Write your answers on a separate piece of paper.

Know The Facts

1. Organisms with bioluminescence are found in the __?__. (open ocean, deep ocean, intertidal zone)
2. Sharp, pointed teeth is an adaptation found in __?__. (carnivores and herbivores, herbivores and omnivores, carnivores and omnivores)
3. A leafless plant covered with spines would most likely be found in the __?__ biome. (desert, grassland, tropical rain forest)
4. Migration and hibernation are adaptations most likely found in animals of the __?__ biome. (desert, deciduous forest, tropical rain forest)
5. Animal niches are most likely to occur in "layers" in the __?__ biome. (tundra, grassland, tropical rain forest)
6. The climate zone with four distinct seasons is the __?__ zone. (temperate, tropic, polar)
7. Two adaptations that prevent water loss in plants are __?__. (spines and no roots, waxy layers and needles, lack of leaves and tallness)
8. Wheat, corn, barley, and clover are the dominant vegetation in the __?__ biome. (grassland, tundra, deciduous forest)
9. As altitude increases, the temperature __?__ in mountain biomes. (increases, stays the same, decreases)
10. The tundra biome has __?__. (much rainfall and a long growing season, little rainfall and a short growing season, permafrost and a long growing season)
11. Instincts are behaviors an animal __?__. (learns, is born with, learns by practicing)
12. __?__ enables animals to blend in with their environment. (camouflage, migration, bioluminescence)
13. Organisms in the __?__ must withstand large daily changes in the environment. (tundra, tropical rain forest, desert)

Understand The Concepts

14. Describe two food-getting adaptations in animals.
15. Define climate. Name three climate zones.
16. Name a biome in each of the three climate zones.
17. Describe some of the adaptations that help desert plants and animals survive.
18. What is instinct? Name one instinctive behavior that helps an organism survive.
19. Describe three protective adaptations in plants or animals.
20. Explain how climate and biome are related.
21. What is hibernation?
22. Explain how hibernation and migration are adaptations.
23. What factors cause a mountain to have different biomes from the base to the top?

24. Compare an alpine tundra and an Arctic tundra. How are they alike? How are they different?
25. What adaptations help organisms survive the changing conditions in the intertidal zone?

Challenge Your Understanding

26. Which biome is likely to have a greater variety of niches, a desert or a tropical rain forest? Explain.
27. A biome and an ecosystem both contain communities of living things. How does a biome differ from an ecosystem?
28. Compare the tundra biome to the desert biome. How are they alike? How do they differ?
29. What would happen if you moved a tundra organism to the desert? Explain.
30. What are some adaptations that deep-sea organisms need in order to survive the special conditions there? Explain.

Projects

1. Use an old aquarium or a wide-mouthed jar to set up a model of the desert biome. Obtain several cacti and other desert organisms for your terrarium. Find out how to care for them.
2. Insect mouth parts are adaptations that help insects get specific kinds of food. Find out what kinds of food different insects eat. Make a poster showing a close-up of each insect and its mouth parts. Explain how the mouth parts help each insect get food.
3. Prepare a report on insect-eating plants and their food-getting adaptations. Put together a bulletin-board display.
4. Read magazines and newspapers to find out about the effects of mining and agriculture on the Amazon tropical rain forest biome in South America. Report back to your class.
5. Monarch butterflies undergo a long-distance migration each year. Using library reference materials, trace the annual migration route of the monarch. Draw a map of the migration route. Label it to show where the monarch is during each season. Find out where the monarch mates and lays eggs. How does the monarch sustain itself during the migration?
6. Find out what precautions were taken to prevent the destruction of the North American tundra biome during the construction of the Alaskan oil pipeline.
7. Read the journals of explorers and naturalists who first crossed the North American continent, such as Powell, Peary, Muir, Sacagawea and Lewis and Clark, and Henson. What were North American biomes like at the time of their explorations? How has the continent changed?
8. Visit a local science or natural-history museum to view models of different biomes.

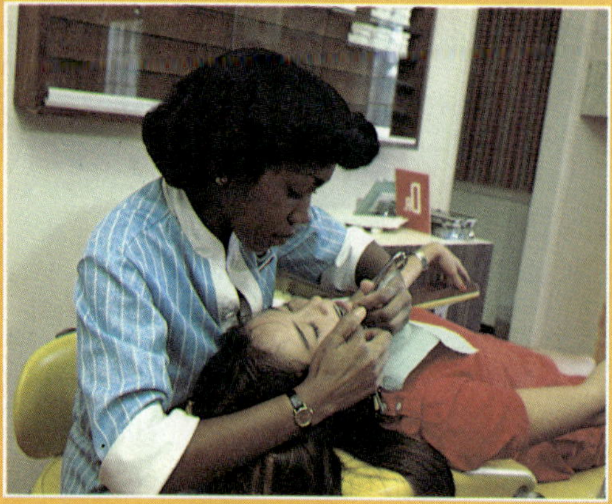

Dental Assistant

Community clinic seeks dental assistants to help examine and treat patients. Assistants put patients at ease and assist dentists by preparing equipment, suctioning patients' mouths, processing x rays, and instructing patients in oral hygiene. In addition, some assistants may be asked to help manage the office by arranging appointments, processing bills, and ordering supplies.

A high school diploma is required. A two-year college degree is preferred. Courses in biology, chemistry, health, typing, and office practice would be helpful. Applicants should be able to handle expensive equipment and work in a standing position. They also should be friendly and supportive with patients. As needed, some on-the-job training will be offered.

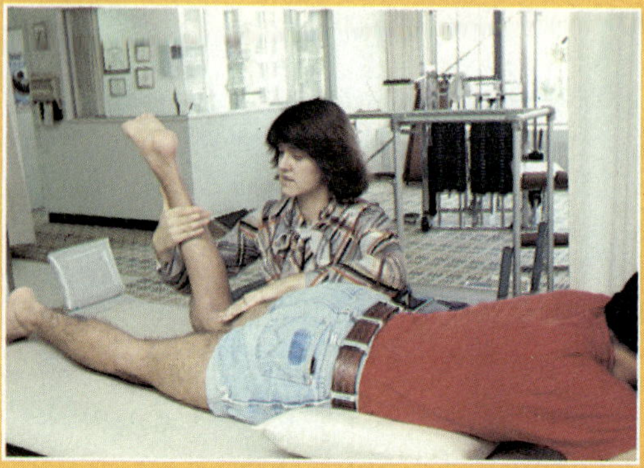

Physical Therapist

Rehabilitation center seeks physical therapists to treat disabled persons. Therapists work with the patient's physician to establish a program of therapy that will restore as much as possible the normal body functions of the patient. Therapists also help patients and their families adjust to limitations and continue treatment at home.

This job requires patience, tact, and understanding. Therapists must be able to care for the same patients for long periods of time. The work can be tiring, since therapists may have to lift or support patients.

Applicants must have a college degree in physical therapy with course work in health, biology, chemistry, physics, and mathematics. Prior experience with the disabled would be helpful.

Help Wanted

Forestry Technician

Federal government needs forestry technicians to help foresters protect national forest and park resources. Technicians plant trees, thin overgrown areas, supervise road-building, and oversee timber-cutting. Technicians also check areas for disease, fire, and erosion. Most of the work is done outdoors and is physically demanding.

Completion of a one-year or a two-year forestry technician's program is preferred. Knowledge of mathematics and biology plus experience in fighting forest fires or working in a tree nursery would be helpful. Successful applicants will begin as supervised trainees. Forestry technicians should expect to work in all kinds of weather and for long hours during emergencies, such as fires and floods.

Emergency Medical Technician

Hospital seeks emergency medical technicians (EMT's). EMT's drive ambulances to the site of an emergency and give immediate treatment to victims. EMT's must be able to administer first aid, restore breathing, and assist in emergency births.

Applicants must have completed a training course for EMT's. Applicants also must be able to think quickly under pressure and to withstand stress. Courses in driver education, health, and the sciences plus knowledge of local roads would be helpful.

EMT's must be ready to treat victims indoors and outdoors in all types of weather. EMT's also should be able to lift heavy loads. All EMT's must work some night and weekend shifts.

Understanding The Unit

1

Main Review

Write your answers on a separate piece of paper.

Know The Facts

1. The jelly-like substance inside the membrane of the cell is ? . (chlorophyll, diffusion, cytoplasm)
2. The energy for the life activities of a cell comes from ? . (ribosomes, mitochondria, vacuoles)
3. The movement of ? propel paramecia. (cilia, cell membranes, vacuoles)
4. An organ is made up of two or more different types of ? . (systems, tissues, cells)
5. A waste product of photosynthesis is ? . (carbon dioxide, glucose, oxygen)
6. The method of asexual reproduction where the parent cell undergoes cell division that results in two identical daughter cells is called ? . (hybrid, fission, meiosis)
7. A kind of cell division that results in half the number of chromosomes in the daughter cells is ? . (meiosis, mitosis, fission)
8. In flowering plants, sperm nuclei are contained within the ? . (pistils, pollen grains, fruits)
9. Colorblindness and hemophilia are ? traits. (hybrid, dominant, sex-linked)
10. A change in a pair of bases in DNA results in a ? . (recessive trait, mutation, dominant trait)
11. A ? gene is only expressed when the ? is absent. (dominant/recessive, recessive/dominant, recessive/pure trait)
12. ? cushions the bone ends and enables them to move smoothly against one another. (ligament, cartilage, muscle)
13. The muscle that is capable of rapid contractions for your entire life is ? . (striated, smooth, cardiac)
14. Blood is pumped from the right ventricle to the ? . (aorta, left atrium, lungs)
15. Digestive juices contain ? that speed the breakdown of food into simpler substances. (wastes, enzymes, villi)
16. Muscle coordination and balance are controlled by the ? . (cerebellum, cerebrum, medulla)
17. A place where an organism lives within an ecosystem is called a ? . (community, habitat, food web)
18. Green plants in a food chain or web are called ? . (consumers, decomposers, producers)
19. Animals that eat both plants and other animals are called ? . (omnivores, carnivores, herbivores)
20. Earth's three climate zones are polar, ? , and tropical. (forest, grassland, temperate)
21. A large area of land with similar

communities and climate is a(n) _?_. (biome, ecosystem, habitat)

22. A tree that sheds its leaves each autumn is called _?_. (coniferous, deciduous, evergreen)

Understand The Concepts

23. Describe how paramecia respond to conditions in their environment.
24. List the seven levels of the classification system.
25. What is diffusion?
26. When plants produce excess glucose by photosynthesis, how is it stored?
27. Name the nineteenth-century Austrian monk who studied inherited traits and summarize his findings.
28. Is the recessive or the dominant gene expressed in a hybrid?
29. Name three parts of DNA.
30. What causes albinism?
31. Name two involuntary muscles.
32. Identify three types of joints that are found in your body.
33. Where in your body are red blood cells produced?
34. Where are carbon dioxide and oxygen exchanged in your body?
35. Describe the pathway of a stimulus through a nerve cell.
36. Why is the hypothalamus sometimes called the "master gland"?
37. What is the role of nitrogen-fixing bacteria in the nitrogen cycle?
38. What percent of the energy stored in food is available to the organism in the next level of the food chain?
39. Identify four factors that can cause a population to change in size.
40. Name some natural events that can cause the destruction of habitats.
41. Compare a deciduous forest biome with an evergreen forest biome.
42. What are the main producers in the open-ocean zone?

Challenge Your Understanding

43. Compare the feeding activities of the paramecium and the amoeba.
44. How are the members of the Moneran Kingdom different from all other organisms?
45. Describe a method of asexual reproduction that starfish and planaria have in common.
46. Predict what might happen to life on Earth if green plants were suddenly unable to photosynthesize food.
47. Why is the radiation produced by the explosion of a nuclear bomb considered as dangerous as the actual explosion?
48. What might happen to a population of animals if all the natural predators were removed?
49. A farmer grows corn that is used to feed a chicken. What proportion of the energy contained within the corn does the farmer obtain by eating the chicken?
50. Explain the difference between the voluntary and the involuntary muscles in your body.
51. Describe what happens to an impulse at a synapse.
52. Why are there no green plants below a depth of about 20 meters in the ocean?
53. If you were given a desert-dwelling mouse as a pet, what conditions would you provide to give it a desert-like home?

Unit 2 · Chemistry

Chemists make measurements with many different tools such as those shown here.

250
PYREX®
100ml
KIMAX
50ml
KIMAX
30ml
KIMAX

Chapter

6

Properties of Matter

This photograph, taken in Yosemite National Park, shows several forms of matter.

6-1

Mass, Volume, and Density

Goals

1. To explain that mass and volume are properties of matter and to describe ways of measuring them.
2. To define density.

You can tell just by the color of a glass of milk that it is not water or orange juice. Even if you could not see it, you could probably tell it was milk by its taste. Color and taste are two properties that help you tell things apart. Generally, a **property** is any quality that can be useful in identifying something.

property

Milk, water, and orange juice are just three of the many kinds of matter. All matter has two properties in common. **Matter** is anything that takes up space and has mass. **Mass** is a measure of the amount of material in an object. Mass is normally measured using an equal-arm balance. An object of unknown mass is placed on the left side of the balance. Objects of known mass, or *standard masses*, are placed on the right side until both sides just balance. (See Figure 6-1.)

matter

mass

Standard masses are based on the standard **kilogram** (kg), a block of platinum metal that is kept in France. A number of exact copies have been made for use all over the world. One kilogram is equal to 1000 *grams* (g). The mass of

kilogram (KIL uh gram)

Figure 6-1 *An equal-arm balance is used for finding the mass of an object.*

this book is about 1 kg, or 1000 g. Two raisins have a mass of about 1 g.

volume

The space taken up by an object is its **volume**. The volume of a cube or a rectangular solid can be found by multiplying length × width × height. (See Figure 6-2.) Most volumes cannot be measured using a ruler. You would need a measuring cup to find the volume of water or milk, for example. Suppose you filled a cube that measured 10 cm × 10 cm × 10 cm with water. The volume of the water would be the same as that of the container, 1000 *cubic centimeters* (cm^3). A volume of 1000 cm^3 is called a **liter** (L). One liter is equal to 1000 *milliliters* (mL). (See Figure 6-3.) A can of motor oil has a volume of about 1 L. The volume of a raindrop is about 0.1 mL.

liter
(LEE tur)

You can measure out any volume of water that you might want. Changing the volume (or mass) does not change the basic nature of a substance. Some properties of a substance are not so easily changed, however. Suppose you measured out 10 mL of water, for example, and found its mass. Then suppose you measured out 50 mL of water and found its mass. You would find the mass of the second sample to be *five* times the mass of the first. The ratio of the mass of a sample of matter to its volume is the **density** of the sample.

density

$$\text{density} = \frac{\text{mass}}{\text{volume}} \quad \text{or} \quad d = \frac{m}{v}$$

Figure 6-2 *The volume of a rectangular solid can be found by multiplying length times width times height.*

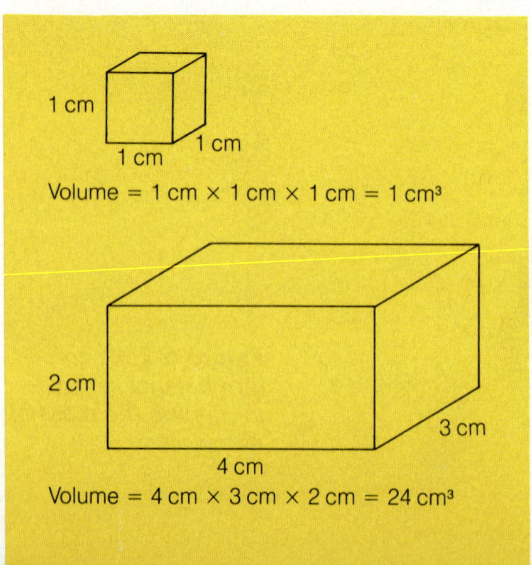

Figure 6-3 *A liter is a little more than a quart. Note that a milliliter and a cubic centimeter are the same volume.*

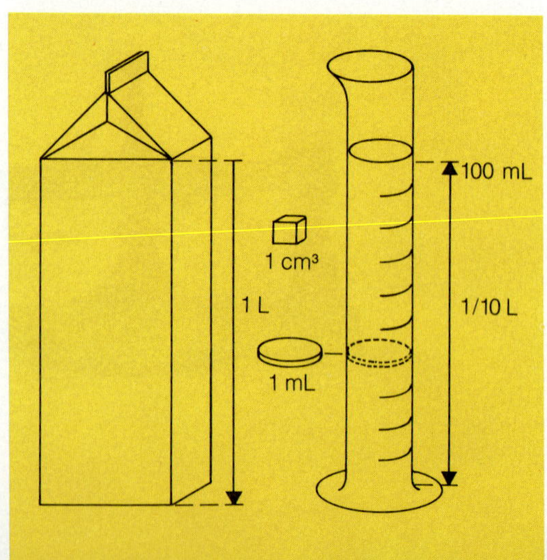

EXAMPLE 1

A 10-mL sample of water has a mass of 10 g. What is its density?

$$d = \frac{m}{v}$$

$$d = \frac{10\text{ g}}{10\text{ mL}} = 1.0\ \text{g/mL}$$

EXAMPLE 2

A sample of gold has a volume of 10.0 cm^3 and a mass of 193 g. What is its density?

$$d = \frac{m}{v}$$

$$d = \frac{193\text{ g}}{10.0\text{ cm}^3} = 19.3\text{ g/cm}^3$$

The density of a substance does not depend on how much of it there is. A small piece of gold has the same density as a large piece. A small sample of water has the same density as a large amount. The density of a substance does not change. Therefore, density can be used to identify a substance. Suppose you find a shiny, yellow object and wonder if it is gold. You measure its mass and volume and find that its density is 5.0 g/cm^3. Since gold has a density of 19.3 g/cm^3, the object cannot be gold. Table 6-1 lists the densities of several substances.

Table 6-1: Densities of Some Substances (at 20°C)

Substance	*Density* (g/cm^3)	*Substance*	*Density (g/mL)*
ice	0.91	helium	0.00018
magnesium	1.74	air	0.00123
glass	2.4–2.8	carbon dioxide	0.00196
aluminum	2.7	alcohol (ethanol)	0.79
iron	7.9	gasoline	0.7
copper	8.9	water	1.0
silver	10.5	mercury	13.6
gold	19.3		
platinum	21.5		

Figure 6-4 *Icebergs in the ocean float because liquid water is more dense than ice. These icebergs were photographed near Carey Island, Greenland.*

Notice in Table 6-1 that ice is less dense than water. This is why ice floats in water. (See Figure 6-4.) Copper is denser than water, so a piece of copper wire sinks in water. But copper floats in mercury, a very dense liquid. Helium, a gas less dense than air, is used in balloons. Metals with low densities, such as aluminum, are used in airplanes. A lump of iron sinks in water, but an iron ship floats. This is because the ship is a container. Most of its volume is occupied by air. The density of the ship is less than the density of water.

TRY THIS

Hold your hands in front of you, with the palms up. Place a piece of paper on each hand. Then close your eyes and have someone place a metal teaspoon on one hand and a plastic teaspoon on the other. Can you tell which spoon is which? Which spoon has more mass? Which is denser?

Checkpoint

1. What is matter?
2. Name two units of mass. How are they related?
3. Describe a method for measuring mass.
4. Which is more, a mL or a cm^3?
5. Water has a density of 1.0 g/cm^3. Solid X has a density of 2.7 g/cm^3. Solid Y has a density of 0.8 g/cm^3. Which of these solids will float in water?

6-2

States of Matter

Goals

1. To describe the three states of matter.
2. To define melting point and boiling point.
3. To predict the state of a substance at a certain temperature, given its melting and boiling points.

states of matter

solid

liquid

gas

melting point

boiling point

evaporation

condensation

Water can be a solid (ice), a liquid, or a gas. Solid, liquid, and gas are the three **states of matter**. A **solid** has a volume and shape that do not change. The shape of a **liquid** changes, but the volume does not. You can change the shape of a liquid simply by pouring it into a different container. A **gas** does not have a definite shape or volume. It fills whatever container it may be in. The densities of gases are much less than those of solids or liquids. (Look again at Table 6-1.)

The state of a substance can be changed. An ice cube in a glass of lemonade melts as it absorbs heat from the lemonade. The change from solid to liquid happens at a temperature called the **melting point**. The melting point of water (ice) is 0°C. At 0°C, liquid water freezes to solid water (ice). The *freezing point*, the temperature at which a liquid changes to a solid, is the same temperature as the melting point of the solid.

At sea level liquid water boils and changes to a gas at 100°C. The temperature at which a liquid changes to a gas, when bubbles form in the liquid, is its **boiling point**.

At lower temperatures, water can change from a liquid to a gas without boiling. This change is called **evaporation**. Water evaporates from puddles on sunny days, and from clothes hung up to dry. The liquid water changes into a gas that mixes with air. Water in the gas state is called *steam* if it is hotter than 100°C. At lower temperatures it is called *water vapor*.

Figure 6-5 *A dripping icicle is a good example of a change of state.*

The change from a gas to a liquid is called **condensation**. Water vapor condenses when it cools. The drops of water that appear on a bathroom mirror when you take a shower have condensed from the air. So have the water drops that form on the outside of a cold drinking glass. The cloud above a pan of boiling water is sometimes called

Figure 6-6 *Water gains or loses heat as it changes state.*

"steam," but the cloud is made up of tiny drops of water that have condensed from the invisible steam. Figure 6-6 summarizes water's changes of state.

The melting and boiling points of a substance can be used to identify it. Table 6-2 lists the melting and boiling points of some substances.

Table 6-2: Melting and Boiling Points

Substance	*Melting point* (°C)	*Boiling point* (°C)
helium	− 271	− 268
oxygen	− 219	− 184
nitrogen	− 210	− 195
mercury	− 39	357
water	0	100
moth flakes	53	174
lead	327	1525
gold	1063	2500
iron	1530	2450

On the Celsius scale, room temperature is about 20°C. At room temperature all substances with a boiling point less than 20°C are gases. Oxygen and nitrogen, which make up more than 99% of the air, are gases. Helium, which is less dense than air, is also a gas. Substances with boiling points above 20°C and melting points below 20°C are liquids at room temperature. Of the substances in Table 6-2, only water and mercury are liquids at 20°C. At 60°C, moth flakes would be a liquid, too.

Checkpoint

1. Describe the properties of each of the three states of matter.
2. What is a melting point? a boiling point?
3. What is evaporation? condensation? Give an example to illustrate each one.
4. In what state (solid, liquid, gas) are the following at 300°C: (a) water, (b) nitrogen, (c) lead, (d) iron, (e) mercury?

TRY THIS

Fill a sink or large bowl with water. Turn a drinking glass upside down and lower it into the water. Why doesn't water enter the glass?

Storing Heat

The sun is being used as a source for heating an increasing number of homes. Such homes must have some way of storing heat for use at night and on cloudy days. Heat can be stored in heavy walls or slabs made of stone or concrete, or in large containers of water. (In the photograph, the sun shines on solar collectors on a monastary in New Mexico.)

Storage space can be a problem, however. One 1940's solar house had a collector that covered half the roof and a hot-water storage tank that filled the whole basement!

Maria Telkes, a chemical engineer, decided that using large tanks of water or tons of rock to store heat was not practical. She found what she thought would be a perfect material for storing the sun's heat. It is Glauber's salt, which melts above room temperature, at about 32°C. When the salt melts, it absorbs heat. When it freezes, it gives off heat. It can store seven times as much heat as the same volume of water.

Telkes found a woman from Boston, Amelia Peabody, who agreed to furnish the money to build a new kind of solar house. Then Telkes hired an architect, Eleanor Raymond, to design the house, which was built in 1948. The walls between the rooms contained cans of Glauber's salt. When the sun shone, air was blown across the solar plates, then past the cans of salt. The warm air melted the salt.

When the temperature of the air around the walls decreased to 32°C, the liquid salt began to freeze and released the heat it had absorbed when it was melting. As long as there was any liquid salt, heat was available to warm the house. Even after all the Glauber's salt had frozen, it was still warmer than room temperature. Cool air blown over the warm salt was heated. The warmed air could be circulated throughout the house.

Unfortunately, the Glauber's salt worked well for only two winters. During the third New England winter, the salt separated into solid and liquid layers. The salt would no longer release the heat it absorbed. However, later experimenters solved this problem. Maria Telkes continued studying methods for using solar energy for many years.

6-3
Solutions

Goals

1. To discuss the properties of mixtures and solutions.
2. To define solubility and describe how it can be used to identify a substance.
3. To describe the effect of temperature on solubility.

mixture

A **mixture** is a physical combination of two or more substances. Often you can tell that something is a mixture just by looking at it. A salad is clearly a mixture. Other mixtures are harder to recognize. Salt water is a mixture even though it looks like pure water. Salt water does not look different from pure water because the tiniest particles of salt are distributed evenly throughout the water.

solution

A mixture in which the smallest particles of two or more substances are evenly distributed is a **solution**. The most familiar solutions are those made by dissolving a solid in a liquid. However, solutions may include any of the three states of matter. Carbonated drinks, such as soda water, are solutions of a gas (carbon dioxide) in water. One popular drink is a solution containing water, cranberry juice, corn syrup, and vitamin C (a solid). Common solid solutions, or *alloys*, include steel, brass, and bronze. (See Figure 6-7.)

A substance in which another substance dissolves is called a *solvent*. The substance that dissolves in a solvent is called a *solute*. Water is a good solvent for many substances. It is a poor solvent for others. Water does not dissolve tar or plastics, for example. This is why solvents other than water are used to dry-clean clothes.

The largest solution of solid in liquid is the ocean. Its volume is 1.3 billion cubic kilometers. It contains enough salt to cover all the land on Earth with a layer of salt 150 meters deep.

There is a limit to how much salt can be dissolved in a given amount of water at a given temperature. If you continue to add salt to water after that limit has been reached, the salt does not dissolve. It simply sinks to the bottom of the container. A solution that contains the maximum amount of a substance that will dissolve at a given temperature is a *saturated solution*.

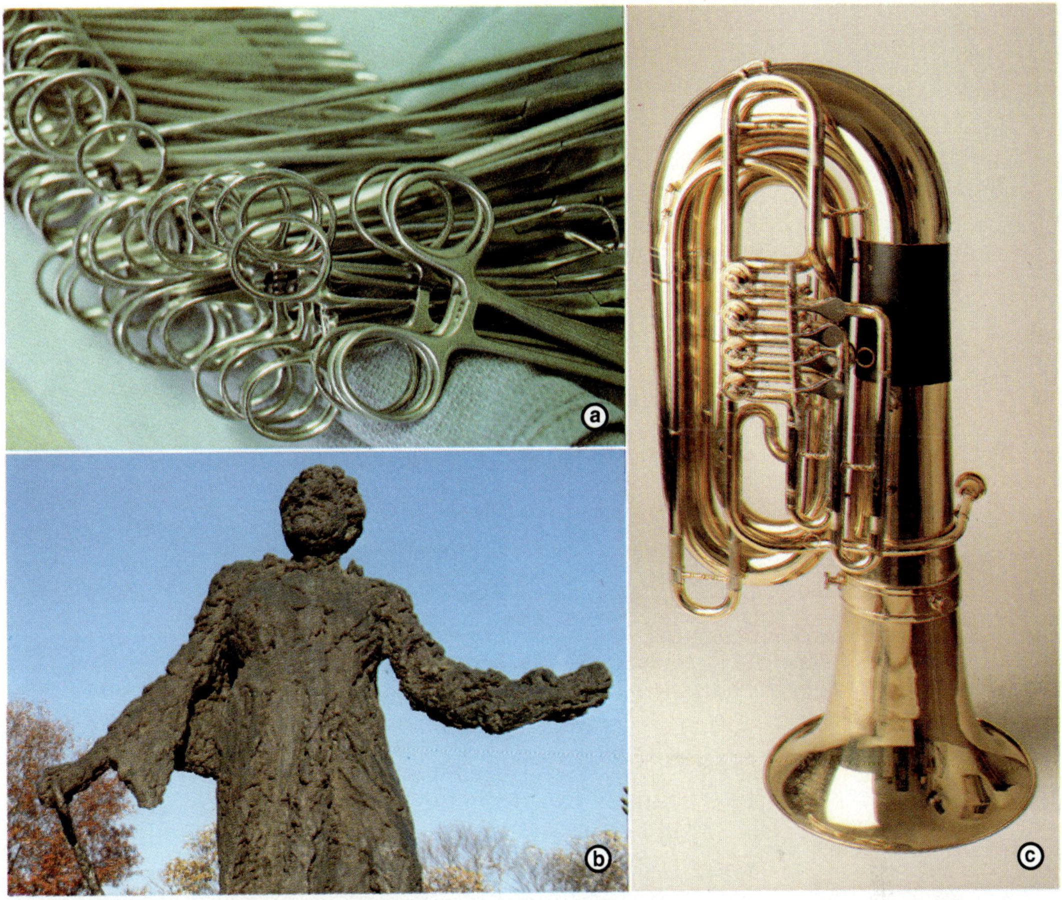

Figure 6-7 *Uses of alloys: (a) stainless steel surgical instruments, (b) a bronze statue of Mary McLeod Bethune, (c) a brass tuba.*

The **solubility** of a substance is the maximum amount of that substance that will dissolve in a given amount of solvent at a given temperature. The solubility of common salt (sodium chloride) in water at 20°C is 37 g per 100 g of water. Solubility is one property that can be used to identify a substance. For example, sodium nitrate looks like sodium chloride. But the solubility of sodium nitrate is 85 g per 100 g of water at 20°C.

solubility
(sahl yuh BIL uh tee)

A solution that is saturated at 20°C may not be saturated at higher temperatures. At 100°C a saturated solution of sodium chloride has 40 g of sodium chloride in 100 g of water. The solubility of sodium chloride increases by only 3 g per 100 g of water when the temperature rises from 20°C to 100°C. But the solubility of sodium nitrate more than doubles with the same rise in temperature. It goes from 85 g per 100 g of water at 20°C to 180 g per 100 g of water at 100°C.

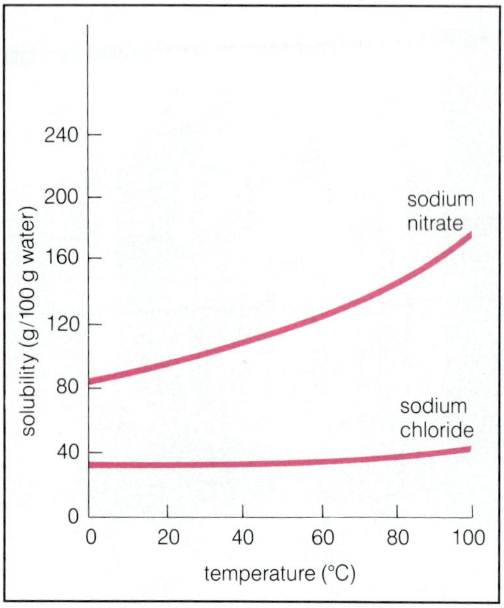

Figure 6-8 *As the temperature increases, the solubility of sodium nitrate changes much more than that of sodium chloride.*

Figure 6-9 *Hot, mineral-rich solutions erupt from cracks in the ocean floor. The minerals precipitate in the cold water.*

(See Figure 6-8.) The change in solubility as the temperature changes can be used to identify a substance.

Hot water generally dissolves more of a solid than cold water. As a hot solution cools, the excess solid *precipitates*, or falls out of the solution. (See Figure 6-9.)

Cold water generally holds more gas in solution than hot water. If a glass of carbonated drink is warm, the dissolved carbon dioxide gas escapes faster.

Fish take in oxygen dissolved in water through their gills. Sometimes the water in a lake or stream becomes very warm because of hot weather or because hot water is dumped into it from a factory or power plant. Fish may die because less oxygen can dissolve in the warm water.

TRY THIS

Add sugar, a teaspoon at a time, to a cup of cold water. How many teaspoons of sugar dissolve? Repeat the same procedure using hot tap water. What effect does the temperature of the water have on the solubility of the sugar?

Checkpoint

1. How do solutions differ from other mixtures?
2. Define solubility. Describe how this property can be used to identify a substance.
3. Describe the effect of temperature on the solubility in water of a typical (a) solid, (b) gas.

6-4

Separating Mixtures

Goals

1. To distinguish between mixtures and pure substances.
2. To give examples of the ways different types of mixtures can be separated.

Mixtures can be separated into the substances of which they are made. When a substance is no longer part of any mixture, it is known as a *pure substance*. A pure substance has a set of properties that identify it. Water is a pure substance. It freezes at 0°C. It boils at 100°C. Its density is 1.0 g/mL. It is colorless, odorless, and tasteless. One hundred grams of water will dissolve 37 g of sodium chloride at 20°C. These properties of water will not change no matter how many times it is frozen, boiled, or separated from mixtures.

If water is mixed with other substances, the properties of the mixture are different from those of water alone. For example, the freezing point of a solution of sodium chloride in water is less than 0°C and its boiling point is more than 100°C. Antifreeze is mixed with the water in a car's cooling system to keep the water from freezing. The solution that is made when antifreeze is added to the water does not freeze until the temperature is well below 0°C.

Some mixtures are easier to separate than others. The easiest ones to separate contain substances that are clearly different. The monkeys in Figure 6-10 separate a mixture of grain and sand by putting it in water. The sand sinks to the bottom, and the grain floats on top.

Figure 6-10 *Japanese macaque monkeys have been seen separating a mixture of grain and sand by putting the mixture in water.*

In a mixture of oil and vinegar you can see the droplets of oil. If you let the mixture stand for a while it separates itself. The less dense oil forms a layer on top of the denser vinegar. A mixture that clearly does not have the same properties throughout is called a *heterogeneous* (het ur uh JEE nee us) mixture.

Most heterogeneous mixtures are easy to separate. To separate an oil-and-vinegar mixture, you can simply let it stand for a while and then pour off the layer of oil. Another way to separate a heterogeneous mixture is to dissolve one or more of the substances in water. A mixture of salt and sand can be separated in this way. The salt will dissolve in water and the sand will settle out. Then the solution of salt and water can be poured off.

A mixture that has the same properties throughout is called a *homogeneous* (hoh muh JEE nee us) mixture. Solutions are homogeneous mixtures. In general, homogeneous mixtures are more difficult to separate than heterogeneous ones. But it is not very difficult to separate the substances in a mixture of salt and water. If there is no need to recover the water, simply let it evaporate away, leaving the salt behind.

Sometimes the water is the substance that people want to recover from salt water. In places where drinking water is scarce and the ocean is nearby, *solar stills* may be used for this purpose. (See Figure 6-11.) As the sun heats it, the water evaporates. Pure water then condenses on the cooler plastic cover.

Another way to separate water from salt water is by freezing. Normally when salt water freezes, the ice that forms contains only water. Tests are being made to see whether towing icebergs from cold regions to places that are

Figure 6-11 *A solar still is a large container partly filled with seawater. The sun evaporates the water, which collects on the plastic cover. These stills are at the University of California.*

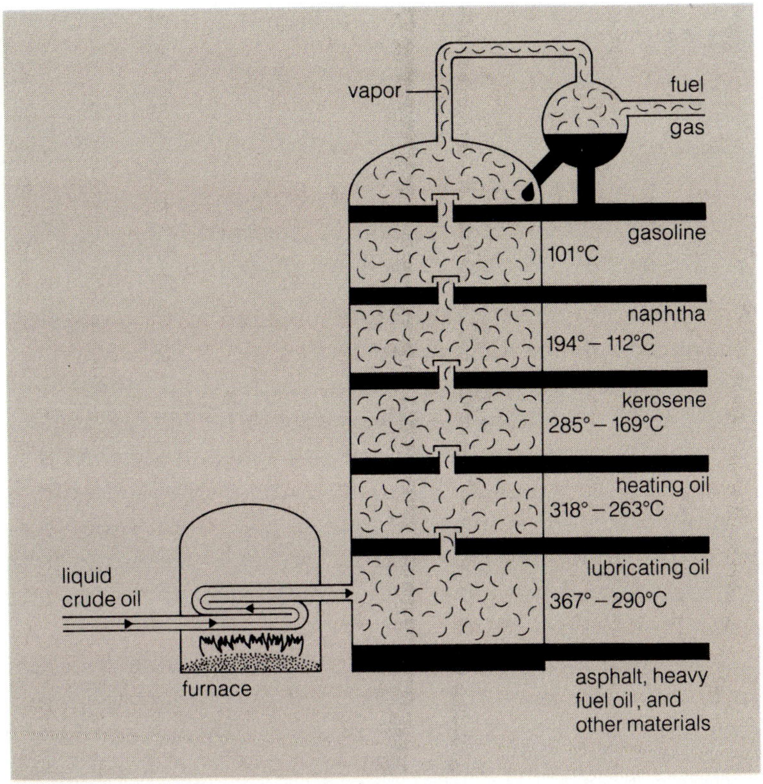

Figure 6-12 *A diagram of a fractionating tower at an oil refinery. As the hot vapor from the furnace rises and cools, different substances condense at different temperatures.*

short of drinking water is economical.

Petroleum, the thick liquid pumped from oil wells, is a mixture of many different substances. To be useful, petroleum must first be separated into simpler mixtures such as gasoline, kerosene, and asphalt. The petroleum is heated in a giant *fractionating column*. (See Figure 6-12.) Different heights in the column are kept at different temperatures. Each substance in petroleum condenses at a different temperature. So groups of substances, or fractions, condense at different heights in the column and flow into separate pipes.

Checkpoint

1. How does a pure substance differ from a mixture?
2. What is a heterogeneous mixture? a homogeneous mixture? Give an example of each.
3. Explain how you could separate (a) a mixture of salt and water, (b) a mixture of salt and sand.

6-5

Elements and Compounds

Goals

1. To explain how chemical and physical changes differ.
2. To define elements and compounds.
3. To state the law of definite proportions.

physical changes

The methods used to separate mixtures involve **physical changes**, in which no new substances are formed. When the water evaporates from salt water, leaving solid salt behind, there is no change in the salt or water. The salt left behind has the same properties as the salt that was dissolved in water to form the solution. The water that evaporated is simply water in a different state. If the water vapor condenses, it will have the same properties as the water from which the solution was made. Physical changes include changes of state, shape, and size.

chemical changes

In **chemical changes**, new substances, with different properties, are formed. If electricity is passed through water, the water is changed into two new substances. (See Figure 6-13.) The new substances, hydrogen and oxygen, are gases. They have properties that are quite different from those of water.

If hydrogen and oxygen gas are mixed together, they do not produce water. However, if a spark is added to the mixture, an explosion takes place as hydrogen and oxygen

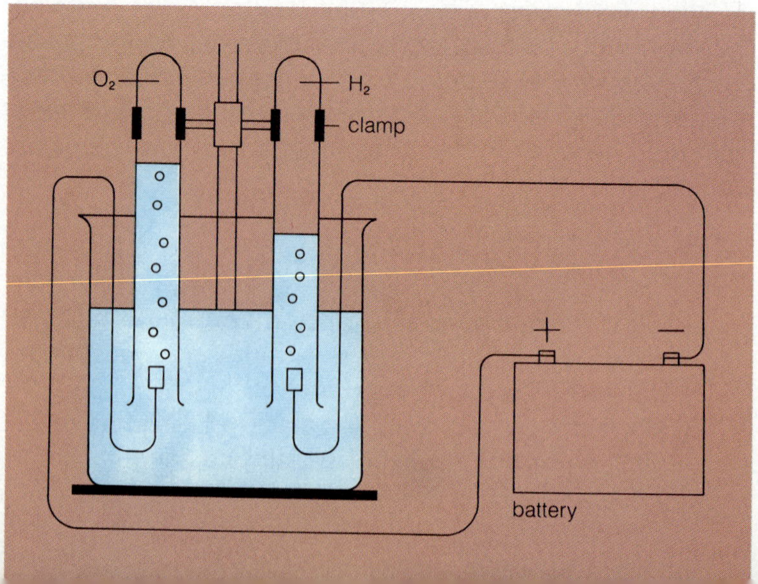

Figure 6-13 *When electricity goes through water, the water breaks into two new substances — hydrogen and oxygen.*

react, forming water. This is a chemical change, also. Chemical changes often produce noticeable heat or light, or both. Whenever something burns, chemical changes are taking place.

Chemical changes can be slow and unspectacular, too. The rusting of iron is a chemical change. As iron combines with oxygen to form rust, there is a color change. Other chemical changes involve color changes. The ripening of fruit and the fading of clothing are two examples.

Sometimes bubbles are a sign of a chemical change. When vinegar is added to baking soda, the carbon dioxide gas that is formed bubbles out of the mixture. When two liquid mixtures are poured together, a solid sometimes precipitates out. This happens when vinegar is added to milk. (The milk is said to "curdle.") Physical changes can produce bubbles and precipitates, too. But with physical changes the bubbles and precipitates usually appear during heating or cooling, not during mixing. Figure 6-14 summarizes some common signs of chemical changes.

Figure 6-14 *Some chemical changes: (a) a seltzer tablet in water, (b) curds forming in milk during the making of cheese, (c) a fire, (d) ripening blueberries.*

elements

Neither hydrogen nor oxygen can be broken down into simpler substances. They are two of more than 100 such pure substances, which are called **elements**. Elements do not break down into simpler substances in physical or chemical changes. Figure 6-15 shows some elements.

compounds

Pure substances that are a combination of two or more elements are called **compounds**. Most pure substances are compounds. The properties of a compound are very different from the properties of the elements from which it was made. Sodium is a metal that reacts violently with water. Chlorine is a poisonous green gas. These two elements combine vigorously, forming sodium chloride, or common salt. (See Figure 6-16.)

Figure 6-15 *Some elements: (a) copper, (b) mercury, (c) carbon, (d) iodine, (e) gold, (f) oxygen.*

Figure 6-16 (a) Sodium is stored in kerosene to keep it from reacting with oxygen and water in the air. (b) A tank of chlorine gas. (c) The combination — table salt, sodium chloride.

When elements combine to form a compound, they always combine in the same percentages by mass. Water is always found to be 89% oxygen and 11% hydrogen. For example, the amount of oxygen and hydrogen in a 10-g sample of water are 8.9 g and 1.1 g, respectively.

$$\text{grams of oxygen} = (0.89)(10\text{ g}) = 8.9\text{ g}$$
$$\text{grams of hydrogen} = (0.11)(10\text{ g}) = 1.1\text{ g}$$

Sodium chloride is 39% sodium and 61% chlorine by mass. Table sugar (sucrose) is 42.1% carbon, 51.5% oxygen, and 6.4% hydrogen.

Thousands of experiments have shown that each compound contains a definite percentage by mass of the elements that make it up. The results of these experiments are stated as the **law of definite proportions**: A compound consists of two or more elements combined in definite percentages by mass.

law of definite proportions

Checkpoint

1. How does a chemical change differ from a physical change?
2. What are two common signs of chemical change?
3. What is an element? a compound?
4. How do you know that water is a compound?
5. State the law of definite proportions.

6-6

ENRICHMENT

Combustion

Goals

1. To define and give examples of combustion.
2. To discuss the dangers of fire and some fire safety rules.

Many substances react with the oxygen in air so rapidly that noticeable amounts of heat and light are produced. This happens when wood or kerosene burns. When a substance reacts with oxygen, releasing heat and light, the process is called *combustion*. Matter that will burn is said to be *combustible* or *flammable*. Combustion is a chemical change. Oxygen and the combustible substance form other substances in the process.

Because wood burns, it is used as a fuel. But wood is also used in construction. Any wooden building is combustible. Curtains, furniture, and other household items will burn, too. Some materials, including many plastic products, release poisonous gases when they burn.

Scientists are studying how fires start and spread. (See Figure 6-17.) These studies will help engineers design safer materials and find ways to prevent and control fires.

Figure 6-17 *This drapery fabric in a testing laboratory turned out to be flammable.*

The production of heat and smoke and the consumption of oxygen are three reasons why fires are dangerous. Thousands of people die in fires every year. Most of them die from smoke inhalation. The tiny drops of liquid and solid particles that make up smoke fill their lungs, taking the place of oxygen.

Deaths from fires could be reduced if smoke detectors were installed in every home. These sensitive detectors signal the presence of smoke, allowing people to get out of burning buildings quickly and safely.

A fire needs fuel, oxygen, and heat. It can be put out by removing any one of these. Water will cool a fire and also keep oxygen from the fuel. Fires can be smothered by adding a dense gas, such as carbon dioxide, that will not support combustion. Small fires can be smothered by simply covering the flames to keep oxygen from the fuel. Table 6-3 provides some do's and don't's for fire safety.

Table 6-3: Some Fire-safety Tips

If	Do	Don't
Clothing catches fire.	Lie down and roll; wrap body in blanket.	Run.
Oil is the fuel.	Smother by covering; use chemical extinguisher.	Put water on the fire.
Wood is the fuel.	Add water.	Breathe the smoke.
Flammable matter is in your home.	Read labels and warnings.	Allow flames or sparks near the substance.

Checkpoint

1. What is combustion? Give two examples of it.
2. What are three reasons why fires are dangerous?
3. What can you do if someone's clothing catches fire?
4. What can you do if an oil fire starts on a stove?

TRY THIS

Light a birthday candle and place it on a can lid in the sink. Mix some vinegar with baking soda in a glass. "Pour" the gas produced by the mixture onto the flame as shown in Figure 6-18. The gas is carbon dioxide. Why is carbon dioxide used in fire extinguishers?

Figure 6-18 *Carbon dioxide gas will put out the flame.*

Understanding The Chapter 6

Main Ideas

1. Matter is anything that has mass and takes up space.
2. Density is the ratio of mass to volume.
3. Matter exists in three states: solid, liquid, and gas.
4. Substances melt and boil at characteristic temperatures called their melting and boiling points.
5. Properties such as density, melting point, boiling point, color, and solubility can be useful in identifying substances.
6. Matter exists in mixtures or as pure substances.
7. Mixtures may be heterogeneous or homogeneous. Solutions are homogeneous mixtures.
8. Pure substances are either elements or compounds.
9. Changes in matter may be either physical or chemical. There are some key signs that indicate chemical change: noticeable heat and light; a color change; the formation of bubbles and precipitates during mixing.
10. A compound consists of two or more elements combined in definite percentages by mass. This statement is known as the law of definite proportions.
11. Combustion is a chemical change in which a substance reacts with oxygen, producing heat and light.

Vocabulary Review

From the following list, choose the term that best completes each of the statements. Write your answers on a separate piece of paper.

boiling point	liter
chemical change	mass
compound	matter
condensation	melting point
density	mixture
element	physical change
evaporation	property
gas	solid
kilogram	solubility
law of definite proportions	solution
liquid	states of matter
	volume

1. The state of matter that has no definite shape or volume is a(n) __?__.
2. The __?__ is a unit of volume.
3. A(n) __?__ is a pure substance that cannot be changed into a simpler pure substance.
4. The temperature at which a substance freezes is also its __?__.
5. The __?__ of water is 1 g/mL.
6. The __?__ of a certain substance is 32 g per 100 g of water at 20°C.
7. When steam comes into contact with a cold surface __?__ occurs.
8. A(n) __?__ is a pure substance that can be broken down into two or more simpler pure substances.
9. The rusting of iron is an example of a(n) __?__.
10. The freezing of water is an example of a(n) __?__.

11. The change from a liquid to a gas without boiling is __?__.
12. A salad is obviously a(n) __?__.
13. __?__ is defined as anything that takes up space and has mass.
14. A mixture in which the smallest particles of two or more substances are evenly distributed is a(n) __?__.

Chapter Review

Write your answers on a separate piece of paper.

Know The Facts

1. The temperature at which water changes to steam is its __?__. (melting point, freezing point, boiling point)
2. The standard unit of mass is the __?__. (liter, kilogram, cubic centimeter)
3. A __?__ has a definite volume, but no definite shape. (solid, liquid, gas)
4. The solubility of oxygen in water __?__ when the water temperature increases. (increases, decreases, stays the same)
5. An equal-arm balance is normally used to measure __?__. (mass, volume, density)
6. The solubility of a solid in water usually __?__ as the temperature of the water increases. (increases, decreases, stays the same)
7. Ice floats on water because it has a lower __?__ than water. (mass, volume, density)
8. A solid may precipitate from a solution when the solution is __?__. (warmed, cooled, stirred)
9. A mixture that has the same properties throughout is __?__. (homogeneous, heterogeneous, impossible)
10. A __?__ is made up of two or more elements combined in definite percentages by mass. (mixture, solution, compound)
11. A fire can be put out by adding __?__. (oxygen, carbon dioxide, heat)

Understand The Concepts

12. A substance has a density of 0.00134 g/mL. Is the substance a solid, a liquid, or a gas?
13. Some sugar is completely dissolved in water. Is the mixture homogeneous or heterogeneous? Explain.
14. List four key signs that a chemical change may be taking place.
15. Is water an element or a compound? How do you know?
16. Give two examples of a physical change.
17. List four properties that can be used in identifying a substance.
18. Salt (sodium chloride) is added to water at 80°C until the solution is saturated. What will happen as the solution cools down to room temperature?
19. What is the volume of a plastic box that is 8 cm × 4 cm × 5 cm? How many mL of water will just fill the box?
20. Give two examples of a chemical change.
21. People cannot drink salt water, yet icebergs that form in salt water could be used for drinking water. How do you explain this?

22. An object of unknown mass is placed on the left pan of an equal-arm balance. It is just balanced when a 500-g mass and a 10-g mass are placed on the right pan. What is the mass of the object in (a) grams, (b) kilograms?
23. What three things are needed to support combustion? Describe two ways in which these things can be removed in order to put out a fire.

Challenge Your Understanding

24. Describe how you could prepare a saturated solution of sugar in water. Then describe how you would determine the solubility of sugar in water at room temperature.
25. Explain what you could do to a lump of clay in order to make it float on water.
26. A block of metal has the dimensions $2 \text{ cm} \times 5 \text{ cm} \times 10 \text{ cm}$. The mass of the block is 890 g. What is the density of the block? Of what metal may the block be made?
27. Is air a pure substance, a heterogeneous mixture, or a homogeneous mixture? Explain your answer.
28. The solubility of potassium nitrate in water at 90°C is 200 g/100 g water. Its solubility at 20°C is 20 g/100 g water. If a potassium nitrate solution that is saturated at 90°C is cooled to 20°C, what mass of potassium nitrate will precipitate out of the solution?
29. Why do you think it is dangerous to add water to an oil fire?

Projects

1. Prepare a mixture of crushed ice and water or snow and water. What is the temperature of the melting ice or snow? Add salt to the mixture and stir. What happens to the temperature?
2. Determine the solubility of sugar in water at several different temperatures. Organize your results in a data table. Then plot a graph showing the number of grams that dissolve at different temperatures in degrees Celsius.
3. Prepare some ice "cubes" that have the same volume but different surface areas. How does the surface area affect the time it takes the ice to melt? (To speed up the melting you could put the ice in a pan of water.)
4. Find out what methods are being used to make salt water drinkable. Make a scale model of one of the plants currently operating.
5. Investigate the uses of *chromatography* to separate mixtures.
6. Visit a firehouse. Describe the function of each major piece of equipment that is used in fighting fires.
7. Prepare a poster on fire safety that will help others to prevent fires and react properly to a fire should they encounter one.
8. The purity of food and drugs is a concern to all consumers. Use the library or interviews to prepare a report on some aspect of food and drug inspection.

Chapter

7

Atoms and Molecules

Imagine the surface of this silicon crystal made up of regularly arranged particles called atoms.

Figure 7-1 *Nineteenth-century scientist John Dalton proposed an atomic model of matter.*

7-1

Dalton's Atomic Model

Goals

1. To explain what is meant by a scientific model.
2. To describe Dalton's atomic model.
3. To define atoms and molecules.

A scientific hypothesis is an educated guess. If you noticed that something had been eating the lettuce in your garden, you might guess that a rabbit had done it. Other indirect evidence, such as rabbitlike footprints, would support your guess. But you would not know for certain unless you saw a rabbit eating the lettuce.

Sometimes a scientific hypothesis is a guess about something that has never been seen or that cannot be seen. Such a hypothesis must be based on indirect evidence only. Suppose that the footprints near the lettuce were not rabbitlike, nor like those of any known animal. You might try to imagine some kind of creature that would eat lettuce and leave the kind of footprints that you observed. Your imaginary creature would be a **scientific model**. A scientific model explains something that is not familiar or that cannot be seen in terms of something that is familiar or visible.

scientific model

Over 2000 years ago some Greek thinkers, or philosophers, devised a model of matter. They said that all matter is made up of invisible particles so small that they cannot be divided any further. They called these particles **atoms**, which means "indivisible." In 1808, John Dalton, an English schoolteacher, described matter in a strikingly similar way. Dalton's model, unlike that of the Greek philosophers, was based on the results of scientific experiments.

atoms

Dalton said that all elements are made up of small, invisible, dense particles that cannot be divided into anything smaller. The particles were like very tiny billiard balls or marbles. Like the Greek philosophers, Dalton called the particles atoms. According to Dalton's model, atoms of a particular element are alike in mass and every other property.

Since Dalton's time, the atomic model has undergone many changes. However, many of its features are still valid. In today's model, for example, atoms are the smallest particles of individual elements. Also, different atoms have different properties. For example, the mass of an oxygen atom is about 16 times as great as that of a hydrogen atom.

Dalton also assumed that the atoms of different elements have different properties. He thought that in a chemical change, atoms are neither created nor destroyed too. Instead, they are only rearranged. When oxygen and hydrogen combine, two hydrogen atoms can combine with one oxygen atom, forming a water **molecule**. (See Figure 7-2a.) In general, a molecule is a particle of matter made up of chemically joined or bonded atoms. Molecules are extremely small. There are many trillions of water molecules in the tiniest drop of water, for example.

molecule
(MAHL uh kyool)

Oxygen and hydrogen can form another compound, called hydrogen peroxide. When hydrogen peroxide is dissolved in water, the resulting solution can be used as a bleach or an antiseptic. Hydrogen peroxide molecules each contain two hydrogen atoms and two oxygen atoms, as shown in Figure 7-2b.

Dalton's atomic model explains what is called the *law of definite proportions*. Suppose one atom of element A combines

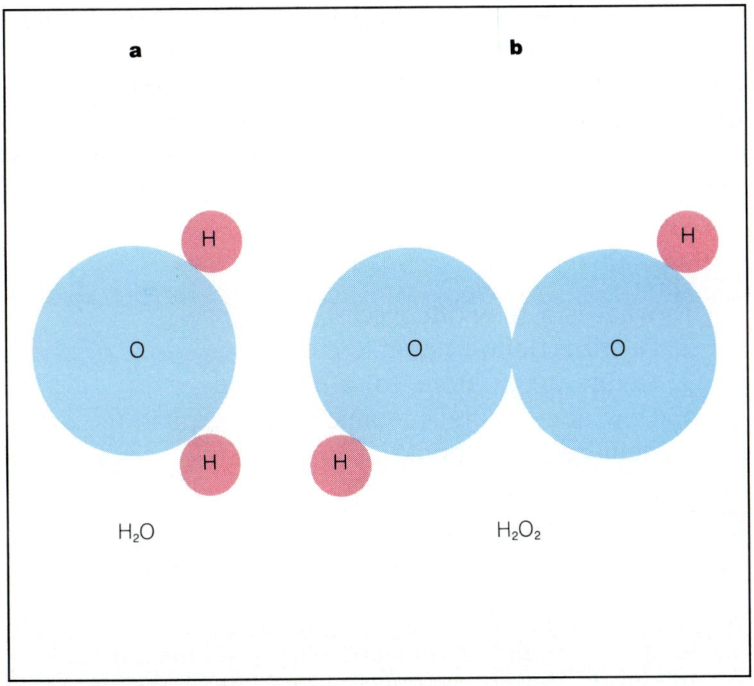

Figure 7-2 *Hydrogen and oxygen can combine to form two different kinds of molecules: (a) water, (b) hydrogen peroxide.*

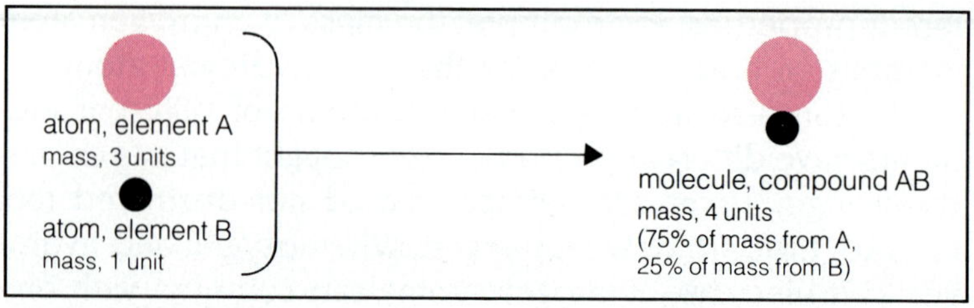

Figure 7-3 *According to the law of definite proportions, compound AB will always have the same percentages by mass of A and B.*

with one atom of element B, forming one molecule of compound AB. (See Figure 7-3.) Suppose, further, that one atom of A has a mass three times as great as that of an atom of B. If we call the mass of a B atom 1 unit, the mass of an A atom must be 3 units. Then 3/4 of the mass of every molecule of AB must come from A and 1/4 must come from B. AB is 75% A and 25% B, by mass.

$$\frac{\text{mass of A}}{\text{mass of AB}} = \frac{3}{4} = 75\% \qquad \frac{\text{mass of B}}{\text{mass of AB}} = \frac{1}{4} = 25\%$$

All molecules of AB contain these percentages by mass because all of them contain exactly one A atom and one B atom, and cannot contain fractions of these atoms. The law of definite proportions is a consequence of the fact that only whole atoms combine. Atoms are indivisible in chemical changes.

Checkpoint

1. What is a scientific model?
2. Decribe Dalton's atomic model.
3. Define (a) an atom, (b) a molecule.
4. How does Dalton's atomic model explain the law of definite proportions?

TRY THIS

Scientific models can be developed by imagining and experimenting. Get a sealed container that has been prepared by your teacher or another student. Try to develop a model of the inside of the box and its contents. Since you cannot actually see the inside, develop your model by gathering indirect evidence. Then sketch the inside of the box and its contents, based on the evidence you have collected.

7-2

Symbols and Formulas

Goals

1. To describe how chemical symbols are used.
2. To define a chemical formula.
3. To determine the number of atoms of each element in a chemical formula.

A *symbol* is a picture or sign that stands for something else. Highway signs contain symbols, as shown in Figure 7-4. If you play a musical instrument, you read symbols, called notes, that stand for certain musical tones. If you take shorthand, you write symbols that stand for words or phrases. Each time you solve a math problem, you use symbols—numerals and signs—that stand for certain quantities and mathematical operations.

chemical symbols

As a convenience, scientists use **chemical symbols** to stand for the names of the elements. Each chemical symbol is made up of one or two letters, the first of which is capitalized. Many of the chemical symbols are abbreviations for the English names of the elements. For example, S stands for sulfur and He stands for helium. Other symbols come from Latin names. For example, the symbol for lead (Pb) comes from the Latin word *plumbum*. Lead is a soft metal that was used by the Romans to make pipes. The English word *plumber* ("one who makes pipes") comes from the Latin word for lead. The symbol for gold is Au, from the Latin word for

Figure 7-4 *Highway signs, such as these, contain symbols that stand for certain objects.*

gold, *aurum*. Mercury's symbol, Hg, comes from the Latin word *hydrargyrum*, which means "silver water." Table 7-1 presents the symbols for some of the most common and useful elements.

Chemical symbols serve as more than abbreviations for the elements. Each symbol can stand for one atom of a particular element. Thus, Au stands for one atom of gold, and Hg stands for one atom of mercury. Si stands for an atom of silicon and W for an atom of tungsten.

Chemical symbols can also be used to show the number of atoms in a molecule. A molecule of water is made up of two atoms of hydrogen (H) and one atom of oxygen (O). Its symbol is H_2O. The subscript, or small number written to the right of H, shows that there are two H atoms. The symbol for oxygen has no subscript. Since it represents a single atom, the subscript 1 is understood. The number and types of atoms that are combined in a molecule of a substance are given by its **chemical formula**. The chemical formula for a molecule of ordinary table sugar is $C_{12}H_{22}O_{11}$. The formula shows that there are 12 atoms of carbon (C), 22 atoms of hydrogen, and 11 atoms of oxygen in each molecule of ordinary table sugar.

chemical formula

So far, we have looked only at single molecules of a substance. Suppose you want to represent three molecules of water. You know that the chemical formula for one molecule of water is H_2O. To represent three molecules, the coefficient 3 would be written in front of the chemical formula for water. Three molecules of water are then represented as $3H_2O$. To find the total number of atoms of each kind in the three molecules, multiply the coefficient times the number of each kind of atom shown in the formula. In three molecules of water, there are 3×2, or 6, atoms of hydrogen and 3×1, or 3, atoms of oxygen.

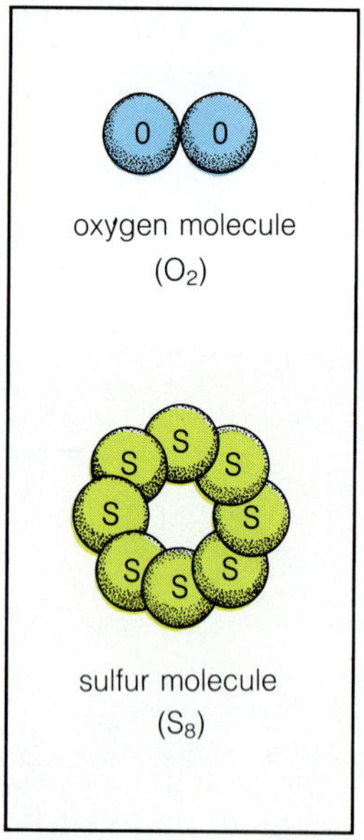

Figure 7-5 *Different elements may be made up of molecules that have different numbers of atoms.*

Scientists have found that in nature, atoms of hydrogen, oxygen, and several other elements usually exist in pairs. For this reason, molecules of these elements are called *diatomic*, or two-atom, molecules. Thus, the formula for the naturally occurring element hydrogen is H_2, and the formula for oxygen is O_2. Bromine (Br_2), iodine (I_2), nitrogen (N_2), chlorine (Cl_2), and fluorine (F_2) are also diatomic molecules. The formula for sulfur (S), a yellow solid, is S_8. The formula shows that each sulfur molecule contains eight sulfur atoms. (See Figure 7-5.)

Table 7-1: Some Elements and Their Symbols

Name	Symbol
aluminum	Al
calcium	Ca
carbon	C
chlorine	Cl
copper (Latin name: *cuprum*)	Cu
fluorine	F
gold (*aurum*)	Au
helium	He
hydrogen	H
iodine	I
iron (*ferrum*)	Fe
lead (*plumbum*)	Pb
mercury (*hydrargyrum*)	Hg
neon	Ne
nickel	Ni
nitrogen	N
oxygen	O
phosphorus	P
platinum	Pt
potassium (*kalium*)	K
silicon	Si
silver (*argentum*)	Ag
sodium (*natrium*)	Na
sulfur	S
tin (*stannum*)	Sn
uranium	U
zinc	Zn

Checkpoint

1. In what two ways are chemical symbols used?
2. What information does a chemical formula provide?
3. List the number of atoms of each kind in the following: (a) H_2SO_4, (b) $2CO_2$, (c) $3CH_4$.
4. Which of the following are diatomic molecules: $C_{12}H_{22}O_{11}$, Br_2, CO_2, S_8, H_2O, Cl_2?

***Figure 7-6** A washer-paper clips model of a molecule of water (H_2O).*

TRY THIS Let paper clips represent hydrogen atoms; washers, oxygen atoms; and bolts, carbon atoms. Use these "atoms" to represent "molecules" of carbon monoxide (CO), carbon dioxide (CO_2), and hydrogen peroxide (H_2O_2).

7 3

Chemical Equations

Goals

1. To explain what happens during a chemical reaction.
2. To identify the parts of a chemical equation.
3. To state the law of conservation of mass.
4. To tell whether a chemical equation is balanced.

chemical reaction

chemical equation

Have you ever watched a burning match or baked a loaf of bread? If so, you have witnessed a **chemical reaction**, a process in which chemical change takes place. The changes that take place in a chemical reaction can be represented by a set of symbols called a **chemical equation**. An arrow is placed between the substances that change chemically, the *reactants* (ree AK tunts), and the substances formed by the chemical change, the *products*. The arrow is a symbol for the word *forms*, or yields. Reactants undergo chemical change to form products. The reaction in which elements A and B combine, forming compound AB, can be written as follows:

$$\underset{\text{reactants}}{A + B} \longrightarrow \underset{\text{product}}{AB}$$

A chemical reaction occurs when the elements copper (Cu) and sulfur (S) are heated together. The two elements combine, forming the compound copper sulfide (CuS). Suppose that this reaction is carried out as shown in Figure 7-7. The photo at the left shows a crucible, a small container used to heat chemicals. The crucible, which contains copper and sulfur, is placed on a device used to measure mass. The mass of the crucible and its contents is recorded. To start the reaction, the mixture of copper and sulfur is then heated, as shown in the middle photo. After the reaction has stopped, the crucible and its contents are allowed to cool. Their mass is measured again, as shown at the right. The mass of the crucible and its contents is the same as it was before the reaction.

law of conservation of mass

Experiments such as this had led to the discovery of a scientific law before Dalton published his atomic model of matter. The law, called the **law of conservation of mass**, states that mass is neither created nor destroyed during a chemical reaction. Dalton's model explains why this law holds. He assumed that atoms are neither created nor destroyed in a chemical reaction, but only rearranged.

A chemical equation, then, must reflect the law of conservation of mass. It must show that there is no change in the number of atoms involved. A chemical equation in which the reactants and products contain the same number of each kind of atom is called a **balanced equation**. The following examples illustrate a step-by-step procedure for writing and balancing equations.

balanced equation

EXAMPLE 1

Write a balanced equation for the reaction between copper and sulfur that produces copper sulfide, CuS.

Step 1 Write the correct formulas for the reactants and product.

$$Cu + S_8 \longrightarrow CuS$$

Note that the equation is *not* balanced. The number of sulfur atoms on each side are unequal.

Figure 7-7 *When a mixture of copper and sulfur is heated, copper sulfide is formed. As this sequence shows, the mass remains the same throughout the process. This illustrates the law of conservation of mass.*

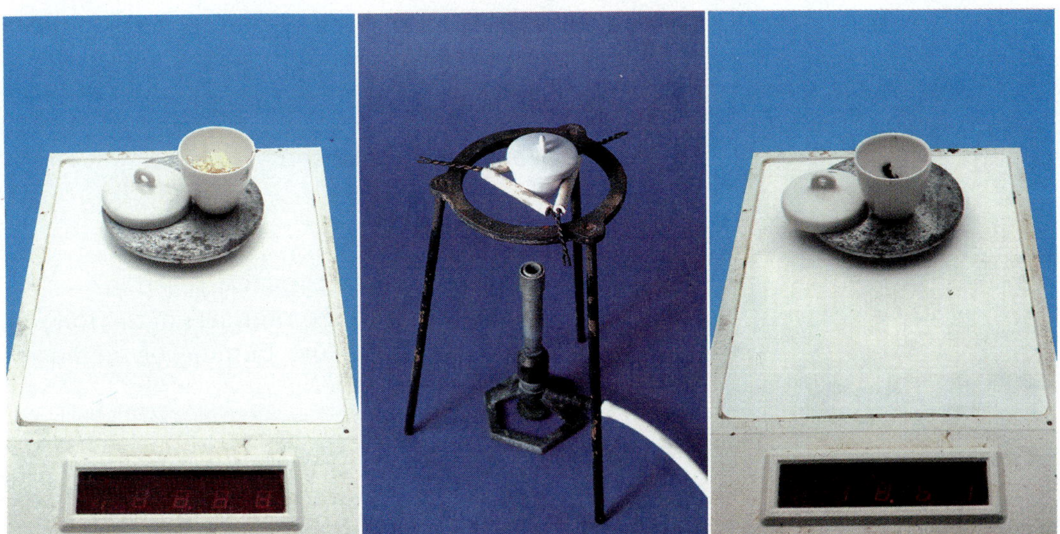

Step 2 Since eight atoms of sulfur react, there must be eight atoms of sulfur in the product. We cannot change the chemical formulas, but we can add coefficients.

$$Cu + S_8 \longrightarrow 8CuS$$

Step 3 Since the product side now contains a total of eight copper atoms, eight copper atoms must have reacted.

$$8Cu + S_8 \longrightarrow 8CuS$$

The equation is now balanced. Note that a coefficient of 1 is understood in the case of S_8.

EXAMPLE 2

When heated, a certain oxide of mercury (HgO) decomposes, forming mercury and oxygen. Write the balanced equation for the reaction.

Step 1 Write the correct formulas for the reactant and products.

$$HgO \longrightarrow Hg + O_2$$

Step 2 Since two atoms of oxygen are produced, two atoms of oxygen must be present in the reactant.

$$2HgO \longrightarrow Hg + O_2$$

Step 3 Since two atoms of mercury react, two must be produced.

$$2HgO \longrightarrow 2Hg + O_2$$

The equation is now balanced.

Checkpoint

1. What takes place during a chemical reaction?
2. Identify the reactants and the products in the following chemical equation for the burning of methane, CH_4, found in "natural gas."

$$CH_4 + O_2 \longrightarrow CO_2 + H_2O$$

3. Balance the equation in Question 2.
4. State the law of conservation of mass.

7-4

The Atomic Model Is Modified

Goals

1. To describe Thomson's model of the atom.
2. To describe Rutherford's model of the atom.
3. To define electrons, neutrons, and protons.

Scientific models are useful—for example, in predicting and explaining scientific laws. But all models have some limitations. As scientists do more experiments, models must often be altered to account for new observations. Dalton's billiard-ball model of an atom, for example, helped explain why elements combine in definite proportions. One of its limitations was that it did not explain *how* atoms combine as they do. That is, it did not explain what caused the "billiard-ball atoms" to unite chemically.

In 1897, almost 100 years after Dalton proposed his atomic model, J. J. Thomson, an English physicist, suggested a new model of the atom. A new model was needed to explain the results of experiments that Thomson had done. These experiments had shown interesting facts about what is called the *electric charge* of matter.

Scientists had known for some time that there are two kinds of electric charge—positive charge and negative charge. Two objects with the same charge repel each other. Two objects with opposite charges attract each other. In a series of experiments, Thomson observed that when atoms are strongly heated, they give off particles that are even smaller than atoms. This was a surprising fact, considering that atoms had been thought to be indivisible. The particles were attracted to objects having a positive electric charge. Since oppositely charged objects attract, Thomson reasoned that the particles being given off must be negatively charged. He called the tiny negatively charged particles **electrons**.

electrons (ih LEK trahnz)

Ordinary matter is *neutral*. That is, it has no electric charge. If matter contains electrons (negatively charged particles) and is neutral, there must be an equal amount of positive charge to *balance* the negative charge of the electrons. Thomson modified the Dalton atom to include positive and negative charge. He proposed that the electrons

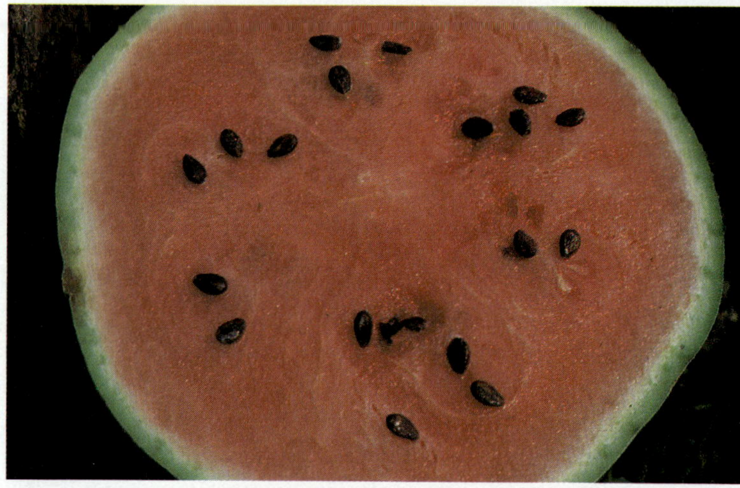

Figure 7-8 *The inner structure of a watermelon, with its seeds embedded throughout its red flesh, is similar to that of the atom proposed by J. J. Thomson. The "seeds" in Thomson's model are the negatively charged electrons, and the "flesh" is the positive part of the atom.*

were embedded within the atoms. The positive charge was spread uniformly throughout the atoms. He thought this explained why electrons are readily removed from atoms, but the positive portion is not. You might picture Thomson's atom as being somewhat like a watermelon in structure. (See Figure 7-8.) The seeds stand for the electrons. The red fleshy part of the melon stands for the positive part of the atom.

In 1911, Ernest Rutherford, a physicist from New Zealand who worked with Thomson, tested Thomson's model. He tested it by aiming a beam of fast-moving, positively charged particles at a very thin sheet of gold foil. (Even a thin sheet of gold foil is hundreds of atoms thick, and the atoms are packed closely together). Rutherford expected that most of the positive particles would pass through the gold foil with a very slight change in direction. This was because the positive charge in the atoms was thought to be thinly spread out, as Thomson had suggested. It would therefore have little effect on the bulletlike particles being fired through it.

The results of the experiment were very different from what Rutherford had expected. In the first place, most of the particles passed through the foil with no change of direction, just as though they were not passing through anything at all. However, a few of the particles swerved greatly from their straight-line path. Some even came straight back toward the source of the beam. Rutherford said that this was as surprising as if he had fired a cannon ball at a piece of tissue paper and it had come back and hit him!

To explain these strange results, Rutherford suggested yet another model of the atom. In Rutherford's model, an atom is mostly empty space. At the atom's center is a tiny, positively charged **nucleus**. (The plural of nucleus is *nuclei.*) All the atom's positive charge and almost all its mass are located in the nucleus, which is much more massive than the very light electrons found outside the nucleus. Rutherford knew that the atom's nucleus must be very small because most of the positively charged particles fired at the atoms passed right through the gold foil. To get an idea of the size of the nucleus compared to that of the entire atom, picture a tiny bug in the center of a large athletic stadium. The bug represents the nucleus. The stadium represents the entire atom. (See Figure 7-9.)

nucleus
(NOO klee us)

Rutherford knew that the nucleus was positively charged because the few positive particles that swerved or came straight back must have been repelled, rather than attracted, by the nucleus. (Remember, like charges repel

Figure 7-9 *Compared to the atom, the nucleus is about as small as an insect would be when compared to a sports stadium.*

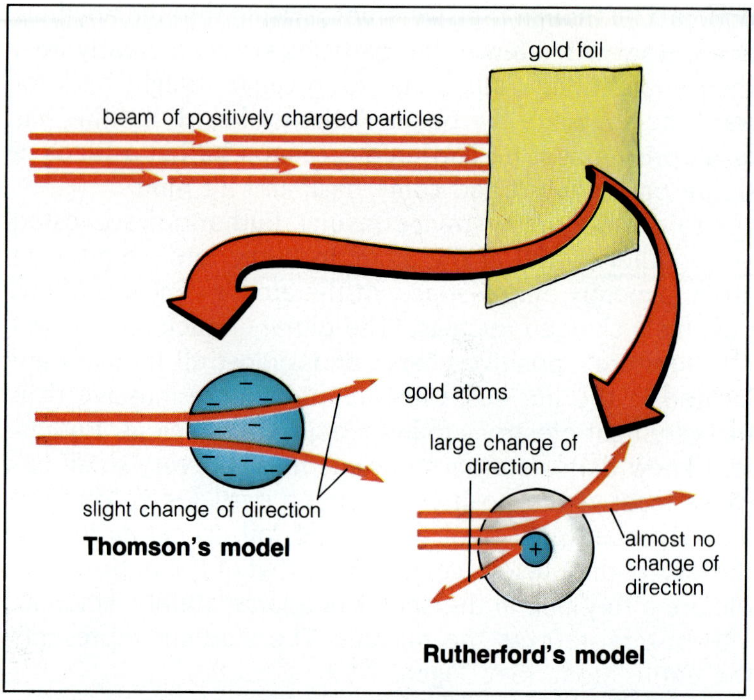

Figure 7-10 The atoms proposed by Thomson and by Rutherford would behave differently when beams of positively charged particles are passed through them. Rutherford's idea that the atom has a nucleus turned out to be correct.

each other.) Figure 7-10 shows how Rutherford's model of the atom accounts for the observed results.

Experiments with metals other than gold showed that the positive charge of the nucleus varies from element to element. Nuclei of heavier atoms carry more charge than do nuclei of lighter atoms. This led to a model of the atom in which the nucleus includes small positively charged particles called **protons**. The more protons an atom contains, the greater the charge on its nucleus, and, in general, the heavier the atom.

protons

However, careful measurements showed that the number of protons in a nucleus alone did not determine the mass of the nucleus. Rutherford's model therefore had to be modified. In addition to protons, the nucleus also contains particles called **neutrons**. Neutrons have roughly the same mass as protons but they carry no charge. They are neutral. The number of protons plus the number of neutrons is the **atomic mass**.

neutrons

atomic mass

Table 7-2: The Charges and Masses of Protons, Neutrons, and Electrons

Particle	*Charge*	*Approximate atomic mass (amu)*
	(proton or electron = 1 unit)	(proton or neutron = 1 unit)
proton	+1	1
neutron	0	1
electron	−1	1/1840

Protons and neutrons each have a mass that is more than 1840 times that of an electron. A proton is assigned 1 unit of atomic mass and 1 unit of positive charge. A neutron is assigned 1 *atomic mass unit* (or 1 amu) and 0 units of charge. Thus, a helium nucleus with two protons and two neutrons has a mass of 4 and a charge of +2. A gold nucleus contains 79 protons and 118 neutrons, giving it a mass of 197 units and a charge of +79. The number of protons in a nucleus is called the **atomic number**. The atomic number of helium is 2. That of gold is 79.

atomic number

We may sum up by saying that atoms were once thought to be the tiniest particles of matter. Later, evidence from experiments led scientists to believe that atoms were made up of even smaller particles. Electrons were eventually pictured by Rutherford as tiny, negatively charged particles that existed in the space surrounding the tiny, heavy, positively charged nucleus.

Checkpoint

1. What are the two kinds of electric charge and how do the kinds of charge affect one another?
2. Describe Thomson's model of the atom and tell why Thomson's model is somewhat like a watermelon in structure.
3. Describe Rutherford's model of the atom. What evidence led to Rutherford's model?
4. Compare the masses and charges of protons, neutrons, and electrons.

Figure 7-11 From top to bottom: Lithium compounds give off red light; copper compounds, green light; and sodium compounds, yellow light in a Bunsen burner flame.

7-5

Line Spectra

Goals

1. To explain how elements can be identified by the light they give off.
2. To describe the Bohr model of the atom.

When compounds of certain metals are heated in the flame of a Bunsen burner, they give off light of a characteristic color. The resulting flame color helps chemists identify the atoms present in the compound. Chemists know, for example, that lithium compounds give off red light in the flame. Copper compounds give a green light, and sodium compounds give a yellow light. (See Figure 7-11.)

By looking at a flame through a *spectroscope*, an instrument that spreads light out into a spectrum, or band of colors, a chemist can tell which element is present. The light from a hot, glowing solid, such as tungsten, can produce a *continuous spectrum*, (plural, *spectra*), in which colors blend into each other. (See Figure 7-12a.) The light given off by a glowing gaseous element, such as hydrogen, does not give off a continuous spectrum. Its spectrum is a series of distinct, colored lines. (See Figure 7-12b.) No two elements have the same *line spectrum*. An element's line spectrum can therefore be used to identify the element. You might think of an element's line spectrum as its "fingerprint."

Niels Bohr, a Danish physicist working in Rutherford's laboratory in 1913, wanted to account for the line spectrum produced by hydrogen. Hydrogen is the simplest and lightest element, having only one proton in the nucleus and one electron. Its line spectrum should therefore be the easiest to understand. Bohr reasoned that the closer the negative electron is to the positive nucleus, the less energy it has. It must gain energy to move away from the positively charged nucleus that attracts it. The farther away it moves, the higher its energy must become. (If it gains enough energy, an electron can completely "escape" from an atom, as Thomson's experiments showed.)

In order to explain the line spectrum, Bohr assumed that the electron in a hydrogen atom could have only certain "allowed" energies. Bohr thought that each energy corre-

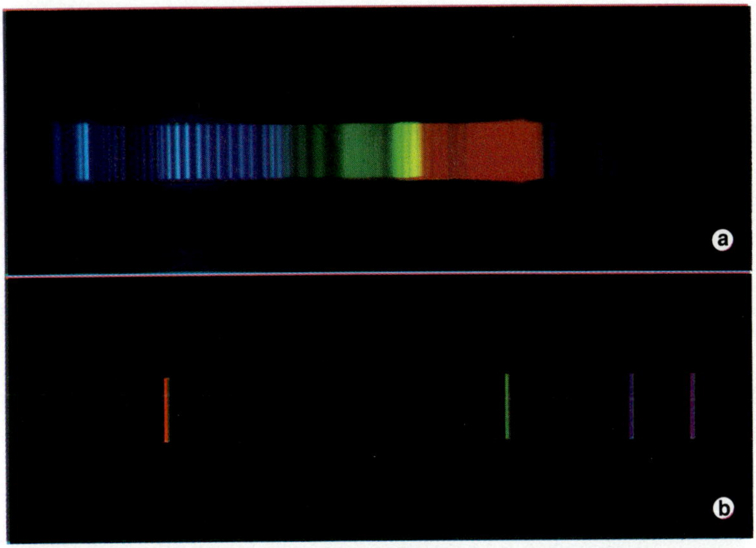

Figure 7-12 *Elements such as tungsten produce a continuous, rainbowlike spectrum, as shown in (a). Elements such as hydrogen produce a spectrum made up of separate, colored lines, as shown in (b).*

sponded to a certain average distance from the nucleus. You might think of the electrons as airplanes in assigned holding patterns above an airport. A plane may circle at the lowest "allowed" level, or altitude, or it may circle at one of the higher "allowed" levels. A plane gains energy in order to move from a lower level to a higher level. It loses energy in moving from a higher to a lower level.

energy levels

The energies that an electron can have in an atom are called **energy levels**. Energy levels were pictured as orbits, or (in three dimensions) as shells, by Bohr. (See Figure 7-13.) Normally an electron occupies the lowest available energy level. When it gains energy—by being heated, for example—it moves to a higher energy level.

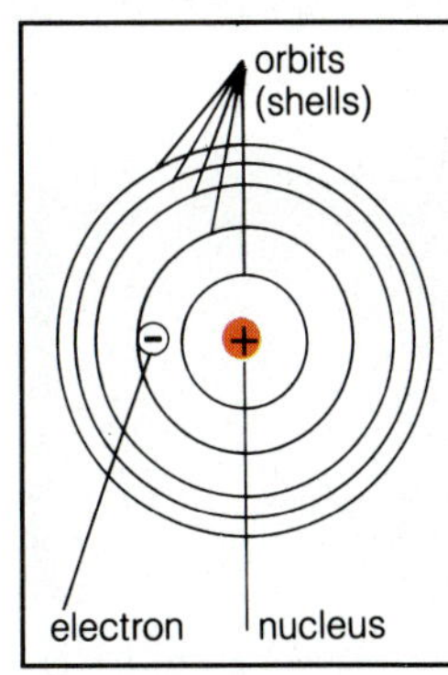

Figure 7-13 *In Bohr's model of the atom, an electron could only be found in certain definite energy levels, which Bohr pictured as orbits or shells.*

A ball at the top of a hill tends to roll back down if there is nothing to prevent it from doing so. Similarly, an electron at a higher-than-normal energy level tends to move back to a lower level. When an electron moves from a higher to a lower energy level, it gives off energy that it had originally absorbed. The energy is given off in the form of light. The color of the light depends on the amount of energy lost by the electron. For example, blue light has more energy than does red light. Electrons giving off blue light are therefore losing more energy than those giving off red light.

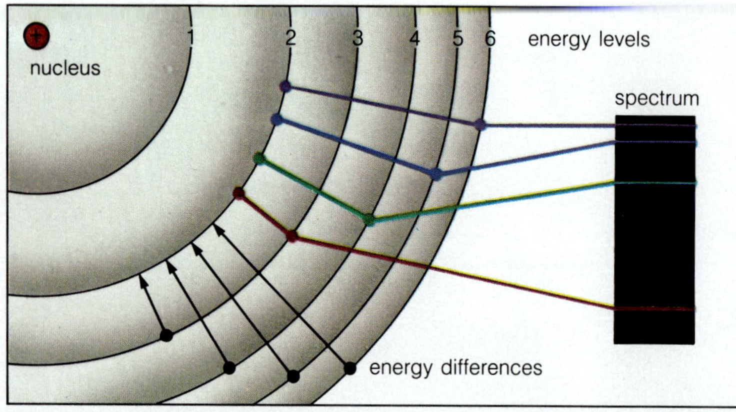

Figure 7-14 *The colored lines of the hydrogen spectrum are produced when electrons move down from various higher energy levels.*

Each line in the hydrogen spectrum corresponds to the energy difference between two energy levels. Figure 7-14 shows the energy differences for the visible lines. Note that these result from the electron's movement from higher levels to lower levels.

Bohr's explanation worked well in accounting for the hydrogen spectrum. It formed the starting point for the understanding of other atoms. Eventually, scientists were able to describe the arrangements of electrons in more complicated atoms. They found that each energy level could be occupied by only a certain number of electrons. The arrangement of electrons in the atom is what determines the chemical properties of a given element.

Checkpoint

1. How can the light that elements give off be used to identify the elements?
2. Describe the Bohr model of the atom.
3. How did Bohr explain the hydrogen line spectrum?

TRY THIS

Light a fluorescent lamp tube in the following way, without plugging it in. On a dry day, darken a room. Gently rub the tube up and down several times on your sleeve. (CAUTION: Be careful not to break the tube.) This rubbing removes electrons from the glass, and the resulting charge on the glass produces electricity inside the tube.

7-6

The Current Model of the Atom

ENRICHMENT

Goal

To compare the current model of the atom with the Bohr model.

The atomic model used by scientists today includes features of the Bohr model. Electrons in an atom are still thought to exist outside a tiny central nucleus. But they are not thought to be moving in definite "orbits." Werner Heisenberg, an atomic physicist, first concluded that it is impossible to know both the exact location and direction of movement of an electron in an atom, and that definite paths could not be assigned to electrons. His conclusion is often referred to as Heisenberg's uncertainty principle. Scientists can calculate only the *probability*, or likelihood, of finding an electron at a certain distance from the nucleus. Figure 7-15 shows that the electron in a hydrogen atom is most likely to be found fairly close to the nucleus.

Scientists can calculate the energies that electrons have in atoms. As Bohr assumed, the electrons can have only certain "allowed" energies. In moving from one energy level to another, an electron must gain or release an exact amount, or "packet," of energy. Each packet of energy is called a *quantum* (plural, *quanta*). The current atomic model is generally referred to as the *quantum model*. It is a mathematical model that gives no very clear "picture" of electron motion. The predictions that scientists have made based on the quantum model have been borne out by experiments, however. Therefore, that model is the accepted one, even though it is very difficult to visualize.

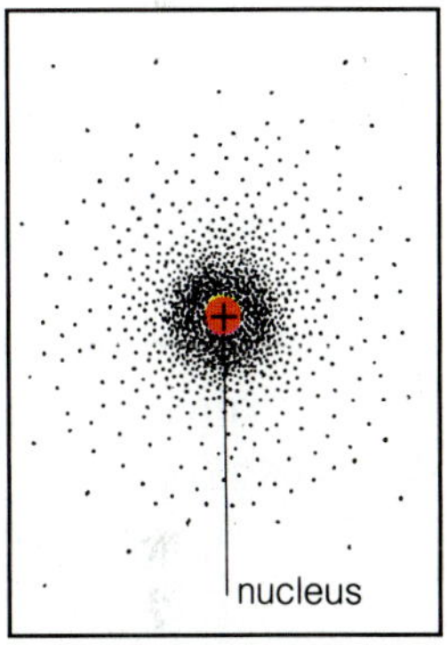

Figure 7-15 *According to the current model of the atom, the electron in a hydrogen atom is most likely to be found in the more dense regions of dots shown here.*

Checkpoint

1. In what ways is the quantum model of the atom like the Bohr model? In what ways is it different?
2. What information does the quantum model of the atom give concerning the location of an electron in an atom?
3. Why is the quantum model accepted by scientists?

FEATURE ARTICLE

Cold Light

Nearly all natural light comes from the sun. You can feel the heat it produces. Light from artificial sources is produced in several ways. Materials may be heated until they glow, or become *incandescent*, as in an ordinary electric light bulb. If you put your hand next to a lighted bulb, you can feel its warmth. Fluorescent lights also produce heat, but not as much. Light without heat is unusual.

Imagine that you are standing in a dark room, holding a slim plastic tube. When you bend the tube, an eerie green light suddenly fills the room. You might think the tube is some kind of electric light bulb. However, there are no wires leading to the tube. The tube also remains cool to your touch, even as it releases light.

Before it was bent, the tube contained a liquid and a thin-walled capsule filled with a second liquid. When the tube was bent, the capsule was broken, releasing the second liquid. The two liquids reacted chemically, forming a substance that gives off light as a result of the reaction.

Some of the atoms of which the light-giving substance is made contain electrons that have moved to higher-than-normal energy levels. The electrons absorbed energy during the chemical reaction and moved up to higher levels. After the reaction, the electrons give off the absorbed energy as they fall back to their normal energy levels. This loss of energy appears as light. The light from this particular substance is green, but other substances release different colors of light.

Cold light sources are used as emergency lights in boats and around explosive materials, where flames or electric sparks would be dangerous. Cold light sources are now expensive compared to conventional light sources. However, cold light sources may eventually be even less expensive and more useful than fluorescent lights.

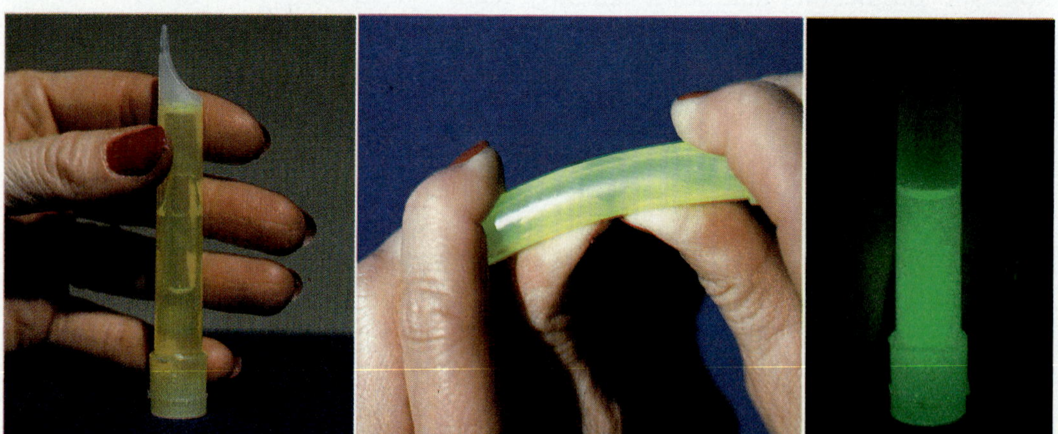

Understanding The Chapter 7

Main Ideas

1. A scientific model explains something unfamiliar or invisible in terms of something that is familiar or visible.
2. An atom is the smallest particle of an element. A molecule is a particle of matter made up of chemically joined atoms.
3. Chemical symbols stand for the names of the elements and also stand for one atom of an element.
4. A chemical formula gives the number and types of atoms that are combined in a molecule.
5. The changes that take place in a chemical reaction can be represented by a chemical equation.
6. During a chemical reaction, mass is neither created nor destroyed.
7. The nucleus of an atom contains positively charged protons and uncharged neutrons, and its mass accounts for most of the mass of the atom.
8. Each element has a unique line spectrum that can be used to identify it.
9. Electrons exist outside the nucleus in certain energy levels.
10. It is impossible to determine both the exact location and direction of movement of an electron in an atom.
11. Electrons in an atom can gain or release energy only in certain exact amounts called quanta.

Vocabulary Review

From the following list, choose the term that best completes each of the statements. Write your answers on a separate piece of paper.

atom	law of conservation of mass
atomic mass	molecule
atomic number	neutron
balanced equation	nucleus
chemical equation	products
chemical formula	proton
chemical reaction	reactants
chemical symbol	scientific model
electron	spectroscope
energy level	

1. The __?__ for carbon dioxide is CO_2.
2. A(n) __?__ is found in the nucleus of an atom, and has a charge of +1.
3. A(n) __?__ has very little mass and a charge of −1.
4. A(n) __?__ of water contains two atoms of hydrogen combined chemically with one atom of oxygen.
5. In chemical formulas, each kind of atom is represented by a(n) __?__ that is different from those of all other elements.
6. When hydrogen and oxygen react, forming water, hydrogen is one of the __?__.
7. The reactants in a chemical reaction change into the __?__.
8. A(n) __?__ has a mass equal to that of a proton, and carries no charge.

9. The changes that take place in a chemical reaction can be represented by a(n) _?_.
10. A chemical equation in which the reactants and products contain the same number of each type of atom is a(n) _?_.
11. The _?_ of a gold atom has a greater charge than that of a hydrogen atom because it contains more protons.
12. Normally, an electron in an atom occupies the lowest available _?_.

Chapter Review

Write your answers on a separate piece of paper.

Know The Facts

1. A chemical _?_ is used to represent a chemical reaction. (equation, formula, product)
2. _?_ is an example of a diatomic molecule. (CO_2, H_2O, O_2)
3. Atoms of oxygen and carbon combine chemically, forming _?_ of carbon dioxide. (nuclei, atoms, molecules)
4. _?_ atom(s) of sulfur is (are) represented in the formula H_2SO_4. (3, 2, 1)
5. Au is the symbol for _?_. (gold, mercury, lead)
6. Most of an atom's mass is in its _?_. (electrons, protons, nucleus)
7. In Thomson's model of the atom, electrons are embedded throughout a _?_ sphere that makes up the rest of the atom. (positive, negative, neutral)
8. A scientific _?_ can be used to explain and predict scientific laws and observations. (method, model, investigation)
9. The total number of atoms represented by "$2CO_2$" is _?_. (3, 6, 8)
10. $2Fe +$ _?_ $Cl_2 \rightarrow 2FeCL_3$. (2, 3, 6)
11. The number of carbon atoms in a molecule of sugar ($C_{12}H_{22}O_{11}$) is _?_. (11, 12, 22)
12. _?_ correspond to the energy differences between energy levels in atoms. (electrons, nuclei, quanta)

Understand The Concepts

13. According to Dalton's atomic theory, how do atoms of the same element compare with one another?
14. What did Rutherford's experiments show about the size and charge of the nucleus?
15. How can Dalton's atomic model be used to explain (a) the law of conservation of mass, (b) the law of definite proportions?
16. What evidence led Bohr to propose his atomic model? How did his model explain this evidence?
17. What evidence is there for the presence of electric charges in matter?
18. Using the balanced equation
$2KClO_3 \rightarrow 2KCl + 3O_2$,
show that mass is conserved in this reaction.
19. Compare the masses and charges of protons, neutrons, and electrons.
20. Why can an element's line spectrum be used to identify the element?
21. A neutral gold atom has 79 protons and 118 neutrons in its nucleus. How many electrons does the atom have? How do you know?

22. Why does the quantum model of the atom give only the probability of an electron's location in the atom?

Challenge Your Understanding

23. Balance the following equations:
    (a) $H_2 + O_2 \rightarrow H_2O$
    (b) $H_2 + O_2 \rightarrow H_2O_2$
    (c) $C + O_2 \rightarrow CO$
    (d) $N_2 + H_2 \rightarrow NH_3$
    (e) $H_2 + Cl_2 \rightarrow HCl$
    (f) $Na + Cl_2 \rightarrow NaCl$
24. A compound is found to consist of 1 part by mass hydrogen (atomic mass = 1 unit) and 16 parts by mass sulfur (atomic mass = 32 units). What is the simplest formula for the compound?
25. How does the quantum model differ from the Bohr model of the atom? What features do the two models have in common?

Projects

1. Write a brief biography of one of the following scientists, emphasizing his or her contribution to science: John Dalton, Ernest Rutherford, Niels Bohr, Maria Goeppert Mayer.
2. At a library, find out about the history of the symbols used to represent elements.
3. Use styrofoam balls and spray paint to prepare some atomic and molecular models. Use your models to represent the chemical reactants and products of certain chemical reactions. Write the reactions on a card to be displayed beneath the models.
4. Build a series of models to represent the various atomic models that were developed before the 1930's.
5. Obtain three identical tubes that will produce cold light. Bend all three tubes in a warm, dark room to produce light. Then leave one tube in the room, place one tube in a refrigerator, and place the third tube in a freezer. What differences do you observe among the tubes during the next few hours?
6. Obtain a diffraction grating, and use it to view light from various sources. The grating spreads light out into a spectrum. View (a) the straight vertical filament of a clear (unfrosted) light bulb (continuous spectrum), (b) an upright fluorescent bulb (mercury line spectrum superimposed on a continuous spectrum), (c) other light sources, such as neon signs, sodium lamps, and mercury vapor lamps (line spectra). You might also try looking at a white light source through a grating, with a colored filter placed between the grating and the source.

Chapter

8

Chemical Elements

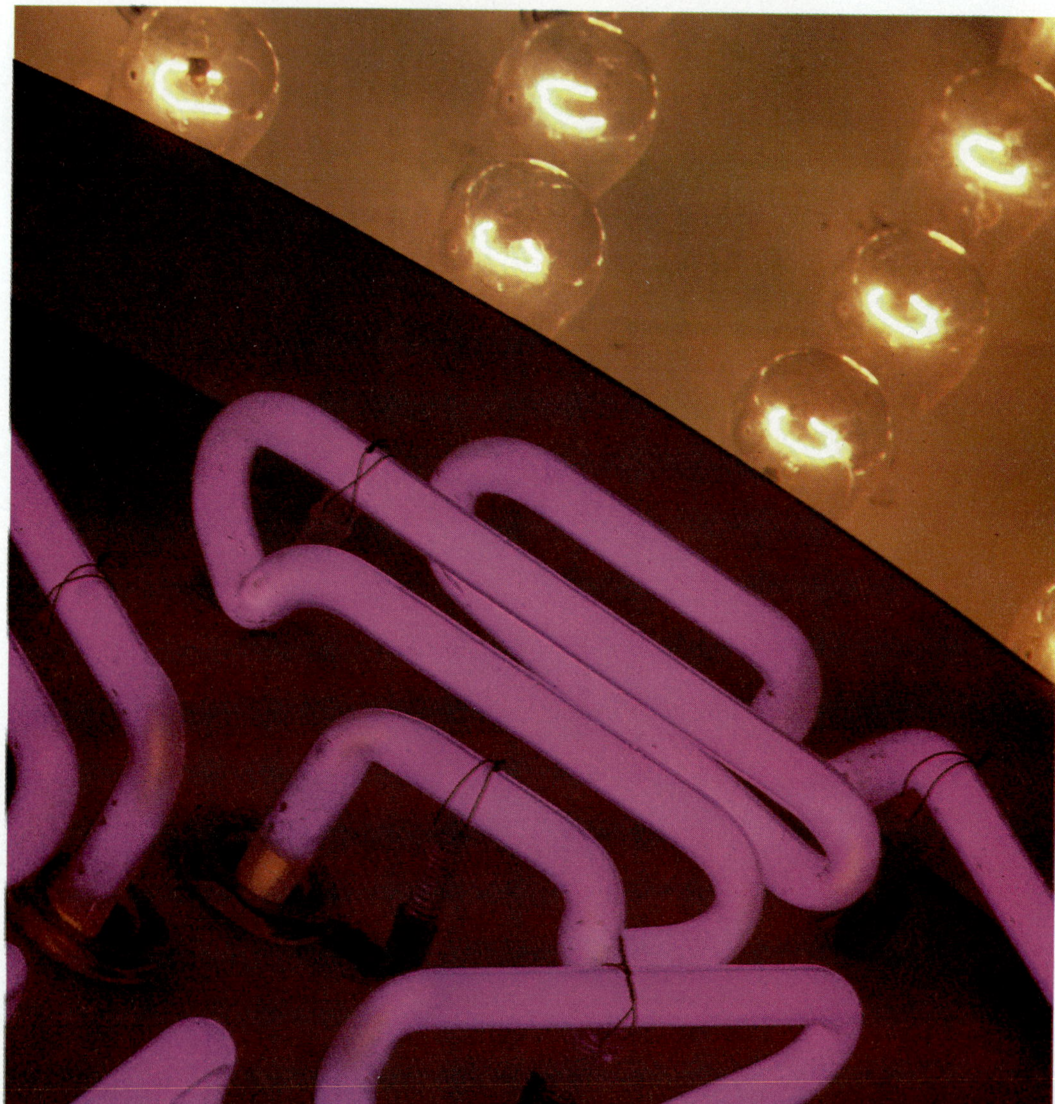

The chemical element neon is used for signs.

8-1

Metals and Nonmetals

Goals

1. To describe typical physical properties of metals and nonmetals.
2. To give examples of metals and nonmetals and state some of their uses.
3. To define and give examples of metalloids.

There are more than 100 chemical elements. Each element is unique, with its own set of properties. For example, hydrogen, a *nonmetal*, is a very light gas that will explode if mixed with oxygen and ignited with a spark. Gold, on the other hand, is a very dense, soft *metal* that can be hammered into sheets so thin that light will pass through them. Sheets like these can be used for lettering on glass, buildings, and leather book bindings.

Metals have many different uses. Copper and aluminum can be stretched into wires for electrical appliances, industrial machines, telephones, and cables. Iron can be melted and mixed with small amounts of carbon and other elements to make the steel used in automobiles, bridges, skyscrapers, and cooking utensils. Lead is a soft, dense metal that melts at 328°C.

Figure 8-1 *The metallic properties of gold make it possible to hammer the element into thin sheets, as this goldsmith is doing.*

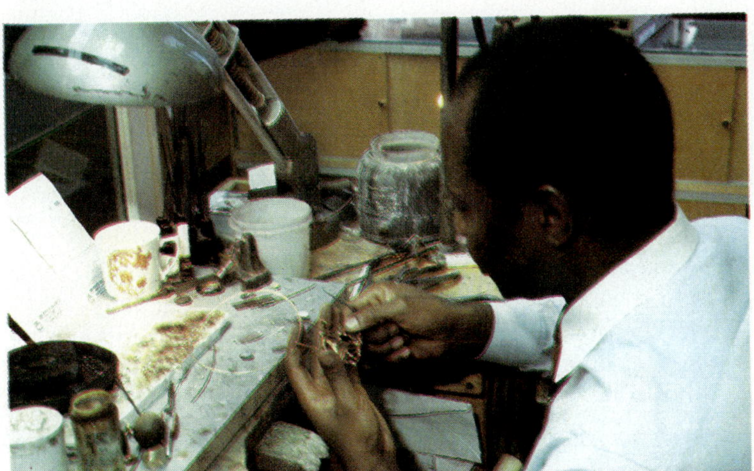

Figure 8-2 *Sulfur is a nonmetallic yellow element that is mined for a variety of uses.*

Oxygen, the world's most abundant element (by mass), is a nonmetal. Oxygen compounds account for nearly nine tenths of the mass of Earth's water and nearly half the mass of its minerals. Elemental oxygen accounts for more than one fifth of the mass of Earth's air. Most living organisms need oxygen. The element sulfur (S) is a yellow nonmetallic solid that is used in producing sulfuric acid (H_2SO_4), which is widely used in industry. The element phosphorus (P) is yet another nonmetal. Ordinary phosphorus is a white, waxy solid that bursts into flame when exposed to air. (See Figure 8-3.) For this reason, it is stored under water.

metals

nonmetals

Some elements share enough similar properties that they can be grouped together. The oldest way of classifying elements is to divide them into the two groups described here: **metals** and **nonmetals**. Such a division is based on differences in physical properties that are easy to observe. For example, samples of metals tend to be heavy for their size. That is, they tend to have a higher density than do nonmetals. The physical properties of metals and nonmetals are listed in Table 8-1. Notice that metals are good conductors of heat and electricity. Nonmetals are poor conductors. Metals and nonmetals also differ in their chemical properties, or the way they react.

A few elements, the *metalloids*, have properties that are intermediate between those of metals and nonmetals. The metalloids include silicon, germanium, arsenic, and antimony. Though not good conductors of electricity, their crystals are better conductors than nonmetals. Because of their

Figure 8-3 *White phosphorus is a nonmetal that bursts into flame when it comes into contact with air.*

Table 8-1: Physical Properties of Metals and Nonmetals

Metals	*Nonmetals*
1. Good conductors of heat and electricity	1. Mostly poor conductors
2. Shiny surface when polished	2. Dull
3. Almost all solid at 20°C	3. Many are liquids or gases at 20°C
4. Can be hammered, stretched, bent, or molded into sheets, wires, or other shapes	4. Solids tend to be brittle
5. Tend to be relatively dense	5. Tend to have low densities

Figure 8-4 *Crystals of the metalloid element silicon are often used in the electronics industry.*

moderate ability to conduct electricity, metalloids are called *semiconductors,* and are often used in the electronics industry. Such semiconductor crystals are found in transistors and in solar cells that convert sunlight to electricity. A *crystal* is a regular, repeated arrangement of the particles (such as atoms) of which it is made. (See Figure 8-4.)

Checkpoint

1. How do metals and nonmetals differ?
2. Give two examples each of metals and nonmetals. What are some of their uses?
3. What are metalloids? Give two examples.

TRY THIS

Using a hand lens, look closely at the backs of some pieces of real gold and silver jewelry. What words or numbers do you see? Do research in the library to find out what sterling silver is. What is the difference between 14-carat gold and 18-carat gold?

Figure 8-5 *The welding shown in this photograph is called heliarc welding. The noble gas helium is used in the process.*

8-2

The Noble Gases

Goals

1. To identify similarities in properties among the noble-gas elements.
2. To describe some uses of the noble gases.
3. To explain why the noble gases are unreactive.

The metals and nonmetals make up two large groups of elements. The elements can also be classified in other ways. They can be placed in small groups, called *families*. Elements in a family have similar properties.

A member of one important family of nonmetals was discovered in the nineteenth century. During an eclipse of the sun in 1868, a French astronomer took photographs through an instrument called a spectroscope. The photographs showed the spectral lines given off by glowing gases in the sun. A few of the lines had never been seen before. In 1871, scientists proposed that the lines belonged to a new element, which they named helium. (*Helios* in Greek means "sun.")

Helium does not react with other substances. At ordinary temperatures, it is a gas that is less dense than air. That is why helium balloons rise. Helium is also used in blimps and weather balloons. The most important industrial use of helium is in heliarc welding, a type of electric-arc welding shown in Figure 8-5. Liquid helium is used as a cooling agent. It has the lowest boiling point (−269°C) of any substance known. When substances are placed in liquid helium, they become very cold. At the very low temperatures produced in this way, many metals become *superconductors*. These are substances that offer practically no resistance to the flow of electricity. Such a property is very useful in large computers, where currents must flow rapidly through many circuits.

A gas similar to helium was detected in 1892 when the British physicist Lord Rayleigh removed all the oxygen, carbon dioxide, nitrogen, and water from an air sample. The small amount of gas that remained would not react with anything. The gas was thought to be a new element and was named argon from the Greek word meaning "lazy." Like he-

Figure 8-6 *The noble gas krypton is used in airport runway lights.*

lium, argon does not react with other elements. Because of this low reactivity, argon is used as the gas in light bulbs.

During the next few years, other stable, unreactive elements of the same family as helium and argon were found in the atmosphere. These elements are krypton, xenon, and radon. They are all gases under normal conditions.

Chemists group all these unreactive elements in a family called the **noble gases**. They are listed in Table 8-2, along with some of their properties.

noble gases

Neon signs contain gaseous neon, which emits reddish-orange light when electricity is passed through it. The light from krypton penetrates fog so well that krypton lamps are used to illuminate airport runways, as shown in Figure 8-6.

Table 8-2: Some Properties of the Noble Gases

Noble gas	*Atomic number (number of protons)*	*Atomic mass (amu)*	*Melting point (°C)*	*Boiling point (°C)*
helium (He)	2	4.0	−272	−269
neon (Ne)	10	20.2	−249	−246
argon (Ar)	18	39.9	−189	−186
krypton (Kr)	36	83.8	−157	−152
xenon (Xe)	54	131.3	−112	−107
radon (Rn)	86	222.0	−71	−62

Table 8-3: Electron Arrangement in the Noble Gases

Noble gas	*Total number of electrons*	*Electrons in each shell*	*Electrons in outermost shell*
helium	2	2	2 (filled)
neon	10	2-8	8 (filled)
argon	18	2-8-8	8 (filled)
krypton	36	2-8-18-8	8 (filled)
xenon	54	2-8-18-18-8	8 (filled)
radon	86	2-8-18-32-18-8	8 (filled)

Because xenon emits a very intense light, it is used in flash lamps made for high-speed photography. Radon, like radium, emits radiation and is used to treat cancer patients.

The arrangement of the electrons in the outermost energy level, or shell, determines the chemical properties of an element. Those electrons are sometimes called *valence* (VAY luns) electrons. The outermost electron shell of any atom can hold a maximum of 8 electrons (except for the first shell which can only hold 2). If this shell is full, the atom is unreactive. It does not tend to gain or lose electrons or combine with other atoms, even atoms of its own kind.

Table 8-3 indicates the total number of electrons in each noble-gas atom, the number in each shell of the atom, and the number in the outermost shell. Note that helium's outermost shell (the first) is filled. It contains two electrons. Each of the other noble-gas atoms has eight electrons in its outermost shell. These shells are also filled. It is the filled outermost shell that gives each noble gas its special stability, or unreactivity.

Checkpoint

1. What are the noble gases? What properties do they have in common?
2. How were helium and argon discovered?
3. List a use for each of the noble gases.
4. Define valence electrons.
5. What property of their atoms makes the noble gases unreactive?

8-3

The Halogens

Goals

1. To identify similarities in properties among the family of elements called the halogens.
2. To describe some uses for halogens and halides.
3. To explain why the atoms of halogens tend to gain or share electrons.

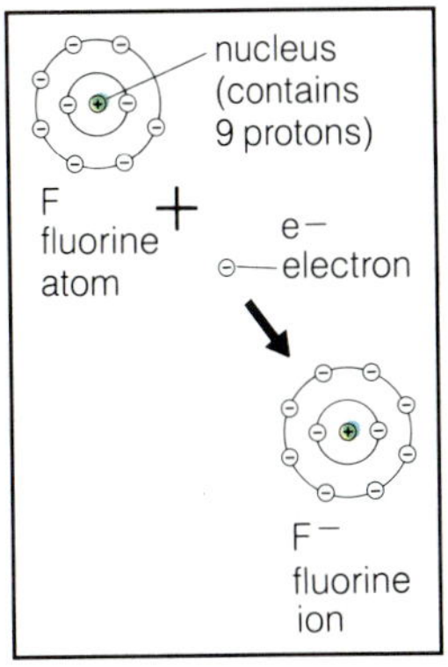

Figure 8-8 *By gaining one electron, a neutral fluorine atom becomes a stable fluoride ion, with a filled outermost shell.*

Fluorine, chlorine, bromine, and iodine make up a family of elements called the **halogens**. (See Figure 8-7.) Only traces of a fifth family member, astatine (At), occur naturally. Some physical properties of the halogens are listed in Table 8-4 on page 200.

The halogens are highly reactive. This is because each halogen atom has seven electrons in its outermost shell. That is one less electron than the noble-gas atoms have. By gaining one electron, each halogen atom can fill its outermost shell, becoming more stable. An atom (or group of atoms) that has gained or lost one or more electrons is called an **ion.** Since the number of electrons is no longer equal to the number of protons, an ion has an electric *charge*. When a fluorine atom gains an electron it becomes a fluoride ion, F^-, with 8 electrons in its outermost shell and a charge of −1. (See Figure 8-8.) Generally, when a halogen atom gains an electron, it becomes a hal*ide* ion.

halogens
(HAL uh junz)

ion
(EYE un)

Figure 8-7 *The halogen elements include yellow-green gaseous chlorine, reddish-brown liquid bromine, and blue-black solid iodine.*

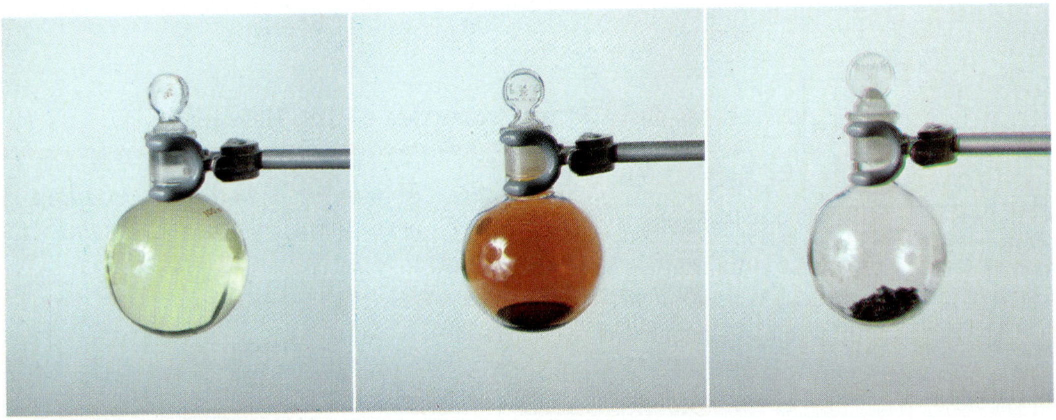

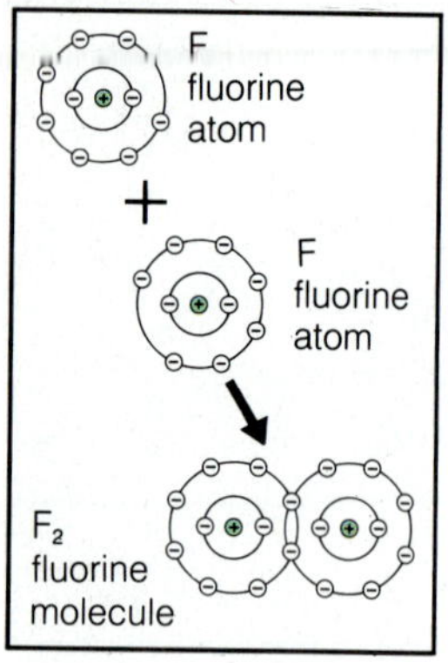

Figure 8-9 By combining to form a diatomic molecule, two halogen atoms, such as fluorine (F), share electrons and achieve filled outermost shells.

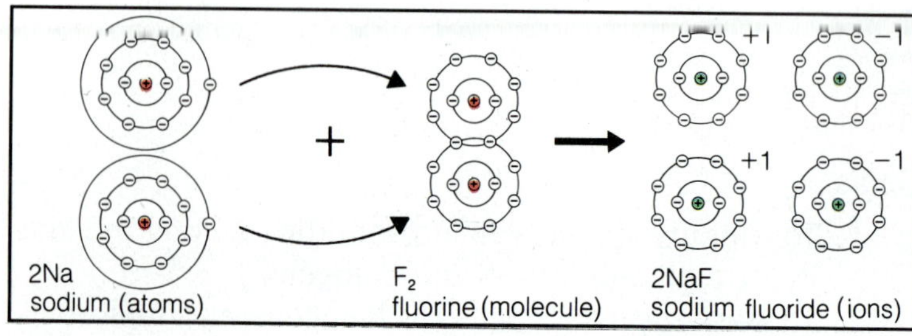

Figure 8-10 Sodium atoms can combine with a halogen molecule, such as fluorine, to produce stable sodium and halide ions.

Another way for halogen atoms to become more stable is to *share* electrons. Under ordinary conditions, the halogens exist as diatomic, or two-atom, molecules. They are more stable in this form than as individual atoms. By sharing electrons, the halogen atoms achieve filled outermost shells, as shown in Figure 8-9. You will learn more about such sharing and transfer of electrons later in this chapter.

Any compound of a halogen and a metallic element is called a hal*ide*. Thus, NaCl is sodium chlor*ide* and NaI is sodium iod*ide*. The sodium halides are solid *salts*. The sodium salts are made up of ions. In a reaction between sodium and a halogen, each halogen atom gains an electron from a sodium atom. (See Figure 8-10.) A typical equation is:

$$2Na + F_2 \longrightarrow 2NaF$$

The halogens react with hydrogen to form hydrogen halides. (See Figure 8-11.) A typical equation is:

$$H_2 + F_2 \longrightarrow 2HF$$

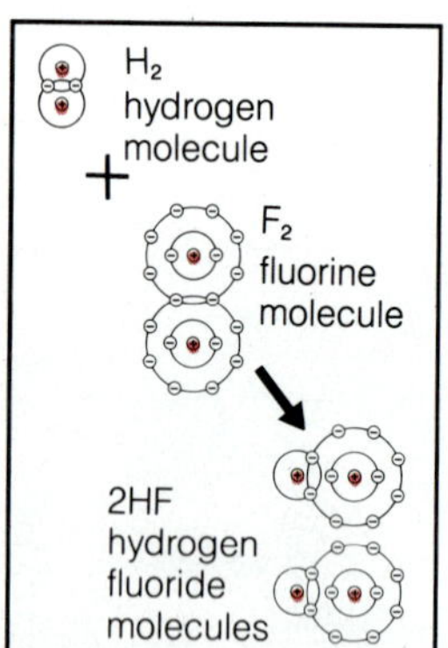

Figure 8-11 A hydrogen molecule can combine with a halogen molecule, such as fluorine, to produce stable hydrogen halide molecules.

Table 8-4: Some Properties of the Halogens

Halogen	*Atomic number*	*Atomic mass (amu)*	*Melting point (°C)*	*Boiling point (°C)*
fluorine (F)	9	19.0	−220	−188
chlorine (Cl)	17	35.5	−101	−35
bromine (Br)	35	80.0	−7	59
iodine (I)	53	127.0	114	184

Figure 8-12 *A photographic negative contains dark silver deposits that result from the breakdown of silver bromide by light. The resulting positive is light in those areas that correspond to the dark areas of the negative.*

The hydrogen halides are gases that form substances called *acids* when they dissolve in water.

All the halogen elements are poisonous. However, many halogen compounds, including table salt (NaCl), are harmless or useful. Small amounts of fluoride salts are added to toothpaste because they strengthen tooth enamel and reduce tooth decay. Small amounts of chlorine are added to swimming pools and drinking water to kill bacteria that could cause disease or infections. Iodide salts, often added to table salt, help the thyroid gland to function properly.

Photographic film (and paper) are coated with a thin layer of gelatin containing tiny crystals of light-sensitive silver bromide (AgBr). In those places where light strikes the gelatin layer, the compound changes to silver and bromine. The silver deposits form the picture on the negative. To make prints from the film, light is shone through the negative onto paper that is coated with a light-sensitive layer. The places that were darkest on the negative are lightest on the print, and vice versa. (See Figure 8-12.)

Checkpoint

1. Why are the halogens grouped together in a family?
2. Why are halide ions stable?
3. Using a diagram, explain how hydrogen and fluorine bond to form HF.
4. List five uses for halogens and halides.

8-4

The Alkali Metals

Goals

1. To identify similarities in properties among the family of elements called the alkali metals.
2. To describe some uses for the alkali metals.
3. To explain why the atoms of alkali metals tend to lose one electron.

alkali metal
(AL kuh ly)

Each atom of the **alkali metal** family has only one electron in its outermost shell. The alkali metals include lithium, sodium, potassium, rubidium, and cesium. Only traces of a sixth alkali metal, francium (Fr), occur naturally. Some properties of the alkali-metal elements are given in Table 8-5. The alkali metals are very soft and can be cut with a knife.

An alkali-metal atom attains the stability of the noble-gas electron arrangement by losing its single outermost electron. It is thus left with a filled shell and becomes a positive ion. (See Figure 8-13.) Very little energy is required

Table 8-5: Some Properties of the Alkali Metals

Alkali Metal	*Atomic number*	*Atomic mass (amu)*	*Melting point* (°C)	*Boiling point* (°C)
lithium (Li)	3	6.9	181	1342
sodium (Na)	11	23.0	98	892
potassium (K)	19	39.1	64	774
rubidium (Rb)	37	85.5	39	686
cesium (Cs)	55	133.0	28.7	669

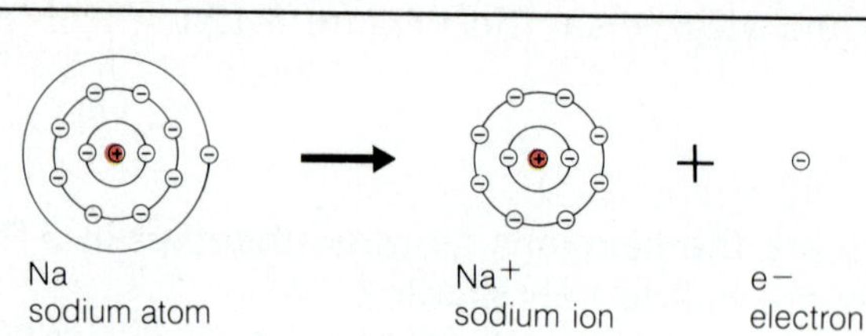

Figure 8-13 *By losing one electron, a neutral sodium atom becomes a stable sodium ion, with a filled outermost shell.*

Figure 8-14 *When exposed to the air, alkali metals react vigorously with the oxygen and water vapor in the air.*

to remove the outermost electron from an alkali-metal atom. This is another way of saying that alkali metals are very reactive. They can react, sometimes explosively, with halogens to form halides, and with oxygen to form oxides. Alkali metals react with water, releasing hydrogen gas. The energy given off in the reaction can ignite the hydrogen and produce a dangerous explosion. To prevent alkali metals from reacting with air or moisture in the air, they are usually stored submerged in kerosene.

Many alkali-metal compounds, such as sodium chloride, sodium hydroxide (NaOH), sodium carbonate (Na_2CO_3), sodium bicarbonate ($NaHCO_3$), and potassium nitrate (KNO_3), are abundant and widely used in homes and industry. For example, sodium bicarbonate—baking soda—is used in cooking. The uncombined elements themselves have limited use. Sodium's excellent conductivity has led to its use in underground cables. Sodium metal is also used as a cooling agent in nuclear reactors. It has a low melting point, a high boiling point, and a good capacity for absorbing the large amounts of heat released in such reactors.

TRY THIS

Look at the compounds in your kitchen and bathroom cabinets. Which compounds contain alkali metals? What are the compounds used for?

Checkpoint

1. What elements are in the alkali-metal family? What properties do they have in common?
2. Why are alkali metals found in compounds instead of by themselves?
3. What gives alkali-metal ions their stability?

8-5

The Periodic Table

Goals

1. To explain the use of the periodic table.
2. To describe how ionization energy varies.

For many years chemists tried different ways of classifying elements on the basis of their properties. In 1871, the Russian chemist Dmitri Mendeleev arranged the elements in order of increasing atomic mass. When he did, he found that similar properties occurred over and over again at fairly regular intervals. The table he set up was the first **periodic table**, an arrangement in which elements that have similar properties are grouped together. At that time, some of the elements known today had not been discovered. For example, there was no known element with an atomic mass of 72. So Mendeleev left spaces for that and other missing elements in his table. He even correctly predicted what the properties of the missing elements would be.

periodic table

In the modern periodic table, the elements are ordered by increasing atomic number, rather than atomic mass. The vertical columns in the table are called **groups**, or *families*. The alkali metals, with one electron in their outermost shell, are in group I. The halogen family, with seven electrons in their outermost shell, are in group VII. The noble gases, with their stable eight-electron arrangement, make up group VIII.

groups

Each horizontal row in the periodic table is a **period**. From left to right along a period, the properties of the elements change dramatically. The first period contains only hydrogen and helium. The second period goes from lithium to neon. Neon atoms, with 10 electrons, have filled outermost energy levels. Sodium, the next element, has 11 electrons and lies below lithium, which it resembles in many ways. Magnesium, with properties similar to beryllium, is the eighth element after beryllium, just as sodium is the eighth element after lithium. This progression continues across the third period up to the noble gas argon, the eighth element after neon. It is the repetition of similar properties that gives the periodic table its name. The lengths of the periods depend on the number of electrons that can occupy each shell.

period

The Periodic Table

atomic number — 12; average atomic mass — 24.3; symbol of element — Mg; element name — Magnesium

columns → I … VIII; rows ↓ 1 … 7

rows	I	II											III	IV	V	VI	VII	VIII
1	1 1.0 H Hydrogen																	2 4.0 He Helium
2	3 6.9 Li Lithium	4 9.0 Be Beryllium											5 10.8 B Boron	6 12.0 C Carbon	7 14.0 N Nitrogen	8 16.0 O Oxygen	9 19.0 F Fluorine	10 20.2 Ne Neon
3	11 23.0 Na Sodium	12 24.3 Mg Magnesium											13 27.0 Al Aluminum	14 28.1 Si Silicon	15 31.0 P Phosphorus	16 32.1 S Sulfur	17 35.5 Cl Chlorine	18 39.9 Ar Argon
4	19 39.1 K Potassium	20 40.1 Ca Calcium	21 45.0 Sc Scandium	22 47.9 Ti Titanium	23 50.9 V Vanadium	24 52.0 Cr Chromium	25 54.9 Mn Manganese	26 55.8 Fe Iron	27 58.9 Co Cobalt	28 58.7 Ni Nickel	29 63.5 Cu Copper	30 65.4 Zn Zinc	31 69.7 Ga Gallium	32 72.6 Ge Germanium	33 74.9 As Arsenic	34 79.0 Se Selenium	35 79.9 Br Bromine	36 83.8 Kr Krypton
5	37 85.5 Rb Rubidium	38 87.6 Sr Strontium	39 88.9 Y Yttrium	40 91.2 Zr Zirconium	41 92.9 Nb Niobium	42 95.9 Mo Molybdenum	43 99 Tc Technetium	44 101.1 Ru Ruthenium	45 102.9 Rh Rhodium	46 106.4 Pd Palladium	47 107.9 Ag Silver	48 112.4 Cd Cadmium	49 114.8 In Indium	50 118.7 Sn Tin	51 121.8 Sb Antimony	52 127.6 Te Tellurium	53 126.9 I Iodine	54 131.3 Xe Xenon
6	55 132.9 Cs Cesium	56 137.3 Ba Barium	57–71* see below	72 178.5 Hf Hafnium	73 180.9 Ta Tantalum	74 183.9 W Tungsten	75 186.2 Re Rhenium	76 190.2 Os Osmium	77 192.2 Ir Iridium	78 195.1 Pt Platinum	79 197.0 Au Gold	80 200.6 Hg Mercury	81 204.4 Tl Thallium	82 207.2 Pb Lead	83 209.0 Bi Bismuth	84 209 Po Polonium	85 210 At Astatine	86 222 Rn Radon
7	87 223 Fr Francium	88 226 Ra Radium	89–103** see below	104 (unnamed)	105 (unnamed)	106 (unnamed)												

*Lanthanide series

57 138.9 La Lanthanum	58 140.1 Ce Cerium	59 140.9 Pr Praseodymium	60 144.2 Nd Neodymium	61 147 Pm Promethium	62 150.4 Sm Samarium	63 152 Eu Europium	64 157.3 Gd Gadolinium	65 158.9 Tb Terbium	66 162.5 Dy Dysprosium	67 164.9 Ho Holmium	68 167.3 Er Erbium	69 168.9 Tm Thulium	70 173.0 Yb Ytterbium	71 175.0 Lu Lutetium

**Actinide series

89 227 Ac Actinium	90 232.0 Th Thorium	91 231 Pa Protactinium	92 238.0 U Uranium	93 237 Np Neptunium	94 242 Pu Plutonium	95 243 Am Americium	96 247 Cm Curium	97 249 Bk Berkelium	98 251 Cf Californium	99 254 Es Einsteinium	100 253 Fm Fermium	101 256 Md Mendelevium	102 259 No Nobelium	103 257 Lr Lawrencium

Figure 8-15 *The elements are arranged according to atomic number in the modern periodic table.*

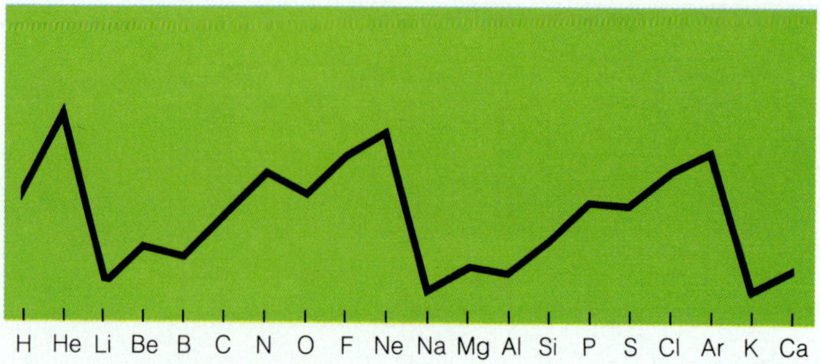

Figure 8-16 *The ionization energies for the first 20 elements of the periodic table are plotted in this graph.*

Although Mendeleev did not know it, the periodic arrangement of the elements corresponds to their electron arrangements. Elements in a group, such as the alkali metals or halogens, have similar chemical properties because they have similar electron arrangements. The changes in chemical properties across a period correspond to the changes in the elements' electron arrangements.

ionization energy (eye uh nuh ZAY shuhn)

The energy required to remove an electron from an atom is called the **ionization energy**. A graph of the ionization energies for the first 20 elements is shown in Figure 8-16. Notice that the energy required to remove an electron tends to increase as you go from lithium to neon or from sodium to argon. All the alkali metals have very low ionization energies. This explains why they are so reactive. It takes very little energy for these metals to lose their one outermost electron, forming an ion with a charge of +1. Elements in group II tend to lose two electrons, and so form ions with a charge of +2. Those in group VI have six electrons in the outermost shell of their atoms. Rather than losing six electrons (which would require a great deal of energy), they tend to gain two, forming ions with a charge of −2.

Checkpoint

1. How are elements ordered in the periodic table?
2. What is (a) a period, (b) a group?
3. Identify three chemical families in the periodic table.
4. What is ionization energy? How does it vary across a period in the periodic table?

8-6

Bonding of Elements

Goals

1. To explain how ionic bonding occurs.
2. To state how covalent bonding occurs.
3. To describe polar molecules.

An understanding of the electron arrangements of the elements can lead to an understanding of how the atoms react to form compounds. Recall the reaction that produces sodium chloride, discussed earlier. In reaching the stable noble-gas arrangement, sodium gives up an electron and chlorine receives one. After the electron transfer takes place, sodium has a positive charge and chlorine a negative charge. Oppositely charged particles attract each other. This attraction holds the ions together in **ionic bonds**. Many such ions may come together in a regular repeating arrangement to form a crystal. Figure 8-17 shows that in sodium chloride crystals, each ion is surrounded by ions of opposite charge. Ionic bonds usually form between atoms that have very different ionization energies. One kind of atom tends to lose electrons, and the other kind tends to gain electrons.

ionic bonds

Atoms sometimes form chemical bonds by sharing electrons rather than by gaining or losing them. Such bonds formed by sharing valence electrons are called **covalent bonds**. You have already seen an example of covalent bonding in the case of the fluorine (F_2) molecule. By sharing electrons, atoms may acquire enough to fill their outermost shells. Covalent bonds usually form between atoms that have similar ionization energies. Neither atom in the molecule tends to give up electrons to the other atom.

covalent bonds (koh VAY lunt)

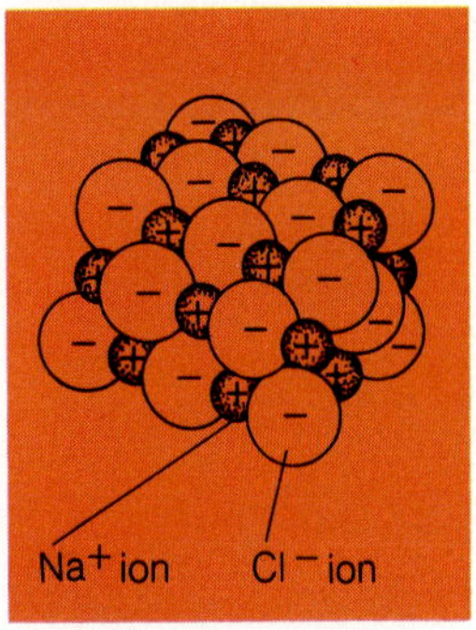

Figure 8-17 *In a sodium chloride crystal, ionic bonds hold together particles of opposite charge.*

When atoms in a compound have ionization energies that are neither very similar nor very different, an unequal sharing of electrons may result. Such a bond is called a *polar covalent bond*. The electrons in such bonds are held more closely to one atom than to another. One side of the bond is slightly positive, while the other is slightly negative. The bond between hydrogen and chlorine in the hydrogen chloride (HCl) molecule, for example, is a polar covalent bond. The chlorine has a greater attraction for electrons than hydrogen does. Thus the chlorine end of the HCl bond is

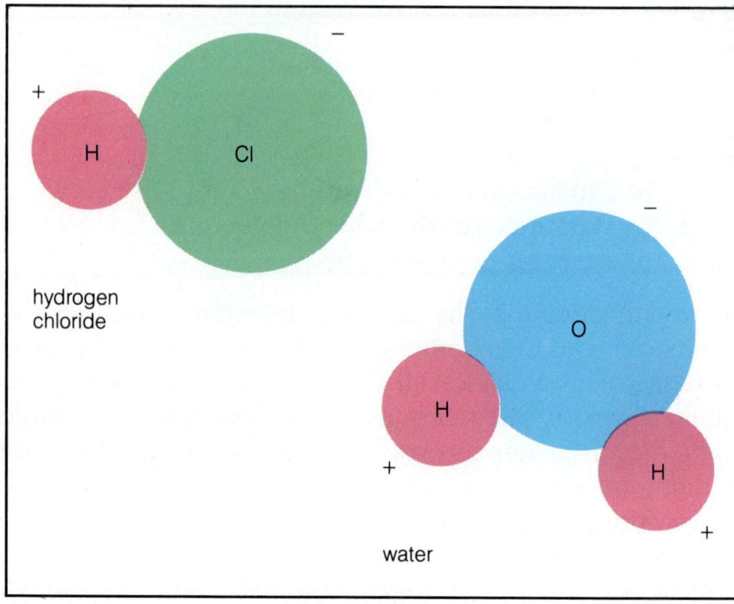

Figure 8-18 *Hydrogen chloride and water are examples of polar molecules.*

slightly negative and the hydrogen end slightly positive.

polar molecules

Molecules such as hydrogen chloride are called **polar molecules**. Water molecules are also polar. The two hydrogen atoms in water are slightly positive, and the oxygen is slightly negative. Note that the atoms in water are not in a line, as is shown in Figure 8-18. If all three atoms in H_2O were in a line, the two polar bonds would "cancel out" and the molecule would not be polar. Thus, the polarity of a molecule depends upon its shape as well as its bonds.

An effect of the polarity of water molecules can easily be observed. A charged rubber rod held near a thin stream of water causes the stream to bend. The negatively charged rod attracts the positive ends of the water molecules, and the molecules are pulled sideways.

Checkpoint

1. What kind of chemical bond holds sodium chloride crystals together?
2. What are covalent bonds? Give some examples.
3. What are polar covalent bonds? Give at least two examples.

8-7

Carbon: Element Six

ENRICHMENT

Goals

1. To explain why there are so many carbon compounds.
2. To tell what a structural formula represents.

The element carbon can occur in the form of graphite or diamonds. Carbon forms many more different compounds than any other element. Carbon compounds make up fossil fuels—coal, petroleum, and natural gas. The energy in these fuels is used to heat homes, to drive turbines, and to turn the wheels of automobiles. Coal and petroleum also serve as the basic materials for making many plastics, fertilizers, perfumes, synthetic fabrics, medicines, and other substances. (See Figure 8-19.)

The large number of carbon compounds is primarily due to the ability of carbon atoms to form covalent bonds with other carbon atoms. The linking of carbon atoms can produce chains of carbon atoms of almost any length.

Figure 8-19 *Fossil fuels serve as the basic materials for making compounds found in a wide variety of products.*

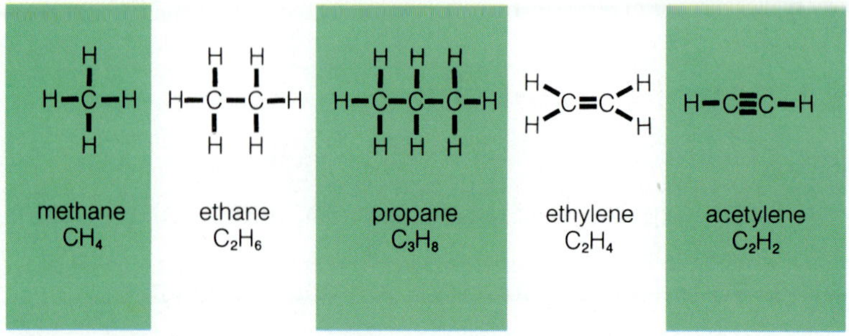

Figure 8-20 *Single lines represent single bonds, double lines double bonds, and triple lines represent triple bonds in the structural formulas of the hydrocarbons shown here.*

hydrocarbons

structural formulas

Compounds that contain only hydrogen and carbon are called **hydrocarbons**. Methane, ethane, and propane are three simple hydrocarbons. They are gases that are commonly used as fuels. Their molecular formulas are CH_4, C_2H_6, and C_3H_8, respectively. *Molecular formulas* show only the type and number of atoms in a molecule. **Structural formulas** are used to show the arrangement of atoms in a molecule. In a structural formula, each pair of shared electrons is shown as a single line. Such a pair of electrons is called a *single bond*. (See Figure 8-20.) Each carbon atom can form four single bonds. Therefore, each carbon atom in a structural formula must be surrounded by four lines.

Atoms may sometimes share more than a single pair of electrons. If atoms share two pairs of electrons, they form a *double bond*. If they share three pairs of electrons, they form a *triple bond*. In structural formulas, double bonds are represented by two lines, and triple bonds by three.

The simplest hydrocarbon having a double bond is ethene (C_2H_4), or ethylene. Ethylene molecules can react with each other, forming very long chains, or *polymers*, such as polyethylene, a common plastic. The simplest hydrocarbon having a triple bond is ethyne (C_2H_2), or acetylene. This compound is commonly burned in oxygen, producing a very hot flame for welding.

Checkpoint

1. Why are there so many carbon compounds?
2. What information does a structural formula give that a molecular formula does not give?

Photoelectric Cells

Have you ever reached out to push open a supermarket door when, suddenly, as if by magic, the door springs open? The device that makes this happen is often a photoelectric cell, sometimes called an electric "eye." It is the same device used to open garage doors, to turn on lights automatically after sunset, and to set off burglar alarms.

Electrons can be made to flow from a negatively charged *cathode* to a positively charged *anode*, within a tube from which most of the air has been removed. Such a tube is called a *vacuum tube*. Modern photoelectric cells may be vacuum tubes. When light shines on the cathode, electrons are released. (See Figure 1.) The cathode is often coated with a thin layer of an alkali metal, such as cesium or potassium. These metals release electrons easily.

Ultraviolet light, an invisible high-energy light, shining on the cathode in a burglar alarm system causes a constant flow of electrons from the cathode to the anode. Ultraviolet light, sometimes called black light, is used because it cannot be seen by a burglar. If the burglar crosses the ultraviolet light beam, electrons stop flowing because there is no light to drive the electrons out of the alkali-metal atoms on the cathode. This causes the alarm to go off. The diagram in Figure 2 shows a burglar-alarm system. Some smoke alarms work in much the same way. A tiny ray of light inside is interrupted by smoke particles and the alarm sounds.

The solar, or photovoltaic, cells that convert sunlight to electricity are also photoelectric cells. Solar cells are often seen in a group on the roofs of houses. Solar cells may also power your watch or calculator. The electrons usually come from a metalloid semiconductor rather than from an alkali-metal coating on a vacuum-tube cathode.

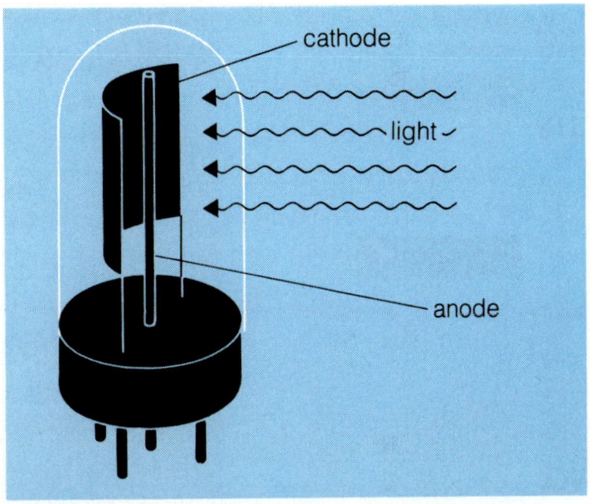

Figure 1

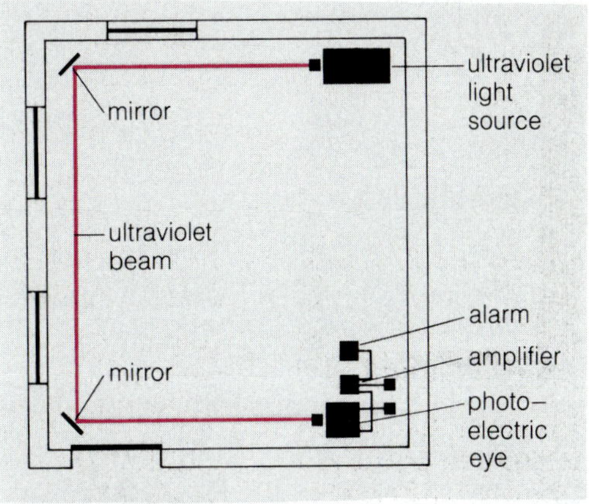

Figure 2

Understanding The Chapter 8

Main Ideas

1. Elements differ greatly in physical and chemical properties.
2. Elements can be classified as metals, nonmetals, and metalloids on the basis of their properties.
3. The noble gases are a group of very unreactive elements.
4. The unreactive, stable nature of the noble gases is due to filled outermost electron shells.
5. The halogens are a very reactive family of elements that have seven electrons in their outermost shell. Halogen atoms can become halide ions by gaining one electron, giving them the stable electron arrangement of the noble-gas atoms.
6. Alkali metals are a very reactive family of elements that have one electron in their outermost electron shell. They attain the noble-gas electron arrangement by losing an electron.
7. Elements are arranged in the periodic table according to their atomic number. The vertical columns of the table contain families of elements that have similar properties.
8. Some compounds form when electrons are transferred from the atoms of one element to the atoms of another. Ionic bonds result.
9. Some compounds form when atoms of different elements share electrons, to form covalent bonds.
10. If one atom in a covalent bond has a greater attraction for electrons than the other, a polar covalent bond results.
11. Carbon atoms form four covalent bonds with one another as well as with other elements.

Vocabulary Review

From the following list, choose the term that best completes each of the statements. Write your answers on a separate piece of paper.

alkali metal	ionization energy
covalent bond	metal
crystal	metalloid
family	noble gas
group	nonmetal
halide	period
halogen	periodic table
hydrocarbon	polar covalent
ion	polar molecule
ionic bond	structural formula

1. Elements arranged vertically in the periodic table constitute a(n) __?__, or __?__, of elements.
2. The bond that forms when an alkali metal combines with a halogen to form a compound is a(n) __?__.
3. A(n) __?__ is a compound made from a halogen and another element.
4. Bromine is a(n) __?__.
5. Most __?__ elements have atoms with eight electrons in the outermost shell.

6. __?__ elements are poor conductors of heat and electricity.
7. __?__ elements have atoms with one electron in the outermost shell.
8. The water molecule is an example of a(n) __?__.
9. A(n) __?__ forms when atoms share electrons.
10. A horizontal row in the periodic table is called a(n) __?__.
11. __?__ elements are good conductors of heat and electricity.
12. If shared electrons are attracted more strongly by one atom than another, the bond is said to be __?__.
13. A(n) __?__ is a regular, repeated arrangement of the particles of which it is made.
14. An atom, or group of atoms, that has gained or lost one or more electrons is a(n) __?__.

Chapter Review

Write your answers on a separate piece of paper.

Know The Facts

1. __?__ is a nonmetal. (lead, sodium, fluorine)
2. The elements lithium, carbon, and fluorine are all found in the same __?__ of the periodic table. (period, column, group)
3. __?__ is a metalloid. (sodium, fluorine, silicon)
4. Metals tend to release electrons __?__ nonmetals. (more readily than, less readily than, about as readily as)
5. __?__ atoms have eight electrons in their outermost shell. (helium, chlorine, argon)
6. Atoms of the __?__ elements have seven electrons in their outermost shell. (noble-gas, halogen, second-period)
7. Halogens are not found in the __?__ period of the periodic table. (first, second, fourth)
8. When a halogen atom gains an electron it becomes a(n) __?__. (noble gas, alkali metal, ion)
9. The noble gases have __?__ ionization energies. (high, low, intermediate)
10. __?__ formulas show the arrangement of atoms in a molecule. (chemical, molecular, structural)

Understand The Concepts

11. Explain why argon is commonly used in light bulbs.
12. All the halogen elements are poisonous. Describe three halogen compounds that are not.
13. List the alkali metals in terms of their (a) increasing atomic mass, (b) increasing melting point, (c) increasing boiling point.
14. Explain how the reaction of an alkali metal with a halogen produces ions with the same outermost electron arrangement as in a noble-gas atom.
15. Why is liquid helium used in preparing very low-temperature environments?
16. List five ways in which the properties of metals and nonmetals are different.

17. List three practical uses for the halogens and halides.
18. Explain the difference between a covalent bond and an ionic bond.
19. Give two examples of compounds that have (a) ionic bonds, (b) covalent bonds.
20. List four products made from carbon compounds.

Challenge Your Understanding

21. Would you expect the bonds between halogen atoms in the molecule Cl_2 to be ionic, covalent, or polar covalent? Explain your answer.
22. Write the chemical equations that describe the reactions between (a) potassium and chlorine, (b) hydrogen and bromine, (c) sodium and water.
23. Why do you think the ionization energy shown in Figure 8-16 increases across a period from lithium to neon or sodium to argon?
24. Draw the structural formula for a three-carbon hydrocarbon having (a) all single bonds, (b) a double bond, (c) a triple bond.

Projects

1. Visit a machine shop, a wire-making factory, a sculpture studio, or some other place where metal is being worked. Write a two-page report summarizing what you found.
2. Visit a steel plant or do some research in your library to find out how steel is made.
3. Go to the supermarket and look at the lists of ingredients on various packages of salt. How does kosher salt differ from other table salt? What is "lite salt"? What is "no salt"?
4. The alkali metals in group I of the periodic table will react with water to form hydrogen. Will metals of group II react with it to form hydrogen? To find out, add small chips of magnesium to a test tube of cold water, with your teacher's supervision. Try the same thing using hot water. Do you see any evidence of a reaction? Repeat the experiment, using calcium. What do you find? How do you explain your results?
5. Investigate the controversy regarding the addition of fluoride compounds to public water supplies. What are the positions involved? What is your position after your investigation?
6. Compounds that have the same molecular formulas but different structural formulas and different properties are known as *isomers*. Do some library research to find examples of isomers. Make a poster showing the names, molecular formulas, and structural formulas of these isomers.

Chapter 9

Chemical Reactions

A chemist working at American University in Washington, D.C.

9-1

Energy and Changes of State

Goals

1. To contrast the particle motions in the three states of matter.
2. To describe the energy changes that occur when matter changes state.
3. To explain what happens to its particles when matter changes state.

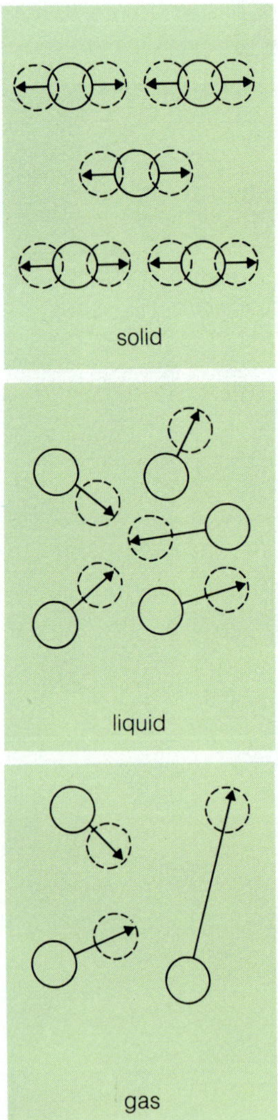

Figure 9-1 *The motions of particles in solids, liquids, and gases are represented above, using arrows to suggest movement.*

The smallest particles of elements and compounds are constantly in motion. The molecules of water in a glass of water, the sodium and chloride ions in a shaker of salt, and the neon atoms in a neon sign are all moving. In the solid state, particles only vibrate back and forth about fixed positions. In a liquid, the particles are so close together that they do not move very far before colliding with other particles. In a gas, the particles are so far apart that they move nearly independently of each other. In Figure 9-1, you can compare the motions of particles in the three states of matter.

The *temperature* of a substance is a measure of the average energy of motion of its particles. The higher the temperature of a substance, the greater the average energy and average speed. When heat is applied to a substance, however, change in temperature does not always result. For example, as a container full of water is heated, a thermometer suspended in the water shows an increase in temperature only until the water starts to boil. At that point, the temperature stops rising. It remains at 100°C as long as the water continues to boil. If the temperature of the steam given off by the boiling water is measured, it is found to be the same as that of the boiling water, 100°C. At the boiling point, all of the energy being supplied to the water is used to change the water from liquid to gas.

Water molecules attract each other strongly because they contain polar bonds. Although a water molecule is neutral, the charge within the molecule is not distributed evenly. The hydrogen atoms have a slightly positive charge and the oxygen atom has a slightly negative charge. As a result, the hydrogen atoms in a water molecule are attracted to the oxygen atoms in nearby water molecules, and vice

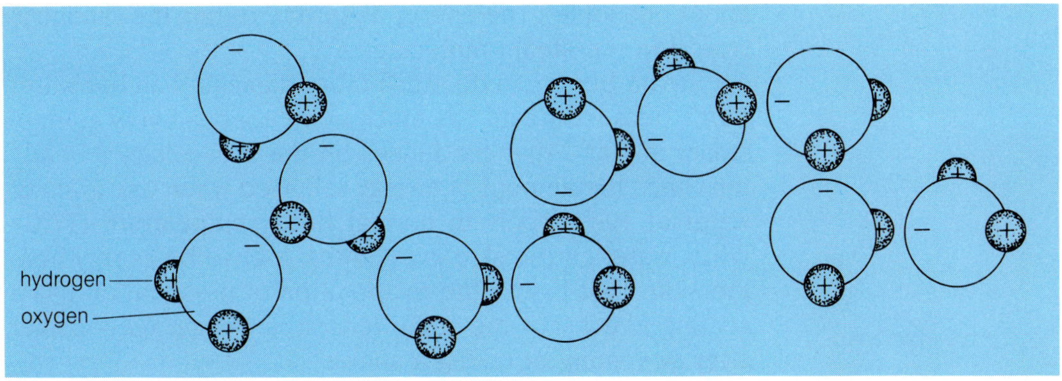

Figure 9-2 *The hydrogen atoms in a water molecule have a slightly positive charge. They attract the slightly negative oxygen atoms in other water molecules.*

versa. (See Figure 9-2.) The closer two oppositely charged objects are, the stronger the attraction between them.

Water molecules are much closer together in liquid water than in steam. When water changes to steam, the volume becomes more than 1500 times as great. Since water molecules attract each other strongly in liquid water, energy is needed to move them apart.

More energy is needed to change water into steam, or *vaporize* it, than is needed to vaporize many other substances. The energy needed to change one gram of a liquid to a gas at the boiling point is the *heat of vaporization* (vay pur uh ZAY shuhn). Of the substances listed in Table 9-1, water has the greatest heat of vaporization. Each gram of water that changes to steam absorbs 2300 joules of energy. The *joule* (J) is a unit of energy, or heat, used by scientists. (The energy needed to lift an apple a distance of one meter is

Table 9-1: Some Heats of Vaporization

Substance	*Heat of vaporization* (J)	*Boiling point* (°C)
water	2300	100
air	214	−191
carbon dioxide	365	−56
grain alcohol	860	78
helium	25	−269
oxygen	214	−184
phosphorus	546	200
sulfur dioxide	400	−10
wood alcohol	1100	65

about one joule.) The energy absorbed during the change is stored in the steam molecules.

Steam burns can do much more damage than hot-water burns because steam has much more energy. Every gram of steam at 100°C that condenses on the skin releases nearly ten times the amount of energy released when one gram of water at 100°C cools to normal body temperature (37°C). When water on the skin evaporates, it absorbs energy from the skin. That is why the evaporation of perspiration has a cooling effect. For the same reason, a person feels cooler after swimming or taking a shower.

When one gram of steam changes back to liquid water, or condenses, it releases 2300 J. The energy gained by the water molecules when they moved apart to form the steam is given up when they move together again, forming liquid water. The *heat of condensation* (kahn dun SAY shuhn), the energy given up when a gas condenses to a liquid at its boiling point, is equal to the liquid's heat of vaporization.

When a solid melts, energy is absorbed. When a gram of ice melts at 0°C, 335 J of energy are absorbed. The energy needed to melt ice is much less than that needed to boil the same amount of water. When a liquid freezes, the particles lose the same amount of energy that they would gain on melting. Each gram of water that freezes at 0°C releases 335 J of energy.

Checkpoint

1. Contrast the particle motions in the three states of matter.
2. What energy change occurs when water (a) boils, (b) condenses?
3. What energy change occurs when (a) ice melts, (b) water freezes?
4. What happens to water molecules when water (a) boils, (b) condenses? How does your answer explain the observed energy change in each case?

TRY THIS

Sprinkle some warm water on your palm. Then wave your hand through the air. Why does this cause your hand to feel cool?

9-2

Energy and Chemical Reactions

Goals

1. To explain why the energy associated with chemical changes is, in general, greater than that associated with physical changes.
2. To explain the difference between exothermic and endothermic reactions.

The energy required to *decompose,* or to break down, one gram of water into the elements hydrogen and oxygen is 15,900 J. Since energy is absorbed in this process, you might guess that it is "stored" in the molecules of hydrogen and oxygen that are formed. To test this idea, elemental hydrogen and oxygen can be ignited to form water. When this is done, energy is released. For every gram of water that is formed; 15,900 J of energy are released. The energy that was stored in the hydrogen and oxygen molecules is released when they combine, forming water.

The energy involved in the decomposition or the formation of water is nearly seven times as large as the heat of vaporization of water. Table 9-2 shows that there is a large difference between these values for other substances, as well. In general, more energy is released or absorbed when a compound is formed or decomposed than when it changes state. In other words, chemical changes tend to involve more energy than do physical changes.

The energy associated with a physical change involves the physical forces of attraction between particles that are not chemically bonded. When water boils, the molecules

Table 9-2: Energies of Some Physical and Chemical Changes

Substance	Heat of vaporization (J/g)	Heat released in formation of compound (J/g)	Heat needed to decompose compound (J/g)
water	2300	15,900	15,900
carbon dioxide	365	8900	8900
sulfur dioxide	400	4600	4600
hydrogen chloride	416	4500	4500

gain enough energy to overcome the forces of attraction between them. As a result, they move farther apart. When steam condenses, the energy stored in the molecules is released, and the molecules move closer together.

The energy associated with a chemical change involves the breaking of chemical bonds. When water is decomposed, the bonds between the hydrogen and oxygen atoms in the water molecules are broken. The energy associated with the breaking or forming of chemical bonds within molecules is generally much greater than that needed to overcome physical forces of attraction between molecules.

In a chemical reaction, the reactants may either gain or lose energy. When hydrogen and oxygen molecules react, forming water, energy is released. The product molecules therefore have less energy stored in them than did the reactant molecules. Similarly, the burning of sodium in chlorine releases energy. The reaction produces 18,000 J per gram of sodium. The combustion of fuel in a furnace releases about 46,000 J per gram of fuel. (See Figure 9-3.) All such chemical reactions, which release energy, are called **exothermic** reactions.

exothermic
(ek soh THUR mik)

endothermic
(en doh THUR mik)

Chemical reactions that require energy in order to take place are called **endothermic** reactions. The decomposition of water is an example of an endothermic reaction. In such reactions, the product molecules have more energy stored in them than did the reactant molecules. Thus, in the decomposition of water, molecules of hydrogen and oxygen have more energy than the reactant molecules of water.

Chemical equations often include an energy term that makes it clear whether the reaction is endothermic or exothermic. For example, the first equation below shows that the decomposition of water is endothermic. In endothermic reactions, the energy term appears with the reactants on the lefthand side of the equation. The second equation shows that the burning of hydrogen in oxygen is exothermic. In exothermic reactions, the energy term appears next to the product molecules on the righthand side of the equation.

$$\text{energy} + 2H_2O \longrightarrow 2H_2 + O_2$$

$$2H_2 + O_2 \longrightarrow 2HO + \text{energy}$$

Figure 9-3 *When fuels burn, they release energy. Energy-releasing reactions are called exothermic reactions.*

Checkpoint

1. Why is the energy associated with chemical changes generally larger than that associated with physical changes?
2. Which has the greater amount of stored energy: (a) two water molecules or (b) two hydrogen molecules plus one oxygen molecule? Explain your answer.
3. Define exothermic and endothermic reactions.

TRY THIS

Add some vinegar to a small glass. Then use a thermometer to find the temperature of the liquid. Add a teaspoon of baking soda. The chemical reaction that occurs releases carbon dioxide gas. Is the reaction exothermic or endothermic? How can you tell?

9-3

Oxidation and Reduction

Goals

1. To define oxidation and reduction.
2. To give examples of oxidation-reduction reactions.

oxidation
(ahk suh DAY shuhn)

reduction
(rih DUHK shuhn)

The burning of fossil fuels is an example of an exothermic reaction. Because fuels combine with oxygen when they burn, burning is sometimes called *oxidation*. To a chemist, oxidation is a much broader term. In chemistry, **oxidation** is defined as the loss of electrons by an atom or ion. When sodium burns in chlorine, sodium atoms lose electrons to chlorine atoms, as you learned in Chapter 8. The sodium is oxidized. The chlorine atoms are said to be reduced. **Reduction** is the gain of electrons by an atom or ion. If one substance is oxidized, another must be reduced. The electrons lost by one substance must be taken up by another. Reactions involving oxidation and reduction are called *oxidation-reduction reactions* or redox reactions.

Figure 9-4 *The reaction between zinc and copper sulfate is an oxidation-reduction reaction. (a) At the start of the reaction, gray zinc metal is shown in the blue copper sulfate solution. (b) During the reaction, the zinc starts dissolving and the blue copper sulfate solution becomes paler, indicating that solid copper metal is being formed.*

An oxidation-reduction reaction takes place when zinc (Zn) is placed in a solution of copper sulfate ($CuSO_4$). A copper sulfate solution contains copper ions (Cu^{2+}), which have a double positive charge, and sulfate ions (SO_4^{2-}), which have a double negative charge. Zinc dissolves in the solution as zinc atoms lose negatively charged electrons to copper ions. When copper ions gain electrons they become copper atoms. Each copper ion accepts two electrons and becomes a neutral atom. As a result, solid copper precipitates, or separates out, from the solution. (See Figure 9-4.) The sulfate ions do not participate in the reaction. We may write the following set of chemical equations to symbolize the reaction.

Figure 9-5 The rusting of iron is an example of an oxidation-reduction reaction. The iron loses electrons to oxygen.

Oxidation $$Zn \longrightarrow Zn^{2+} + 2e^-$$
Reduction $$Cu^{2+} + 2e^- \longrightarrow Cu$$
Overall reaction $$Zn + Cu^{2+} \longrightarrow Zn^{2+} + Cu$$

The rusting of iron is a familiar example of a process called *corrosion*. Rusting and many other forms of corrosion are oxidation-reduction reactions. Such reactions are sometimes described as "slow combustion" reactions. Iron rusts by reacting with oxygen to form iron oxides. (See Figure 9-5.) In the process, iron gives up electrons to the oxygen, which results in formation of positive iron ions and negative oxide ions. Metals other than iron can also be oxidized. When aluminum is oxidized, the white aluminum oxide that is produced forms a coating that protects the metal from further attack.

Checkpoint

1. Define (a) oxidation, (b) reduction.
2. Give an example of an oxidation-reduction reaction.

TRY THIS

Clean a nail with steel wool or fine sandpaper. Place it in a test tube of vinegar, so that the lower half of the nail is under the liquid. Examine the nail a day or two later. Is there any evidence of corrosion?

Figure 9-6 *(a) A "D"-cell is an example of an electrochemical dry cell. (b) A lead storage battery contains several electrochemical cells.*

9-4

Electrochemical Cells

Goals

1. To describe how electrochemical cells work.
2. To explain how unwanted electrochemical reactions can cause corrosion.
3. To describe some ways of preventing corrosion.

Electrochemical (ih lek troh KEM ih kul) *cells* produce electricity that results from oxidation-reduction reactions within the cells. The dry cells shown in Figure 9-6 are electrochemical cells. The battery that is shown is made up of several electrochemical cells connected together. An electrochemical cell called the Daniell cell is shown in Figure 9-7. It produces electricity by means of the oxidation-reduction reaction between zinc and copper sulfate, which was described earlier.

Every electrochemical cell includes *electrodes* (ih LEK trohdz), typically plates through which electrons can flow into and out of the cell and at which oxidation or reduction occurs. In the Daniell cell, one electrode is a piece of copper. It is placed in a porous cup full of copper sulfate solution. The other electrode is a piece of zinc. It is placed in a zinc sulfate solution that surrounds the porous cup. The porous cup prevents the zinc metal and the copper sulfate solution from coming into direct contact. Ions can move slowly through the cup. If a wire connects the electrodes, electrons flow through the wire from the zinc electrode to the copper electrode.

At the copper electrode, copper ions pick up electrons and are reduced to copper atoms, which collect on the electrode. Atoms in the zinc electrode give up electrons and go into solution as zinc ions. Because zinc is the source of electrons, it is the negative electrode. The overall reaction is exactly the same as the reaction of zinc and copper sulfate.

$$\begin{array}{lrcl} \text{Oxidation} & Zn & \longrightarrow & Zn^{2+} + 2e^- \\ \text{Reduction} & Cu^{2+} + 2e^- & \longrightarrow & Cu \\ \hline \text{Overall reaction} & Zn + Cu^{2+} & \longrightarrow & Zn^{2+} + Cu \end{array}$$

The difference between this setup and that described in the last section is that the zinc is not now in direct contact with copper ions. The only path by which the electrons released by the zinc can reach the copper ions is the wire connecting the electrodes. Like any electrochemical cell, the Daniell cell provides a means of using the energy of an oxidation-reduction reaction to obtain electricity for use outside the cell.

Figure 9-7 *A Daniell cell produces electricity through use of the oxidation-reduction reaction of zinc and copper sulfate.*

Different metals have different tendencies to release electrons to ions present in solution. Figure 9-8 shows that lead is oxidized by copper ions and that zinc is oxidized by lead ions. Furthermore, it shows that neither lead nor copper is oxidized by zinc ions.

By means of experiments like those illustrated in Figure 9-8, chemists have ranked metals in order of their tendency to release electrons. This ranking is commonly shown as a list called the *electromotive series*. Table 9-3 on page 226 shows part of this series. Hydrogen is included in the list as a reference point. Metals that have a greater tendency than hydrogen to release electrons are called *active metals* and occur above hydrogen in the table. Any metal in the table is able to be oxidized in the presence of ions of elements below it

Figure 9-8 *Metals differ in their tendencies to be oxidized, or to give up electrons. (a) Lead metal will oxidize in blue copper sulfate solution. (b) Lead metal is not oxidized in zinc nitrate solution. No reaction occurs. (c) Zinc metal is oxidized in lead nitrate solution. Lead metal precipitates from the solution. (d) Copper metal is not oxidized in lead nitrate solution. No reaction occurs.*

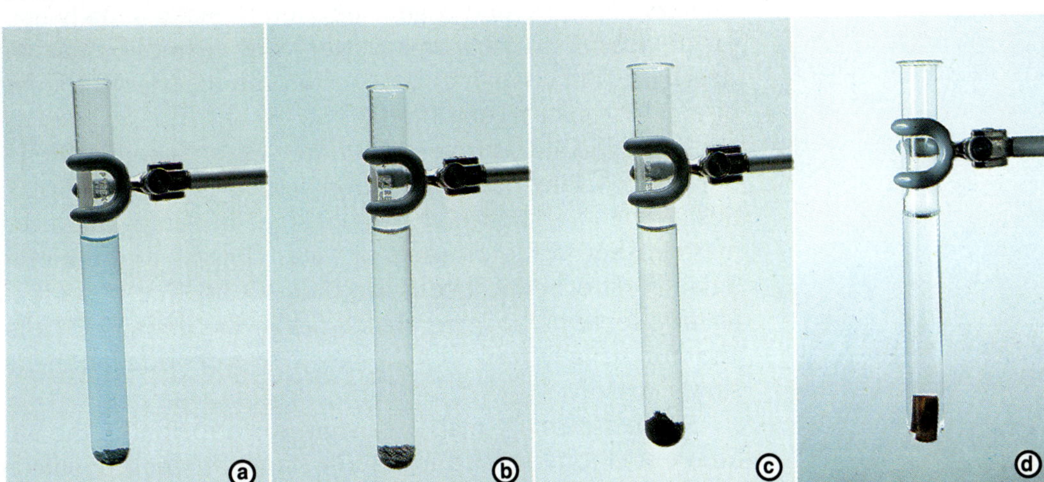

Table 9-3: Electromotive Series

Element	Reduced form		Oxidized form
potassium	K	$\longrightarrow$	$K^+ + e^-$
sodium	Na	$\longrightarrow$	$Na^+ + e^-$
magnesium	Mg	$\longrightarrow$	$Mg^{2+} + 2e^-$
aluminum	Al	$\longrightarrow$	$Al^{3+} + 3e^-$
zinc	Zn	$\longrightarrow$	$Zn^{2+} + 2e^-$
iron	Fe	$\longrightarrow$	$Fe^{2+} + 2e^-$
lead	Pb	$\longrightarrow$	$Pb^{2+} + 2e^-$
hydrogen	H	$\longrightarrow$	$H^+ + e^-$
copper	Cu	$\longrightarrow$	$Cu^{2+} + 2e^-$
silver	Ag	$\longrightarrow$	$Ag^+ + e^-$
mercury	Hg	$\longrightarrow$	$Hg^{2+} + 2e^-$
gold	Au	$\longrightarrow$	$Au^+ + e^-$

in the table. For example, zinc metal is oxidized to zinc ions in a solution such as lead nitrate, which contains lead ions. The lead ions would accept the electrons lost by the zinc, and would become neutral lead atoms.

With the use of Table 9-3, a number of electrochemical cells can be set up. The negative electrode must be a more active metal than the one used for the positive electrode. The positive electrode should be surrounded by a solution containing ions that will accept electrons from the negative electrode. Electrochemical cells are basically quite simple to make, but careful design is needed to make a really useful device. A dry cell or a battery, for example, must be designed well enough to be able to produce a large amount of electricity for a long time.

It is difficult to prevent unwanted electrochemical reactions from taking place. Most iron and steel contain impurities. When the surface of iron or steel is damp, there are ions present in the droplets of water clinging to the metal. Small electrochemical cells are thus set up. The impurities usually serve as positive electrodes in the unwanted cells. The iron usually provides the negative electrode, and so it slowly dissolves.

Some steels are made resistant to corrosion by the addition of chromium and nickel (in stainless steel) or silicon (in duriron) to the iron. To reduce corrosion, metal surfaces can also be covered with paint, with a ceramic enamel (such as those usually found on sinks, stoves, and refrigerators), or with an oxide coating. (See Figure 9-9.) A thick protective coating is often sprayed on the undersides of cars to prevent corrosion. Without such treatment, the undersides of cars may be corroded by unwanted electrochemical reactions, especially if the cars are driven on wet or salt-covered roads.

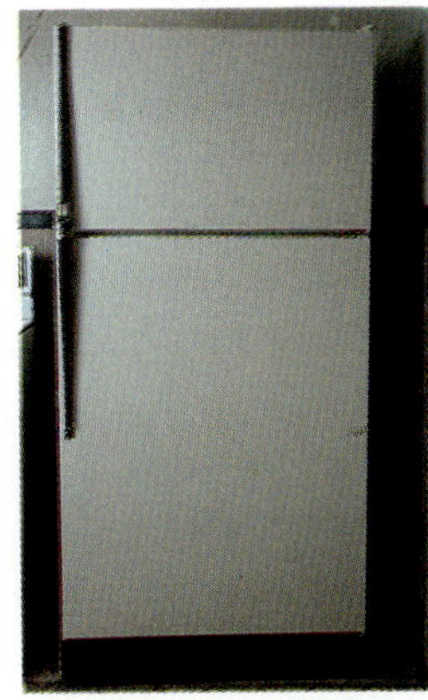

Figure 9-9 *Refrigerators are usually coated with ceramic enamel in order to prevent unwanted electrochemical reactions that would cause corrosion.*

Another way to reduce corrosion involves the use of "sacrificial electrodes." For example, an underground steel tank that needs protection against corrosion can be made the positive electrode of a cell. Zinc, magnesium, or another active metal is used as the negative electrode. Electrons flow from the negative electrode to the positive electrode. As a result, the active metal (the sacrificial electrode) dissolves—not the tank. Hulls of steel seagoing ships are often protected in the same way.

Iron or steel pipes and steel sheet metal that need to be protected against corrosion are sometimes coated with a thin layer of zinc. The zinc serves as more than simply a protective cover. If scratches expose the iron, in the zinc-iron cells that develop, the zinc—not the iron—will act as the negative electrode and will corrode. Iron or steel coated with zinc is sometimes said to be *galvanized* (GAL vuh nyzd). The term comes from the name of an eighteenth-century Italian scientist, Luigi Galvani. Galvani was a pioneer in the development of electrochemical cells.

Checkpoint

1. Explain how a Daniell cell works.
2. Arrange the following metals in order of decreasing tendency to be oxidized (place the one most easily oxidized at the top, then the next-most-easily-oxidized, and so on): zinc, copper, silver, lead, sodium.
3. Describe two methods for protecting metals from corrosion, and explain how the methods work.
4. An electrochemical cell has zinc and silver electrodes immersed in zinc and silver ions, respectively. (a) Make a diagram of the cell. (b) What is oxidized? (c) What is reduced? (d) Which metal is the negative electrode? (e) Which is the positive electrode?

9-5

Acids and Bases

Goals

1. To state the properties of acids and bases.
2. To describe what happens when an acid and a base react.
3. To explain what is meant by the *p*H of a solution.

acids

If you have ever tasted lemon juice, you know that it is sour. A sour taste is one characteristic property of an acid. Generally, **acids** in solution (a) taste sour, (b) conduct electricity, (c) turn blue litmus pink, (d) contain hydrogen, and (e) react with active metals, producing hydrogen gas. (CAUTION: Never taste a substance unless you know it is safe to do so.) Some common acids are listed in Table 9-4. Note that they all contain hydrogen. In solutions, acids produce hydrogen ions (H^+). The hydrogen ions are responsible for the characteristic acid properties.

You have already learned that the gas hydrogen chloride (HCl) consists of polar molecules. When this gas dissolves in water, the molecules break up into hydrogen ions and chloride ions. The polar water molecules interact with the HCl molecules to bring about the ion formation. A solution of HCl in water is called hydrochloric acid. The H^+ ion in hydrochloric acid reacts vigorously with active metals, such as zinc, producing hydrogen gas:

$$Zn + 2H^+ \longrightarrow Zn^{2+} + H_2$$

Table 9-4: Some Common Acids

Acid	Formula
hydrochloric acid	HCl
sulfuric acid	H_2SO_4
nitric acid	HNO_3
acetic acid	$H_4C_2O_2$
citric acid	$H_8C_6O_7$
phosphoric acid	H_3PO_4
carbonic acid	H_2CO_3
boric acid	H_3BO_3

Table 9-5: Some Common Bases

Base	Formula
sodium hydroxide	NaOH
potassium hydroxide	KOH
calcium hydroxide	$Ca(OH)_2$
aluminum hydroxide	$Al(OH)_3$
ammonia	NH_3
sodium carbonate	Na_2CO_3
magnesium hydroxide (milk of magnesia)	$Mg(OH)_2$

Acids can be classified as either strong or weak. A property of strong acids is that they are good conductors of electricity. In general, a *strong acid* is one that produces a large percentage of hydrogen ions when dissolved in water. Nearly all of the HCl molecules in a water solution break up into ions, and so HCl is a strong acid.

indicator

bases

Vinegar is a dilute solution of acetic acid in water. Acetic acid is an example of a weak acid. It reacts with zinc, producing hydrogen, but the reaction takes place very slowly. Weak acids are poor conductors of electricity. In general, a *weak acid* is one that produces a small percentage of hydrogen ions when dissolved in water. Acetic acid produces only a few hydrogen ions per thousand molecules.

Hydrogen ions in solution can be detected with an indicator such as litmus paper. An **indicator** is a substance that changes color in the presence of certain other substances. Blue litmus, for example, turns pink in an acid, as shown in Figure 9-10.

Figure 9-10 *Litmus is an acid-base indicator.*

(a) Blue litmus paper turns pink in an acid solution.

(b) Red litmus paper turns blue in a basic solution.

Red litmus placed in a solution of sodium hydroxide turns blue. This color change indicates the presence of a base. **Bases** are compounds that (a) taste bitter, (b) conduct electricity, (c) turn red litmus blue, (d) feel slippery. (CAUTION: Never taste or touch a substance unless you know it is safe to do so.) Some common bases are listed in Table 9-5.

All of the bases listed in Table 9-5 contain hydroxide ions (OH^-) except ammonia and sodium carbonate. However, all of the bases produce hydroxide ions in water.

$$NH_3 + H_2O \longrightarrow NH^+ + OH^-$$

ammonia → ammonium ion + hydroxide ion

$$2Na^+ + CO_3^{2-} + H_2O \longrightarrow HCO_3^- + OH^- + 2Na^+$$

sodium carbonate → bicarbonate ion + hydroxide ion

When an acid and a base are mixed together, the hydrogen and hydroxide ions combine, forming water:

$$H^+ + OH^- \longrightarrow H_2O$$

The acid and base are said to *neutralize* (NOO truh lyz) each other. When hydrochloric acid and sodium hydroxide are mixed together, they react as follows:

$$H^+Cl^- + Na^+OH^- \longrightarrow Na^+Cl^- + H_2O$$

The resulting solution contains common salt, sodium chloride. Actually the sodium and chloride ions are not really involved in the reaction and can be left out of the equation. In that case the reaction is identical to one given before.

$$H^+ + OH^- \longrightarrow H_2O$$

salt

The ionic compound formed when *any* acid and base react is called a **salt**. In general:

$$\text{acid} + \text{base} \longrightarrow \text{salt} + \text{water}$$

neutralization (noo truh luh ZAY shuhn)

A reaction in which an acid and a base neutralize each other, forming a salt and water, is called a **neutralization** reaction.

Acids and bases are very commonly occurring substances. (See Figure 9-11.) Hydrochloric acid is secreted in the stomachs of human beings where it helps to break down food. Sulfuric acid is widely used in making plastics, fertilizers, and dyes. It is used in the lead storage batteries found in cars, also. The production level of sulfuric acid is considered to be a measure of a nation's industrial development.

Figure 9-11 *(a) Some commonly used acids. (b) Some commonly used bases.*

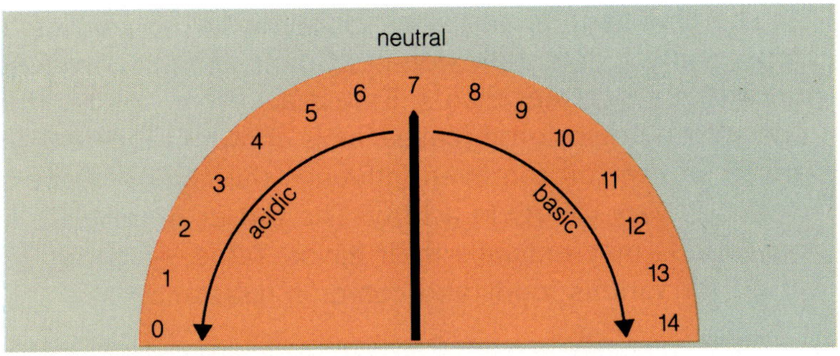

Figure 9-12 *The* pH *scale ranges from 0 to 14. Neutral solutions have a* pH *of 7.*

Sodium hydroxide (lye) is used in making soaps and detergents, cellophane, and paper. Potassium hydroxide is vital in producing fertilizers and dyes. Ammonia is used in the manufacture of explosives, synthetic fibers, medicines, and cleaning agents.

Acids and bases are commonly used to neutralize each other. Farmers use lime (CaO) to reduce the acidity of soil. (When mixed with water, the lime forms calcium hydroxide, $Ca(OH)_2$.) People take antacids to neutralize excess acid in the stomach.

***p*H scale**

It is often important to know how acidic or basic a solution is. For example, the chlorine that is added to a swimming pool makes the water acidic. Testing the acidity of the water can reveal whether the correct amount of chlorine has been added. Chemists have devised a ***p*H scale** to measure how acidic or basic a solution is. The symbol "*p*H" is an abbreviation for "power of hydrogen ion." For dilute solutions, the range of the *p*H scale is 0–14. (See Figure 9-12.) A *p*H of 7 indicates a *neutral* solution—one with equal numbers of H^+ and OH^- ions. A *p*H below 7 indicates an acidic solution. A *p*H above 7 indicates a basic solution. Between each two successive numbers on the scale there is a tenfold difference. A solution with a *p*H of 6 has ten times as many H^+ ions as the same volume of a neutral solution. A solution with a *p*H of 8 has one tenth as many H^+ ions, or ten times as many OH^- ions, as the same volume of a neutral solution. A mixture of dyes has been devised that changes color gradually over the entire range of the *p*H scale. This universal indicator is available in the form of *p*H paper. (See Figure 9-13.)

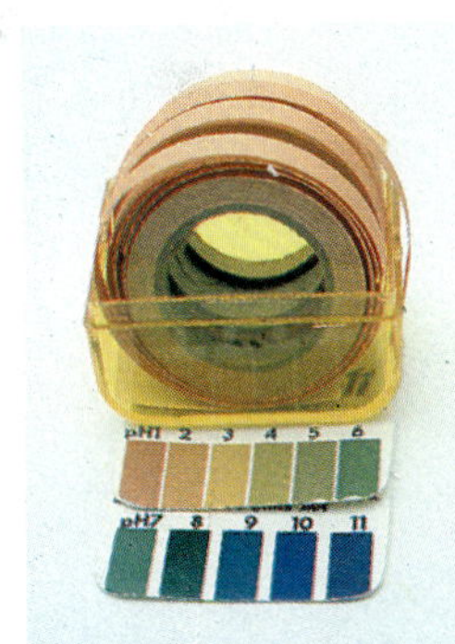

Figure 9-13 *The mixture of dyes in* pH *paper can indicate* pH *over the entire range. The rainbowlike colors range from orange (highly acidic, low* pH*) through yellow-green (neutral) to dark blue (highly basic, high* pH*).*

Figure 9-14 *The discoloration of the leaves on this plant is due to excess soil acidity.*

The *p*H of soil is an important factor in the growing of crops. Some plants grow well in acidic soil. Others require soil with a *p*H greater than 7. Excess acidity in soil is often indicated by discolored leaves, as is shown in Figure 9-14. The *p*H of the soil can even influence the color of flowers.

Foods vary widely in acidity. The tart or bitter taste of some foods may enhance their flavor. Table 9-6 shows *p*H values for various foods and common substances.

Table 9-6: *p*H of Various Solutions

Substance	***p*H**
stomach acid	1.0
vinegar	2.4
cola drink	2.5–3.0
grapefruit juice	3.2
tomato juice	4.3
orange juice	2.6–4.4
boric acid solution	5.0
saliva	6.6
milk	6.6–6.9
human blood	7.4
seawater	8.0
egg white	8.0
household ammonia	11–12

Checkpoint

1. What are the characteristic properties of (a) acids, (b) bases?
2. What ions are always present in solutions of (a) acids, (b) bases?
3. What happens when an acid and a base are mixed?
4. What is the *p*H scale? What does it measure?

TRY THIS

Using litmus or *p*H paper, find out whether each of the following is acidic, basic, or neutral: tap water, grape juice, and orange juice. Repeat the test on solutions made by dissolving each of the following in water: tea, coffee, aspirin, salt, sugar, soap, and baking soda (sodium bicarbonate).

9-6

Rates of Reaction

ENRICHMENT

Goals

1. To state the factors that affect the rate of a chemical reaction.
2. To explain why these factors have their effects.

The rate of a chemical reaction is the speed at which it occurs. This can be measured in terms of how much product is formed per unit of time. The rates of chemical reactions vary greatly. A mixture of hydrogen and oxygen reacts with explosive speed when ignited with a spark. A piece of iron may take years to react completely with oxygen, forming rust. By learning what factors affect reaction rates, people can control chemical reactions. They can make reactions more safe, efficient, and economical.

The rate of a reaction depends on (a) the temperature, (b) the amounts of reactants and products in a given volume (their *concentrations*), (c) the size and surface area of the reactants, and (d) the presence of a catalyst. A **catalyst** is a substance that changes the rate of a reaction without being used up in the reaction.

catalyst (KAT ul ist)

As a rule of thumb, a 10°C increase in temperature tends to double the rate of a reaction. As the temperature rises, molecules move faster, collide more often, and hit each other harder. This causes more of them per second to break apart and form new chemical bonds. The cooking of food involves chemical reactions. By regulating the temperature, people can control the rate at which foods cook.

If the concentration of the reactant molecules increases, the number of molecular collisions per second will also, in general, increase. This increases the rate of the reaction. If a small amount of copper sulfate is dissolved in water, the solution will react slowly with a piece of zinc. If a large amount of copper sulfate is dissolved in the same amount of water, the reaction will take place more quickly. The concentration of a solution can be increased by dissolving more of the solute in the same amount of water. The concentration of a gas can be increased by compressing it. This forces the molecules of the gas into a smaller volume, causing them to collide more often. In an automobile engine, the

Figure 9-15 *Reaction rate depends on the amount of exposed surface area, as well as on other factors. (a) A large piece of wood tends to burn slowly in air. (b) The same amount of wood in the form of wood chips has a greater surface area and combines more rapidly with oxygen in the burning process.*

fuel-air mixture that is ignited is highly compressed. It reacts very rapidly when a spark is applied, providing power to move the car.

The surface area of a substance can be increased by breaking or grinding it into smaller particles. When this is done, the substance reacts more rapidly. This happens because more molecules of one reactant are exposed to the other reactant. A log of wood burns at a slow, steady rate. The same amount of wood in the form of chips can burn rapidly when ignited. (See Figure 9-15.)

A catalyst may provide a more efficient "pathway" for a reaction, increasing the reaction rate. Substances called *enzymes* act as catalysts within the human body. They enable reactions to take place more rapidly than they normally would at body temperature.

Platinum is often used as a catalyst in reactions between gases. Many gases attach themselves to a platinum surface, enabling the gas molecules to react more rapidly. Platinum is used as a catalyst for the reaction between oxygen and sulfur dioxide. These gases react to form another gas, sulfur trioxide. The sulfur trioxide is then made to react with water, forming sulfuric acid. Catalysts can be used to slow down, as well as speed up, reactions. Such catalysts are sometimes called "negative catalysts."

When the products of a reaction do not escape from the reaction area, their concentrations increase. The products may re-form into reactants. The rate of this *reverse reaction*—products re-forming into reactants—tends to increase as the concentration of products increases. There is a limit to the "completeness" of the *forward reaction* under such conditions. That is, there may be reactants as well as products present at the end of the reaction. When products escape or are removed from the reaction area, a reaction can go more nearly to completion. When zinc reacts with hydrochloric acid, producing hydrogen as a product, the gas escapes. This escaped hydrogen is no longer present to re-form into the reactant hydrochloric acid. Therefore, there is no chance of a significant reverse reaction. The forward reaction tends to go to completion.

Checkpoint

1. What factors affect the rate of a chemical reaction?
2. Explain how, at the molecular level, each factor affects the reaction rate.
3. Why does compressing a gas increase its concentration?

TRY THIS

Add a seltzer tablet to half a glass of hot water. At the same time add an identical tablet to half a glass of cold water. In which case is the reaction faster?

To half a glass of cold water add a whole seltzer tablet. At the same time add a crushed seltzer tablet to an identical amount of water at the same temperature. In which glass is the reaction faster? Why?

Internal Combustion Engines

Most automobile engines are internal combustion engines. The fuel that provides the energy to move the car burns inside the engine. The energy causes pistons to move, which cause a crankshaft to turn, which then causes the wheels to turn. (See Figure 1.) The fuel is usually gasoline, a mixture of hydrocarbon compounds. When the fuel burns, the carbon and hydrogen combine with oxygen. Energy is given off when the fuel burns.

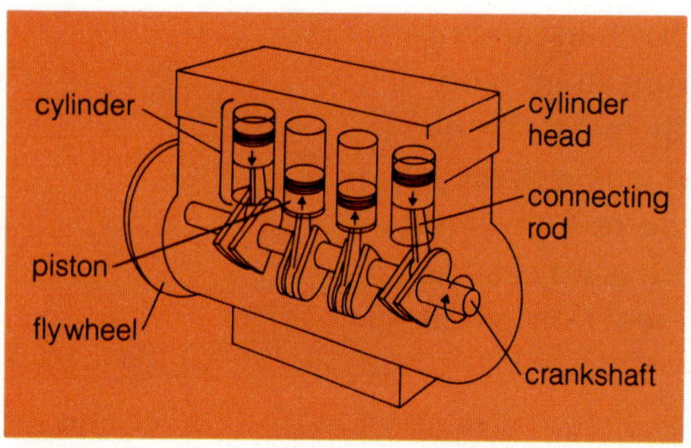

Figure 1

In an automobile, the engine is normally started by turning the ignition key. This connects the starter motor to the battery. Powered by the battery, the motor turns the crankshaft, which causes at least one piston to move downward. At the same time a mechanism opens the inlet valve of the cylinder in which the piston is located and closes the exhaust valve. As the air in the cylinder expands, its pressure decreases. As a result, a mixture of air and gasoline vapor is drawn in from the carburetor (the chamber in which the air and fuel are normally mixed). When the pressure inside the cylinder equals the air pressure outside, the inlet valve closes.

As the crankshaft completes its first full turn, the ascending piston compresses the fuel-air mixture, heating the gases in the process. At the top of the stroke, the spark plug gives off a spark. The gases ignite and burn. The high temperature produced by the combustion causes the gaseous products (mostly carbon monoxide, carbon dioxide, and water vapor) to expand. This drives the piston downward, providing power to the crankshaft. As the crankshaft completes its second turn, the ascending piston this time forces the gaseous waste products through the open exhaust valve. (See Figure 2.)

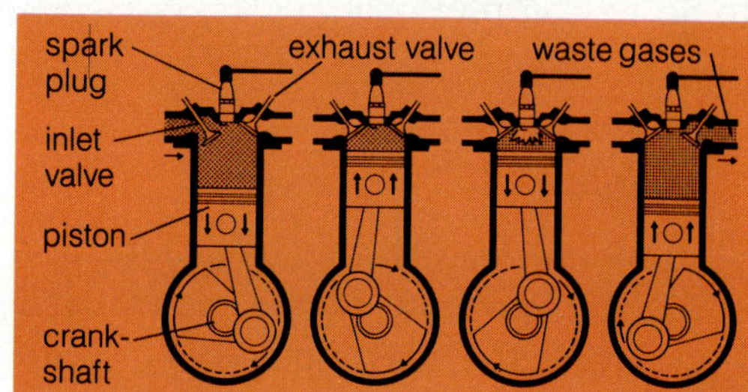

Figure 2

In most cars, there is a minimum of four cylinders. They fire so that the crankshaft gets a push (power stroke) every half turn. Each piston goes through four complete cycles between explosions within its cylinder. Once started, the cycles continue as long as fuel, air, and electricity are provided.

Understanding The Chapter 9

Main Ideas

1. Energy is needed to separate the particles of a liquid when it changes to a gas. The energy is given up when a gas changes into a liquid.
2. Chemical changes generally involve more energy than do physical changes. In chemical changes, chemical bonds are broken.
3. Exothermic reactions release energy. Endothermic reactions absorb energy.
4. Oxidation is the loss of electrons by an atom or ion in a chemical reaction. Reduction is the gain of electrons.
5. Electrochemical cells are based on oxidation-reduction reactions.
6. Corrosion often results from unwanted electrochemical reactions.
7. Active metals are those that give up electrons readily.
8. In solutions, acids produce hydrogen ions and bases produce hydroxide ions.
9. Acids have a *p*H less than 7, and bases have a *p*H greater than 7.
10. A neutralization reaction is one in which an acid and a base react with each other, forming a salt and water.
11. The rate of a chemical reaction depends on the temperature, the concentrations of reactants and products, the surface area of the reactants, and the presence of a catalyst.

Vocabulary Review

From the following list, choose the term that best completes each of the statements. Write your answers on a separate piece of paper.

acid	heat of condensation
active metal	heat of vaporization
base	indicator
catalyst	neutralization
electrochemical cell	oxidation
electrode	*p*H scale
endothermic	reduction
exothermic	salt

1. Litmus paper is a(n) __?__.
2. A(n) __?__ has a sour taste.
3. __?__ is the loss of electrons by an atom or ion.
4. A(n) __?__ reaction is one in which heat is absorbed.
5. A reading of 9 on the __?__ indicates that a substance is basic.
6. __?__ occurs when an atom or ion gains electrons.
7. A(n) __?__ produces OH^- ions in solution.
8. The burning of hydrogen is an example of a(n) __?__ reaction because heat is released.
9. The energy needed to transform a liquid into a gas at its boiling point is its __?__.
10. A reaction in which an acid and a base react to form a salt and water is called a __?__ reaction.

11. A(n) _?_ can be a convenient source of electricity.
12. A(n) _?_ is a substance that can change the rate of a chemical reaction without itself being used up.

Chapter Review

Write your answers on a separate piece of paper.

Know The Facts

1. The energy released when a gram of gas changes to liquid at the boiling point is the heat of _?_. (reaction, condensation, vaporization)
2. When water that is being heated starts to boil, the temperature _?_. (slowly rises, slowly falls, stays at about 100°C)
3. The forces of attraction between water molecules are _?_ compared to those acting between molecules of most other substances. (large, small, average)
4. Energy changes in chemical reactions are generally _?_ those in physical changes. (smaller than, equal to, greater than)
5. The forces between atoms *within* molecules are generally _?_ the forces *between* molecules. (less than, equal to, greater than)
6. In a Daniell cell _?_ are reduced. (zinc atoms, zinc ions, copper ions)
7. Lead is a more active metal than is _?_. (zinc, iron, copper)
8. The formation of zinc ions when zinc reacts with an acid indicates that zinc has been _?_. (reduced, oxidized, neutralized)
9. Acetic acid, which is found in vinegar, is a _?_ acid. (strong, weak, active)
10. _?_ is found in all acids. (oxygen, nitrogen, hydrogen)
11. Ammonia, sodium carbonate, and lye are all _?_. (acids, bases, salts)
12. The rate of a chemical reaction can be increased by _?_. (lowering the temperature, increasing the concentration of reactants, increasing the concentration of the products)

Understand The Concepts

13. It takes 400 J of energy to convert one gram of liquid sulfur dioxide (SO_2) to vapor at the boiling point (−10°C). How much energy is released when one gram of SO_2 gas condenses at −10°C?
14. How do you explain the fact that the heat of vaporization of water is so much higher than that of most other liquids?
15. In each of the following reactions, what is oxidized and what is reduced?

    (a) $2H^+ + Mg \longrightarrow Mg^{2+} + H_2$
    (b) $Fe + Pb^{2+} \longrightarrow Fe^{2+} + Pb$
    (c) $4Al + 3O_2 \longrightarrow 4Al^{3+} + 6O^{2-}$

16. Explain how you could build an electrochemical cell using lead and silver electrodes, lead nitrate and silver nitrate solutions, a beaker, a porous cup, and wires. Draw a sketch of the cell. Which metal is the negative electrode?
17. Describe at least two ways in which metals can be protected from corrosion.

18. How would you tell whether a liquid is (a) an acid, (b) a base?

19. What is the difference between strong and weak acids?

20. Explain why ammonia and sodium carbonate are bases even though they contain no OH^- ions.

21. Some seltzers come in tablet form and others come as powders. Which form would you expect to react faster in water? Why?

22. What increase in the rate of a reaction might you expect if the temperature were to rise by 20°C?

Challenge Your Understanding

23. How will the number of hydrogen ions per unit volume of an acid with a *p*H of 2 compare with one that has a *p*H of 6?

24. Boric acid is sometimes used as an eye rinse. Why is it safe to use boric acid this way, when other acids, such as sulfuric acid, could cause permanent blindness?

25. Explain how the connection of a piece of magnesium to a ship's iron hull can protect the hull from corrosion.

26. What would you expect to be the difference between strong and weak bases?

27. If you were to heat a solution of sodium chloride in water, you would find that the temperature would continue to rise after the solution has started to boil. How might you explain this?

28. Which of the following ions would you expect to be easiest to reduce: Cu^{2+}, Zn^{2+}, Mg^{2+}? Explain your answer.

Projects

1. Study the effects of corrosion on steel, with and without a "sacrificial electrode." Place a piece of steel in seawater (or a solution made by adding 30 g of salt to a liter of water). In a second container of salt water, place a similar piece of steel that you have connected to a piece of clean zinc by means of a wire. (A current-measuring device called an ammeter can be used to measure the current flowing between the two metals.) Allow the setup to remain for several weeks, adding water to replace any lost by evaporation. Note which piece of steel corrodes more.

2. Pay a visit to your school or community swimming pool, if there is one. Find out how the *p*H of the water is tested. Also, find out how the *p*H of the water is adjusted, when necessary.

3. Talk to a farmer or your local farm bureau agent about the effect of lime on soil. Find out which crops grow best in acidic soils and which grow best in basic soils. Find out how the *p*H of soils is changed.

4. At a library, do research on acid rain. What is acid rain? What causes it? What are its effects? What can be done about it?

5. At a library, do research on the history of electric-powered vehicles. What vehicles now in use are powered by electrochemical cells? Do you think an all-purpose electric-powered automobile is likely to be developed in the near future? Why or why not?

Chapter

10

Nuclear Reactions

Aerial view of the Stanford University linear accelerator in California.

10-1

Radioactivity

Goals

1. To describe the discovery of radioactive elements.
2. To list the types of radiation given off by radioactive nuclei.
3. To define transmutation.

In 1896, Henri Becquerel made an amazing discovery. Becquerel had left a piece of rock that contained uranium on top of a sheet of unexposed photographic film that was covered with paper. Some time later, when Becquerel developed the film, he was surprised by the image he saw. The image had the same size and shape as the rock that had been on top of the film, as is shown in Figure 10-1. Something in the rock had given off invisible rays, or *radiation* (ray dee AY shuhn), that could pass through the film cover. It was found that the uranium in the rock was the source of the radiation. Further, the radiation was given off by the *nuclei* of uranium atoms. Nuclei that give off radiation are said to be *radioactive nuclei.*

Two years later, Marie Curie, one of Becquerel's associates, began working with pitchblende. It is a mineral even more radioactive than uranium-containing minerals. After long, painstaking work, she found that pitchblende contained two previously unknown radioactive elements. She

Figure 10-1 *In the photograph at the left, a rock that contains radioactive nuclei is resting on a piece of photographic film. Later, when the film has been developed, the rock's image is seen, as shown at the right.*

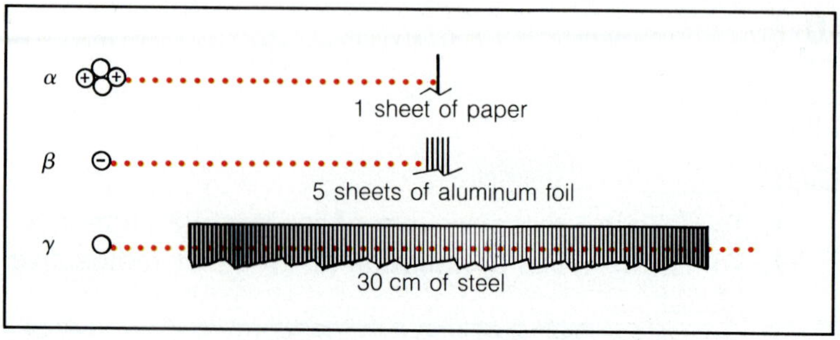

***Figure 10-2** Alpha (α), beta (β), and gamma (γ) radiation differ in ability to pass through various thicknesses of substances.*

named one of the elements polonium (Po) in honor of Poland, her native country. She named the other element radium (Ra) because of the intense radiation it gave off. From tons of pitchblende, Marie Curie was able to extract only a tenth of a gram of pure radium.

Three types of radiation are given off by radioactive nuclei. They are named after the first three letters of the Greek alphabet: *alpha* (α), *beta* (β), and *gamma* (γ). **Alpha radiation** consists of helium nuclei, which are made up of two protons and two neutrons. Because alpha particles move at relatively low speeds, they can be stopped by a sheet of ordinary writing paper. (See Figure 10-2.) **Beta radiation** consists of electrons moving at extremely high speeds. Beta particles, depending on their speeds, can be stopped by aluminum foil, 2 to 10 sheets thick. **Gamma radiation** consists of light rays that have more energy and greater penetrating power than do x rays. Some gamma rays can penetrate steel that is 30 cm thick.

alpha radiation

beta radiation

gamma radiation

When a nucleus gives off an alpha or a beta particle, the number of protons in the nucleus changes. A nucleus that has given off an alpha particle has lost two protons and two neutrons. A nucleus that has given off a beta particle has lost a neutron and gained a proton. (Actually, a neutron turns into a proton and an electron, which escapes from the nucleus.) When the atoms of an element gain or lose protons, a new element is formed. This is called **transmutation**.

transmutation

Checkpoint

1. How was radioactivity discovered?
2. What types of radiation are given off by radioactive nuclei?
3. What is transmutation?

10-2

Isotopes

Goals

1. To define isotopes.
2. To explain why isotopes of a given element have the same chemical properties.
3. To compare nuclear forces with those that act between charged objects.

Each atom of a given element has the same number of protons. However, the atoms may have different numbers of neutrons and, therefore, different masses. Atoms of the same element that have different masses are called **isotopes**. For example, most naturally occurring hydrogen atoms have a single proton in their nucleus. Their atomic mass is 1 amu. However, some hydrogen atoms have a nucleus that consists of a proton and a neutron. (See Figure 10-3.) These atoms have a mass of 2 amu and are sometimes referred to as atoms of *heavy hydrogen* or *deuterium* (doo TEER ee um). Only two out of every 1000 naturally occurring hydrogen atoms are deuterium. (See Table 10-1.) The mass of hydrogen is given in tables as 1.008. This represents the

isotopes
(EYE suh tohps)

Table 10-1: Naturally Occurring Isotopes of Several Elements

Element	Atomic number	Masses of isotopes (amu)	Percent in nature
hydrogen	1	1	99.8
	1	2	0.2
carbon	6	12	98.89
	6	13	1.11
neon	10	20	90.92
	10	21	0.26
	10	22	8.82
chlorine	17	35	75.5
	17	37	24.5
uranium	92	234	0.01
	92	235	0.72
	92	238	99.27

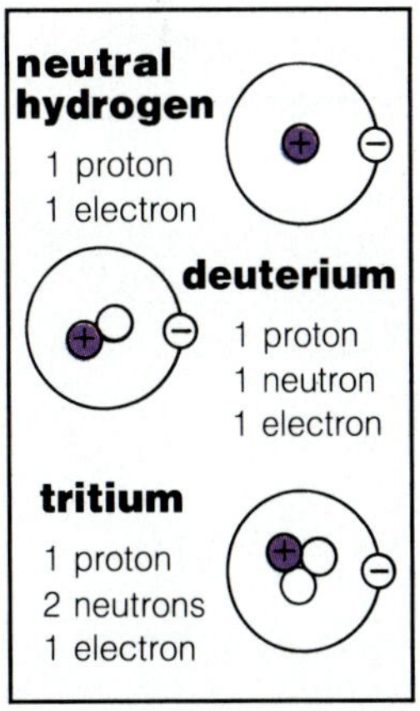

Figure 10-3 *The three isotopes of hydrogen differ in the number of neutrons present in the nucleus.*

mass of the naturally occurring *mixture* of hydrogen isotopes, 99.8% of which have a mass of 1 amu and 0.2% of which have a mass of 2 amu.

There is a third isotope of hydrogen, called *tritium*, which can be produced in nuclear experiments. Its nuclei contain one proton and two neutrons. Unlike the naturally occurring isotopes of hydrogen, tritium is radioactive.

Each element has several known isotopes, but each of the isotopes shows the same chemical behavior. This is because the chemical properties of an element depend on the number of electrons in its atoms. The number of electrons in a neutral atom depends, in turn, on the number of protons in the nucleus of the atom. The number of neutrons in the nucleus does not significantly affect the chemical behavior.

Each of the hydrogen isotopes has one electron, so the isotopes undergo the same chemical reactions. For example, all three isotopes react with oxygen, forming water. The water formed by causing naturally occurring hydrogen to react with oxygen includes minute amounts of *heavy water*, D_2O. ("D" stands for deuterium.) Heavy water can be separated from ordinary water, since it has a greater mass. Heavy water has properties that are useful in nuclear reactors.

radioactive nuclei

In a nucleus, the particles are held together by attractive forces. The force of attraction within a stable nucleus is much stronger than the force of repulsion between protons, which have the same charge. If it were not stronger, protons in a nucleus would push away from each other, destroying the nucleus. **Radioactive nuclei** are unstable nuclei that break down, releasing particles and also large amounts of energy.

Some nuclei are more stable than others because the number of protons and neutrons they contain fit together in an especially stable arrangement. For example, the helium nucleus that consists of two protons and two neutrons is a very stable unit. Other combinations of protons and neutrons form less-stable arrangements. For instance, the tritium nucleus (1 proton and 2 neutrons) is unstable. Tritium nuclei give off beta particles, becoming more stable as a result.

The effect of the arrangement of protons and neutrons on the stability of a nucleus is similar to the effect of the arrangement of electrons on the stability of an atom. There

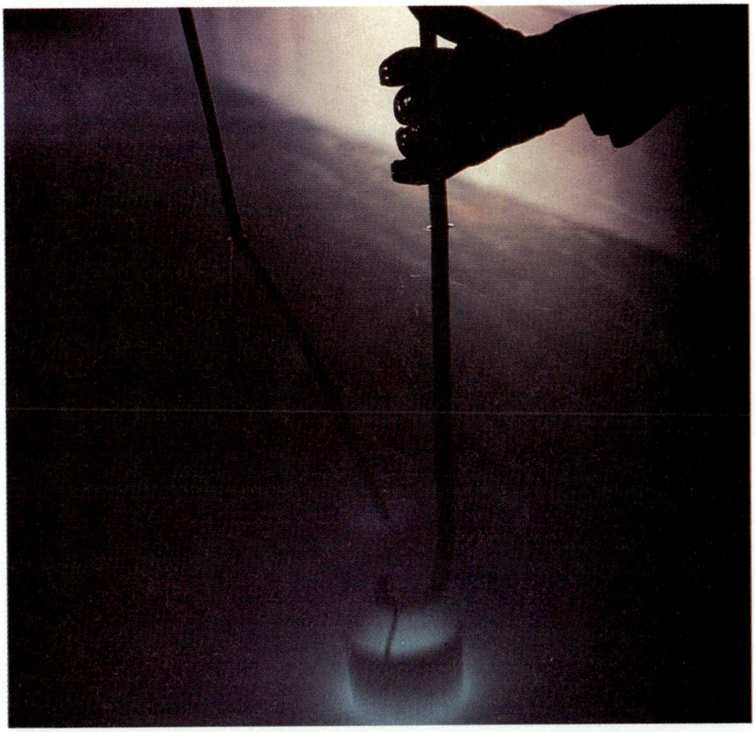

Figure 10-4 *This submerged sample of an isotope of cobalt is so radioactive it makes the water glow.*

are important differences, however. For one thing, nuclear forces are much stronger than the forces involved in electron arrangements. That is why nuclear reactions involve far more energy than do chemical reactions. For another thing, the arrangements of protons and neutrons in nuclei are more complicated than are electron arrangements in atoms. The effect of the arrangement of nuclear particles on the properties of nuclei is less well understood than is the effect of electron arrangements on chemical properties.

Checkpoint

1. What are isotopes?
2. Why do the isotopes of a given element show the same chemical behavior?
3. Use Table 10-1 to find the number of neutrons in the naturally occurring isotopes of: (a) hydrogen, (b) neon, (c) carbon, (d) uranium.

10-3

Radioactive Decay

Goals

1. To describe radioactive decay.
2. To define the half-life of a radioactive isotope.
3. To interpret nuclear equations.
4. To explain how stable isotopes can be converted into radioactive isotopes.

radioactive decay

half-life

You have seen that when a radioactive isotope gives off an alpha or a beta particle, it becomes a different element. This change of a radioactive isotope into a different element is called **radioactive decay**. Tritium, the isotope of hydrogen with a mass of 3 amu, gives off beta radiation. As a result, it undergoes transmutation, becoming the isotope of helium with a mass of 3 amu. (See Figure 10-5.) Using symbols, scientists say that H-3 decays into He-3. The number 3 stands for the total number of protons and neutrons. This number, which is nearly equal to the mass in amu, is called the *mass number*. A sample of tritium that has a mass of one gram will decay to half a gram in 12.3 years. Tritium is said to have a *half-life* of 12.3 years. The **half-life** of a radioactive isotope is the time it takes for half of its atoms to decay. Half-lives range from less than a second to billions of years, as is shown in Table 10-2.

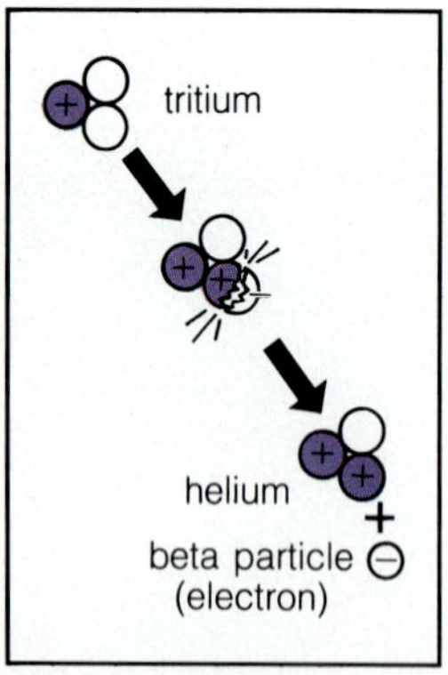

Figure 10-5 *A radioactive tritium nucleus gives off a beta particle and becomes a helium nucleus.*

Table 10-2: Half-lives of Some Radioactive Isotopes

Isotope	*Half-life*
H-3	12.3 years
He-6	0.8 seconds
C-14	5570 years
P-32	14.3 days
Co-60	5.27 years
Sr-90	28 years
I-131	8.05 days
Po-210	138 days
Po-218	3.05 minutes
Ra-226	1662 years
Th-234	24.1 days
U-238	4.5 billion years

The uranium isotope with a mass of 238 amu (U-238) has a half-life of 4.5 billion years. That is, it takes 4.5 billion years for half of the atoms in a sample of U-238 to decay. A U-238 atom gives off an alpha particle (He-4), becoming an atom of thorium (Th-234). (See Figure 10-6.)

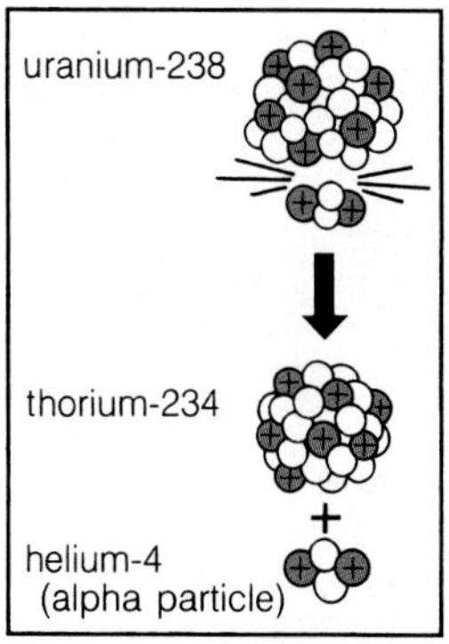

Figure 10-6 *A nucleus of uranium-238 can decay, thereby producing a nucleus of thorium-234 and an alpha particle.*

Nuclear reactions, such as the decay of a radioactive element, can be shown by means of a **nuclear equation**. The decay of U-238 into Th-234 is shown as follows:

nuclear equation

$$^{238}_{92}U \longrightarrow ^{234}_{90}Th + ^{4}_{2}He$$

The number at the upper left of each symbol in the equation indicates the mass number, or total number of protons and neutrons in the nucleus. The lower number is the charge of the nucleus. A charge of 1 unit is the amount of charge on a proton or an electron. The charge is positive unless there is a minus sign in front of the charge number. In the case of an isotope, the charge number is also the number of protons, or the *atomic number*.

In balancing nuclear equations, as in chemical equations, mass and charge are assumed to be conserved. In the equation above, the mass number to the left of the arrow is equal to the sum of the mass numbers to the right: $238 = 234 + 4$. The charge on the left is equal to the sum of the charges on the right: $92 = 90 + 2$.

The decay of U-238 to Th-234 is just the first in a long series of radioactive decays. The Th-234 nucleus is unstable, also. When Th-234 decays, it produces yet another unstable nucleus. The series of decays continues until a stable nucleus results. The U-238 decay series ends with a stable isotope of lead (Pb-206).

A stable isotope can become unstable when it is hit by a high-energy particle. High-energy particles from outer space called **cosmic rays** are constantly showering Earth. These particles include neutrons. Some cosmic-ray neutrons collide with nitrogen atoms, which are the most numerous in the atmosphere. The reaction between nitrogen-14 and a high-energy neutron produces carbon-14 and a proton, or hydrogen nucleus.

cosmic rays

$$^{14}_{7}N + ^{1}_{0}n \longrightarrow ^{14}_{6}C + ^{1}_{1}H$$

Figure 10-7 *Particle accelerators are used to produce nuclear reactions. New elements can be made in this way.*

The carbon-14 that is produced is unstable. It decays, with a half-life of 5570 years, producing nitrogen-14 and a beta particle, or electron.

$$^{14}_{6}C \longrightarrow {}^{14}_{7}N + {}^{0}_{-1}e$$

The mass of an electron is shown as zero in a nuclear equation because it is very small compared to the mass of a proton or neutron.

Reactions similar to those produced by cosmic rays can occur when stable isotopes are bombarded with radiation given off by unstable isotopes. Neutrons were first detected as the result of such a reaction. The reaction was produced by the bombardment of beryllium-9.

$$^{9}_{4}Be + {}^{4}_{2}He \longrightarrow {}^{12}_{6}C + {}^{1}_{0}n$$

Particles can be accelerated to high speeds in devices called *accelerators*. (See Figure 10-7.) Particles traveling at high speeds have the high energies needed to produce nuclear reactions. Such reactions have provided a great deal of information about the nucleus. They have also produced many very useful radioactive isotopes that do not occur naturally.

Particle accelerators have been used to produce elements that do not occur naturally on Earth. Before 1940, uranium was the element with the largest atomic number (92). Since then, more than a dozen new elements have been produced by bombarding other elements with high-speed particles. Elements with atomic numbers greater than 92 are sometimes called *transuranium elements*, since they are "beyond" uranium in the periodic table.

TRY THIS

Build a 1-m-long pendulum. Place a ruler under the bob, and note its rest position. Pull the bob 20 cm to one side. Release the bob and record the time. Calculate the pendulum's "half-life," which is the amount of time it takes for the bob to swing 10 cm from its rest position. Find out how long it will take the bob to swing 5 cm from its rest position.

Checkpoint

1. Define radioactive decay.
2. What is meant by the half-life of a radioactive isotope?
3. $^{214}_{84}Po \longrightarrow {}^{210}_{82}Pb + \underline{\ ?\ }$. Complete the equation.
4. Describe two ways in which a stable isotope may be converted into a radioactive one.

10-4

Uses of Radioisotopes

ENRICHMENT

Goal

To describe some practical uses of radioactive isotopes.

Radioactive isotopes, or *radioisotopes*, can be put to practical use. There are important uses in medical treatment. Radiation can destroy or harm body cells. It can cause cancer, also. However, the ability to destroy body cells can be used to *treat* cancer. When directed at a cancer, radiation can destroy the cancer cells, while doing little damage to normal cells nearby. (See Figure 10-8.) Cobalt-60 is a radioisotope used for this purpose.

Different kinds of radioisotopes may be used to treat different cancers. For example, if cancerous tissue absorbs phosphorus, the radioisotope phosphorus-32 can be used to destroy the cancer. Similarly, since iodine is absorbed in the thyroid gland, the radioisotope iodine-131 can be used to destroy thyroid cancer. Both P-32 and I-131 have half-lives of less than 15 days. As a result, these radioisotopes do not remain in the body for long.

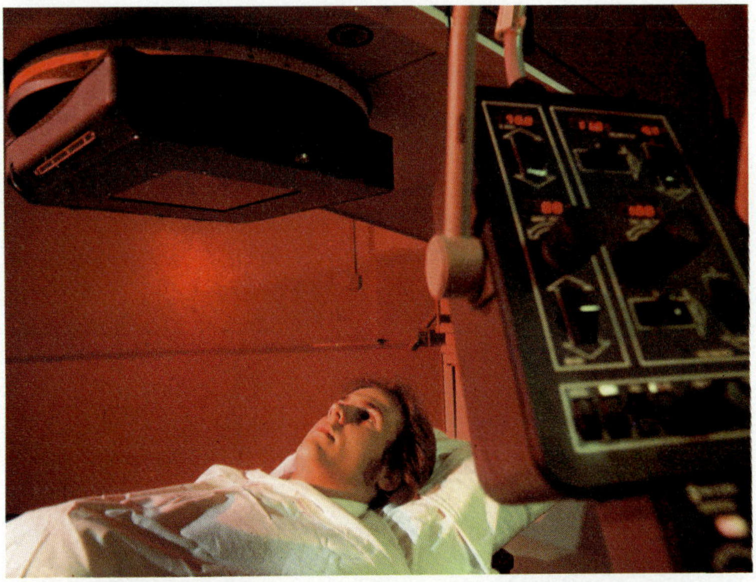

Figure 10-8 *Radioisotopes can be used to treat certain cancers. One way patients are treated is by external radiation.*

Figure 10-9 *The thickness of metal sheets can be determined through the use of radioisotopes.*

Radioisotopes behave basically the same way in chemical reactions as do nonradioactive isotopes of the same element. Therefore, radioisotopes can be used to find out how chemical reactions work. A good example involves the photosynthesis reaction, which enables green plants to make food:

$$CO_2 + H_2O \xrightarrow{\text{sunlight}} \text{food} + O_2$$

In order to explain how this important reaction works, scientists wanted to know where the oxygen that was released came from. Did it come from the CO_2, from the H_2O, or from both? When they caused ordinary CO_2 and water that contained a radioisotope of oxygen to react, they found that the O_2 given off was radioactive. This was not the case when they "tagged" the CO_2 with a radioisotope of oxygen. This showed that the oxygen released by the photosynthesis reaction comes from the water. Radioisotopes used in this way are often called **tracers**, because scientists can trace a course of events by detecting the path of a radioisotope. Tracers can be used in many ways. For example, the path of oil in a pipeline or the flow of water beneath the ground can be followed through the use of tracers.

tracers

Radiation can be stopped by a metal sheet, depending on the thickness of the metal and the energy of the radiation. A radioisotope can be placed beneath a sheet of metal that is being produced. A radiation detector above the metal can then be used to measure how much radiation gets through the metal sheet. This setup can be used to measure the thickness of the sheet. If the metal is too thick or thin, automatic controls linked to the detector can adjust the thickness.

TRY THIS

Think of a way to use a radioisotope (a) to find out how water circulates in a plant, (b) to detect cracks in a piece of metal.

Checkpoint

1. Describe how radioisotopes can be used in treating cancer.
2. What is a radioactive tracer? Give an example of the use of a tracer.
3. Describe how radioisotopes can be used to measure the thickness of metal sheets.

10-5

Nuclear Energy

Goals

1. To define binding energy.
2. To identify the source of binding energy.
3. To explain, in words, what $E = mc^2$ means.

The mass of a proton or a neutron is approximately 1 amu. However, to account for nuclear reactions, more precise values are needed. The mass of a proton is 1.00728 amu. That of a neutron is 1.00867 amu. A helium nucleus formed in a nuclear reaction contains two protons and two neutrons. You might expect that its mass would be:

$$(2 \times 1.00728) + (2 \times 1.00867) = 4.0319 \text{ amu}$$

However, its measured mass is only 4.0015 amu. It seems that 0.0304 amu is missing from the helium.

The explanation for the missing mass is that it was converted into energy when the nucleus was formed. The energy given off when a nucleus is formed from protons and neutrons is called the **binding energy**. If the nucleus breaks apart into individual protons and neutrons, energy equal to the binding energy is required. Each nucleus has its own binding energy.

binding energy

The amount of energy involved in forming or in breaking apart a nucleus is extremely great. Suppose 400 g of neutrons and protons were to combine, forming helium. The energy given off in the process could cause a lake averaging 1.6 km in diameter and 20 m in depth to boil!

Scientists no longer find it surprising that mass can be converted into energy, and vice versa. Nor do they think it strange that a small amount of mass can be transformed into an extremely large amount of energy. However, these notions seemed astounding to most people when Albert Einstein first proposed them. In 1905, Einstein published his well-known equation that shows the relationship between energy and mass: $E = mc^2$. E stands for energy, m is mass, and c is the speed of light. The equation shows that a mass (m) can be transformed into an amount of energy (E) equal to the product of the mass and the square of the speed of

Figure 10-10 *Scientist Albert Einstein first proposed that mass could be changed into energy.*

light (c^2). The speed of light is a very large number. When this number is squared, a much larger number is obtained. Thus a tiny amount of mass can be converted to a very large amount of energy.

Checkpoint

1. Define binding energy.
2. What is the source of the binding energy of a nucleus?
3. Write the equation $E = mc^2$ in words.

10-6

Nuclear Fission

Goals

1. To define nuclear fission.
2. To explain what is meant by the critical mass of a fissionable material.
3. To describe what happens in a chain reaction.

When the radioactive isotope uranium-235 is struck by a neutron, it may split into two smaller nuclei. The splitting of a nucleus into two smaller nuclei is called **nuclear fission**. A reaction for the fission of U-235 is:

nuclear fission (FISH un)

$$^{1}_{0}n + ^{235}_{92}U \longrightarrow ^{138}_{56}Ba + ^{95}_{36}Kr + 3^{1}_{0}n + \text{energy}$$

Notice that three neutrons are produced in this reaction ($3\,^{1}_{0}n$). Because neutrons are produced by the fission of U-235, a single U-235 fission reaction can give rise to as many as three others. (See Figure 10-11.) Each of these three reactions can then give rise to three others, and so on. When the products of a reaction begin new reactions, the reaction rate can increase, and a **chain reaction** can occur. A nuclear chain reaction can produce a vast amount of energy in a very short period of time.

chain reaction

Uranium compounds are present at many locations in Earth's crust. The uranium in these compounds consists of several radioactive isotopes. U-235 makes up only a small percentage of the isotopes. Since the other isotopes do not undergo fission as readily as does U-235, nuclear chain reac-

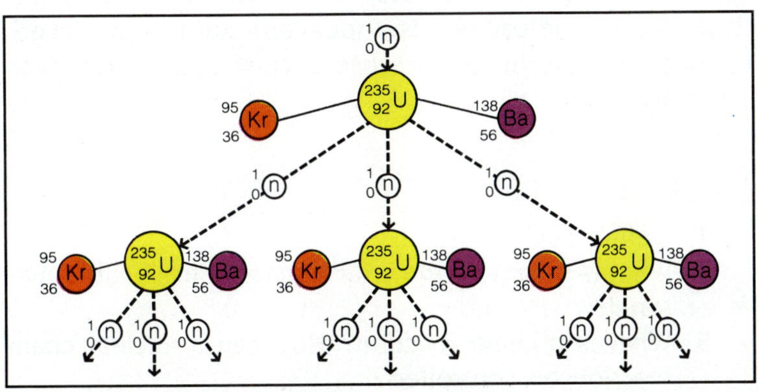

Figure 10-11 *A single fission reaction involving uranium-235 can produce many other fission reactions, in a so-called chain reaction.*

Figure 10-12 In the fission reactions involved in an atomic-bomb explosion, very large amounts of energy are released.

tions do not occur naturally. In order to produce a chain reaction, uranium must contain a greater fraction of U-235 than is present in naturally occurring uranium. Even in a uranium sample that has been "enriched" with U-235, a chain reaction will not necessarily occur, however. If the mass of uranium is too small, many neutrons can escape without hitting U-235 nuclei. If the mass is made larger, fewer neutrons can escape without causing other U-235 nuclei to fission. If the mass is made large enough, a chain reaction takes place. The smallest mass of fissionable material that will support a chain reaction is called the **critical mass**.

critical mass

In a fission (or "atom") bomb, two or more masses that are smaller than the critical mass are pushed together to produce a critical mass. The chain reaction that results produces an awesome explosion, as shown in Figure 10-12. By controlling the number of neutrons available to produce fission, a fission reaction can be made to proceed at a steady rate. The energy given off by the controlled reaction can be harnessed for power production. Whether fission is controlled, as in nuclear reactors, or uncontrolled, as in a nuclear bomb, radioactive isotopes are always produced. These fission products can cause severe injury or death to living organisms.

TRY THIS

Arrange a set of dominoes so that each domino will knock over two other dominoes when it falls. Then push the first domino and watch the "chain reaction." How is this chain reaction like a nuclear chain reaction? How is it different?

Checkpoint

1. Define nuclear fission.
2. What is meant by the critical mass of fissionable material?
3. What is a chain reaction? How can a nuclear chain reaction be controlled?

10-7

Nuclear Reactors

Goals

1. To list the parts of a nuclear reactor and explain their functions.
2. To explain what nuclear wastes are and to describe how they might be safely stored.

nuclear reactors

Controlled nuclear reactions are carried out in **nuclear reactors**. Figure 10-13 shows the parts of a nuclear reactor. Refer to the diagram as each part is described.

The *fuel elements* in a reactor (usually uranium-235 or plutonium-239) are the materials that undergo fission. The fuel is the source of neutrons, also. The fuel elements are surrounded by a substance called a *moderator*, which slows down neutrons. The moderator is often graphite (a form of carbon) or heavy water. Fast-moving neutrons are slowed down by bouncing off the small nuclei in the moderator.

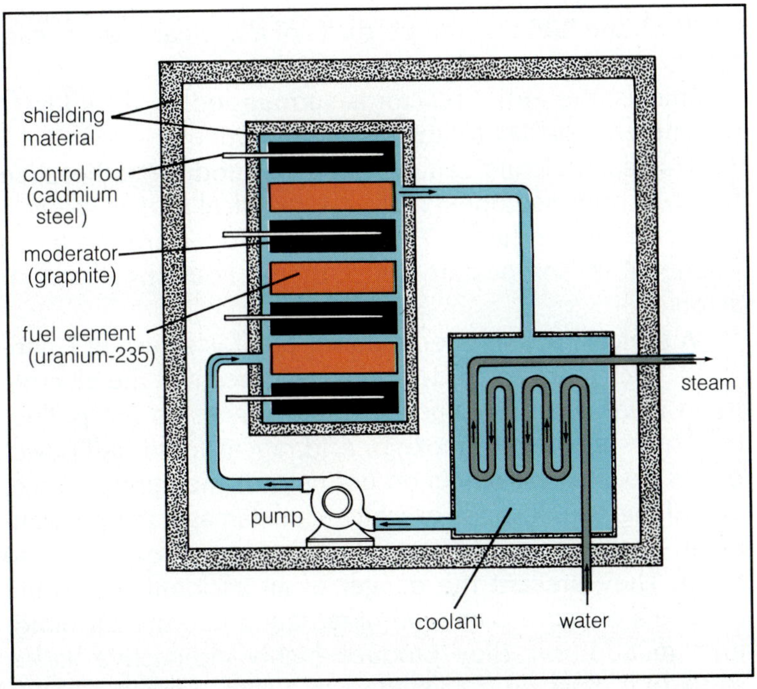

Figure 10-13 *This diagram shows the principal parts of a nuclear-fission reactor.*

Slow neutrons are more easily "captured" by the nuclei in the fuel than are fast neutrons. Therefore, fission occurs more readily with slow neutrons.

In order to control the rate of a nuclear reaction, the number of neutrons must be controlled. *Control rods* are used to absorb neutrons and thus control the reaction rate. These rods are made of substances such as the elements boron or cadmium that absorb neutrons strongly. When control rods are pushed into the fuel area, they absorb neutrons, thereby slowing the reaction. If the rods are pushed in far enough, they stop the reaction. When control rods are pulled out, more neutrons are available to react, and the reaction speeds up.

A *coolant* is pumped through the reactor vessel to absorb heat released by the fission reaction. The coolant is usually water or liquid sodium. After absorbing heat from the reaction, the coolant flows into the *heat exchanger*. There it gives up heat to the surrounding water, which causes the water to boil. The steam produced by the boiling water is used to drive turbines. The turning turbines run electric generators, which produce electricity.

Both the fuel and the products of a nuclear-fission reaction are radioactive. Since this radiation is harmful to living organisms, the entire reactor is surrounded by thick layers of concrete *shielding* to absorb the radiation.

Nuclear reactors can be used to produce radioactive isotopes (radioisotopes) that have practical uses in science, medicine, and industry. These radioisotopes are produced when stable isotopes are struck by the neutrons in the reactor.

A typical nuclear-energy power plant is shown in Figure 10-14. Such plants provide about ten percent of the electricity in North America. Nuclear power plants do not pollute the atmosphere with smoke or acid rain as fossil-fuel power plants do. Nuclear plants do not require the mining of coal or drilling for oil, both of which can damage the environment. However, nuclear power plants are very expensive to build. They present the danger of an accident that could release a dangerous amount of radiation into the surroundings. In addition, they produce highly radioactive *nuclear wastes* that must be disposed of in a safe manner.

Most of the radioactive isotopes in nuclear wastes decay

Figure 10-14 *This nuclear power plant provides energy by controlling fission reactions.*

rapidly. However, a small percentage of the wastes takes much longer to decay to a safe level. Some of these wastes take thousands of years to decay. Nuclear wastes can be buried in locations in which they will not be disturbed for many years. A major danger is that buried nuclear wastes might leak into underground water systems. This contaminated water might then mix with a supply of drinking water.

A number of methods have been proposed for safe disposal of nuclear wastes. A popular proposal involves the "four-barrier approach." The first step of this approach is to embed the waste in a nonradioactive ceramic material. (Ceramic materials include glass, porcelain, earthenware, and other nonmetallic substances.) The ceramic material can hold the wastes securely within it, so that they cannot easily leak out. The waste-containing ceramic material is then sealed into a ceramic or metal container. Next, an absorbent mineral filling is packed around the container. Finally, everything is securely buried underground in an area that is free from the danger of earthquakes or some other natural

Figure 10-15 *According to some scientists, a salt mine, such as this, might provide a suitable burial site for nuclear wastes.*

disaster. (See Figure 10-15.) However, many people still feel that there is simply no way in which to dispose of nuclear wastes safely.

Once a disposal site has been selected, wastes must be transported to it from reactors. There are several hazards in transporting such wastes. One is that people might be exposed to dangerous amounts of radioactivity. Another is that the wastes might form a critical mass. These two hazards can be controlled by using adequate shielding. Other hazards include the possibility of fires and other accidents. These hazards can be reduced by using a combination of specially designed shipping casks. Such casks can withstand exposure to fire, immersion in water, and other rigorous conditions without leaking. It is possible that some people might try to steal radioactive waste materials in order to make a weapon of some kind. Since such an attempt is most likely to occur during transport, special care would have to be taken to guard against this possibility. Once again, the dangers involved have led many people to the conclusion that the use of fission reactions as an energy source is unwise.

Checkpoint

1. List the major parts of a nuclear reactor.
2. Explain the function of each of the major parts.
3. What are nuclear wastes? Why are they hazardous?
4. Describe a way in which nuclear wastes might, according to some people, be safely stored.

10-8

Nuclear Fusion

Goals

1. To define nuclear fusion.
2. To state the source of the energy given off by nuclear fusion.
3. To describe the major obstacle to producing a controlled fusion reaction.

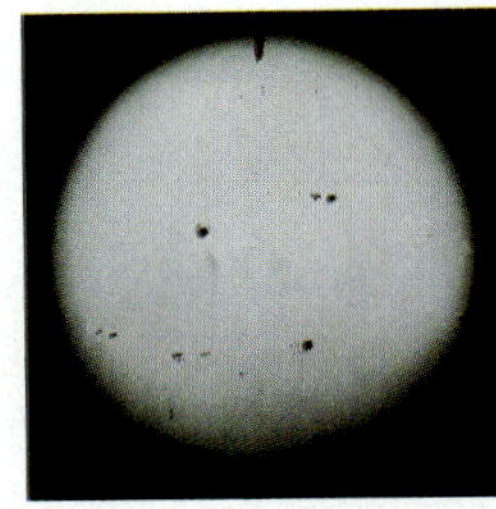

Figure 10-16 *Nuclear-fusion reactions are the source of the sun's energy.*

Combining hydrogen isotopes to form helium produces energy, just as the splitting of uranium atoms does. Indeed, the difference between the binding energies of reactants and products is much greater for such a reaction than for a fission reaction. Therefore the energy given off is much greater.

A nuclear reaction in which lighter nuclei combine to form heavier nuclei is called a **nuclear fusion** reaction. Nuclear fusion is the source of the sun's energy. In the sun, 650 million tons of hydrogen are converted to helium each second. In the process, several tons of mass are converted into energy. Hydrogen fusion will occur only at very high temperatures (about 50 million °C).

nuclear fusion (FYOO zhun)

Scientists are trying to produce a controlled fusion reaction. This involves reaching very high temperatures without melting the containing vessel. One way of meeting this requirement is to carry out the reaction in a properly designed magnetic field that would contain the high-energy material. A device built to do this, called the *tokamak*, has produced some promising results. (See Figure 10-17.) However, a practical fusion reactor may still be decades away.

If scientists succeed in producing a practical fusion reactor, there will be plenty of fuel available for it. Earth's ocean can provide a nearly endless supply of hydrogen.

Figure 10-17 *The tokamak device can be used to produce a controlled nuclear-fusion reaction.*

Checkpoint

1. What is nuclear fusion?
2. What is the source of the energy given off by a nuclear-fusion reaction?
3. What is the major problem in producing controlled nuclear fusion?

Carbon-14 Dating

The carbon-14 produced by cosmic rays can be used to find the age of carbon-containing materials, such as wood and cloth. Despite the decay of C-14 to N-14, there is always a small amount of C-14 in the atmosphere due to the cosmic rays. The rate at which C-14 is produced by cosmic rays and the rate at which C-14 decays balance, keeping the amount of atmospheric C-14 constant.

Living plants use carbon dioxide (CO_2) from the atmosphere to make food by photosynthesis. A small fraction of the carbon atoms in this CO_2 is C-14. Most of the other carbon atoms in the CO_2 are C-12, and a few are C-13. Neither C-12 nor C-13 is radioactive. When a plant dies, it stops absorbing more carbon dioxide from the atmosphere. The C-14 atoms already in the plant continue to decay. The longer a plant has been dead, the less C-14 it contains. The half-life of C-14 is 5570 years. Therefore, only half of the C-14 present when a plant dies is still present after 5570 years.

A wooden object or a piece of cloth or paper made from plant fibers loses C-14 as it decays into N-14. The ages of such objects can be estimated by finding out how much C-14 is present, compared to the amount of C-12 and C-13. Dates obtained from C-14 analysis have been confirmed by other methods. For example, the age of a tree can be estimated by counting the number of annual rings. Estimates of the ages of old trees obtained by C-14 analysis are close to those obtained by counting the tree rings.

Understanding The Chapter 10

Main Ideas

1. Radioactive nuclei give off radiation in the form of alpha and beta particles and gamma rays.
2. In transmutation, an element gives off alpha or beta radiation and becomes a new element.
3. Isotopes are atoms of the same element that have different masses.
4. The half-life of a radioactive isotope is the time it takes for half the atoms in a sample to decay into another isotope.
5. Nuclear equations can be used to represent nuclear reactions.
6. Radioactive isotopes have uses in science, medicine, and industry.
7. Binding energy is the energy given off when a nucleus is formed. It results from the transformation of mass into energy.
8. Nuclear fission is the splitting of a nucleus into two smaller nuclei.
9. A chain reaction can occur during nuclear fission because the fission reaction produces neutrons. The smallest mass of fissionable material that will produce a chain reaction is the critical mass.
10. In nuclear reactors, the number of neutrons that can produce fission is controlled in order to produce a steady reaction rate.
11. Nuclear reactors produce dangerous radioactive wastes.
12. In nuclear fusion, light nuclei combine, forming heavier nuclei.

Vocabulary Review

From the following list, choose the term that best completes each of the statements. Write your answers on a separate piece of paper.

alpha radiation	isotopes
beta radiation	nuclear equation
binding energy	nuclear fission
chain reaction	nuclear fusion
cosmic rays	nuclear reactor
critical mass	radioactive decay
gamma radiation	tracers
half-life	transmutation

1. The __?__ of a radioactive isotope is the time required for half its atoms to decay to another isotope.
2. For a fission chain reaction to take place, there must be a(n) __?__.
3. The decrease in mass that occurs when protons and neutrons combine, forming a nucleus, appears as __?__.
4. The helium nuclei released when some radioactive isotopes decay are called __?__.
5. To increase its atomic number by one, a radioactive isotope must release __?__.
6. The change of one element to another when the element emits alpha particles is an example of __?__.
7. __?__ occurs when hydrogen nuclei combine to form helium nuclei.
8. Chemists use __?__ to follow the path of an element in a reaction.

9. High-energy particles that shower Earth from outer space are called _?_.
10. The rate of a fission reaction is controlled in a _?_.
11. Atoms of the same element that have different masses are called _?_.
12. The splitting of a nucleus into two smaller nuclei is called _?_.

Chapter Review

Write your answers on a separate piece of paper.

Know The Facts

1. The purpose of the _?_ in a nuclear reactor is to slow down neutrons. (moderator, coolant, shielding)
2. The _?_ of hydrogen nuclei provides the energy released by the sun. (fission, fusion, radioactive decay)
3. Adjusting the _?_ changes the rate of the fission reaction in a nuclear reactor. (shielding, turbines, control rods)
4. The most common isotope of hydrogen is _?_. (H-1, H-2, H-3)
5. _?_ radiation is like x rays, but has more energy. (alpha, beta, gamma)
6. All isotopes of an element have basically the same _?_. (half-life, chemical properties, tendency to undergo fission)
7. The amount of energy involved in a nuclear reaction is _?_ that involved in a chemical reaction. (greater than, less than, about the same as)
8. Thorium-234 nuclei contain 90 protons and _?_ neutrons. (90, 100, 144)
9. A radioactive decay series always ends with _?_. (a stable isotope, an unstable isotope, lead)
10. The fraction of U-235 in naturally occurring uranium is _?_ the fraction of U-235 in the uranium used in a nuclear reactor. (greater than, less than, about the same as)
11. A _?_ cannot be used to produce radioactive isotopes. (nuclear reactor, particle accelerator, chemical reaction)

Understand The Concepts

12. Explain what happens to the atomic mass and atomic number of an isotope that gives off alpha particles.
13. What could you use to stop (a) alpha particles, (b) gamma rays?
14. Complete the following nuclear equations:

(a) $^{234}_{91}Pa \longrightarrow\, ^{234}_{92}U + $ _?_

(b) $^{218}_{84}Po \longrightarrow\, ^{214}_{82}Pb + $ _?_

(c) $^{210}_{82}Pb \longrightarrow$ _?_ $+\, ^{0}_{-1}e$

15. What happens to the atomic mass and atomic number of an isotope that gives off a beta particle?
16. What are some of the advantages and disadvantages of using nuclear reactors to produce electricity?
17. What might happen to a nuclear reactor if it had no coolant?
18. Describe the "four-barrier" approach to the disposal of nuclear waste.

19. What is the difference between nuclear fusion and nuclear fission?
20. What major problem is involved in producing a controlled fusion reaction?
21. Why did scientists prior to 1940 think that there could be only 92 elements?
22. Explain why a critical mass is necessary in order for a nuclear chain reaction to occur.

Challenge Your Understanding

23. An element consists of two isotopes. Isotope A has a mass of 24 amu. Isotope B has a mass of 25 amu. If 70% of the atoms of the element are A and 30% are B, what is the atomic mass of the element?
24. Strontium-90 has a half-life of 28 years. How long will it take for 100 g of strontium-90 to decay to 25 g?
25. Explain why the sum of the masses of the protons and neutrons in a nucleus is less than the sum of their masses when they exist as separate particles.
26. Determine the ratio of neutrons to protons in the stable isotopes of a wide range of elements. Then determine the same ratio for some unstable isotopes of the same elements. What can you conclude about the effect of the neutron-proton ratio on the stability of a nucleus?
27. Why might scientists who wanted to trace the flow of water under the ground choose radioisotopes with short half-lives?

Projects

1. Get information from power companies that own nuclear reactors and from organizations that oppose nuclear power. Additional information may be obtained from your school or local library. Arrange a debate on nuclear power. Let your classmates choose which side (pro or con) they wish to be on.
2. Obtain some radioactive sources from your school's physics or chemistry teacher. Under your teacher's supervision, place several sources at different points on a pack of black and white film. After a few days, remove the radioactive materials and have the film developed. What do you find?
3. Use your library to write a report on one of the following topics: (a) the threat of nuclear war, (b) useful isotopes, (c) cancer and isotopes, (d) radioactive tracers, (e) the development of nuclear reactors, (f) the breeder reactor.
4. Make a chart showing the radioactive decay series that begins with U-238 and ends with Pb-206.
5. Do some research to find out how carbon-14 has been used as a tool by archeologists.
6. Make a scale model of a nuclear reactor.

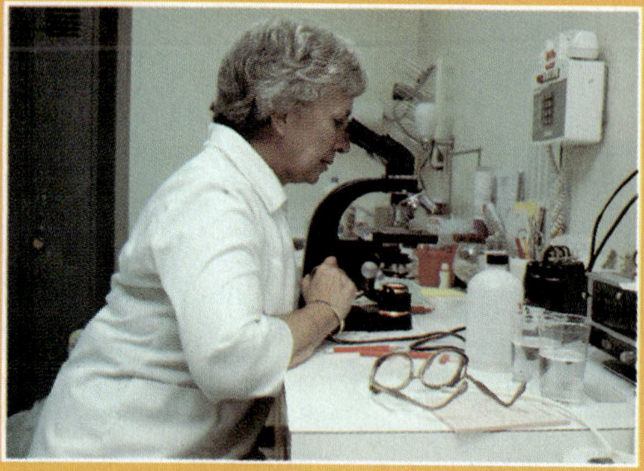

Medical Technologist

Hospital seeks medical technologists to analyze tissue and fluid samples. Technologists perform tests and gather data to assist physicians in making diagnoses. Tasks include analyzing the levels of substances in the blood and identifying bacteria in tissue samples. Technologists also may be asked to develop new tests, teach, or manage some laboratory activities.

Applicants should have a degree from a four-year college with courses in chemistry, biology, and mathematics and special training in sterile laboratory technique. Ability to perform tests precisely and accurately is essential. Technologists should expect to work some evening and weekend shifts.

Wastewater Treatment Plant Operator

Wastewater treatment plant seeks persons to train as equipment operators. Operators control the machinery that removes biological wastes and chemical pollutants from wastewater.

Applicants should have a background in chemistry, biology, and mathematics. High school graduates with some training in a technical school or a community college are preferred. Those hired will receive on-the-job training. Those who wish to become supervisors must obtain a degree in science or engineering from a four-year college.

Some outdoor work is required. Operators also should expect some night, holiday, and weekend shifts. Hazards include noise from the machinery and unpleasant odors.

Help Wanted

Lithographer

Off-set printing company seeks high school graduates to be trained as lithographers. Lithographers help prepare the metal plates used in offset-lithography, the process of printing from a flat surface. Trainees will specialize in one of several jobs: camera operator, lithographic artist, assembler, or platemaker. *Camera operators* take photographs of the material to be printed. *Lithographic artists* sharpen and reshape the images on film. *Assemblers* cut out the film and mount it on special sheets. *Platemakers* use chemicals to transfer the images from film to the metal plate.

Applicants should have a background in chemistry, optics, and photography. Artistic skill and knowledge of electronics would also be helpful. On-the-job training will be provided.

Darkroom Technician

Portrait studio has two positions open for darkroom technicians. One is for an *all-around darkroom technician* who processes both black-and-white and color film. The other position is for a *color laboratory technician*, someone who processes only color film. Applicants with prior studio experience may take photographs as well as process film.

Applicants must be high school graduates with courses in chemistry and mathematics. Good color perception and vision are important. Applicants also should be good at working with their hands. Film-processing experience as part of a hobby would be helpful. The technician's job requires doing one task over and over at a rapid pace. On-the-job training will be offered.

Understanding The Unit 2

Main Review

Write your answers on a separate piece of paper.

Know The Facts

1. Matter takes up space and has _?_. (weight, mass, shape)
2. An equal-arm balance is used to measure the _?_ of matter. (weight, mass, volume)
3. If a cube that measures 1 cm on a side is filled with water, the volume of the water would be _?_. (1 L, 1 mL, 1 cm)
4. Salt dissolved in water is an example of a(n) _?_. (element, compound, mixture)
5. As the temperature of a certain volume of water increases, the mass of the gas required to form a saturated solution _?_. (increases, decreases, remains the same)
6. A _?_ mixture, such as oil and water, does not have the same properties throughout. (homogeneous, heterogeneous, uniform)
7. _?_ cannot be broken down into simpler pure substances. (elements, compounds, mixtures)
8. The electron in a hydrogen atom is found only in certain "allowed" _?_. (energy levels, compounds, molecules)
9. A chemical equation is used to represent a _?_. (chemical reaction, symbol, physical change)
10. An atom's nucleus contains almost all the atom's _?_. (charge, mass, electrons)
11. K, Au, Ag, Na, Pt, Pb, and Mg are the symbols of _?_. (metals, nonmetals, metalloids)
12. Elements in a family have similar _?_. (chemical properties, color, physical properties)
13. Elements with low-ionization energies are located _?_ of the periodic table. (at the left side, in the middle, at the right side)
14. Atoms form _?_ bonds by sharing electrons equally. (ionic, covalent, polar covalent)
15. The energy released when a gram of liquid changes to a gram of gas is called the heat of _?_. (reaction, fusion, vaporization)
16. The rate of a chemical reaction can be reduced by _?_. (lowering the temperature, increasing the concentration of the reactants, decreasing the concentration of the products)
17. U-238 atoms have 92 protons and _?_ neutrons. (146, 238, 92)
18. The process of _?_ provides the energy that comes from the sun. (fission, fusion, combustion)
19. The energy required to break molecules apart is generally _?_ the energy required to separate molecules from one another. (less than, greater than, equal to)
20. The decomposition of water into hydrogen and oxygen is an exam-

ple of an ? reaction. (endothermic, exothermic, explosive)

Understand The Concepts

21. What is the difference between (a) strong and weak acids, (b) strong and weak bases?
22. Convert 1550 g into kilograms.
23. What is the density of a liter of salt water with a mass of 1150 g?
24. A piece of gold has a mass of 193 g and a density of 19.3 g/cm^3. What is the volume of the gold?
25. A 10-cm^3 piece of metal has a mass of 78 g. What will be the mass of a 50-cm^3 piece of the same metal?
26. 3 $C_{12}H_{22}O_{11}$ represents how many (a) molecules, (b) atoms?
27. Helium atoms have two protons and two neutrons in their nuclei. How many electrons are found in each neutral helium atom?
28. Balance the equations below:
    (a) $C + O_2 \rightarrow CO$
    (b) $C_6H_{12}O_6 + O_2 \rightarrow CO_2 + H_2O$
    (c) $NaClO_3 \rightarrow NaCl + O_2$
    (d) $K + Cl_2 \rightarrow KCl$
    (e) $HCl \rightarrow H_2 + Cl_2$
29. Compare the properties of metals and nonmetals.
30. Why do some companies that sell table salt add KI to the salt?
31. How do the following families of elements attain the stable electron configuration of the noble gases: (a) halogens, (b) alkali metals?
32. Why do many steel structures need to be coated with paint?
33. Why do you feel more uncomfortable on a muggy day when the temperature is 30°C than on a dry day when the temperature is 31°C?
34. What are isotopes?
35. Tritium produces beta radiation. What is the other product of this reaction?
36. In the equation, $E = mc^2$, what do the letters represent?

Challenge Your Understanding

37. If a solid has the dimensions 10 cm × 5 cm × 20 cm and a mass of 7.8 kg, what is the density of the solid (a) in g/cm^3, (b) in kg/cm^3?
38. A 10-mL volume of mercury has a mass of 136 g. What volume of mercury has a mass of 2.72 kg?
39. Carbon atoms have a mass of 12 amu. Hydrogen atoms have a mass of 1 amu. Methane molecules have the formula CH_4. What percentage of methane is carbon? What percentage is hydrogen?
40. Pieces of red and blue litmus paper are placed in a beaker of water. A piece of potassium is added to the water. What will happen to the litmus paper? Why?
41. Why must there be more than one neutron per fission in order for a fission reaction to sustain itself?
42. Compare a power plant that uses nuclear fuel to one that uses coal or oil to produce electricity.
43. A saturated solution of sodium nitrate contains 85 g of the solid in 100 g of water at 20°C. How much of the solid will be required to saturate a liter of water at the same temperature?
44. Discuss the advantages and disadvantages of nuclear power.
45. As fossil fuels become more scarce, how will energy be obtained?

Unit 3 · Earth and Space Science

The water that covers three fourths of planet Earth can strike the shore with tremendous force.

Chapter 11

Earth Materials

This geologist is studying mineral samples at the Smithsonian Institution in Washington, D.C.

11-1

Minerals

Goal

To explain what minerals are and how minerals can be identified.

When you hear the word *rock* you probably think of the pebbles in a driveway or rock garden, or the flat stones you have tried to make skip across the water. Rocks, stones, and pebbles are part of Earth's outermost solid layer, its **crust**. The crust includes the continents and the ocean bottoms. **crust**

Everything on Earth is made from less than 100 elements. *Elements* are substances that cannot be broken down into anything simpler by chemical means. A few solid elements are found in the crust in a pure state. For example, gold can be found as individual lumps of pure gold. Most of the elements of the crust, however, are combined in *compounds*. Naturally-occurring solid elements and compounds are called **minerals**. Most rocks are mixtures of minerals. **minerals**

The most abundant elements in Earth's crust are oxygen and silicon (SIL ih kun). (See Table 11-1.) They are the main ingredients of minerals called *silicates* (SIL ih kaytz). Silicates, sometimes called "rock-forming minerals," make up 87 percent of all minerals. Granite (GRAN it), for instance, is a rock made up of several different silicate minerals. Granite is chiefly biotite, feldspar, quartz, and hornblende.

Table 11-1: Most Abundant Elements in Earth's Crust

Element	Symbol	Percent of crust (by mass)
oxygen	O	46.60
silicon	Si	27.72
aluminum	Al	8.13
iron	Fe	5.00
calcium	Ca	3.63
sodium	Na	2.83
potassium	K	2.59
magnesium	Mg	2.09

Most of the elements present on Earth are *metals*. Familiar metals include iron, copper, aluminum, and silver. In Table 11-1, all the elements listed are metals, except for oxygen. Metals, and other elements, are often obtained from rocks called *ores*. The mineral hematite is an important source of iron. Cinnabar is a source of mercury, and malachite (MAL uh kyt) is a source of copper. (See Figure 11-1.)

crystals

In order to identify minerals, you need to look for certain physical properties. For example, many minerals form as regularly-shaped solids, or **crystals**. A crystal's shape results from the regular pattern in which its basic units line up. The *shape* of its crystals is a clue that can help to identify a mineral. For example, halite (table salt) crystals are cubic in shape. (See Figure 11-2.) All crystals of halite, no matter how large or small, have the same shape.

Many minerals have identifying *colors*. Halite, for instance, is white (or colorless), malachite is green, and cinnabar is red. Some minerals come in more than one color. An example is orthoclase feldspar, which is a common mineral in some kinds of rock. Orthoclase can be white, gray, pinkish, or yellowish.

The color of a mineral sample can't always be used for identification. Impurities in a mineral can change its color. The *streak test* is a better indicator of a mineral's true color. The streak is a sample of the powdered mineral. To observe

Figure 11-1 *(a) Malachite is a compound of copper and oxygen. (b) The red mineral is cinnabar, containing mercury. (c) Hematite is iron oxide.*

Figure 11-2 *Table salt is halite, or sodium chloride. Its crystals are always cubic.*

Figure 11-3 *Calcite cleaves in three directions that are not at right angles to each other.*

the color of the streak, scratch the mineral on an unpolished porcelain tile. The streak may not be the same color as the mineral. A mineral's streak color is just about the same for all samples of that mineral.

If you test a large variety of mineral samples, you will find that some have a colorless streak. Minerals that are very hard (see the next paragraph) leave no streak at all. A diamond is one of those.

Hardness is another reliable feature of a mineral. To test hardness, you can scratch one mineral with another. A harder mineral always scratches a softer mineral. On a scale of hardness, the higher the number, the harder the mineral.

Because of their crystal structure, most minerals break in a particular pattern. If a mineral breaks along flat surfaces, it is said to show *cleavage*. (See Figure 11-3.) This property can be used to help identify a mineral. The sample in Figure 11-3 has cleaved in three directions. The three directions are *not* at right angles to each other. A mineral that cleaves in three directions that *are* at right angles to each other will form cubic shapes. Galena (guh LEE nuh) is an example. The mineral mica (MY kuh) cleaves in only one direction.

The way a mineral reflects light is called its *luster*. Shiny new coins and polished silverware have a high luster. Luster is divided into two main types: metallic (like metal) and nonmetallic (like glass). The dark gray mineral galena, for

Figure 11-4 *(a) Galena has a beautiful metallic luster. Note the cleavage in three directions at right angles. (b) Quartz shines like glass—it has a glassy luster.*

example, shines like a highly polished metal. It is said to have a metallic luster. Quartz has a nonmetallic, glassy luster. (See Figure 11-4.)

Another test that is used to help identify minerals is their *specific gravity*, or their "heft." If you have two mineral samples of the same size and one of them feels a lot heavier than the other, you are noticing a difference in their specific gravity. Specific gravity is measured by comparing the mass of a sample to the mass of an equal volume of water.

Some rare minerals have special combinations of beauty and hardness. They are the minerals used in jewelry, and are called *gems*. Colorless diamonds can be cleaved, cut, and polished to beautiful sparkling jewels. Emeralds are dark green samples of the mineral beryl, which can be cut and polished. Sapphires are deep blue gems that are pieces of the mineral corundum.

Checkpoint

1. What is a mineral?
2. Name at least five properties of minerals that can be used to help identify them.
3. What is the name for rocks from which metals and other elements are obtained?
4. Why may the color of a mineral and its streak be different?

TRY THIS

Look at a sample of granite through a magnifier. How many different kinds of mineral crystals can you see? Try to identify them.

11-2

Rocks

Goals

1. To identify the three classes of rocks.
2. To describe the rock cycle.

There are three classes of rocks, determined by the way the rocks are formed. Most of the rock that makes up the crust is igneous rock. **Igneous rocks** form from extremely hot, melted rock material, called **magma**. Magma comes from deep within Earth. The word *igneous* means "fire-formed." The differences among igneous rocks are caused by the chemical make-up of the magma and its rate of cooling. Magma may cool and harden very slowly, very rapidly, or at various rates in between.

igneous rocks (IG nee us)

magma

If magma is forced between layers of rock in the crust without ever reaching the surface, it cools very slowly. The more slowly the magma cools, the larger the crystals are. One igneous rock formed by slow cooling is *granite*. Granite, with its crystals of various colors, is frequently used for buildings. It is very hard and does not crumble easily when exposed to wind and rain, heat and cold. (See Figure 11-5a.)

Figure 11-5 *Igneous rocks: (a) granite, (b) basalt, (c) obsidian, and (d) pumice.*

Figure 11-6 *Sedimentary rocks: (a) shale, (b) sandstone, and (c) limestone.*

lava
(LAH vuh)

sedimentary rocks

TRY THIS

Make a small batch of cooked fudge. When it is at the proper temperature, pour a small amount of boiling fudge into a cup of cold water. Pour the same amount of fudge into a small pan to allow slow cooling. When both samples are cool, observe their textures. Describe the difference. Relate it to differences in igneous rock.

Magma that comes to Earth's surface—in a volcanic eruption, for instance—is called **lava**. When lava flows onto Earth's surface, it cools fairly rapidly. Crystals form, but they don't have time to grow very large. In fact, you may need a magnifier in order to see them. *Basalt* (buh SAWLT) is an igneous rock that forms from cooling lava. It is a dark-colored, hard rock with very small crystals. (See Figure 11-5b.)

Sometimes lava cools so fast that no crystals can grow. When this happens the result is a *glass*. One rock that forms in this way is *obsidian*, a glassy rock. (See Figure 11-5c.) Some obsidian jewelry is called "Apache tears" because of its shape. Obsidian was also used for arrow points because of its hardness and sharp edges.

Another common form of igneous rock is *pumice* (PUHM is). Like obsidian, pumice cools too rapidly for crystals to form. Pumice can be considered a glass, also. Pumice is a frothy volcanic glass. It is full of holes, like foam that has become solid. Pumice is so full of holes it floats. (See Figure 11-5d.) Ground-up pumice is used in some hand soaps to help grind out dirt from soiled hands. It can be used as an abrasive also, to remove calluses from hands and feet, and to sharpen knives.

There are many types of igneous rocks besides granite, basalt, obsidian, and pumice. These four examples show how igneous rocks can differ, depending on how rapidly the molten rock cooled.

The second class of rock is the one that may be most familiar to you. **Sedimentary rocks** are rocks made of min-

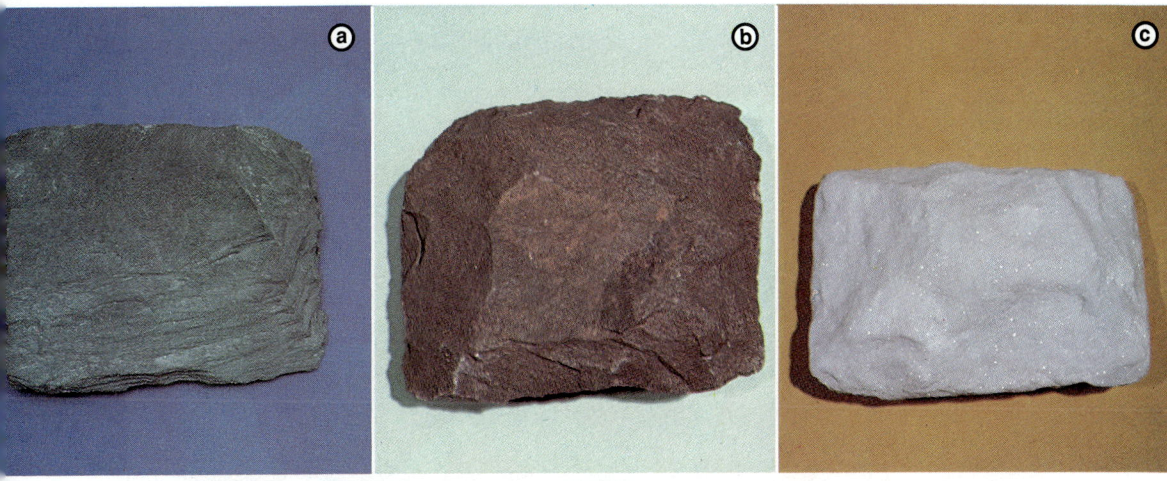

Figure 11-7 Metamorphic rocks: (a) slate, (b) quartzite, (c) marble.

eral grains that become cemented together, under pressure, forming solid rock. They usually form in water, where grains settle out in layers on the bottom of rivers, lakes, or oceans. Sometimes grains are deposited by wind, and form layers of sand in the desert.

Sedimentary rocks are known by their relative softness, their lack of crystals, and their layered appearance. Some common sedimentary rocks are *sandstone*, *shale*, and *limestone*. (See Figure 11-6.) Sandstone is made up of sand grains. Shale is made up of smaller fragments of clay and mud. Limestone is formed from the powdery remains of shells, teeth, and bones. Sedimentary rocks can be used for building materials. They do not, however, hold up under extreme weather conditions as well as igneous rocks do.

metamorphic rocks

The third class of rock making up Earth's crust is called metamorphic rock. This name means "changed" rock. Some rocks become buried deep within Earth. There large amounts of heat and pressure can change them. **Metamorphic rocks** are rocks that have been formed by changes in sedimentary or igneous rocks.

Some examples of metamorphic rock are *slate* (changed from shale), *marble* (changed from limestone), and *quartzite* (changed from sandstone). (See Figure 11-7.) These rocks are very hard and most have crystals. Many show some form of organization, such as the *banding* shown in Figure 11-8. Banding shows that similar minerals have moved and collected with others of the same kind. This happens to a rock when it gets hot enough to flow under extreme pressure.

Figure 11-8 Banding in gneiss, a metamorphic rock.

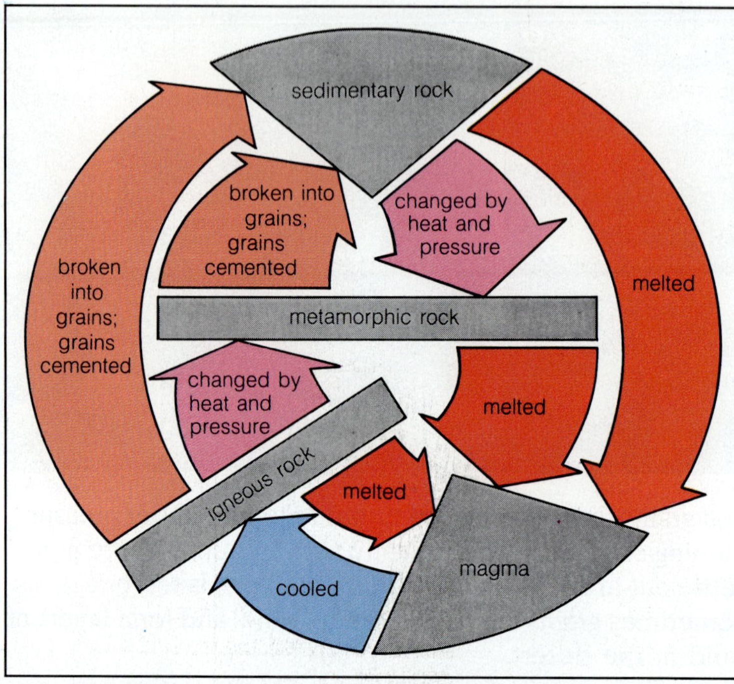

Figure 11-9 *The rock cycle. A grain of quartz from magma could become part of igneous, then sedimentary, then metamorphic rock.*

After rocks have formed, they do not remain unchanged. You have learned that metamorphic rocks can be formed from either igneous or sedimentary rocks. You may have wondered where the grains, or particles, come from that make up sedimentary rocks. They can be from igneous or metamorphic rocks, or even other sedimentary rocks. Igneous rocks can be formed from sedimentary or metamorphic rocks that become buried deep enough in the crust to form magma. The changes from one rock type to another are called the **rock cycle**. It is a never-ending process of change. The next time you pick up a rock, remember that it was not always in the form you see now. (See Figure 11-9.)

rock cycle

TRY THIS

Visit a nearby cemetery. Look for monuments and markers made of various kinds of rock. In what ways can you distinguish between granite and sandstone? limestone and marble?

Checkpoint

1. What is the difference between lava and magma?
2. How can you tell an igneous rock from a sedimentary rock?
3. If an igneous rock is glassy, what was its rate of cooling?
4. What is a metamorphic rock?
5. Describe the rock cycle.

11-3

Fossils

Goals

1. To define fossils.
2. To list conditions for the preservation of fossils and describe likely fossil locations.
3. To list some uses for fossils.

fossils

If you are interested in rock and mineral hunting, knowledge of fossils will add a new excitement to your hobby. **Fossils** are evidence of formerly-living things. Fossils can be the actual bones, teeth, or shells of animals. Or they can be indirect evidence of former life, such as footprints, trails, or imprints of shells in rocks.

Fossils are almost always found in sedimentary rocks. (See Figure 11-10.) Igneous or metamorphic rocks do not contain fossils because the heat that forms these rocks would destroy the fossils. In some rare cases fossils are found in layers of volcanic dust that have hardened to rock.

Figure 11-10 *Fossils: (a) a hardwood tree leaf; (b) a fish; (c) insects.*

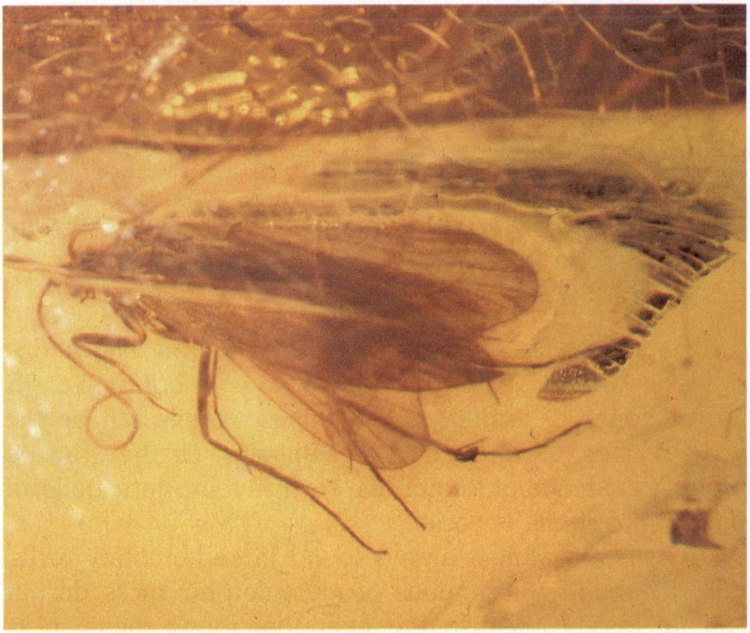

Figure 11-11 *A fossil insect in a piece of amber (hardened pine sap). It was found on the shore of the Baltic Sea and is about 30 million years old.*

There are two general conditions necessary for the formation of fossils. First, the organism should be buried rapidly, to prevent destruction by other organisms or by natural events, such as storms. Second, the organism should have hard parts, such as bones, shells, teeth, or stems, to ensure their holding up through time.

If you are going fossil hunting, look in areas where sedimentary rocks are plentiful: along the banks and cliffs near rivers, streams, or lakes. If there is a quarry nearby, you may be able to get permission from the owners to explore the area. If construction companies have been clearing land and have uncovered rocks or made cuts through sedimentary rock layers, you may find fossils there.

Fossils have been preserved in some unusual places. They have been found in frozen soil or ice, tar pits (Rancho La Brea Tar Pits, Los Angeles), and sap from trees. (See Figure 11-11.) Some organisms have become natural "mummies" from extreme drying conditions.

An unusual fossil you may have seen is *petrified wood,* or wood that has changed to stone. For wood to become like stone, a tree must be under water for a long time. The minerals dissolved in the water fill in the cells of the tree, turning the tree's structure to stone.

You are probably wondering what fossils look like, especially the ones you might find on your first expedition. Depending on your location, you are most likely to find various species of small, shelled animals that lived in water. Plant fossils are more difficult to find because they have such fragile parts. They are less likely to survive the process of being buried.

Fossils can provide valuable resources. All organisms contain carbon and hydrogen. Under certain conditions, the carbon and hydrogen in the remains of organisms combine, forming *hydrocarbons*. Hydrocarbons are the fuels in oil and natural gas. Oil and natural gas are called *fossil fuels* because they are formed from the remains of organisms. Coal is another fossil fuel. It is formed from organisms that were buried in ancient swamps.

Scientists can use fossils to learn about past environments. For instance, if coral fossils are found in a quarry, scientists know that the area must have been covered by a warm, salty sea at the time the corals were living. That is the environment needed for modern coral to survive.

Now that you know what fossils are, keep your eyes open for them. You may find them in a slab of limestone used for a sidewalk or in the foundation of a building. You can most certainly find some fossils in any nearby limestone quarry.

Checkpoint

1. What is a fossil? What conditions are needed for good fossil formation?
2. In what kind of rocks are fossils most often found?
3. How can wood become a fossil?
4. Name a fossil fuel.

TRY THIS

You can produce a fossil-like image by making an imprint of your hand or foot in modeling clay. Mix plaster of Paris and pour it into the imprint in the clay. After the plaster hardens, remove it from the clay. The plaster forms a *cast* of the "fossil" foot or hand. The imprint of the hand or foot in the clay is called a *mold* of the fossil. The cast is a three-dimensional model of the fossil.

11-4

The Water Cycle

Goal

To describe the water cycle and discuss some of the variations of the cycle.

Water is a compound so familiar to all of us that we take it for granted. It covers about three quarters of Earth's surface and is abundant in the atmosphere and under the ground. You, as well as other living things, could not live long without it. In fact, water is what makes our planet unique. No other planet in the solar system has the proper range of temperature to allow water to exist in all three states: solid (ice), liquid (water), and gas (water vapor).

water cycle

The movement of water from the land to the rivers and oceans, to the atmosphere, and back to land again is called the **water cycle**. (See Figure 11-12.) This description of the water cycle is highly simplified. A number of variations of the cycle can occur. Consider, for example, a drop of water in an inland lake. This water drop could *evaporate*, or change from liquid to gas, and become part of the atmosphere as water vapor. It later could *condense*, or change from gas to liquid, as a drop in a cloud. From here the water drop could fall back to the land, and then return to the lake, completing a cycle that doesn't include the ocean.

Next, consider a water drop that evaporates off the leaf of a tree, condenses in a cloud, and falls on the ocean. In this cycle the drop doesn't enter a river before entering the ocean. Over large areas of ocean, a water drop could evaporate into the atmosphere, condense in a cloud, and fall back to the ocean. In this cycle the water drop would never reach land.

Water can remain in one place for a long period of time, thus delaying its movement through the water cycle. Imagine how long a drop of water can be trapped in the ice caps in the polar regions! On the average, a drop of water can be expected to spend about 98 out of every 100 years in the ocean. Of the remaining two years, the drop spends about 20 months as ice. It is present in lakes, ponds, and rivers for only about half a month. It spends less than a week in the atmosphere. Although only a small proportion of Earth's

Figure 11-12 *The water cycle. There are many pathways that a drop of water can follow between air, ground, and water.*

water is in circulation at any given time, there is a very large amount moving through the water cycle. Water on Earth is in constant circulation.

Checkpoint

1. Why is Earth called the "water planet"?
2. What is (a) evaporation, (b) condensation?
3. Describe the simplified water cycle and one of its variations.

TRY THIS

Study the water-cycle diagram (Figure 11-12). List as many other paths as you can for a water drop moving through the water cycle.

11-5

Drinking Water

Goal

To discuss drinking water and its sources.

You might think that with all the ocean water on Earth there would be plenty of water for everyone. However, ocean water is salty, and people cannot safely drink salt water. There are two main sources of drinking water on Earth. One is the large bodies of fresh water on the surface, such as lakes and rivers. The other is water stored in the ground.

In inland areas that are short of water, dams may be built on rivers to form reservoirs (REZ ur vwarz). *Reservoirs* are artificial lakes that supply water for large and, in some cases,

Figure 11-13 *Ground water is present in the saturated zone. The upper boundary of the saturated zone is known as the water table.*

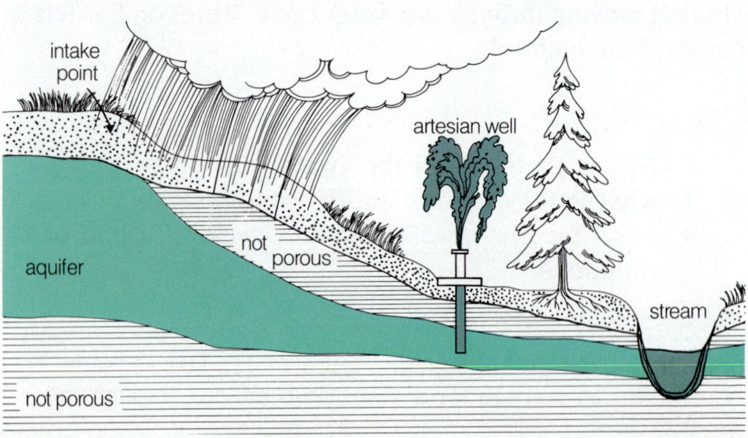

Figure 11-14 *An artesian well works because its top is below the level of the water table farther up the hill.*

distant areas. In some places huge pipes called *aqueducts* (AK wuh duhktz) carry water from reservoirs across mountains to supply water for desert areas or for large cities.

Communities that don't have large bodies of surface water nearby obtain their water from the ground. When it rains, some of the water soaks into the ground, as if the ground were a sponge. This water, called *ground water*, fills the spaces between soil particles and between grains of sedimentary rock. The part of the ground that contains ground water is called the **saturated zone**. The irregular upper boundary of the saturated zone is called the **water table**. The depth of the water table determines how deep a well must be drilled to reach water. (See Figure 11-13.)

saturated zone

water table

Large volumes of ground water are stored in layers of rock called **aquifers**. To be a good aquifer, the rock must have openings, or *pore spaces*, within it. The pore spaces can be filled with air or water. Sandstone is a porous rock that makes a good aquifer.

aquifers
(AK wuh furz)

In some places an aquifer may be sandwiched between two layers of less porous rock. These keep the water trapped between them. If these layers are on a slope, the water in the aquifer builds up pressure near the lower end. When a well is sunk there, the water spurts up naturally, forming an *artesian well* (ar TEE zhun). (See Figure 11-14.)

You may have heard of a stream that is particularly clear and cold because it is "spring-fed." *Springs* (see Figure 11-15) are places where ground water flows naturally out to the

Figure 11-15 *Seven Veils Falls in Yoho National Park, British Columbia. When water flowing through porous material reaches a rock layer that is not porous, the water is forced to the surface, forming a spring.*

TRY THIS

Stir up a concentrated salt solution in an empty paper milk carton. Put it in a freezer. When it has frozen solid, cut away the carton. Where do you find the salt?

surface. They are usually found on the sides of hills, where the saturated zone meets a nonporous rock layer. The water has nowhere to go but out.

Two special kinds of springs are mineral springs and warm springs. *Mineral springs* occur when water dissolves minerals from the ground as it flows through. *Warm springs* occur when ground water passes near hot rock. The hot rock can be far below the ground, or closer to the surface. A good example of the latter case is found in Yellowstone Park. There the hot layer of rock is close to the surface, and there are many warm springs.

Besides surface water and ground water, there is another source of drinking water. Some coastal cities are making use of ocean water. They must remove the salt from the water, however. This process is called desalination or, sometimes, desalinization.

Any water used for drinking purposes must not only be free of salt, but should also be free of foreign matter. Bacteria from sewage, harmful chemicals, such as fertilizers and highway salt, and waste products from industry are examples of foreign matter. Suppose a water supply comes from a sandstone aquifer. If the water has been in the aquifer long enough, it is cleansed by the filtering action of the fine-grained sandstone.

Various government agencies set the standards for drinking water. They conduct frequent tests of water supplies to ensure good quality. Chlorine can be added to water to kill bacteria. In many communities, fluorides are added to protect people from tooth decay. These additions are made by local water departments.

TRY THIS

Stir up a salt solution in a glass jar. Let it stand uncovered for several days until the water has evaporated. Has the salt evaporated?

Checkpoint

1. What are the two main sources of drinking water on Earth?
2. What is the difference between a spring and an artesian well?
3. Define (a) saturation zone, (b) water table.
4. What is the difference between an aquifer and an aqueduct?

FEATURE ARTICLE

Minerals and Modern Technology

Much of modern technology depends on metals extracted from ores. At one time iron and copper were used extensively. But as technology has changed, so has the need for different metals. To increase gasoline mileage, automobile manufacturers need to use structural materials that are lighter in weight than those formerly used. Jet engines and rocket motors need metals that can withstand great heat and stress.

Unfortunately, many of the metals needed for the new technologies are not readily available in North America. Because of their importance they must be imported. These metals—niobium, manganese, tantalum, cobalt, chromium, platinum, and titanium—are known as *strategic metals*. Niobium, manganese, and chromium are used in steel production. Cobalt and titanium are used in jet engines. Strategic metals are needed in the aerospace, automobile, electronic, and chemical industries, also. The prices of these metals are rising at an ever-increasing rate.

A number of solutions to the strategic-metal problem have been offered. For one thing, recycling can be increased, in order to make more use of the strategic metals already purchased. New areas can be mined—the ocean floor, for instance, is a source of minerals. More easily available metals can be substituted for some strategic metals (for example, nickel instead of cobalt).

One of the best ways to reduce dependence on imported metals is to find substitutes. Ceramics (materials made from clay or similar minerals), plastics, and glass are being studied to find new ways of using them. Technology may have to develop in entirely new directions to make use of these materials. Who knows what jet engines of the future will be made of?

11-6

ENRICHMENT

Geologic and Topographic Maps

Goal

To describe features and uses of geologic and topographic maps.

There are many kinds of maps. You have probably seen maps on classroom walls, for example. They generally are *political* maps that show such things as countries, states, provinces, and major cities. They may also show physical features such as rivers, lakes, mountains, and deserts. You have probably seen road maps, too. They show highways, cities, and other information useful to motorists.

geologic map

Earth scientists use special kinds of maps to help them study Earth. A **geologic map** shows the various kinds of rock found at the surface or under the soil in a particular area. (See Figure 11-16.) Geologists can learn a lot about an area from a geologic map. For instance, if there are many fine-grained igneous rocks, there may have been volcanic activ-

Figure 11-16 *The colors on this geologic map represent rocks of different kinds and ages found in northwestern Wyoming.*

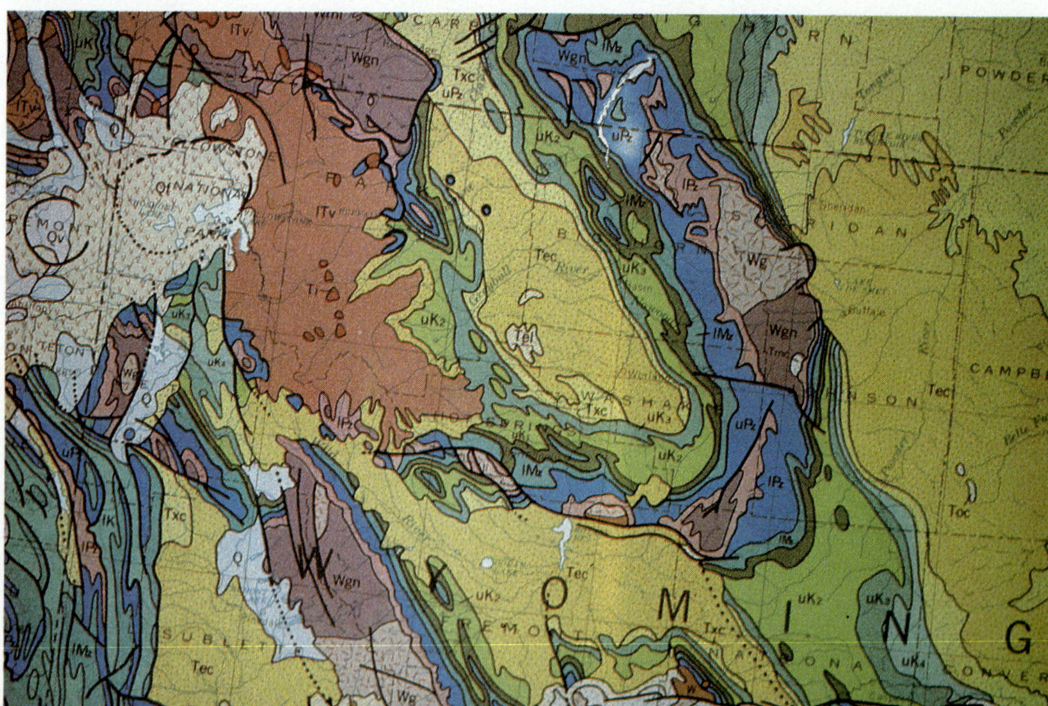

Figure 11-17 *A topographic map of a part of southern California. It shows land contours, altitudes, mountains, lakes, streams, highways, and towns.*

ity close by at some time in the past. If there are layers of limestone, the area must have been covered by an ocean at least once in the past. Also, knowing what kinds of rocks are found in an area is the first step in locating natural resources, such as mineral ores.

topographic maps

Topographic maps have a special pattern of lines to show the shape of the land surface. Points that have the same elevation above sea level are connected by lines called *contour lines.* (See Figure 11-17.) If the contour lines are close together, the land surface rises steeply. If they are far apart, the land is fairly flat. Landmarks such as schools, quarries, and radio towers are identified on topographic maps.

TRY THIS

Using a topographic map for your area, find the elevation of several familiar places. What is the elevation of the highest point shown on the map? of the lowest?

Checkpoint

1. What is the name of the map that shows kinds and ages of rock layers?
2. What kind of map uses contour lines?
3. What information may a geologic map provide?

Understanding The Chapter 11

Main Ideas

1. The outermost solid layer of Earth is made of rocks. This layer is known as Earth's crust.
2. Solid elements and compounds that occur naturally in Earth's crust are known as minerals.
3. Most of the rocks found in Earth's crust are mixtures of minerals.
4. Minerals can be identified from properties such as crystal shape, color, hardness, specific gravity, streak, and luster.
5. Rocks are classified into three groups: igneous, sedimentary, and metamorphic.
6. Fossils are evidence of formerly-living organisms.
7. Fossils can provide valuable clues to ancient climates. They are a source of oil, natural gas, and coal, also.
8. Water is one of the most important substances on Earth.
9. The movement of water from the land to the rivers and oceans, to the atmosphere, and back to land again is called the water cycle.
10. Even though water is abundant on Earth, it is a problem to find enough good drinking water.
11. Drinking water comes from sources on the surface and from under the ground.
12. Geologic and topographic maps are specialized maps that are used by scientists.

Vocabulary Review

From the following list, choose the term that best completes each of the statements. Write your answers on a separate piece of paper.

aquifers	metamorphic rock
crust	minerals
crystals	rock cycle
fossils	saturated zone
geologic map	sedimentary rock
igneous rock	topographic map
lava	water cycle
magma	water table

1. Most rocks are mixtures of __?__.
2. Earth's outermost solid layer is the __?__.
3. The shape of its __?__ is a clue that can help to identify a mineral.
4. Melted rock that is under the surface is called __?__.
5. __?__ has been made by the changing of other kinds of rock under great heat and pressure.
6. Basalt, granite, and obsidian are examples of __?__.
7. The changes in rock from one form to another are called the __?__.
8. __?__ are evidence of formerly-living things.
9. The __?__ describes the way in which a drop of water may move about on Earth.
10. Porous rock layers that store ground water are called __?__.
11. The depth of the __?__ determines how deep a water well must be drilled.

12. A __?__ shows the kinds of rock in an area.
13. The part of the ground that contains ground water is known as the __?__.
14. __?__ is usually formed in bodies of water, where mineral grains settle out in layers on the bottom.

Chapter Review

Write your answers on a separate piece of paper.

Know The Facts

1. Most of the rock that makes up Earth's crust is __?__. (sedimentary, igneous, metamorphic)
2. The size of the crystals in igneous rock depends chiefly on the __?__. (minerals, rocks, cooling rate)
3. Rocks are classified by __?__. (their color, their size, the way they were formed)
4. Luster is a test for a mineral. It describes __?__. (its texture, its hardness, how it reflects light)
5. Fossils are usually found in __?__ rock. (sedimentary, igneous, metamorphic)
6. Probably the most important compound on Earth is __?__. (salt, silica, water)
7. A large pipe that carries water is called a(n) __?__. (aquifer, aqueduct, water table)
8. An artificial lake that supplies water to large areas is a(n) __?__. (reservoir, artesian well, aqueduct)
9. You could find the elevation of your home on a __?__ map. (political, geologic, topographic)
10. When water changes from gas to liquid, the process is called __?__. (condensation, saturation, evaporation)
11. Earth is an unusual planet because it has so much __?__. (atmosphere, water, igneous rock)
12. A chemical used to kill bacteria in water supplies is __?__. (chlorine, alcohol, salt)
13. Yellowstone Park has many bubbling hot springs because __?__. (it has a warm climate, there is sulfur in the ground, hot rock is close to the surface)
14. About __?__ of Earth is covered by water. (¼, ¾, ½)

Understand The Concepts

15. How can igneous rock be changed into (a) sedimentary rock, (b) metamorphic rock?
16. If an igneous rock cools slowly, how will this affect its characteristics?
17. Name two typical characteristics of sedimentary rocks that are different from those of metamorphic rocks.
18. Which would have a higher specific gravity: marble or pumice?
19. Why would you not bother to look for fossils in a granite quarry?
20. How might you use a topographic map to look for a spring?
21. Since the total amount of water on Earth doesn't change, why do you hear of water shortages?
22. Why could icebergs be used as a source of fresh water?
23. Explain how it might be possible for some igneous rocks to contain fossils.
24. Explain why the streak test for a

mineral is a better indicator of the mineral's true color than is its appearance.

25. Explain why granite may be preferable to sandstone as a building material.
26. What is the water table? Why is the water table in a given area important to a person who wants to drill a water well?
27. Why do some minerals fail to produce a streak when they are subjected to a streak test?
28. What special characteristics do gems have that other minerals do not have?

Challenge Your Understanding

29. Why do all crystals of the same mineral have the same shape?
30. Why would you not want a well located on the downhill side of a barnyard where animals are kept?
31. Why don't rivers stop flowing when it stops raining?
32. Oxygen is one of the gases present in the air. It is one of the most abundant elements in Earth's solid crust, as well. What property of oxygen do you think accounts for its abundance in the crust?
33. How does the water cycle benefit all living things?
34. Why do you think a material that forms crystals forms larger ones when it cools slowly?
35. In addition to sedimentary rocks, layers are sometimes found in metamorphic and igneous rocks. In what ways might such layers be formed?

Projects

1. Study the water supply for your community. Where does the water come from? What is done to it before it is sent to the public?
2. Collect rocks and minerals. Put them in a display with proper identification.
3. Collect fossils from your area. Identify them using a key obtained from a library.
4. List all the ways you use or depend on water for recreation.
5. Rent or borrow a metal detector. Where would you expect to find stray coins or pieces of metal in the ground? Why? Make a collection of your findings.
6. Grow crystals using a sugar solution. Get the instructions from your teacher or the library.
7. Make a model of a geyser, which is a spring that throws hot water into the air from time to time.
8. Obtain a geologic map for your area. Write a history of its geologic past. Consult experts, if possible, to check your accuracy.
9. People have increased their use of bottled water. What are people assuming when they buy bottled water? Find out if there are regulations that control the quality of bottled water.
10. Obtain a topographic map for your area. Use it to construct a model landscape out of plaster of Paris, clay, or a similar substance. Reduce the scale of the map. For example, you could let one millimeter in your model represent one meter on the map.

Chapter 12

Earth's Changing Crust

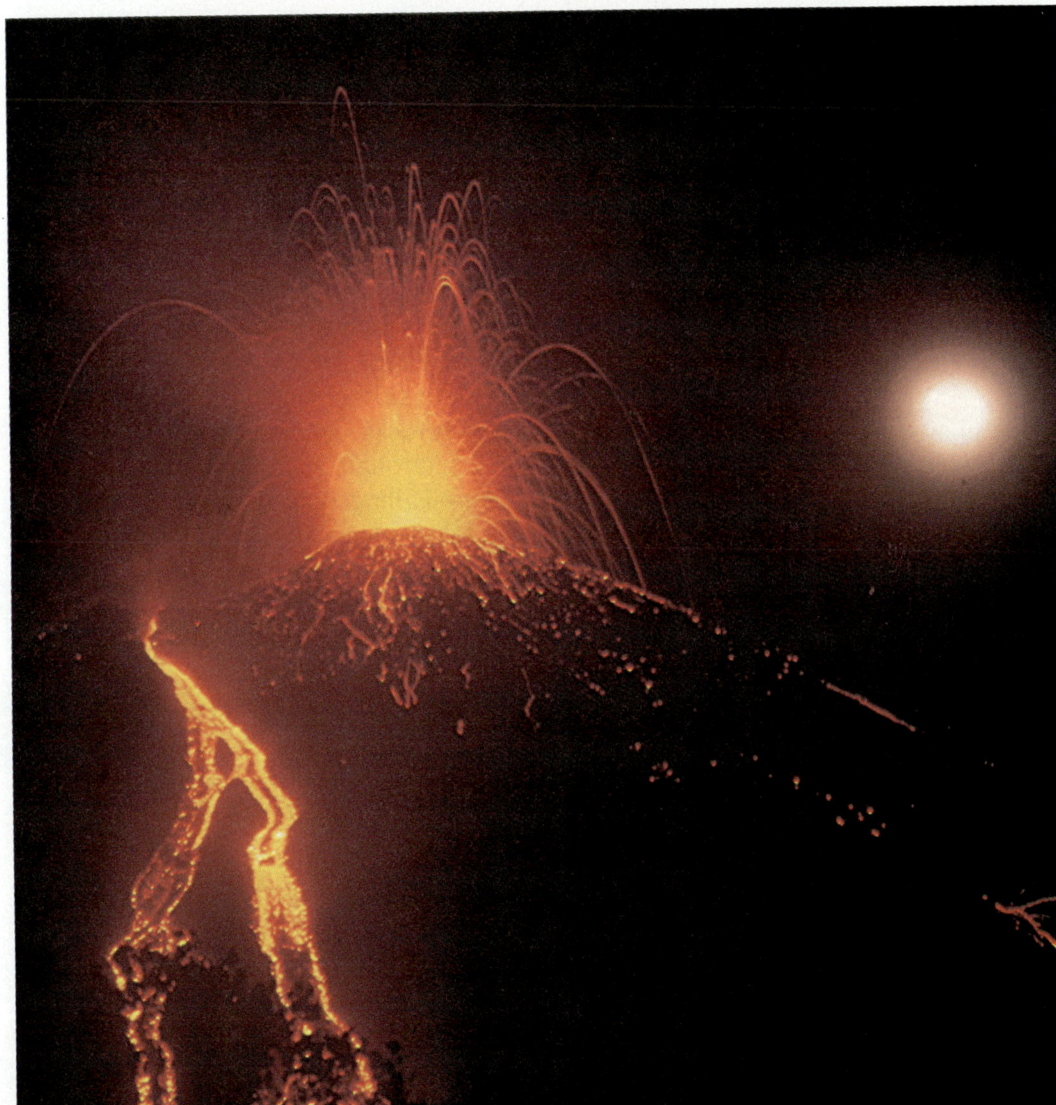

The eruption of Mt. Fuego, Guatemala, is an example of a rapid, dramatic change in Earth's crust.

12-1

Weathering

Goals

1. To define weathering.
2. To list some of the factors that affect weathering.

weathering

All around you are examples of materials breaking down. Sidewalks and driveways crack, paint on buildings peels, and metal on cars and bicycles rusts. The rocks in Earth's crust go through a similar process. When rocks break down, the process is called **weathering**.

There are two types of weathering, physical and chemical. Through *physical weathering*, rocks are broken into smaller chunks. One way this can happen is by alternate freezing and thawing. There is usually some water trapped in the cracks in rocks. Since water expands when it freezes, freezing widens the cracks. This puts pressure on the rock on either side. Eventually the rock breaks into smaller pieces. If this happens high on the side of a cliff, pieces of rock that break off can fall on other rocks, causing them to break, too. You may have traveled through an area where signs are posted saying "Danger—Falling Rocks." (See Figure 12-1.)

Chemical weathering takes place when chemicals in air or water react with rocks to change their composition. For example, carbon dioxide in the air combines with water, forming a weak acid called *carbonic acid*. This acid is strong enough

Figure 12-1 *Physical weathering caused this rock slide in California.*

Figure 12-2 *Physical and chemical weathering combined can ruin a stone statue.*

to dissolve parts of some rocks. (See Figure 12-2.) Chemical weathering also occurs when oxygen in the air reacts with certain minerals in the rocks to form new minerals called *oxides*. Iron oxide, or rust, is a common example. The reddish color of some soils is usually caused by the presence of iron oxide.

Most of the time physical and chemical weathering work together. Physical weathering breaks rocks into smaller pieces. This process exposes more surface area of the rock, allowing chemical weathering to take place more quickly. An increase in surface area can cause physical changes to take place more quickly, too. For example, crushed ice in water melts faster than ice cubes would. With crushed ice, the water comes into contact with more ice surface area, causing it to melt faster.

Several factors determine how rapidly weathering takes place. A large amount of rainfall or extreme temperature variations can increase the rate of weathering. The type of rock in an area has an effect on weathering, also. Some kinds of sedimentary rock, such as soft sandstone, weather faster than igneous rock, such as granite. The type and quantity of vegetation growing in an area are another factor. They determine how moist the ground will be and how much protection leaves may provide for surface rocks. Whatever factors are involved, it usually takes a long time for a large amount of weathering to occur.

Checkpoint

1. Define weathering.
2. Give an example of physical weathering.
3. What is chemical weathering?
4. Name three factors that affect the rate at which weathering occurs.

TRY THIS

Fill two cups or bowls with equal amounts of water. The water in each container should be at the same temperature. Place a cube of sugar in one container and a spoonful of sugar in the other. Stir both at the same time. Which dissolves faster? Why? Relate this to the weathering of rocks.

12-2

ENRICHMENT

Soils

Goals

1. To describe how soil is formed.
2. To list the factors that affect soil development.

soil

One of the most valuable products of the weathering process is soil. **Soil** is a mixture of weathered rock material and decayed *organic* (once-living) material. Soil provides the proper nutrients for plant growth.

The development of soil begins with solid, unweathered rock, called *bedrock*, or *parent material*. The bedrock weathers and breaks up into smaller and smaller pieces. In time, small plants and animals start to live in the layer of weathered rock. Plants help the weathering process and also help prevent the soil from being washed away. As the organisms die, their decayed remains, called *humus* (HYOO muhs), mix with the weathered rock. This forms the uppermost layer of soil, the *topsoil*. Developing soil consists of topsoil above a layer of weathered rock fragments. It takes centuries to develop several centimeters of topsoil.

As weathering of the bedrock continues, the soil develops further. Dissolved minerals are carried downward from the topsoil by ground water. So are particles of *clay*, which is made up of very small, closely packed grains of certain minerals. Eventually a middle layer of soil develops. This middle layer, the *subsoil*, is harder to plow than topsoil because of its clay content. (See Figure 12-3.)

A number of factors affect the development of soil. The

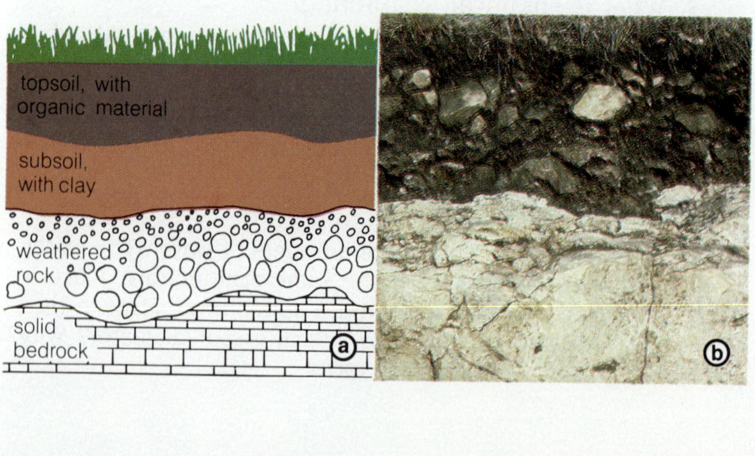

Figure 12-3 *(a) A diagram of soil. (b) The photograph shows topsoil, subsoil, and the less weathered material the soil is forming on.*

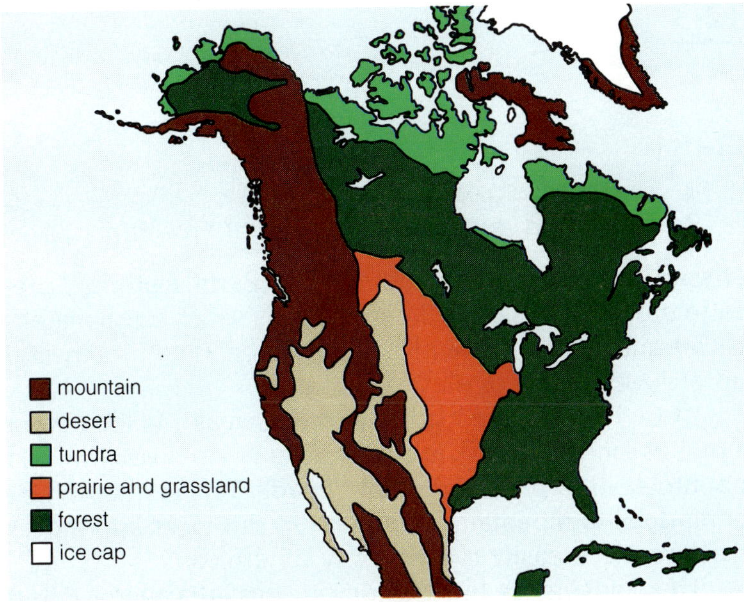

Figure 12-4 *North American soils. Mountain soils are thin and rocky. Forest soils are thin with little humus. Grassland and prairie soils have a thick layer of rich, black topsoil with more humus than forest soils. Desert soils are high in minerals because there is little rain to wash the minerals downward. The lower layers of tundra soil remain frozen, reducing drainage in the top layers.*

type of bedrock in an area affects the soil that develops from it. The kinds of organisms that live in an area affect soil development, also. Since soil needs time to develop, time is another factor. The most important factor in the development of soil, however, is climate. In warm, moist climates rocks weather rapidly, and soil develops quickly. Desert areas, on the other hand, have little rainfall. Rock weathers much more slowly in deserts.

The North American continent has many different types of soil. Soils differ because of where they occur, and what the local climates are. (See Figure 12-4.)

TRY THIS

Find a spot in the schoolyard or at home where you can dig a hole deep enough to expose the depth of the topsoil. How does topsoil differ from the layer beneath it? How thick is the topsoil? Replace the soil when you are finished.

Checkpoint

1. What is soil?
2. How is topsoil formed?
3. How is subsoil formed?
4. What is the major influence on the type of soil that develops in an area?

12-3

Erosion

Goals

1. To define erosion.
2. To list and describe the agents of erosion.

erosion

Physical and chemical weathering of Earth materials precede erosion. **Erosion** is the process by which Earth materials are moved from one place to another. They usually end up at lower levels of elevation.

If you have a large load to move, such as a fallen tree or a pile of sand, your task is much easier if you break the load up into smaller pieces or smaller loads. This is what weathering does to mountains. It breaks up the rocks into pieces that are more easily carried away by erosion.

The "movers" in the erosion process are called *agents of erosion*. The agents of erosion are water, ice, wind, organisms, and gravity. The one that moves the most material is water.

Earth materials can be moved by running water in three different ways. One way is by *pushing* the water along the bottom or sides of a stream. (Have you ever used running water to push dirt off a sidewalk?) Another way that water can carry particles is as a cloud of mud and sand called a *suspension*. River water is usually cloudier than lake water

Figure 12-5 *The dramatic features that occur along the Oregon coast were produced by wave erosion.*

Figure 12-6 *The dark streaks on this Alaskan glacier are rock fragments weathered from valley walls.*

because there are more particles suspended in the moving water. The third way that water can carry materials is in *solution*. When in solution, particles are dissolved in water, just as salt may be dissolved in water. Dissolved particles are not visible.

Waves pounding against a shoreline over long periods of time are powerful agents of erosion. If you live by the ocean or a large lake, you have seen the erosion done by waves. Depending on the shape of the shoreline (smooth or jagged) and on the kind of rock found there (hard or soft), wave erosion can make some interesting patterns. (See Figure 12-5.)

In areas with cold climates, glaciers may act as agents of erosion. A **glacier** is a large, slowly moving mass of ice, many meters to thousands of meters thick. Glaciers form high in mountain valleys. As they grow they move downward, scraping and scouring the sides and bottom of the valley. Glaciers push and carry large quantities of rock debris with them. They also can act as a "conveyor belt." Rocks that fall onto the top of the glacier, are carried along. (See Figure 12-6.) Unlike running water, glaciers move very slowly. Their movement is measured in meters per year.

glacier
(GLAY shur)

Wind is another agent of erosion. Although wind cannot move pieces as large as those carried by water or ice, it can carry sand and dust long distances. In dry regions, where vegetation is scarce, wind does a considerable amount of erosion. (See Figure 12-7.) Farmers are encouraged to leave their fields unplowed until spring because of wind erosion. Plowing loosens the soil. If farmers plow in the fall, strong winter winds may blow away much of the rich topsoil.

Figure 12-7 *A dust storm on the dry plains of New Mexico.*

Figure 12-8 *Flooding of the St. John's River, Maine.*

You might think it strange that organisms are listed as agents of erosion. By stripping the land of vegetation when mining minerals, or by allowing the overgrazing of land by livestock, people have caused a considerable loss of topsoil in some areas. Animals such as cows or sheep can cause erosion on hillsides as they carve out paths where they walk.

Gravity is an agent of erosion because it is the force that causes materials to fall to lower levels of elevation. Gravity helps the other agents of erosion do their work. It causes rivers to flow and rocks to fall and strike other rocks.

When erosion occurs rapidly it makes the headlines. (See Figure 12-8.) Landslides, mudslides, and raging floods are examples of rapid erosion. People can protect themselves from most of these hazards, however. They can avoid building houses on hillsides where mudslides frequently occur or in areas subject to flooding.

TRY THIS

Fill two glasses about ⅔ full of water. Stir a spoonful of salt into one glass and a spoonful of dirt into the other. Describe what you observe. Let the glasses sit undisturbed until all the material has stopped moving. Which one is a suspension?

Checkpoint

1. What is the major difference between erosion and weathering?
2. List several agents of erosion.
3. Which agent of erosion has the greatest effect?
4. Why can people be considered agents of erosion?

12-4

Deposition

Goals

1. To describe various types of depositional features on Earth's surface.
2. To recognize events that caused these features.

Materials moved by erosion and left in another place are called **deposits**. Different agents cause different forms of deposits, or *depositional features* (dep uh ZISH uhn ul).

deposits

Running water leaves behind several recognizable features. One feature is a buildup of sandbars along the inside curves of moving streams. Sandbars form there because the stream slows down on the inside curve as it goes around a bend. (The faster-moving water is on the outside of the curve.) As the water slows down, the heavier material being carried is dropped. The stream continues to move in this way, cutting out the far side of the curve and depositing debris on the inside curves. Eventually, it develops a series of **meanders**, or looping curves. A meander may loop so much that it closes back on itself, forming an *oxbow lake*. (See Figure 12-9.)

meanders (mee AN durz)

Another feature formed by deposits from running water is a delta. **Deltas** are deposits formed at the mouths of rivers as they flow out into lakes, oceans, or bays. The river slows down as it joins a still body of water, and drops the materials it carries. Deltas typically have a triangular shape. (See Figure 12-10.)

deltas

Figure 12-9 *Erosion can cut through the land between meanders, forming an oxbow lake.*

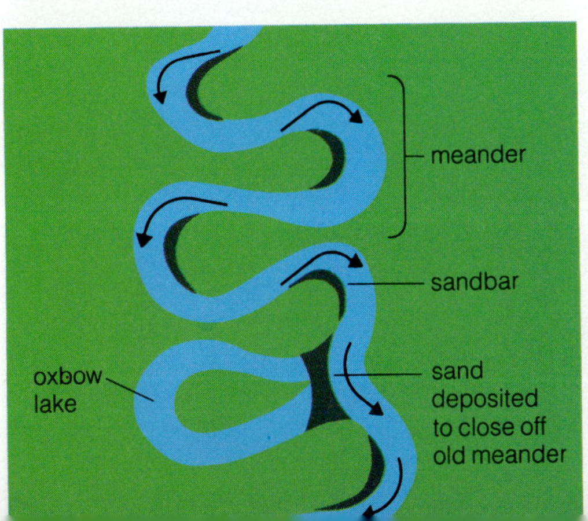

Figure 12-10 *The red triangle is the delta of the Nile River, photographed from space.*

There is a feature common to all materials deposited by running water. As the water slows down and drops its load, the largest particles are dropped first, then the next largest, and so forth. This process is called *sorting*, because the deposits are grouped according to size.

A depositional feature found at the base of mountains and cliffs is a collection of loose rocks and boulders called a *talus slope* (TAY luhs). Talus slopes are caused by gravity pulling down loose material from above.

Glaciers leave a number of recognizable depositional features. Since ice can carry heavier material than running water can, the deposits left by ice are quite different. When a glacier begins to melt, the materials it is carrying—inside the ice, underneath, or on top—are dumped. There is no sorting of this material as there is with running water. Large boulders and small particles are mixed together.

moraines
(muh RAYNZ)

All the unsorted, loose rock material deposited by a glacier is called *glacial drift*. Glacial drift can form many different-shaped features. Deposits of drift dropped in front of a glacier or along its sides are called **moraines**. A *terminal moraine* is formed at the end of a glacier. (See Figure 12-11.) A *lateral moraine* is formed along the sides of a glacier. Moraines form large irregular hills. These hills may display the shape of the glacier that produced them.

Glaciers can carry huge boulders and leave them hundreds of kilometers from their sources. Such boulders are not like the rocks that are most common in the area, and are called *erratics*. (Have you seen any of these rocks in your area?)

Figure 12-11 *The loose material these people are standing on is part of a terminal moraine.*

Figure 12-12 Sand dunes in the Monahans Sandhills of Texas. The tips show that the wind blows from "right" to "left."

Wind also leaves characteristic depositional features. Sand dunes are probably the best example. Desert sand dunes have a typical "boomerang" shape. The tips of the "boomerang" always point away from the wind. (See Figure 12-12.) Snow drifts have a similar shape.

Checkpoint

1. Which agents of erosion sort materials as they deposit them?
2. On which side of a curve do sandbars develop?
3. What is a delta?
4. How do glacial deposits differ from deposits left by running water?
5. What is an erratic?

TRY THIS

Partially fill a long, plastic tube with water. Hold it in an upright position, preferably using a ring stand. Drop spoonfuls of unsorted sand into the water. Allow each spoonful to settle before adding the next one. What do you observe happening to the sediment in the bottom? How does this compare to what happens in a river delta?

12-5

Plate Tectonics

Goals

1. To define plate tectonics.
2. To discuss some evidence in support of the plate theory.
3. To describe the layers of Earth.

In 1912 a scientist named Alfred Wegener (VAY guh nur) came up with a startling idea. He proposed that all the continents had at one time been together, but have been slowly drifting apart for many years. (See Figure 12-13.) He based his hypothesis on the obvious way that the continents could fit together like a jigsaw puzzle, especially South America and Africa. He also discovered matching rock layers on continents separated by the Atlantic Ocean.

mid-ocean ridges

Wegener's idea was given support when scientists learned more about the ocean floor. They discovered a system of underwater mountain chains, or **mid-ocean ridges**, that rise hundreds to thousands of meters above the ocean floor. The youngest ocean floor rocks are near the mid-ocean ridges, and the oldest are near the edges of the ocean basins.

The youngest rocks on the sea floor are near the mid-ocean ridge because the sea floor is spreading apart at the mid-ocean ridges. The ridges are made up of many volcanoes that allow hot, molten rock to flow out onto the ocean floor. Horizontal motion of the sea floor carries the new rock away from the ridge, just as if the floor were on a huge con-

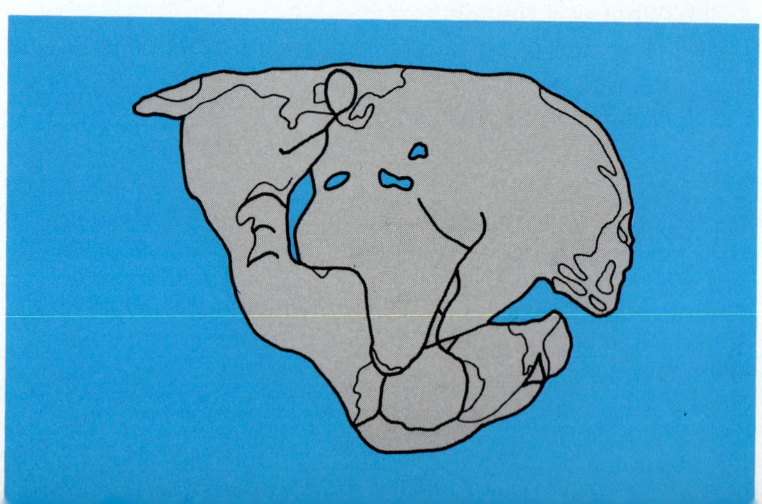

Figure 12-13 *This is Alfred Wegener's map of the land mass he thought broke up into the present continents.*

veyor belt. When the sea floor reaches a continental boundary, it is forced downward beneath the continent. Sea floor *trenches,* long grooves in the ocean floor thousands of meters deep, are formed where the ocean floor is pulled down under the continents.

In order to understand how parts of the crust can move about, it helps to step back and look at Earth as a whole. Earth's interior consists of several layers, each having different properties. (See Figure 12-14.) The **inner core** is believed to be solid, composed mainly of nickel and iron. The **outer core** is a hot liquid. The **mantle** is a thick layer, denser than the crust, that is mostly solid. The outer layer is the crust.

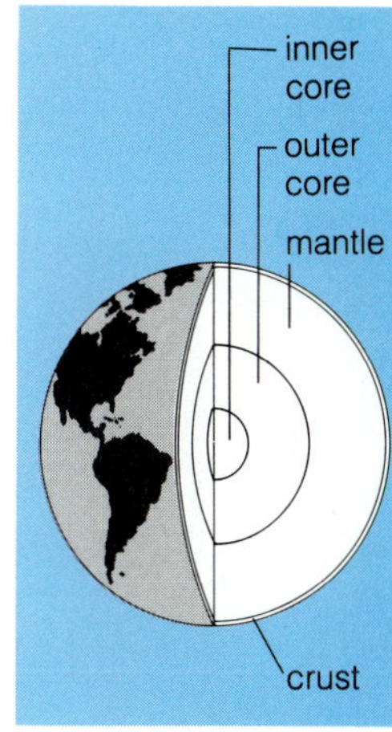

Figure 12-14 *The layers in Earth's interior.*

Since the rocks in the crust are less dense than the material in the mantle, the crust "floats" on the mantle. Where continents are thicker, as in mountain ranges, the crust sinks deeper into the mantle. Where the crust is thinner, as under the ocean, it doesn't sink into the mantle as much. Thinner parts of the crust are similar to an empty ship floating high on the water. The thicker parts, like a ship full of cargo, "float" at a lower level.

inner core

outer core

mantle

plate tectonics
(playt tek TAHN iks)

The crust and the uppermost part of the mantle, with a thickness of about 100 km, are divided into segments called *plates.* (See Figure 12-15.) There are six major plates, and several small ones. According to the theory called **plate tectonics**, these plates move about on Earth, carrying continents and ocean floor with them. Shifting plates account for continental movement and sea floor spreading. When plates move apart, sea floor spreading takes place between them. When plates slip alongside each other, as they do

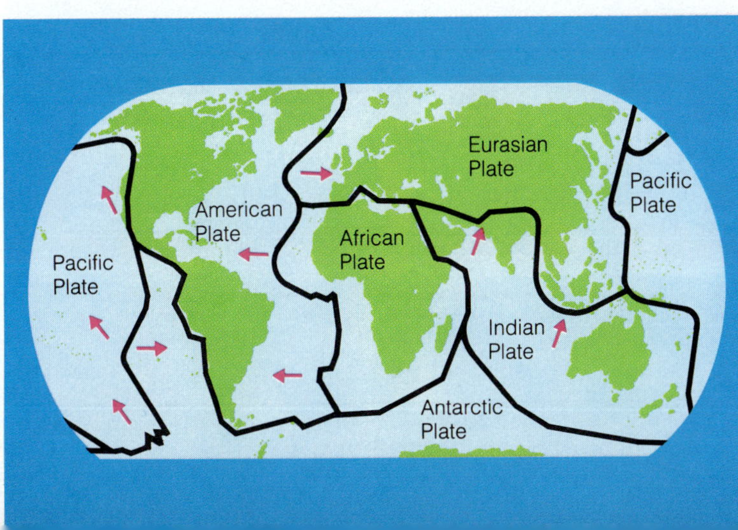

Figure 12-15 *The major plates in Earth's crust. The arrows show directions of crust movement within the plates.*

along the San Andreas Fault in California, earthquakes occur. When one plate slides under another, as they do at continental margins, earthquakes and volcanoes occur. Where plates collide, mountains are pushed up.

Continental motion has been measured at from 1–5 cm per year. As the plates continue to move about, this will produce a slow change in Earth's geography. Each year, for instance, the Atlantic Ocean becomes slightly wider.

Checkpoint

1. Define plate tectonics.
2. Name the layers of Earth, from the inside to the outside.
3. Where are the youngest rocks found on the sea floor? How does this fact provide support for the plate theory?

TRY THIS

Tape two sheets of paper together, making one long sheet. Move two desks or tables together. Fold the paper in half where you have taped it. Place it so that it hangs down between the desks, as in Figure 12-16. The paper represents molten rock before it comes out of the mid-ocean ridge.

Pull the paper up on both sides about 3 or 4 cm and draw a line on the paper at the edge of each desk. Put a "1" on the paper on each side of the gap between the desks. Pull the paper up again, about the same distance as before. Draw another pair of lines and number the "new rock" as "2."

Repeat this until all the paper is on top. Study the pattern that has developed. The lower the number, the older the rock layer. Where do you find the youngest rock?

Figure 12-16 *This model shows why younger rock layers are closer to the mid-ocean ridge than older rock layers.*

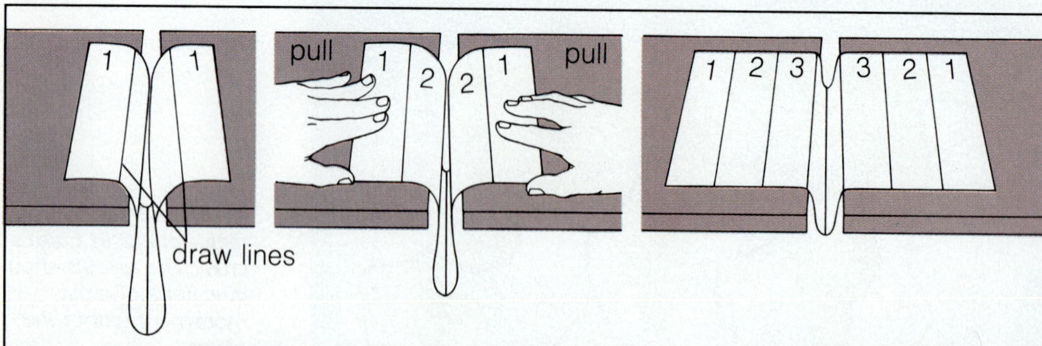

12-6

Earthquakes

Goals

1. To describe earthquakes.
2. To discuss the causes of most earthquakes.
3. To describe earthquake waves.

earthquake

The plates that make up Earth's crust move very slowly and over long periods of time. It is impossible for us to feel this slow overall motion. But sometimes there is a sudden shaking of the crust at a particular place—an **earthquake**. Some earthquakes are so mild that they can be sensed only by the most delicate instruments. Other earthquakes are strong enough to cause widespread damage and destruction. (See Figure 12-17.)

fault

Most earthquakes are associated with faults. A **fault** is a zone of weakness in the crust, along which some movement of rock takes place. Pressure within the crust, often caused by moving plates, causes stress to build up along the fault. When the stress becomes great enough, the rocks slip along the fault. Energy is released and an earthquake occurs.

The sudden release of stress in an earthquake can be compared to the breaking of a tightly wound watch spring. A great deal of stress builds up in a watch spring when it is wound tightly. If the watch is allowed to run down normally, the stress is relieved steadily over time. But if the spring is given one turn past its maximum, the stress becomes too much and the spring suddenly breaks.

Figure 12-17 *The 1971 earthquake in California destroyed parts of the Golden State Freeway.*

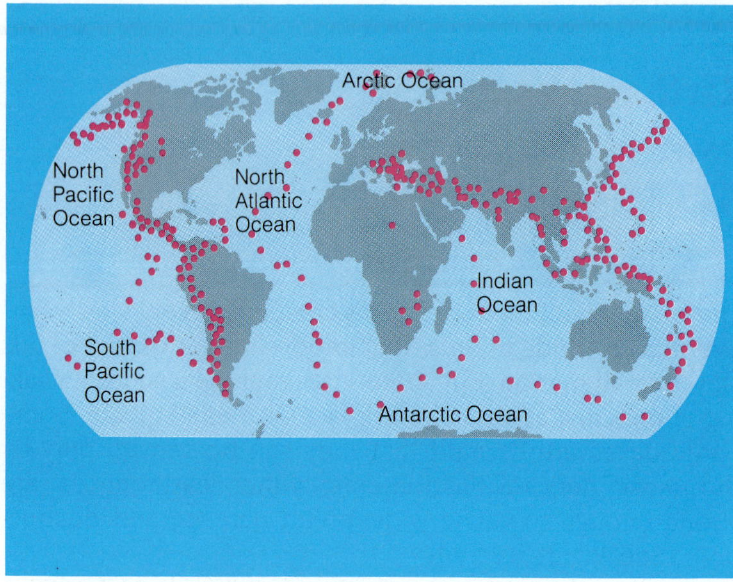

Figure 12-18 *A map of earthquake locations. Compare this map with the one in Figure 12-15.*

More than half of all earthquakes take place along the edges of plates. (See Figure 12-18.) Many earthquakes occur in areas that are not densely populated, so they get little publicity. For instance, there is much activity along the mid-ocean ridges.

focus

The area in the crust where energy is released during an earthquake is called the **focus**. It is the point of origin of the quake. The focus can be shallow (0–70 km below the surface), intermediate (70–300 km), or deep (over 300 km). The deepest earthquakes occur about 700 km below the surface. Most destruction is caused by earthquakes of shallow or intermediate depth.

magnitude

The relative size of an earthquake—the amount of energy released—is called its **magnitude**. Magnitude is expressed in terms of the *Richter scale* (RIK tur). This scale uses numbers from 1 up, with each number indicating a magnitude ten times stronger than the number below. For example, if an earthquake has a rating of 8 on the Richter scale, its magnitude is ten times as great as an earthquake with a magnitude of 7. A magnitude of 7 or higher indicates a major earthquake. If there were an earthquake with a magnitude of 10, it would be felt all over Earth. An earthquake in Chile, in 1960, had the highest magnitude ever recorded, 9.5.

intensity

There is another scale used to describe earthquakes. The *Modified Mercalli Scale* is used to rate **intensity**, the

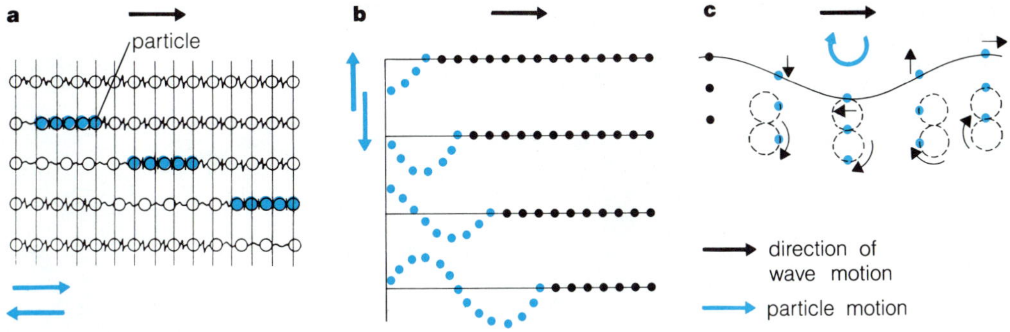

Figure 12-19 *Three kinds of earthquake waves. (a) A compressional wave. (b) A shear wave. (c) A Rayleigh wave. The black arrows show the direction of wave motion; the blue ones show particle motion.*

strength of an earthquake at a particular place. This scale uses Roman numerals from I to XII to describe the damage done by the quake. A rating of XII would indicate total destruction.

An earthquake can have only one magnitude. But the amount of damage it does, its intensity, can vary depending on the location. An earthquake of magnitude 6.3 might cause a lot of damage and have a high intensity at one place, but cause little or no damage 200 km away.

The energy released by an earthquake travels away from the focus in waves. The waves, which travel through and around Earth, are detected by a recording device called a **seismograph**. There are four kinds of waves sent out from the focus of an earthquake. The first waves to arrive at a seismograph are **P waves**, also called primary or *compressional* waves. Compressional waves cause the individual particles in a substance to move back and forth in the direction of the motion of the wave. (See Figure 12-19a.) They can travel through solids, liquids, or gases.

seismograph (SYZ muh graf)

P waves

The second waves to arrive at a seismograph are **S waves**, also called secondary or *shear* waves. Shear waves cause rock particles to move from side to side at right angles to the direction of wave motion. (See Figure 12-19b.) They cannot pass through liquids.

S waves

The third and fourth kinds of earthquake waves, *Love waves* and *Rayleigh waves*, travel along the surface. Love waves involve no vertical movement of the surface. Rayleigh waves involve both horizontal and vertical movement, as indicated in Figure 12-19c. Surface waves cause most of the damage.

The arrival times of P and S waves, recorded on seismographs at three or more different locations, can be used to

Figure 12-20 How to find the epicenter of an earthquake. Each circle represents the distance of the epicenter from a seismograph station.

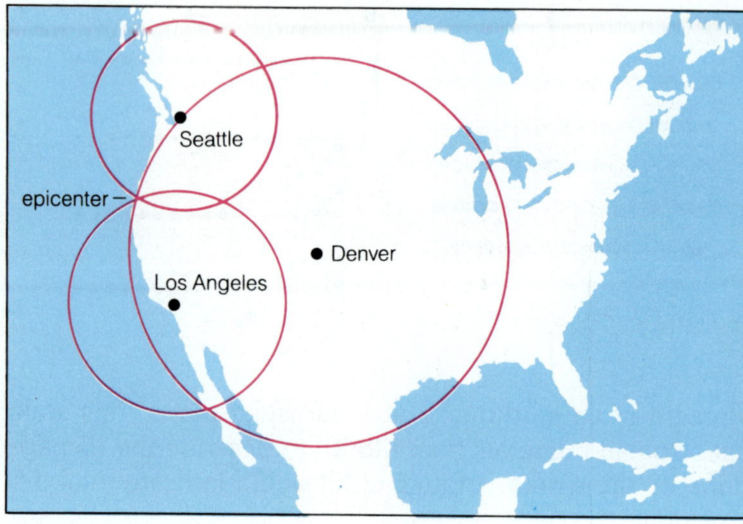

epicenter
(EP uh sen tur)

find the epicenter of an earthquake. (See Figure 12-20.) The **epicenter** is the point on the surface directly above the focus of an earthquake. People at each seismograph station can figure out how far away the epicenter is, but they don't know which direction the waves came from. They draw a circle representing that distance on a map. Then the circles of two other stations can be drawn on the map. The epicenter is where all three circles cross.

When earthquakes occur on the ocean floor, energy is passed through the water, resulting in huge sea waves, or *tsunamis* (tsoo NAH meez). These waves move extremely fast, up to 750 km/h. This is similar to the speed of a jet plane. When a tsunami reaches shallow water near land, a wave as high as 30 m may build up. These waves do great damage to the shorelines where they release their energy.

Checkpoint

1. What is the cause of most earthquakes?
2. How do P waves differ from S waves?
3. What scale is used to measure the magnitude of an earthquake?
4. What scale is used to measure the intensity of an earthquake?
5. What is a tsunami?

Earthquake Prediction

Scientists would like to be able to predict where and when earthquakes will occur, and at what magnitudes. Earthquake-prone areas in which no quakes have occurred for a long time are likely to have major earthquakes fairly soon. Scientists are concentrating their activities in such areas. The San Andreas Fault in California—shown in the photograph—is one example. The last great earthquake along this fault was the San Francisco earthquake of 1906. Since there are many large population centers near this fault, it is being monitored to detect the slightest movement.

Chinese scientists have had some success with earthquake prediction. They have noted changes in water levels in wells prior to a quake. They have observed changes in the amount of a certain gas in the well water, also. They have even noted the behavior of animals, since animals often act in highly unusual ways just before earthquakes. For example, animals that normally live underground, like snakes and mice, leave their burrows up to several days before a quake. Fish that normally stay on the ocean bottom begin to appear near the surface. Dogs, horses, and pheasants become very noisy. Chinese scientists gave advance warning of a major quake in Haicheng, in 1975. As a result, many lives were saved.

In earthquake-prone areas, preparations for a major quake are being made. In California, Civil Defense teams hold regular exercises to prepare them for such a disaster. Many localities have strengthened local building codes to make buildings more quake-proof. New methods of reducing damage to buildings are being tested. One method is to place springs between the walls and the footings of buildings. This method is intended to prevent buildings from breaking off at their foundations when earthquakes occur.

Many of the observations concerning earthquakes seem to be fitting together like pieces of a jigsaw puzzle. They suggest models to explain why earthquakes occur. Only time will tell how successfully such models can predict earthquakes.

12-7

Volcanoes

Goals

1. To describe the causes of volcanoes.
2. To describe the kinds of volcanoes.
3. To identify places where volcanoes can be found.

volcano

A **volcano** is an opening in the crust through which magma from within Earth flows out as lava. Most of the magma that comes to the surface originates in a partly molten "zone of weakness" in the mantle. This zone of weakness is at the base of the plates. The plates put great pressure on the zone of weakness, and this pressure forces magma upward through cracks and other openings in the crust.

The locations of major volcanoes around the world are shown in Figure 12-21. The pattern should look familiar to you. Most volcanoes are near plate boundaries, in the same regions where earthquakes occur. The area around the Pacific Ocean is sometimes called the "Ring of Fire" because so many volcanoes and earthquakes are found there. Plate boundaries are areas of movement in the crust, where most of the cracks and weaknesses in the crust are located. This makes plate boundaries areas where magma is likely to find its way to the surface. There are exceptions to this rule. The Hawaiian Islands, for example, are major volcanoes that are not on a plate boundary.

When lava comes out of a volcano, the pressure on it is released. Gases that were dissolved in the lava are given off. Sometimes a huge cloud of steam, the most common of

Figure 12-21 *Active and recently extinct volcanoes are located on this map. Compare it with Figure 12-18.*

these gases, accompanies a volcanic eruption. Harmful gases, such as carbon monoxide and sulfur dioxide, may be given off also. Besides lava and gases, a volcano can throw out chunks of rock of various sizes. These can range from small particles of volcanic dust to large chunks called *bombs* or *blocks*. Many of these form as the lava hits the cool air, solidifying the molten rock immediately.

The opening at the top of a volcano is called a **crater**. After an eruption is over, the lava settles back down into the crust and a solid cap, or *plug*, forms in the crater. The plug is similar to the hard crust formed on top of warm pudding. A volcano that is quiet or inactive is said to be *dormant*. Before its eruption in 1980, Mount St. Helens was dormant for more than 100 years. A volcano that has not erupted within historic times is called *extinct*.

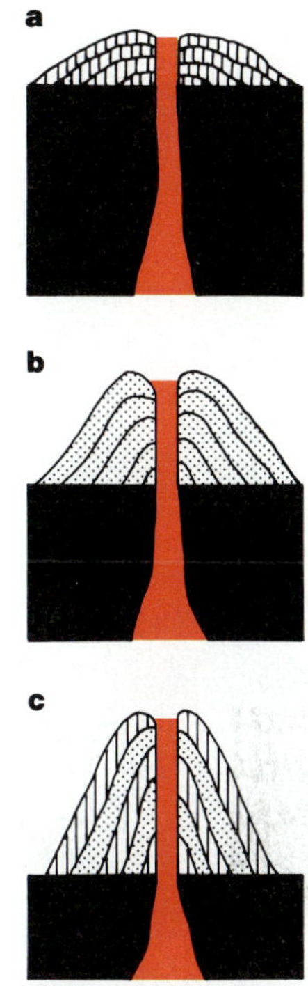

Figure 12-22 *Types of volcanoes: (a) shield cone, (b) cinder cone, (c) composite cone.*

There are three main types of volcanoes, classified by the shape of their cones. The **cone** of a volcano is the pile of material that collects around the opening. The shape of the cone depends on how the lava flows out of the volcano. A *shield cone*, typical of the Hawaiian Island volcanoes, forms when the lava runs out quietly from the opening and forms gently sloping, almost flat layers. A *cinder cone* volcano is typical of an explosive type of eruption and is built up by layers of ash and cinders. It has steeper sides than a shield cone. *Composite cone* volcanoes are built by both quiet and explosive eruptions and have alternating layers of lava and cinders. (See Figure 12-22.)

In the case of Mount St. Helens in 1980 (Figure 12-23), one side of the mountain swelled up before the explosion

crater

cone

Figure 12-23 *Mount St. Helens, before and after erupting.*

that blew off the top of the mountain. Scientists have been watching active volcanic areas for that and other clues to help them predict when a volcano is going to erupt. Many times there is mild earthquake activity in the area around a volcano before an eruption. The plug inside the crater may steam and swell. Predictions of volcanic eruptions are very difficult because there are so many factors that must be considered.

Although volcanoes can cause great destruction to land and life, some of their effects can benefit people. Volcanic ash may spread for hundreds of kilometers around a volcano, creating rich soil for crops. (See Figure 12-24.) In Iceland, which is located on a mid-ocean ridge, hot springs are an important heat source. Twenty-eight towns, including the capital, Reykjavik (RAY kyuh veek), heat their homes directly with water from hot springs. In one town, a bakery uses hot springs to heat its ovens.

Figure 12-24 *Gardening in the rich volcanic soil of Iceland.*

Throughout history there have been some spectacular volcanic eruptions. In A.D. 79 the eruption of Mt. Vesuvius (vuh SOO vee us) buried the city of Pompeii (pahm PAY), Italy. In 1883, the island of Krakatoa near Java was almost completely blown up by a volcanic eruption. The dust from this eruption was carried around the world by winds in the upper atmosphere. One result was extremely colorful sunsets for several years.

By reducing the amount of sunlight reaching Earth's surface, dust and ash from volcanic eruptions can affect weather conditions. The explosion of the Tambora volcano on Java in April, 1815 was one of the largest known. The debris from this explosion is thought to have caused some unusually cold weather. During the summer of 1816, snow fell and temperatures were often below freezing in eastern North America and Northern Europe.

TRY THIS

Using modeling clay of different colors, make a model of each type of volcanic cone. Use powder between the layers of different-colored clay so they can be separated at the end of the activity.

Checkpoint

1. What is a volcano? What causes the occurrence of volcanoes?
2. Where do the majority of volcanoes occur?
3. What are the three main types of volcanoes?
4. What is the most common of the gases given off by volcanoes?

Understanding The Chapter 12

Main Ideas

1. Weathering is a process in which rocks are broken down. Weathering can be physical or chemical.
2. Soil is a mixture of weathered rock material and decayed organic material. The North American continent has many types of soil.
3. Through erosion, weathered materials are carried to lower levels of elevation.
4. Agents of erosion are water, ice, wind, organisms, and gravity.
5. Materials that have been moved through erosion and left in another place are called deposits.
6. Running water, glaciers, and wind are responsible for most of the depositional features on Earth.
7. Earth consists of an inner core, an outer core, a mantle, and a crust.
8. The plate tectonic theory states that Earth's outer layers are divided into six major plates that are moving.
9. Continental drift and sea floor spreading are evidence of the plate theory.
10. Earthquakes and volcanoes are the result of stresses that build up inside Earth.
11. Most earthquakes and volcanoes occur in the same locations on Earth, at the edges of plates.
12. The energy that is released by an earthquake travels in the form of waves.

Vocabulary Review

From the following list, choose the term that best completes each of the statements. Write your answers on a separate sheet of paper.

cone	mantle
crater	meanders
deltas	mid-ocean ridges
deposits	moraines
earthquake	outer core
epicenter	P waves
erosion	plate tectonics
fault	S waves
focus	seismograph
glacier	soil
inner core	talus slope
intensity	volcano
magnitude	weathering

1. __?__ is the process in which rocks break down.
2. Triangular-shaped deposits at the mouths of rivers are called __?__.
3. A(n) __?__ is an opening in Earth's crust from which lava flows.
4. Large hills of material left by a glacier are called __?__.
5. A(n) __?__ is used to detect earthquakes.
6. The point on Earth's surface directly above the origin of an earthquake is called the __?__.
7. The most important agent of __?__ is water.
8. __?__ explains how continents can move.
9. __?__ are mountain ranges on the ocean floor.

10. A(n) __?__ is the opening at the top of a volcano.
11. The relative size of an earthquake is called its __?__.
12. __?__ is a mixture of weathered rock and decayed organic material.

Chapter Review

Write your answers on a separate piece of paper.

Know The Facts

1. Rocks are broken into smaller chunks by means of __?__ weathering. (chemical, physical, organic)
2. The agent of erosion that moves the most material is __?__. (wind, ice, running water)
3. Landslides are an example of __?__. (weathering, erosion, glaciation)
4. An example of a depositional feature is a __?__. (delta, glacier, fault)
5. The kind of earthquake wave that can travel through all kinds of Earth materials is a(n) __?__. (L wave, S wave, P wave)
6. Melted rock still under Earth's surface is called __?__. (lava, magma, dormant)
7. One of the main direct causes of chemical weathering is __?__. (carbonic acid, carbon dioxide, water)
8. A volcanic cone made up of alternate layers of lava and cinders is called a __?__. (shield cone, cinder cone, composite cone)
9. The "Ring of Fire" is another name for __?__. (the boundary of the Mediterranean Sea, the boundary of the Pacific Ocean, the middle of the Atlantic Ocean)
10. Earth's __?__ is composed of solid nickel and iron. (inner core, outer core, mantle)
11. When particles in water are not visible, they are in __?__. (suspension, a mixture, solution)
12. Sea floor spreading seems to be happening because __?__. (continents are drifting, plates are moving apart, plates are colliding)
13. The reddish color in some soils is caused by __?__. (iron oxide, carbon dioxide, nitrogen)
14. An example of chemical weathering is __?__. (ice breaking up rocks, reddish color in rocks, rocks being moved by running water)
15. The Richter scale is used to indicate an earthquake's __?__. (intensity, magnitude, focus)
16. __?__ soil forms where there is little rainfall. (tundra, desert, forest)
17. The layer of soil where water collects and further changes the materials is the __?__. (subsoil, topsoil, parent material)

Understand The Concepts

18. Why would a mountainous area with abundant rainfall tend to have more rockfalls than a mountainous area in a dry climate?
19. How does a thick snow cover in the winter aid in soil conservation?
20. S waves from an earthquake do not pass all the way through Earth. What does this tell you about the interior of Earth?
21. What accounts for the oldest rocks on the sea floor being near the edge of the ocean basin?

22. Why are volcanoes often found in an earthquake zone?
23. How is a talus slope different from a delta? How are the two similar?
24. In what kind of area would you expect wind to be the most important agent of erosion?
25. How does the shape of a sand dune indicate wind direction?
26. How does physical weathering aid chemical weathering?
27. Why do streets and highways show more damage in winter than in summer?
28. Explain how a terminal moraine shows where a glacier stopped moving forward.
29. Why do some kinds of solidified lava have bubblelike holes in them?

Challenge Your Understanding

30. As a glacier moves slowly down a mountain valley, what part of the glacier moves at the fastest rate? Why?
31. Why would a river erode more of the riverbed during floods?
32. How can gravity be said to be the cause of all erosion?
33. Why did scientists conclude that Earth's inner core is solid?
34. Compare a map of the world showing major mountain systems with the map showing the six major plates in Earth's crust. Try to explain how some mountains are formed.
35. Other than at the mouths of rivers, where would sediments be piling up and forcing lower layers down?

Projects

1. Report on Pompeii and the volcano Vesuvius.
2. Research the effect volcanoes have had on Earth's weather.
3. Report on how Iceland has made use of the heat from volcanoes. Make a small model showing how the people use this heat.
4. Report on the geysers and bubbling springs in Yellowstone Park. How do they relate to geothermal energy?
5. Make a model of Earth, showing the layers of the interior. Describe each layer. Explain how scientists have been able to learn about Earth's interior.
6. Make a model of a volcano. Show the structures underneath as well as outside.
7. Use a library to find out where major earthquakes have taken place in the last 30 years. Draw or make a copy of a world map. Mark the earthquake locations on the map. Note whether the locations fall into any noticeable patterns. If they do, try to explain why such patterns may exist.
8. Find out if you live in an earthquake-prone zone. Has your community made any provisions for earthquakes in its building codes? What plans, if any, does it have for coping with an earthquake, if one should occur?
9. Report on Mount St. Helens. What caused it to erupt?
10. Find out whether there are any soil conservation programs in your area. If so, learn how they operate.

Chapter

13

Weather

This meteorologist at Dulles International Airport provides information about atmospheric conditions.

13-1

The Atmosphere

Goals

1. To describe the different layers of the atmosphere.
2. To describe how the atmosphere affects people.

atmosphere

atmospheric pressure

barometer (buh RAHM uh tur)

Earth is surrounded by a blanket of gases about 900 km thick. This is Earth's **atmosphere**. The atmosphere helps to hold in Earth's heat. In addition, it protects Earth's surface from harmful rays of the sun. The atmosphere is difficult to detect on a clear day. You don't sense its presence unless the wind is blowing.

The gases of the atmosphere push down on Earth's surface with great force. Force applied on a given area of surface is called *pressure*. The force exerted by the atmosphere on a given area of Earth's surface is called **atmospheric pressure**, or air pressure. Atmospheric pressure at Earth's surface is sometimes referred to as *one atmosphere* of pressure.

You notice differences in air pressure when it changes rapidly. Think of what happens when you are on a fast-moving elevator. The plugged-up feeling in your ears is caused by a change in air pressure. Air pressure decreases as height above Earth's surface increases. The low pressure at high altitudes makes it necessary for airplanes to have pressurized cabins.

Atmospheric pressure is measured with an instrument called a **barometer**. In the 1600's, Otto von Guericke (GAY rih kuh) made a giant barometer to demonstrate air pressure. (See Figure 13-1.) He attached a 10-m pipe almost full of water to the side of his house. The bottom of the pipe was in a tub of water. When air pressure increased on the water in the tub, the water rose in the pipe. When the pressure decreased, the level in the pipe dropped. Von Guericke tried to predict weather as the water-level rose and fell. Figure 13-2 on the next page shows a modern barometer.

Figure 13-1 *Otto von Guericke's famous barometer.*

Scientists have found that the atmosphere consists of several layers. The layers differ in such physical properties as pressure and the types of gases. The most significant characteristic of each layer, however, is its pattern of temperatures.

troposphere (TROH puh sfeer)

The layer of the atmosphere closest to Earth is called the **troposphere**. (See Figure 13-3.) It contains about 90% of the total mass of the atmosphere. The troposphere is about 16 km thick at the equator. It thins toward the poles. The temperature of the troposphere is highest at Earth's surface. It decreases gradually with height. The lowest temperature in the troposphere is found at its upper boundary—the *tropopause*.

Almost all of Earth's weather occurs in the troposphere. For this reason, it is the part of the atmosphere that has the most noticeable effect on our daily lives.

stratosphere (STRAT uh sfeer)

ozone layer (OH zohn)

The layer above the troposphere is called the **stratosphere**. The stratosphere begins about 16 km above Earth's surface. It is about 32 km thick. In this layer there is a steady increase in temperature due to the absorption of high-energy ultraviolet rays from the sun. (Ultraviolet rays are invisible rays with higher energy than visible light.) You may have heard of the **ozone layer**. This is the part of the stratosphere where absorption of ultraviolet rays occurs. The ozone layer is a broad band of gas that extends through most of the stratosphere. A molecule of oxygen is O_2. When O_2 molecules are hit by ultraviolet rays, they break apart into separate oxygen atoms. Some of these atoms then combine with other O_2 molecules, forming ozone, O_3.

Figure 13-2 *(a) A typical modern barometer contains a box (b) with a very thin top and bottom. Inside is a spring. High pressure squeezes the box; low pressure allows the box to expand.*

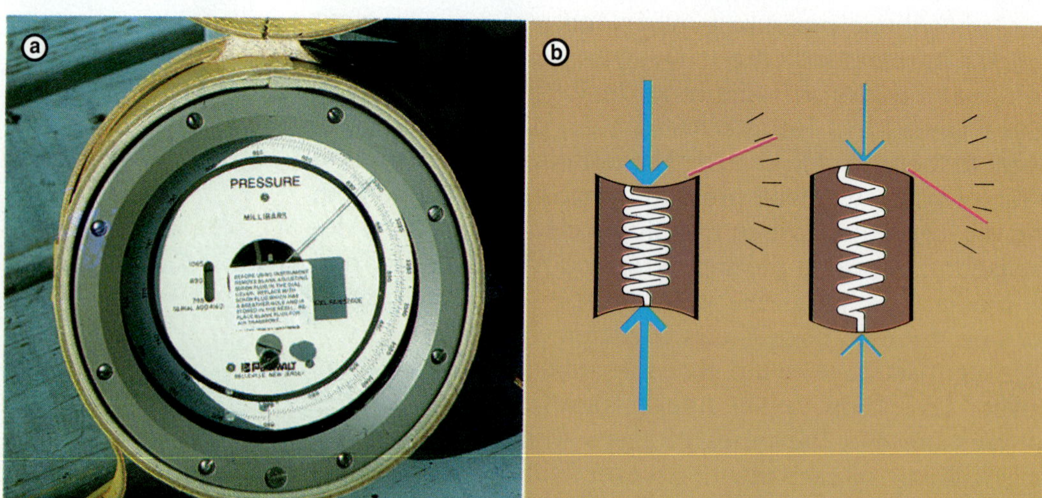

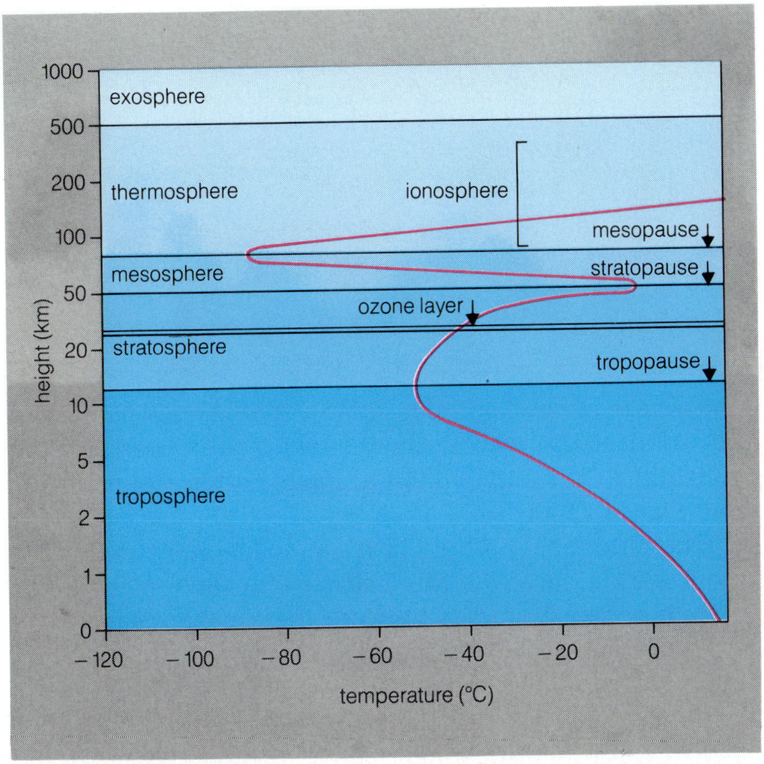

Figure 13-3 *The layers of the atmosphere. The curving line shows how much the temperature varies in the different layers.*

Scientists worry that the ozone layer can be destroyed by some of the gases used in spray cans. Some of these gases react with ozone. This reduces the amount of ozone in the ozone layer, allowing more ultraviolet rays to reach Earth's surface. Ultraviolet rays can be harmful to living things. For example, people who are exposed to large amounts of ultraviolet rays are more likely to get skin cancer. This is one reason people are cautioned to limit exposure to the sun, or to use special lotions that block out ultraviolet rays.

The upper boundary of the stratosphere is called the *stratopause*. Beyond this the temperature decreases again. The layer above the stratosphere is called the *mesosphere*. In the mesosphere the temperature decreases to −100°C. The *thermosphere* lies above the mesosphere. Here the temperature increases to several thousand degrees. Ultraviolet rays from the sun bombard the gases in the thermosphere. This ionizes (charges) them. These ionized gases form a layer within the thermosphere called the **ionosphere**. The ionosphere has electrical properties, and because of these, the

ionosphere
(eye AHN uh sfeer)

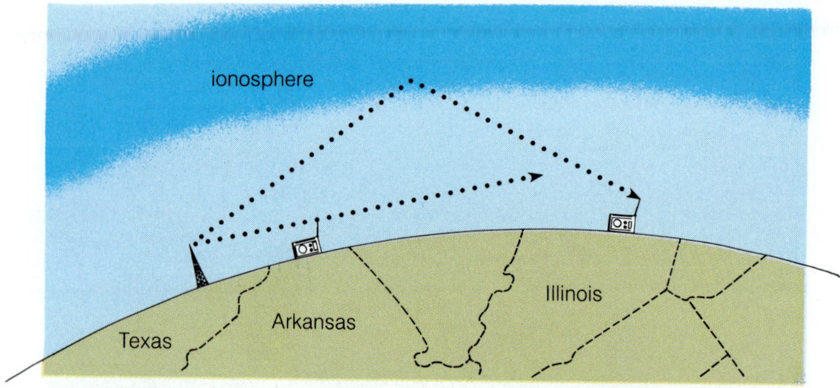

Figure 13-4 *People in Arkansas can listen to the Texas radio station at any time. Because Earth is curved, people in Illinois can't. They can get the radio waves only if the waves bounce off the ionosphere.*

ionosphere is used to relay radio communications around the world. Radio waves can be reflected off parts of the ionosphere. This allows them to travel farther around Earth. (See Figure 13-4.)

The outermost part of Earth's atmosphere extends from about 480 km out to 960 km. This part is called the *exosphere*. There is no real boundary between the exosphere and outer space. The gas particles just get farther and farther apart. Some may be kilometers apart.

The composition of the atmosphere varies. The two lower levels (the troposphere and stratosphere) contain the mixture of gases that we call *air*. The most abundant gases in air (by volume) are nitrogen (78%), oxygen (21%), argon (0.9%), and carbon dioxide (.03%). There are a number of other gases, including water vapor, in air. These gases, however, make up less than 0.1% of the air.

Air contains many solid particles also. Dust-sized bits of matter from industry and volcanic eruptions are examples. In addition, the ocean adds salt particles to the air as water evaporates from ocean spray. Substances that enter the atmosphere may travel far from their sources. They may eventually affect the weather.

Checkpoint

1. Name two ways in which the atmosphere affects people.
2. In what layer of the atmosphere does weather occur?
3. How could a change in the ozone layer affect life on Earth?
4. What is the most abundant gas in air?

13-2

Wind

Goal

To define wind and explain what causes it.

Wind is air in motion. You have probably observed how changeable wind can be. Sometimes there is hardly any wind. At other times, it is difficult to move against it. Wind is something you can't see, but you can see its work.

The major cause of wind is the unequal heating of Earth. Energy from the sun is not received equally all over Earth's surface. The greatest amount of energy is received where the sun's rays striking the surface are *vertical*, that is, at right angles to the surface. A vertical ray delivers more energy per square meter of surface than does a slanting ray. (See Figure 13-5a.) As a result, areas receiving vertical rays are warmer. Because Earth's surface is curved, some places receive vertical rays while other places receive slanting rays. (See Figure 13-5b.)

Other factors too affect how much of the sun's energy is received by Earth. As the sun's energy nears Earth, it is affected by the atmosphere. Some energy is reflected back by clouds. Some is absorbed in the stratosphere or by water vapor in the troposphere. And some is scattered by dust particles in the air. A little less than half of the sun's energy that reaches the atmosphere arrives at Earth's surface.

Figure 13-5 *(a) The same amount of light is spread over a larger area in the oval than in the circle. Thus a given area within the oval gets less light than the same area within the circle. (b) Vertical rays provide more energy to a given area than slanting rays do.*

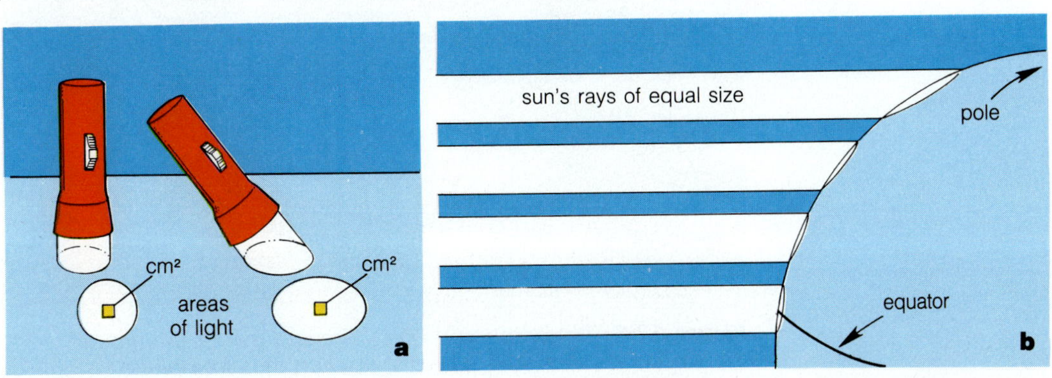

Some heat is added to the atmosphere by the ground. Energy from the sun comes to the surface of Earth in short, high-energy waves. These are absorbed by the ground. Earth, in turn, radiates long, low-energy waves back to the atmosphere. In this way, the ground adds a small amount of energy to the atmosphere. The long-wave energy radiated by Earth is absorbed by carbon dioxide in the atmosphere. This "trapped" energy causes the atmosphere to become warmer. This is called the **greenhouse effect**, because greenhouses work on the same principle. (See Figure 13-6.) The greenhouse effect insulates Earth, keeping it warmer than it would be without an atmosphere.

greenhouse effect

As a result of unequal heating, air is warmer in some places than in others. Warm air expands, becomes less dense, and rises. Cooler air then flows in to replace the warm air. The result is wind. The warmer air cools as it rises, becomes denser, and sinks again. This motion of air is called a *convection current* or *cell*. The principle of convection is used in some heating systems in homes.

Figure 13-6 *The greenhouse effect. The sun warms the ground, and the ground radiates long waves that warm the atmosphere.*

Figure 13-7a *In the daytime, cooler air from the ocean forms an onshore, or sea, breeze.*

Figure 13-7b *At night, cooler air from the land forms an offshore, or land, breeze.*

A common example of a local air convection cell occurs along the seashore. Land heats up faster in the daytime than water does. During the day, warmer air over the land rises and flows toward the water. Cooler air from over the water flows in to replace the warm air. The result is a cool onshore breeze, or *sea breeze*. (See Figure 13-7a.) At night, the reverse takes place. Land cools more rapidly than water. The warmer air over the water rises and moves toward the land. Cooler air from the land flows in to replace the warm air. The result is an offshore breeze, or *land breeze*. (See Figure 13-7b.)

Checkpoint

1. What is wind? What is the major cause of wind?
2. About how much of the sun's energy that reaches Earth's atmosphere actually arrives at Earth's surface?
3. Describe the greenhouse effect.
4. In which direction does the wind on a lakeshore usually blow during a sunny day? Why?

13-3

Wind Patterns

Goal

To describe wind patterns of Earth and their causes.

If Earth were not rotating, air heated at the equator would rise, spread toward each pole, and sink as it cooled. Cool air would then move along the surface toward the equator. The winds at the surface would always be from the north or south. (See Figure 13-8.) However, Earth's rapid rotation from west to east affects air currents. Earth's rotation causes air to be deflected in a clockwise direction in the Northern Hemisphere and in a counterclockwise direction in the Southern Hemisphere. This change in the path of movement caused by Earth's rotation is called the **Coriolis effect**.

Coriolis effect (kor ee OH lis)

In the region between 30°N and 30°S the surface winds are mostly from the east. These are the *trade winds*, or tropical easterlies. Between 30°N and 60°N and 30°S and 60°S, the surface winds are from the west, the so-called *prevailing westerlies*. Finally, between 60°N and 90°N and 60°S and 90°S, the winds are from the east—*polar easterlies*. (See Figure 13-9.) This is the global pattern of winds on Earth. Notice that winds are named according to the direction *from which* they blow.

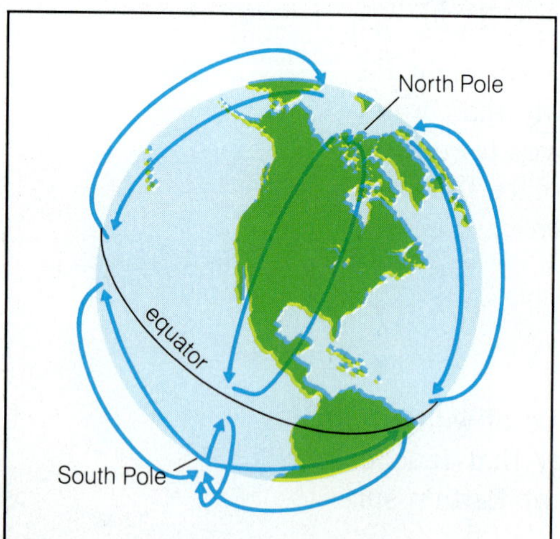

Figure 13-8 *If Earth did not rotate, the air would flow in this simple pattern. Air would rise near the equator and sink at the poles.*

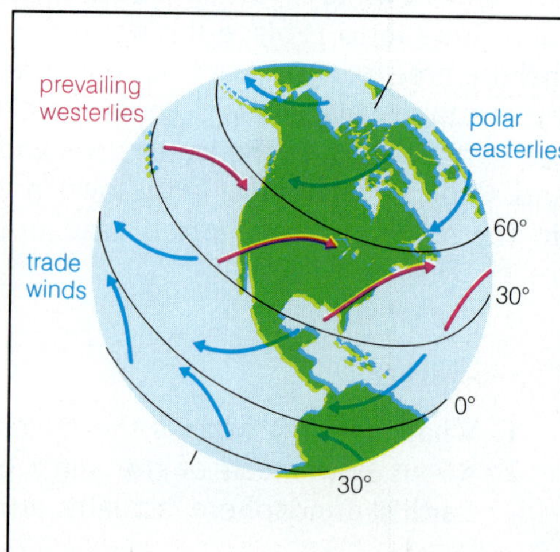

Figure 13-9 *Because Earth rotates, the temperature differences and rotation cause this pattern of easterly and westerly winds.*

Warm air rises because it is less dense than cool air. Cooler air, being more dense, exerts a higher pressure than warm air. Thus, when cool air flows in to replace rising warm air, the air flow is from a region of higher pressure to a region of lower pressure. Regions of high and low pressure, caused by unequal heating of the atmosphere, determine which way the wind will blow.

Wind blowing outward from a high pressure area in the Northern Hemisphere is deflected by the Earth's rotation. The wind moves clockwise. (See Figure 13-10.) Wind blowing inward around a low pressure area in the Northern Hemisphere is deflected in a counterclockwise direction. (In the Southern Hemisphere, these directions are reversed.) The greater the difference in pressure and the closer these systems are together, the faster the winds will blow.

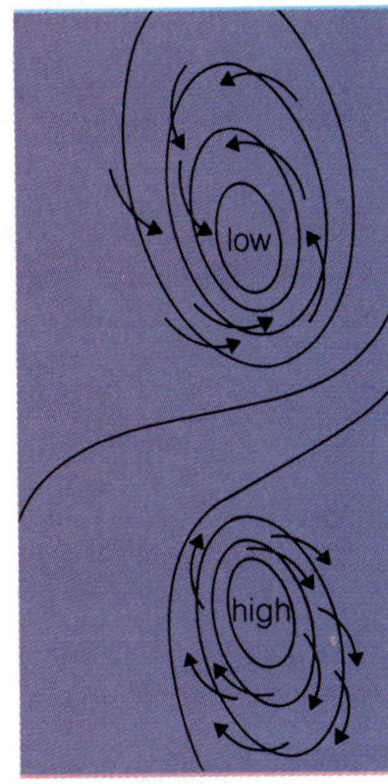

Figure 13-10 *Winds move outward and clockwise around a Northern Hemisphere high, and inward and counterclockwise around a low.*

High above Earth's surface are upper-level winds called **jet streams**. These are about 10 km above Earth's surface in the middle latitudes. The jet streams are caused by extreme differences in pressure at this level of the atmosphere. The major jet streams in the Northern Hemisphere blow from west to east.

High-level winds in the atmosphere are not slowed down by contact with Earth's surface. The air is less dense at this altitude, so winds can reach very high speeds—as great as 400 to 500 km/h.

jet streams

By observing the path of a jet stream, *meteorologists* (scientists who study the weather) can predict the paths of highs and lows on the surface. High pressure areas, or *highs*, on the ground are usually under highs at upper levels. Low pressure areas, or *lows*, on the ground are under lows in a jet stream. Pressure systems on Earth's surface behave like shadows of the jet streams above them.

Checkpoint

1. What is the name for the change in the path of movement of air caused by Earth's rotation?
2. How does air pressure affect wind?
3. What are jet streams? From which direction do they blow?
4. How does knowing the path of a jet stream help weather prediction?

TRY THIS

Using a child's toy pinwheel, how many different things can you do to make it turn? How does each method relate to winds on Earth?

13-4

Clouds and Precipitation

Goals

1. To describe different kinds of clouds and how they form.
2. To describe different types of precipitation.

dew point

condensation nuclei

Clouds form when warm, moist air rises. As it rises, the warm air expands and cools. As air cools it loses its ability to hold water vapor. As warm, moist air rises and cools, it reaches a temperature called the **dew point**. At this point the water vapor it is holding starts to *condense*, or change to liquid water. Water condenses on a solid surface. (You may have seen this happen on the outside of a glass of cold water.) Water in the atmosphere condenses on dust or salt particles. These are called **condensation nuclei**. Clouds are made of billions of water droplets that have condensed out of the air.

stratus clouds

cumulus clouds (KYOO myuh lus)

cirrus clouds (SIHR us)

There are many different kinds of clouds, but they are classified into three main types. (See Figure 13-11.) **Stratus clouds** are layered clouds that usually cover the whole sky. **Cumulus clouds** are billowy clouds that look like huge heaps of cotton. **Cirrus clouds** are high, thin, wispy clouds.

Figure 13-11 *Cloud types: (a) stratus, (b) cirrus, (c) cumulus.*

Once you know the three main cloud types, a few other terms are all you need to help you name clouds. For example, *nimbus*, which is Latin for "rain," can be added to cloud names. *Cumulonimbus clouds* (kyoo myuh loh NIM bus), also called *thunderheads*, are huge, towering clouds that form when a cumulus cloud expands into a vertical rain cloud. (See Figure 13-12.) Cumulonimbus clouds bring winds, lightning, showers, and sometimes hail. *Nimbostratus clouds* are sheetlike clouds that usually bring rain. The prefix *alto* is often used to identify clouds of intermediate height. *Altostratus* and *altocumulus* are "middle-level" clouds. A special term is used to describe stratus clouds at ground level—*fog*.

Cloud droplets are very tiny. It is estimated that it takes a million cloud droplets to make one raindrop! When enough cloud droplets accumulate, a drop becomes heavy enough to fall as *precipitation*.

The kind of precipitation that falls depends on many factors. The four major factors are the type of cloud from which it falls, the temperature of the cloud, the temperature of the air through which it falls, and the temperature of the ground.

Rain is probably the most common form of precipitation. It can fall from any cumulonimbus or nimbostratus cloud. *Sleet* is rain that has fallen through a cold layer of air and is frozen by the time it hits the ground. Sleet usually originates in high stratus clouds. *Drizzle* is the slow fall of very

Figure 13-12 *Cumulonimbus clouds have strong vertical currents. They cause rainstorms, thunder, and lightning.*

Figure 13-13 *If you cut one of these hailstones open, you would be able to see its layers.*

tiny drops of water. They sometimes stay suspended in the air as a mist. Drizzle forms in low-lying stratus clouds. If the temperature of the ground is below freezing, the drizzle freezes as it hits the ground.

If the clouds are high and cold enough, the water freezes in the cloud and starts out as *snow*. If it melts on the way down, it lands as rain. If the temperatures are low enough all the way down, the precipitation remains snow.

Hail is another form of frozen precipitation. It usually accompanies thunderstorms and comes from cumulonimbus clouds. If the clouds are piled up extremely high, the winds inside the cloud form strong convection cells. In these cells, raindrops are lifted to heights where it is so cold they freeze into hailstones. Hailstones can be tossed up and down within the cloud many times. They pick up layers of ice as they move. When they get heavy enough, they fall. The largest recorded hailstone was 19 cm in diameter, about the size of a volleyball. A hailstorm can do a tremendous amount of damage in a few minutes. Hailstorms often flatten growing crops such as grain.

The amount of precipitation varies around the world. Some places, such as the Atacama Desert in Chile, have never recorded any rainfall. In contrast, Mt. Waialeale in Hawaii receives about 1200 cm each year!

Checkpoint

1. What causes a cloud to form?
2. What are the three main types of clouds?
3. What is another name for a cumulonimbus cloud?
4. What is the difference between sleet and snow?
5. What causes hailstones to vary in size?
6. What kind of cloud produces drizzle?

TRY THIS

Find the dew point of the air in the classroom by slowly adding small pieces of ice to a container of water. Use a shiny can for the container, if possible. Gently stir the mixture with a thermometer. Record the temperature when drops of water begin to condense on the outside of the can. Repeat this several times to make sure you get the correct temperature. This temperature is the dew point.

13-5

Storms

Goal

To describe different kinds of storms.

Storms can be exciting to observe. Victims of powerful storms, however, might not call them exciting. Tremendous amounts of energy, in the form of wind, are released in storms. On seacoasts this wind can cause huge waves that cause damage along the shore. By understanding how storms arise and being able to predict their behavior, people can save many lives.

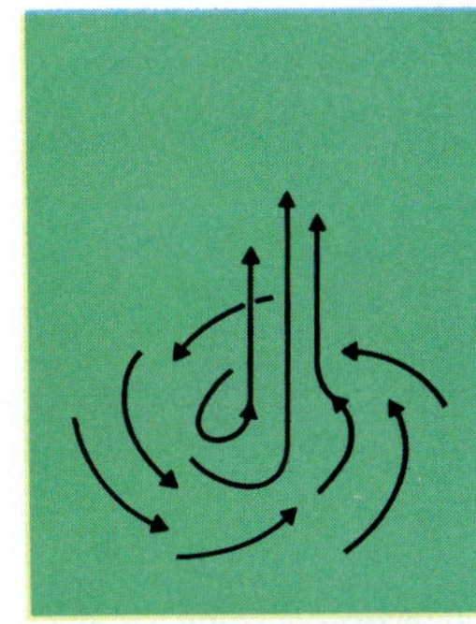

Figure 13-14 *Winds in a cyclone curve inward and upward.*

Most violent storms are caused by circulating winds blowing inward around a low pressure center. These regions of low pressure with their circulating winds are called **cyclones**. In a cyclone, the winds move upward as they circle in toward the center of the low. (See Figure 13-14.) The severity of these storms depends on the speed of the winds.

cyclones
(SY klohnz)

A *tornado* is a small cyclone with a diameter less than a few hundred meters. Tornadoes travel along the ground for only about 25 km. They generally contact the ground for only a few minutes, but often do much damage in that time. They are the most powerful storms on Earth. The winds inside a tornado can reach speeds up to 400 km/h. (See Figure 13-15.)

Figure 13-15 *A tornado touching the central plains of the United States. Tornadoes usually touch ground for only a few minutes.*

Figure 13-16 *Lightning often creates spectacular displays.*

If you are ever caught in tornado weather, keep tuned to radio and television for watches and warnings. If you're inside, stay away from windows and go to the lowest level, preferably to a closet or bathroom. If you're in a car, get out and lie flat in a ditch or ravine. Cars, mobile homes, and other vehicles can be picked up by a tornado. Following these simple rules could save your life. Tornadoes can be predicted, but not as accurately as hurricanes.

Hurricanes are the same kind of storm as tornadoes, but they cover more territory. They occur only over or near water, and they last much longer. A hurricane is shaped like a doughnut, with a "hole" in the middle of the circulating winds. During a hurricane, strong winds will blow for several hours. Then it is quiet as the "eye" or "hole" passes. As the storm moves, winds blow strongly again from the opposite direction. Hurricanes are 480–960 km in diameter, and their winds can approach 320 km/h. (Compare the size of a hurricane with the distance from Chicago to Nashville—about 750 km.) The eye of a hurricane can be 24–40 km across (less than the distance run in a marathon).

Hurricanes begin in tropical oceans. There temperatures are high and large amounts of water are evaporated. During evaporation, energy is stored in the water vapor. When this water vapor condenses in clouds, the stored energy is released, producing wind. When a hurricane reaches land it starts to lose its power because there is much less water vapor to condense.

A more common type of storm is the *thunderstorm*. Thunderstorms originate in cumulonimbus clouds. There the winds are strong, but not as great as the winds in hurricanes or tornadoes. Thunderstorm winds are accompanied by thunder and lightning. (See Figure 13-16.) Lightning can be within parts of the cloud, or between cloud and land, or between two clouds. Thunder is sound produced by the violent expansion of air being heated by the lightning. Heavy rain often falls during a thunderstorm.

To protect yourself from lightning, try to stay indoors during a thunderstorm. Stay away from open windows and doors. Do not touch metal pipes, including faucets. Try not to use electric appliances, such as a television or an iron. Outdoors, stay away from tall trees, poles, or power lines. Never swim during a storm. If you are in a boat, head for land. If you are in a car, stay there, but do not touch any metal parts. The metal of the car will conduct any electricity to the ground.

TRY THIS

Blow a small paper bag up and twist the top so the air will stay inside. Holding the top of the bag in one hand, quickly break the bag between your hands. What did you observe? How is this similar to thunder?

Checkpoint

1. Describe a cyclone.
2. What do cyclones, tornadoes, and hurricanes have in common?
3. Why does a hurricane start to die out when it reaches land?

13-6

Weather Prediction

Goals

1. To discuss some factors that can be used to predict weather.
2. To describe how moving air masses affect the weather.

Farming, aviation, and construction are industries that rely on accurate weather predictions. Weather forecasting is a big business. The movement of weather patterns on Earth's surface is determined by winds high in the atmosphere. Thus, data from upper layers of the atmosphere are collected by weather balloons and satellites. These data, collected in all areas of the country, are sent to the National Weather Service. Forecasts are then sent to newspapers and radio and television stations for local use. In small communities, some individuals collect weather information and send it to area forecasters. Computers have speeded up and improved the collection and interpretation of weather data.

Some data are needed to successfully predict weather. These are temperature, atmospheric pressure, humidity, wind speed and direction, precipitation, cloud type, sky condition, and dew point. Because there are so many factors, weather prediction is a complicated process.

air mass

The first step in weather prediction is identifying the size, location, and movement of an air mass. An **air mass** is a huge volume of air that has the characteristics of the area over which it started out. For instance, an air mass that originates over the ocean contains a lot of water vapor. One that starts out over land is dry. An air mass that originates in the northern latitudes (Northern Hemisphere) is cold. One that starts out in the south is warm. These characteristics can be mixed. Thus, cold-dry, cold-wet, warm-dry, or warm-wet air masses occur.

isotherms (EYE suh thurmz)

Air masses do not stay in one place very long. Earth's rotation sets them in motion. By studying **isotherms**, lines on a weather map that connect areas of equal temperature, the boundaries of air masses can be located.

When one air mass overtakes another, the leading edge

of the approaching air mass is called a *front*. Each type of front has a typical weather pattern associated with it. If a cold air mass moves into a warm air mass, a **cold front** is formed. (See Figure 13-17a.) Because cold air is denser than warm air, cold air in motion has more momentum. The cold air pushes the warm air up. This upward motion causes cumulonimbus clouds to form, and often a short period of precipitation follows. After the cold front passes, the weather becomes clear, dry, and cooler.

cold front

If a warm air mass overtakes a cold air mass moving in the same direction, a **warm front** is formed. (See Figure 13-17b.) The warm air slowly moves up and over the denser cold air. High cirrus clouds announce the approach of a warm front. As time goes by, the clouds get thicker. A long period of precipitation usually occurs. After the front passes, the weather takes the characteristics of the warm air mass.

warm front

Figure 13-17a *A diagram of a cold front, with cold air pushing up warm air.*

Figure 13-17b *A diagram of a warm front, showing warm air sliding over colder air.*

occluded front

An **occluded front** results when a warm air mass is lifted upward as two cold air masses meet. The warm air is "cut off" from the ground by cold air. (*Occluded* means "cut off.") Occluded fronts produce long periods of precipitation. Can you remember a period when it rained for several days? Still another kind of front is a *stationary front*. This is caused by a front that stops moving. It can cause precipitation that lasts as long as the front remains stationary. (Figure 13-18 shows the symbols used on weather maps for these four kinds of fronts.)

relative humidity

Knowing how much water vapor is in the air is useful in predicting weather. An important measure of water vapor is **relative humidity**. The amount of water vapor in the air compared to the maximum amount the air *can* hold at a given temperature is called relative humidity. It is usually expressed in terms of percent. If the relative humidity is 100%, air is said to be *saturated*. It cannot hold any more water vapor at that temperature. If the temperature of the air goes down, water vapor will condense out of the air. If the temperature of the air goes up, the relative humidity will be lower. The warmer air can hold more water vapor.

isobars
(EYE suh barz)

In order to predict wind direction, high and low pressure areas must be located. Lines on a weather map that connect areas of equal pressure are called **isobars**. The pattern of the isobars shows where the centers of the highs and lows are located. (See Figure 13-19.) Remember, in the Northern Hemisphere winds move counterclockwise around a low and clockwise around a high. By knowing the wind direction, you can better predict the type of weather for your area. Wind direction is measured with a *wind vane* and wind speed with an *anemometer*. (See Figure 13-20.)

High pressure areas are characterized by clear skies and drier air. The air at the center of a high is denser than the surrounding air. In addition, it is sinking toward the surface.

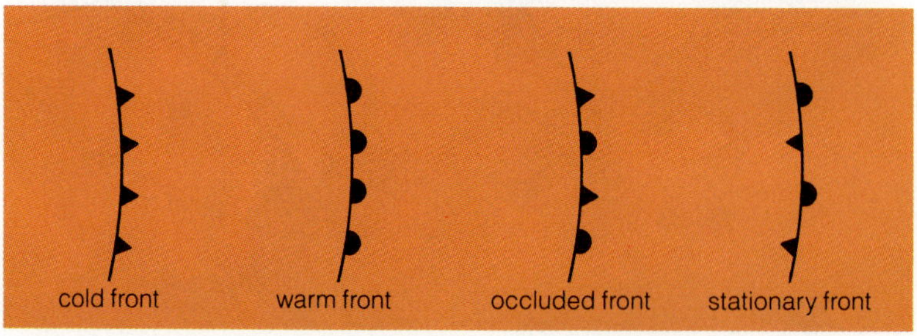

Figure 13-18 *The symbols for weather fronts.*

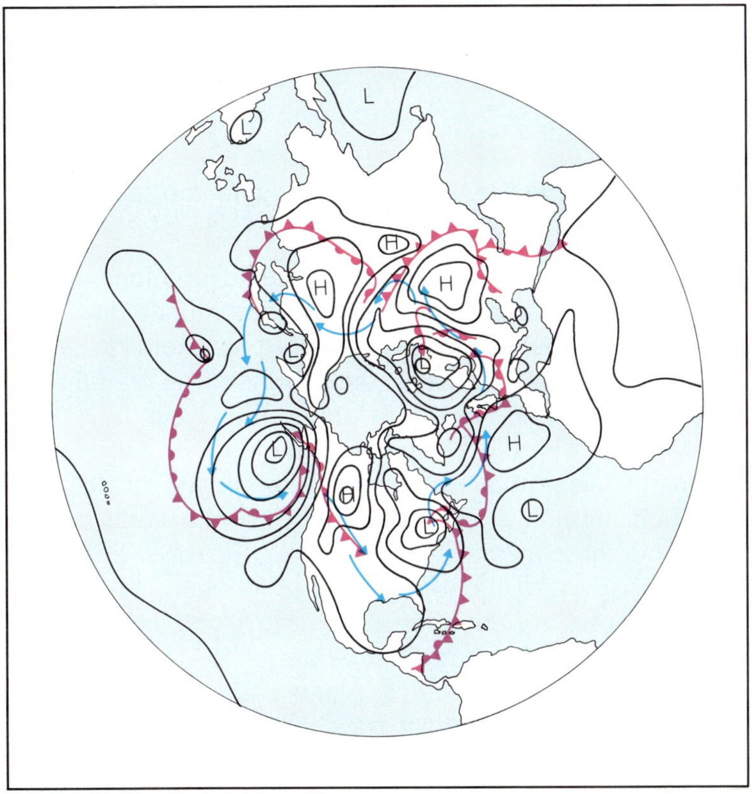

Figure 13-19 *On this weather map, the black isobars outline highs and lows. The blue arrows show upper air currents, which are above the surface weather pattern. Highs and lows tend to move along with the upper air currents. Fronts are shown in red.*

Figure 13-20 *(a) The wind vane shows wind direction.*

(b) Wind turns the cups of the anemometer to register wind speed.

A low pressure region has warmer air and is usually cloudy. When the barometer is steady, you can expect weather to stay the same. When the barometer is either rising or falling, you can expect a change in the weather.

Checkpoint

1. A hot, dry air mass would have its origin over what part of Earth?
2. What causes an air mass to move?
3. What is the name of the lines on a weather map that connect points of (a) equal temperature, (b) equal pressure?
4. What weather pattern is typically associated with (a) a cold front, (b) a warm front?

13-7

ENRICHMENT

Climate

Goal

To define climate, climate zones, and the factors that determine the climate of an area.

climate

Climate, the average weather in an area over a long period of time, affects living things. Climate determines the types of plants and animals that can survive. It also helps to determine the jobs people have, their forms of recreation, the houses they live in, and the clothing they wear.

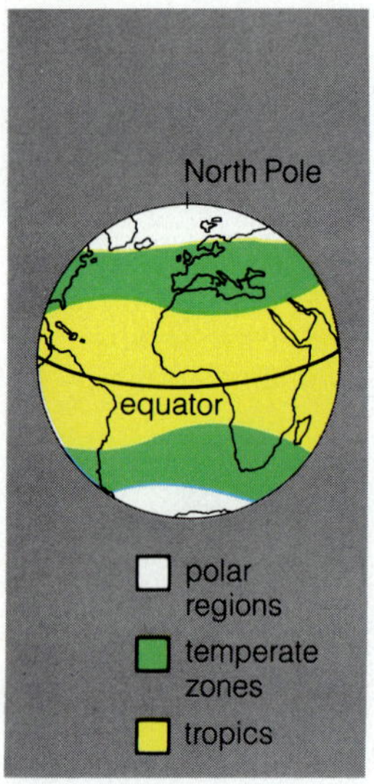

Figure 13-21 *Earth's major climate zones.*

Many factors affect climate. Temperature, however, is the major factor that determines the climate of an area. Based on temperature alone, climates of the world are divided into three major zones: the *tropics*, the *middle latitudes* or *temperate zones*, and the *polar regions*. (See Figure 13-21.)

The tropics are bounded on the north and south by the line of 18°C average daily temperature. This is the lowest temperature at which tropical plants can grow. The tropics do not have much contrast between summer and winter. They receive large amounts of energy from the sun year around.

The middle latitudes, or temperate zones, are regions where hot air masses from the equator meet cold air masses from the poles. The result is a wide variety of weather. Most of the world's population lives in the temperate zones.

The polar regions have long, cold winters. They do not support much life. The average temperature in the warmest month is only 10°C, compared to 22°C for the temperate zones. Little energy is received from the sun, due to the low angle of the sun's rays.

A second factor affecting the climate of an area is *topography* (tuh PAHG ruh fee), the shape of the landscape. For example, the higher the *altitude* (height above sea level) of a place, the cooler the climate. Air at high altitudes is less dense. Therefore, it has less of an insulating effect.

windward

Topography combines with a third factor, precipitation, in determining the climate of an area. If there are mountains, the **windward** side—the side where the wind hits first—usually has more rainfall. The air is forced up on the windward side. Then it flows over the top of the mountains.

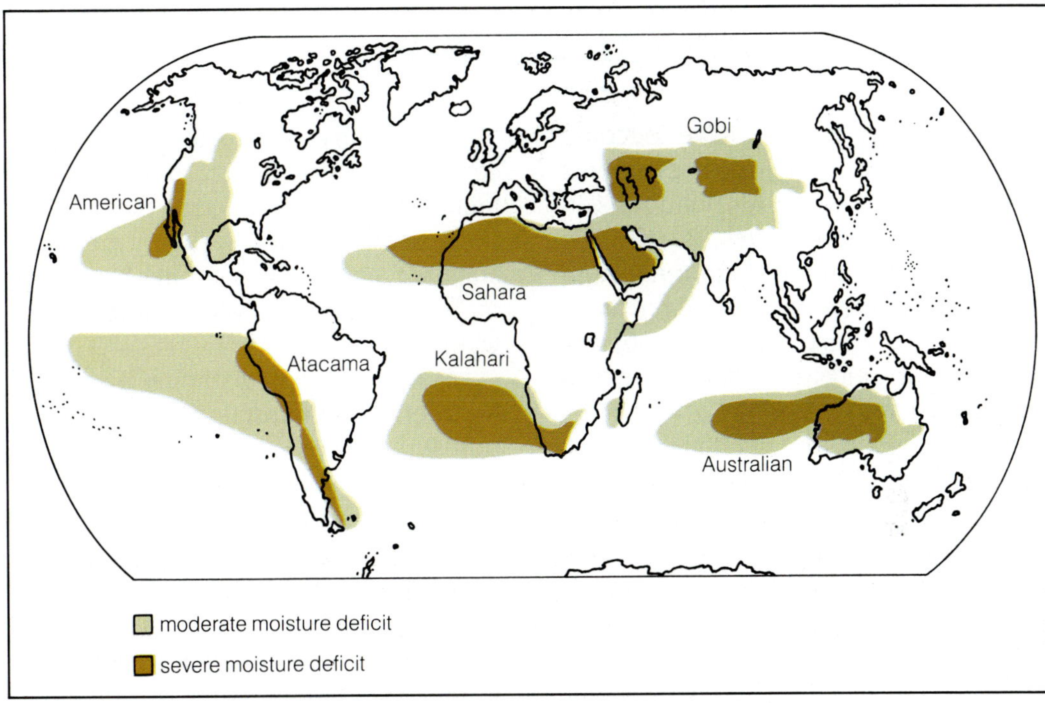

Figure 13-22 *The regions of dry climate around the world. Each of the dry regions contains a desert that is named on the map.*

As warm air containing moisture moves upward, it cools. Cool air can hold less water vapor, so the water condenses and falls as precipitation. On the **leeward** side—the side of the mountain away from the wind—the air dries out as it descends. Deserts are usually found on the leeward sides of mountains. (See Figure 13-22.)

leeward

The interiors of large continents are drier than the coasts. Because of this, there are great temperature differences between summer and winter. Within the interior of the Europe-Asia continent, for example, there is an average annual temperature range of 60°C!

Checkpoint

1. Define climate.
2. Name the three major climate zones of the world.
3. List three factors that influence the climate of an area.
4. In a mountain area, where would the most rain fall?

Air Quality

In recent years people have become increasingly aware of health problems caused by polluted air. Air pollution can cause watery, burning eyes, raw throats, and painful breathing. Extremely high levels of pollution can endanger lives. People with asthma, or other problems involving the lungs or heart, are especially at risk.

According to the Pollution Standards Index (P.S.I.), air quality is classified as good, moderate, or unhealthful. The last category is further subdivided into alert, warning, and emergency conditions. At the emergency level, lives can be in danger.

There are two basic sources of air pollution—natural events and human activities. Erupting volcanoes are natural events. They add particles and poisonous gases, such as carbon monoxide and sulfur dioxide, to the air.

Vehicles and industrial plants are major sources of the air pollution produced by human activities. Pollutants include carbon monoxide, ozone, sulfur dioxide, and carbon dioxide.

Air quality is affected by weather conditions. Pollutants build up to higher levels in air masses that are stationary, or stagnant. A highly polluted, stagnant air mass is most likely to occur on hot, humid days when there is little wind.

People cannot do much about the weather or the natural sources of pollution. But choices can be made about the pollution caused by human activities. In many places people are trying to decrease the amount of pollution produced by their activities.

Pollen-producing plants also can affect air quality, although pollen is not usually considered to be a "pollutant." A lot of pollen in the air (a high pollen count) can cause severe problems for people who have allergies. Pollen counts are usually highest in mid-central North America. They are lowest on the Pacific coast and in southern Florida. Pollen counts are highest during the day—particularly in dry, windy weather in the summer. They are lowest after a good rain.

Understanding The Chapter

13

Main Ideas

1. The atmosphere is a mixture of gases that exerts pressure on Earth's surface.
2. Weather occurs in the troposphere. The ozone layer is in the stratosphere.
3. The major cause of wind is the unequal heating of Earth.
4. Unequal heating of the atmosphere causes high and low pressure air masses to form.
5. Earth's rotation affects air currents and is responsible for the pattern of winds on Earth.
6. The jet streams affect the path of surface winds.
7. Clouds form when water vapor condenses in the atmosphere. The three main types of clouds are stratus, cumulus, and cirrus.
8. The kind of precipitation depends on the type of cloud from which it falls. Other factors are the temperature of (1) the cloud, (2) the air through which the precipitation falls, and (3) the ground.
9. Most severe storms are caused by the winds around a low pressure center.
10. Warm and cold fronts have typical weather characteristics that accompany them.
11. Climate is the average of weather in a region over a long period of time.
12. The major climate zones are the tropics, the temperate zones, and the polar regions.

Vocabulary Review

From the following list, choose the term that best completes each of the statements. Write your answers on a separate piece of paper.

air mass	dew point
atmosphere	greenhouse effect
atmospheric pressure	ionosphere
barometer	isobar
cirrus clouds	isotherm
climate	jet streams
cold front	leeward
condensation nuclei	occluded front
Coriolus effect	ozone layer
cumulonimbus clouds	relative humidity
cumulus clouds	stratosphere
cyclones	stratus clouds
	troposphere
	warm front
	windward

1. The __?__ is the portion of the atmosphere that contains the ozone layer.
2. The change in the path of wind movement caused by Earth's rotation is called the __?__.
3. The instrument used to measure air pressure is called a(n) __?__.
4. Upper-level winds high above Earth's surface are the __?__.
5. The temperature at which water vapor condenses is the __?__.
6. The atmosphere traps heat being radiated from Earth. This is called the __?__.
7. A thunderstorm is most likely to occur if __?__ are overhead.

8. The side of a mountain where deserts are found is the ? side.
9. Weather occurs in the layer of the atmosphere called the ?.
10. Winds circulating around low pressure centers form ?.
11. ? are high, thin, wispy clouds.
12. ? is the average of weather in an area over a long period of time.

Chapter Review

Write your answers on a separate piece of paper.

Know The Facts

1. The atmosphere acts as a blanket for Earth because ?. (it is not very thick, it holds in the heat, it covers the whole Earth)
2. Warm air rises because ?. (it is less dense than cold air, it is more dense than cold air, it can hold less water vapor than cold air)
3. The prevailing winds near the equator are called ?. (prevailing westerlies, trade winds, polar easterlies)
4. Winds always blow ?. (from a low to a high pressure area, back and forth between highs and lows, from a high to a low pressure area)
5. Condensation nuclei are ?. (in the middle of a convection cell, particles in the atmosphere, necessary for a dust storm)
6. The billowy clouds that look like cotton are ?. (cumulus clouds, cirrus clouds, stratus clouds)
7. The primary reason a hurricane loses its strength when it reaches land is that ?. (the winds die down, condensation of water vapor slows down, it cools down)
8. Thunder is caused by ?. (two clouds hitting each other, hail being tossed around in a cloud, rapidly expanding air)
9. The most abundant gas in air by volume is ?. (nitrogen, oxygen, carbon dioxide)
10. Sleet is precipitation that freezes ?. (in a cold cloud, as it falls to the ground, after it hits the ground)
11. An air mass originating over the Gulf of Mexico would be ?. (cold and dry, warm and dry, warm and wet)
12. A line on a weather map that connects areas of equal temperature is called an ?. (isobar, isotherm, isocline)
13. Oceans have a modifying effect on climate because ?. (water is slow to change in temperature, there is a lot of wind, the humidity is low)

Understand The Concepts

14. How do jet pilots make use of the jet streams?
15. What makes warm air less dense than cold air?
16. What would be the characteristics of an air mass that originated over the Arctic Ocean?
17. Why is it so cold at the poles, even though the sun never sets at the poles for part of the year?
18. Why is there such a variety of weather in the middle latitudes?
19. What causes the great range of tem-

perature between summer and winter in central Asia?

20. How does the curved surface of Earth contribute to the unequal heating of the surface?
21. Why would you not expect to find clouds over desert areas?

Challenge Your Understanding

22. Is the coldest month in the Northern Hemisphere the same as the coldest month in the Southern Hemisphere? Explain.
23. How can the relative humidity of the outside air remain high when the temperature is low?
24. Make a rough drawing to show why your ears "pop" when you change your elevation in the mountains, in an elevator, or in an airplane. Why does swallowing make your ears feel normal again?
25. Why is an apparent ring around the moon an indicator of precipitation?
26. Why do pilots avoid flying through large cumulus clouds?
27. How can you account for the fact that some areas on the equator have snow?
28. With what kind of front would drizzle and fog be associated?

Projects

1. Research what is being done in your community to monitor and control air quality.
2. Research the importance of the ionosphere in shortwave radio communications.
3. Learn all you can about tornadoes.
4. Report on the process of cloud seeding. Does it work? What are some arguments for and against seeding?
5. Is it a good idea to be able to control hurricanes? Debate this question or have a panel discussion.
6. Research the effect of jet planes on cloud formation in the upper atmosphere.
7. Make a model showing how the Coriolis effect works.
8. Build a model to demonstrate the greenhouse effect.
9. Report on how satellites are used in meteorology.
10. Research the effect that weather has on human behavior.
11. Ask various people for their weather superstitions. Example: "If the leaves on the trees are turned upside down, it is going to rain." After collecting several, analyze each for its accuracy.

14

Oceans

This is one of the Society Islands in the Pacific Ocean.

14-1

Composition of Seawater

Goals

1. To discuss the composition and temperature of seawater.
2. To describe a method used to study seawater.

Earth is unusual because it is the only planet in our solar system with so much liquid water on its surface. Almost 75% of the planet is covered by water. These large areas of water are called **oceans**. Since water circulates freely throughout the oceans, they are often called simply "the ocean." Smaller areas of the ocean, sometimes partly surrounded by land, are called **seas**.

oceans

seas

Have you ever noticed how easily you float in salt water? Ocean water has a higher salt content than the water in most rivers and lakes. The salty water is denser than fresh water, so it holds you up better. (See Figure 14-1).

Rivers carry dissolved salts picked up by the water as it moves across and through the land. These salts eventually end up in the ocean. The most abundant salt in the ocean is

Figure 14-1 *A vacationer floating in the Dead Sea, Israel.*

sodium chloride, or table salt. The amounts of other salts in ocean water, or seawater, are given in Table 14-1. The amount of salt in seawater is called its **salinity**. The average salinity of seawater by mass is about 35 grams per kilogram of water, or 3.5%. A large pinch of salt in a drinking glass full of water produces a similar salinity.

salinity
(suh LIN uh tee)

The salinity of seawater depends on several factors. If fresh water is added, by precipitation such as rain or snow, for example, the salinity decreases. If water evaporates, the salt is left behind and the salinity increases. Where evaporation is greater than precipitation, the salinity is high. For example, **inland seas**, seas that are almost completely surrounded by land, have higher salinity because evaporation exceeds precipitation. (The Red Sea and the Mediterranean Sea are considered inland seas.) Temperature too can affect salinity. Warm water can hold more salt than cold water. So warmer ocean water has a higher salinity than cold water.

inland seas

Seawater also contains gases, such as nitrogen, oxygen, and carbon dioxide, dissolved from the atmosphere. The gases are mixed with the water by waves at the surface. They are used by plants and animals in the ocean.

The temperature of ocean water varies from place to place. It ranges from 0°C to 30°C (from freezing to quite warm). In general, it is warmest at the surface, where the effect of heat from the sun is greatest, and coldest at the ocean floor. Temperature varies with latitude, also. Ocean water is warmer near the equator than near the poles.

Collecting ocean water samples at various depths and measuring the temperature at those depths was once a diffi-

Table 14-1: Dissolved Salts in Seawater

Salt	g/kg of Water
sodium chloride	27.2
magnesium chloride	3.8
magnesium sulfate	1.7
calcium sulfate	1.3
potassium sulfate	0.9
calcium carbonate	0.1
magnesium bromide	0.1
Total	35.1

Figure 14-2 *This scientist is checking a Nansen bottle that has returned to the surface.*

cult task. Early *oceanographers,* scientists who study oceans, had to lower bottles and thermometers into the water. This task was made much simpler by the invention of the **Nansen bottle**. A thermometer is attached to the outside of each Nansen bottle. Several bottles are hung along a wire and lowered into the water. Each bottle can sample a different depth. When the bottles are in place, a small weight, or *messenger,* is sent down the wire. It trips the first bottle, causing it to turn over. In the process, the valves on the bottle close to collect a water sample. At the same time, the temperature is permanently recorded by the thermometer. As each Nansen bottle is tripped, it releases another messenger that slides down the wire to trip the next bottle. (See Figure 14-2.) Samples from many different places have shown that the composition of ocean water varies around the world.

Nansen bottle

TRY THIS

Dissolve 35 g of table salt in 1 L of water. Fill one small plastic bag with the salt solution and another with the same amount of tap water. Seal the bags with twist ties. Place the bags *gently* on the surface of a bucket or sink full of tap water. Explain what happens.

Checkpoint

1. How much of Earth's surface is covered by water?
2. What causes seawater to be salty?
3. Why does the Red Sea have a high salinity?
4. What are two main reasons for variation in ocean temperatures?
5. What is a Nansen bottle?

14-2

Waves

Goals

1. To identify the parts of waves.
2. To describe how a wave moves.
3. To explain what causes waves.

If you drop a hollow rubber ball into a river or stream, the current carries it along. Suppose you drop the ball into a lake or pond and watch it for a while. You will see that the ball moves up and down as waves pass.

crest

trough (trawf)

Waves can be described by defining the parts of a wave. (See Figure 14-3.) The highest point in a wave is the **crest**. The lowest point is the **trough**. The distance between two crests (and between two troughs) is called the *wavelength*. *Wave height* is the distance between the top of a crest and the bottom of a trough.

The water in a wave does not move forward. The energy of the wave travels forward, but the water moves up and down, "in place." People have studied the motion of water in a wave by hanging objects at various levels in the water. Objects floating on the surface travel in circles. The diameters of the circles equal the wave height. As the depth increases, the circular motion of the objects gets smaller and smaller. At a depth equal to half the wavelength, no motion can be seen. (See Figure 14-4a.)

Figure 14-3 *Locate the crest, the trough, the wavelength, and the wave height in this illustration.*

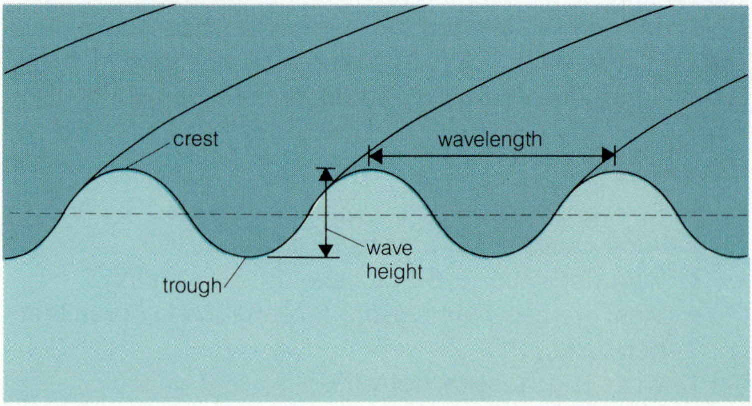

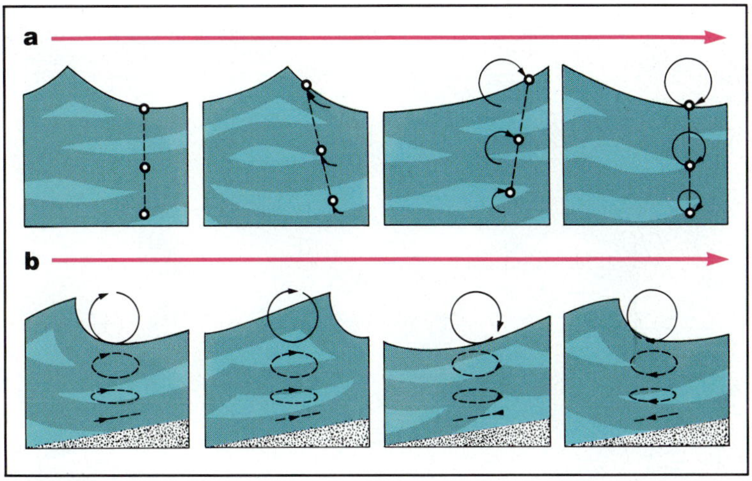

Figure 14-4 *Particle movement in waves. (a) In deep water, the circular paths of particles get smaller with greater depth. (b) In shallow water, the bottom of a wave drags on the ocean floor, causing its circular path to flatten out.*

As a wave approaches the shore, the bottom of the wave "drags" on the sloping beach and causes the circular motion to flatten out. This, in turn, forces the top of the wave to fall forward. This process produces *breakers.* (See Figure 14-4b.)

Waves do much work carving out features on the shoreline. (See Figure 14-5.) You may wonder where waves get all their energy. Most of this energy comes from the wind. Wave size depends on three factors: the speed of the wind, how long the wind has been blowing, and the distance over

Figure 14-5 *Waves can carve out many forms in a rocky shoreline. How many different features do you see?*

Figure 14-6 *Waves become parallel to the shore as they approach.*

which the wind has blown, the *fetch*. Where the wind blows over a large area for long periods of time, waves can get as high as 12 m!

Spectacular and damaging waves called *tsunamis* (tsoo NAH meez) are caused by earthquakes on the ocean floor. A tsunami can travel as fast as 750 km/h (as fast as a jet plane) and have a wavelength as great as 60 km! On a boat in the middle of the ocean, you would not notice a tsunami. But when it hits shore, a tsunami can be over 30 m high.

Have you ever had the opportunity to watch long waves approach a shore? You may have observed that sometimes the waves that are farther out come in at an angle. Yet when they reach the shore, they hit in parallel rows. (See Figure 14-6.) This happens because the part of the wave in shallow water tends to slow down as it drags on the bottom. The part of the wave in deeper water maintains its speed and catches up with the other end. This is like what happens in a game of "crack the whip."

TRY THIS

Build a model shoreline in a stream table and then add water. Make waves in the water by tapping on the surface with your finger. Observe the work done to the shoreline. Try changing the shape of the shoreline. Also try making larger waves.

Checkpoint

1. Define the following terms: crest, trough, wavelength, wave height.
2. Describe the motion of water in a wave.
3. What causes breakers on a shoreline?
4. What is the major cause of waves?

14-3

Tides

Goals

1. To explain tides and their causes.
2. To describe the factors that influence the height of tides.

Tides may have little meaning for people who do not live on a seacoast. If you live there, you know that the level of the sea rises and falls at regular intervals throughout each day. It is important to know the exact times of these changes when you are planning a beach picnic or a hike to find seashells. Knowledge of tidal activity is important when planning buildings and other structures near the shore. The design of docks and harbors also depends on knowledge of tidal behavior.

The tides are caused by the gravitational pull of the moon and sun on the water. The moon has more influence on the tides than the sun because the moon is closer to Earth.

As Earth rotates, the amount of pull by the moon at any one place on Earth changes. The moon's pull is strongest on the side of Earth that faces it. (See Figure 14-7.) The water there bulges outward, and a high tide occurs. At the same time, there is a high tide on the side of Earth farthest away from the moon. The solid part of Earth is between the moon and the water on the side away from the moon. The moon tends to pull Earth away from the water. That causes the water on the side away from the moon to bulge outward, too. High tide occurs about once every 12 hours.

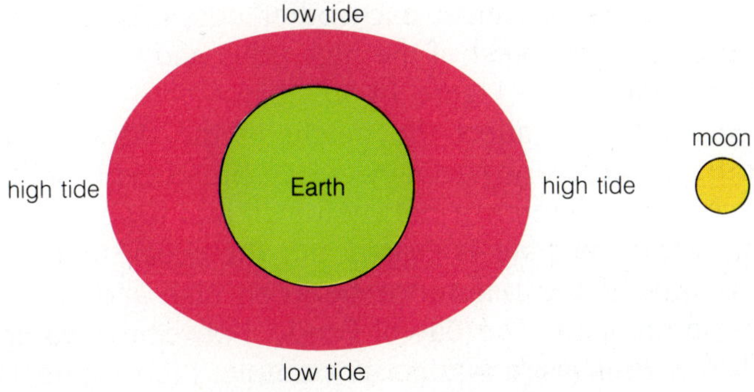

Figure 14-7 *The high tides on Earth are in line with the moon. Low tides occur at a 90° angle to that line.*

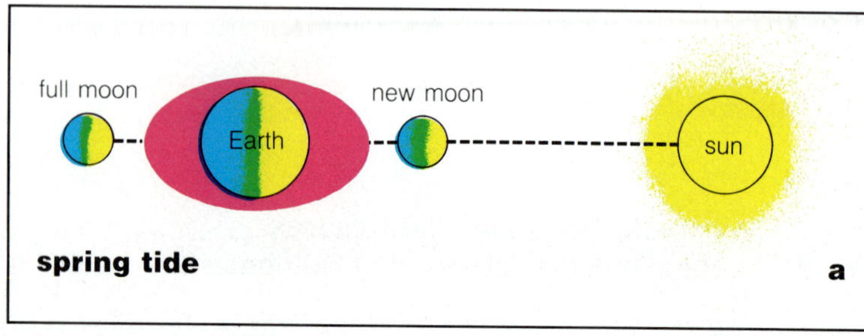

Figure 14-8a *A spring tide occurs when the sun and moon are in line with Earth.*

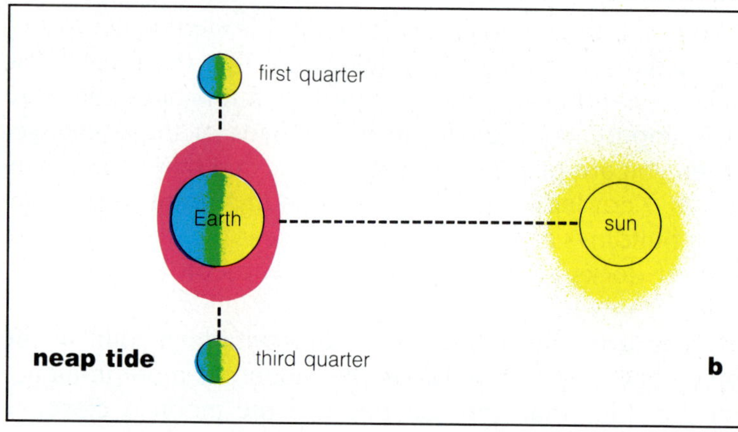

Figure 14-8b *A neap tide occurs when the moon and sun are at 90° angles to Earth.*

The gravitational pull of the moon is weakest on those parts of Earth that are 90° out of line with the moon. These are the places where low tides occur. In a given location, low tides occur about halfway between high tides.

Twice each month the tides are the highest they can be, because the sun and moon are in a direct line with Earth. The sun's pull is added to the moon's pull at these times. (See Figure 14-8a.) These highest high tides are called **spring tides**. On days when the high tides are highest, the low tides are lowest.

spring tides

When the moon and sun are at right angles to Earth, their pulls partly cancel each other out. This causes the **neap tides**, the lowest of the high tides and the highest of the low tides. (See Figure 14-8b.)

neap tides (neep)

Other factors affect the height of the tides. If ocean depth changes very gradually over a broad expanse, there may be little difference between high and low tides. However, if the coastline is jagged and the ocean bottom has deep valleys, the difference between high and low tides may be dramatic. The Bay of Fundy, between Nova Scotia and New Brunswick, is a good example. (See Figure 14-9.)

Figure 14-9 High and low tides in the Bay of Fundy.

Another factor that affects the height of the tides is the distance between the moon and Earth. The moon does not move around Earth in a perfect circle. It is closer to Earth at some times than at others. When it is closer, it pulls more strongly on Earth's ocean water.

Checkpoint

1. What factors cause the tides on Earth?
2. About how much time is there between two high tides in one place?
3. What causes the spring tides? neap tides?
4. How does the shape of the coastline affect tides?

TRY THIS

In a stream table or a plastic tray with sides, use clay to build a model shoreline. Shape it to look like the flat coastline of Florida. Add water for an ocean to one side. Gently tip the table or tray so that the water resembles high tide. (Plan how to tip the table or tray to the same point each time.) Then lower it, making a low tide. Now change the shoreline so that it has deep valleys to resemble the Bay of Fundy. Notice the difference in the heights of high and low tides in each model.

14-4

Currents

Goals

1. To describe the different kinds of currents.
2. To describe the methods by which ocean currents are detected.

ocean currents

Ocean currents are movements of water similar to huge "rivers" running through the oceans. The motion of these currents is caused by several factors. Surface currents are controlled mainly by the prevailing winds. Thus surface current patterns are similar to prevailing wind patterns. The surface currents in the Northern Hemisphere have a clockwise motion. Those in the Southern Hemisphere have a counterclockwise motion. (See Figure 14-10.)

The Gulf Stream on the east coast of North America is a large current flowing north. This warm current is responsible for keeping climate moderate on the east coast. It also causes icebergs from the cold Labrador Current to melt in the north Atlantic. Foggy weather in London, England is due to the warm water in the Gulf Stream meeting cold air at this latitude.

Figure 14-10 *The major surface ocean currents of the world. Note that there is an equatorial counter (opposite) current in the Pacific.*

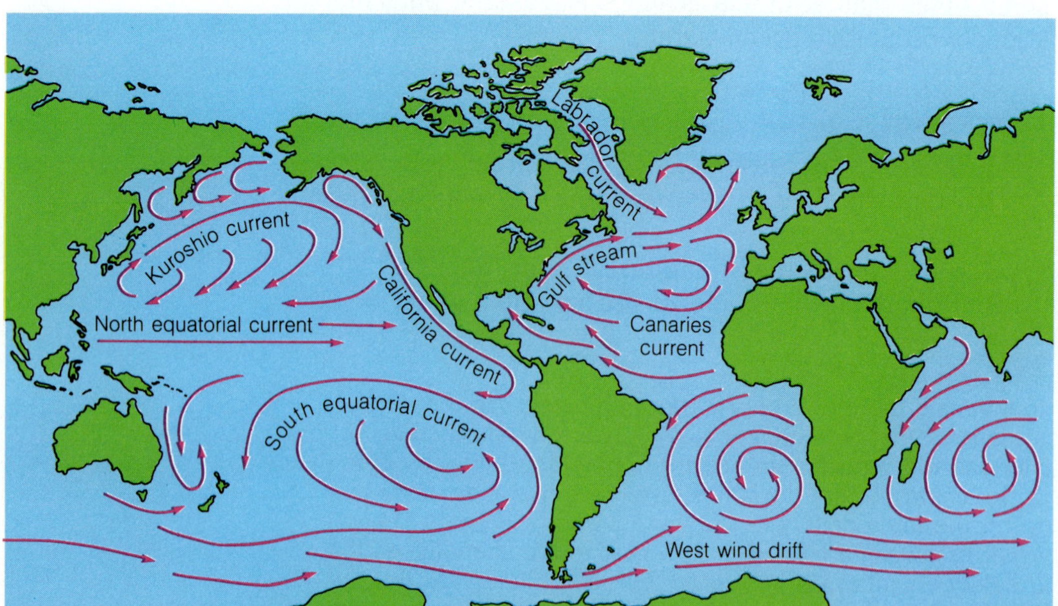

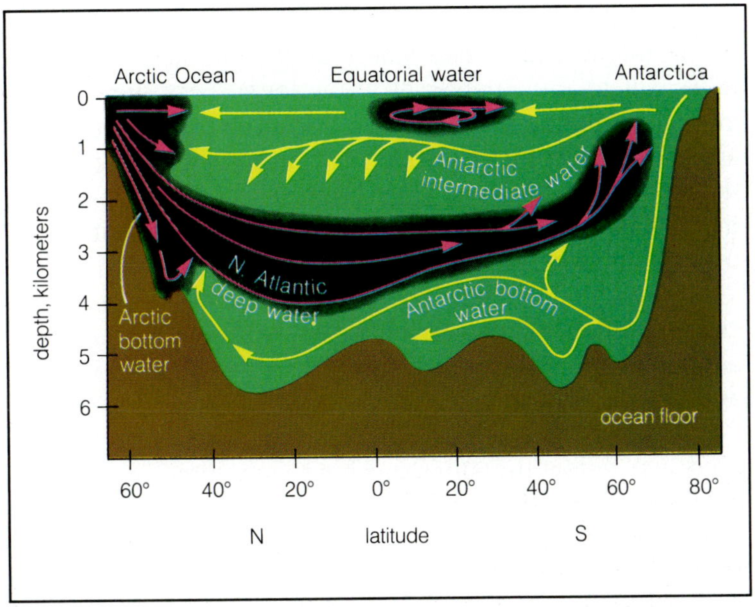

Figure 14-11 *A north-south cross section of the Atlantic Ocean, showing the movement of currents at different depths.*

Around 1750, Benjamin Franklin's cousin, Timothy Folger, was captain of a whaling ship. Sailors on these ships noticed a strong eastward-flowing current in the North Atlantic. The whalers had learned to sail with the current on trips from North America to England. On return trips, they chose a more northerly route to avoid fighting the current. Benjamin Franklin published the first map of the Gulf Stream, based on a map drawn by Captain Folger.

The California Current is a large cold current flowing from the north toward the equator on the west coast of North America. It keeps the west coast cooler than the east coast at the same latitude. The Gulf Stream and the California Current are part of the clockwise pattern in each ocean.

The strong easterly trade winds cause the equatorial currents at the equator. Early navigators in their sailing ships used these powerful currents to get them to their destinations in less time. Modern ocean vessels can use them to help conserve fuel.

Deeper currents in the ocean are caused by density differences. Denser water sinks beneath less dense water, causing a *density current*. It has been found that there are three levels of density currents in the Atlantic Ocean. Two of these come from the Antarctic and one from the Arctic. (See Figure 14-11.)

A good example of density currents occurs at the Straits of Gibraltar. The dense water of the Mediterranean Sea flows out to the Atlantic at a low level. Less dense water from the Atlantic flows into the Mediterranean at a higher level. During World War II, German submarine captains used these density currents. They often shut off the engines of the submarines and floated in and out of the Straits of Gibraltar, undetected.

The motion of ocean currents can be detected with special instruments called *current meters*. A current meter can be anchored in the water at any depth. It measures the speed and direction of flow of the water.

Drift bottles are used to measure surface currents. They float just below the surface, where they aren't affected by winds. Each bottle contains a card with information about where and when the bottle was released, plus instructions to the finder. After the finder returns the card, the direction and speed of the current can be determined.

A *swallow float* anchored in the water is used to measure the flow of deep currents. A swallow float is an electronic device that sends out "pings" that are picked up by a ship's receiver. The "pings" show the speed and direction of the currents.

Checkpoint

1. What is the major cause of surface currents?
2. What causes currents deep in the ocean?
3. How are drift bottles used to measure the speed and direction of surface currents?
4. Name an instrument that is used to measure speed and direction of deep currents.

TRY THIS

Fill a clear plastic shoe or sweater box half full of water at room temperature. Punch some tiny holes in the bottom of a small styrofoam cup and tape the cup into one corner of the box. Only the bottom of the cup should be under the water surface. Fill the cup with ice cubes. Put several drops of food coloring over the ice. Observe what happens. Can you detect a current in the water? What direction is it flowing? Let the water stand overnight. What do you observe the next day?

14-5

Sediments

Goals

1. To describe the types of sediment found on the ocean floor.
2. To describe the methods used for studying the layers of the ocean floor.

If you have ever waded in a river or lake, you have felt the *sediment* on the bottom squeeze up between your toes. This sediment is mostly sand and clay. Sediments are formed from weathered rock material that has settled down to the bottom of bodies of water. The floor of the ocean is covered by sediments, too.

It is much more difficult to study the sediments on the ocean floor because they are so far below the surface. However, oceanographers have divided ocean-floor sediments into two main categories: organic and inorganic. **Organic sediments** come from the remains of organisms that lived in the sea. Bits of shells and other tiny pieces from animals and plants form some organic sediments. Other kinds, called *oozes*, are made up mainly of hard parts of one-celled organisms.

organic sediments

inorganic sediments

nodules (NAHJ oolz)

Inorganic (nonliving) **sediments** come from several different sources. Mineral grains and rock fragments that form clay on the ocean floor are inorganic sediments. So are meteorite dust, volcanic ash, and nodules. **Nodules** are formed when chemicals from the water collect in lumps. Nodules are an excellent source of pure manganese, copper, nickel, and cobalt. In Figure 14-12, you can see manganese nodules on the floor of the Pacific Ocean.

Figure 14-12 *Manganese nodules on the floor of the Blake Plateau, off the southeastern coast of North America.*

Oceanographers have found sediments to be 600 m thick in the Atlantic and 300 m thick in the Pacific Ocean. The reason for the difference is that there are fewer rivers draining into the Pacific Ocean. Rivers carry sediments from the land to the sea. Some sediments brought to the Pacific are deposited in deep *trenches* close to the shorelines. Trenches are deep, narrow valleys on the sea floor. They are the deepest parts of the ocean. The Atlantic Ocean does not have trenches, so more sediments collect on the ocean floor of the Atlantic.

Figure 14-13 *These sediment samples were obtained with a core sampler.*

Samples of sediment layers can be obtained with a *core sampler*. (See Figure 14-13.) A core sampler is a long tube that is forced down into the sediment. When it is brought up, the tube is filled with layers of sediment in the order in which they were formed. (See Figure 14-14.) Much can be learned about the ocean by studying these sediments. The kinds of organisms that once lived in a part of the ocean are clues to the temperature of the water and the available food supply in that place. By observing the layers in sediments, scientists can learn about previous periods of volcanic activity. Certain fossils are clues to the location of oil and other minerals.

Other tools used by oceanographers to gather samples of sediments from the ocean floor are dredges and *box corers*.

Checkpoint

1. What are the two categories of ocean sediment? Give two examples of each.
2. Which ocean seems to have the thickest sediment layers?
3. Which tool can an oceanographer use to get samples of undisturbed layers of sediment?

Figure 14-14 *Coring tubes are used to collect samples of ocean bottom sediments.*

TRY THIS

Place layers of different colored modeling clay in a milk carton with the top cut off. (Use talcum powder between the layers, so they can be separated later.) For a coring tube, use 1.2-cm diameter electrical conduit the same length as the height of the milk carton. Force the coring tube down through the layers of clay and pull the tube out. With a round wood stick to fit, push the "core sample" out on the counter. Observe the layers. They are in the same order as they were put down. How can oceanographers use this information?

14-6

Features of the Ocean Floor

Goals

1. To describe the general landscape and some specific features of the ocean floor.
2. To describe a method used to map the ocean floor.

By looking just at the surface of the ocean, you might assume that the ocean floor is uniformly flat. But this is not the case. The ocean floor has as varied a landscape as the continents do.

continental shelf

As shown in Figure 14-15, the **continental shelf** is an extension of the continent that is covered by water. The shelf can be very narrow or it can extend out about 1000 km from the shoreline. It slopes very gently down to a depth of about 200 m. Next comes the **continental slope**, which drops off more sharply, about as steeply as the aisle in a movie theater. It could be called the wall of the ocean basin. At the bottom of the continental slope is a gently sloping collection of sediments from land called the **continental rise**. Beyond the continental rise is true ocean bottom—the **abyssal plain**. This flat floor of the ocean covers 60% of the ocean floor.

continental slope

continental rise

abyssal plain (uh BIS ul)

mid-ocean ridge

In the middle of the Atlantic Ocean there is a chain of rugged mountains, the **mid-ocean ridge**. This ridge is part of a system of ridges extending through the Indian Ocean into part of the Pacific. These ridges form the longest feature

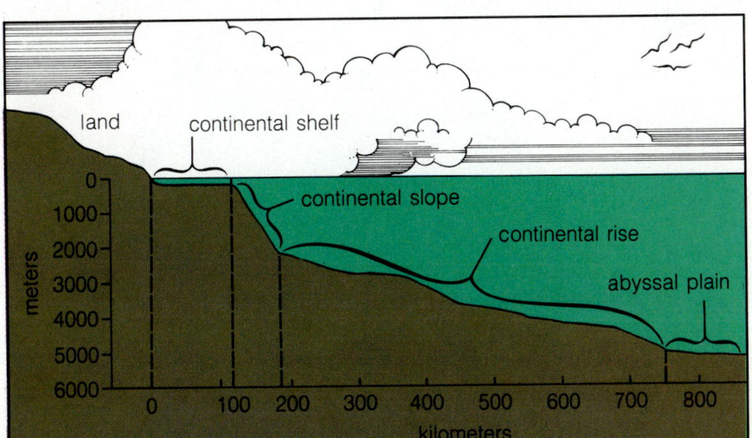

Figure 14-15 *A cross section of a typical ocean floor.*

on Earth—75,000 km long and as much as 1500 km wide in some places. The mountains stand 2 to 4 km high off the ocean floor. This is similar to the height of the Rocky Mountains in western North America. (See Figure 14-16.)

Figure 14-16 *The deep basins, mountain ranges, and zones of movement on the Atlantic Ocean floor.*

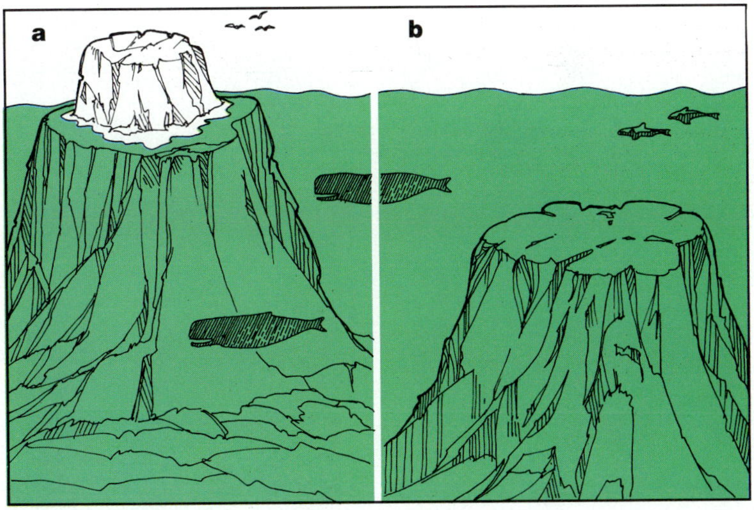

Figure 14-17 *The formation of a guyot. (a) An island is eroded by ocean waves. (b) After the island has been eroded, it becomes submerged.*

All oceans do not have the same pattern of features. Deep sea trenches, for example, are found in the Pacific Ocean but not in the Atlantic. The deepest point in the oceans is about 11.5 km deep, in the Mindanao trench near the Philippine Islands. Compare this depth with the height of the tallest mountain on Earth, Mt. Everest, which is 8.85 km high.

Distributed around the Pacific floor are hundreds of underwater volcanic mountains called *seamounts*. Some seamounts have unusual flat tops. These flat-topped seamounts, called *guyots*, were formed when waves cut across the tops of the mountains. They are evidence that sea level has changed. (See Figure 14-17.) Some underwater volcanoes form mountains high enough to protrude above the surface, forming islands. The Hawaiian Islands are examples of such volcanoes. Other volcanoes occur in groups and form *island arcs*, or chains of islands. The Aleutian Islands and the Philippine Islands are examples of island arcs.

Another feature of the ocean floor is coral reefs. Corals are small animals that live in shallow warm water near the shore. They live in large colonies. *Reefs* are made of coral skeletons. If the shore happens to be on a sinking volcanic island, corals will grow on the skeletons of former corals. So the living corals are always near the water's surface.

There are three kinds of reefs. *Fringing reefs* are on shorelines of land. *Barrier reefs* are found away from land, separated from the land by water. An *atoll* is a complete circle of

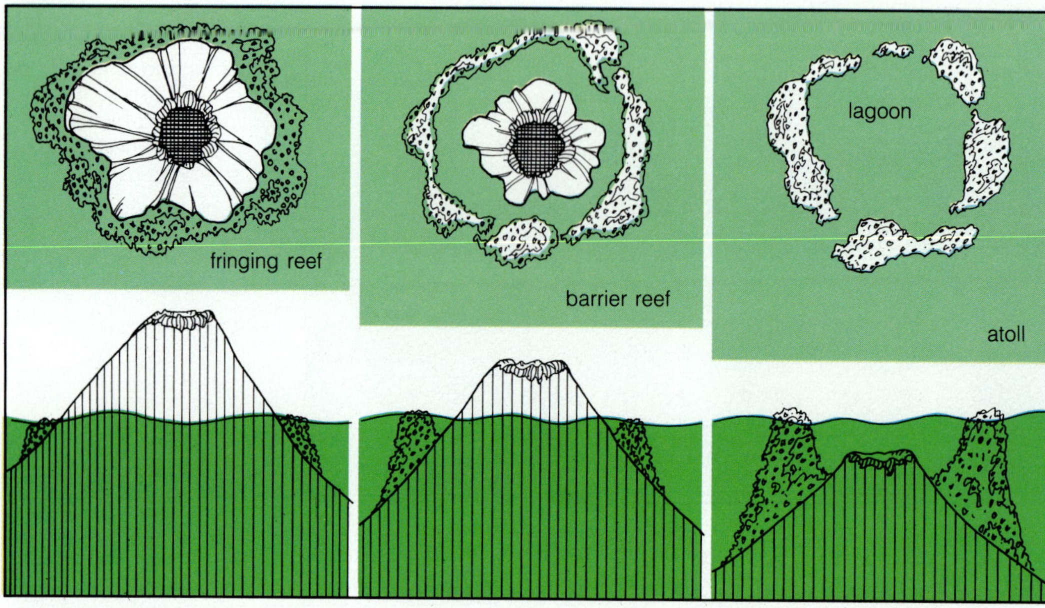

Figure 14-18 *The three stages in the development of an atoll, as an old volcano sinks and the coral reefs grow upward.*

coral with no land visible. Atolls form as the sea level changes relative to the heights of land and coral. Figure 14-18 shows how an atoll develops.

sonar (SOH nar)

It is difficult to explore the ocean depths because of the extreme pressure. Water pressure increases with depth. At 20 m deep the pressure is twice the air pressure we experience at sea level. Imagine what it is at 100 m! Most of the ocean floor has been mapped using **sonar** (**so**und **na**vigation **a**nd **r**anging). Sound waves are sent from a ship to the ocean floor. The time it takes for the sound to go to the bottom and return depends on the depth of the water. The average depth of all the oceans is 3.8 km.

TRY THIS

Fill an empty milk carton with water. Place it on the edge of a sink. With a nail, punch holes in the carton along the sink side, from top to bottom. Does each stream of water shoot out at the same angle? Explain.

Checkpoint

1. What ocean feature serves as the wall of the ocean basin?
2. What is the name for the flat floor of the ocean?
3. Name an ocean feature found in the Pacific Ocean but not in the Atlantic.
4. What is the difference between a seamount and a guyot?
5. Name a method used to map the ocean floor.
6. How is an atoll formed?

14-7

Resources of the Ocean

ENRICHMENT

Goal

To discuss the natural resources of the ocean.

The ocean, or "sea," has been a source of fish and other seafoods for centuries. Recently, there has been progress made in more efficient farming of the sea. Seafood is an excellent source of protein. By concentrating food for fish in small areas, large numbers of fish can be raised. The same thing has been done for oyster beds. (See Figure 14-19.) This farming of the sea is called **mariculture**. The prefix "mari" comes from *mare*, the Latin word for sea.

mariculture (MA ruh kuhl chur)

Salt, used in preparing and preserving food, has long been taken from the sea. By evaporating water from seawater, large amounts of salt are obtained. A cubic kilometer of seawater contains enough salt to supply the world's needs for nine years!

Kelp, another food being cultivated, is a form of seaweed. It can be used for food because it is rich in minerals and other nutrients. Kelp can be ground up into flour to make bread or it can be eaten as a vegetable. Kelp also is used in the production of certain drugs and cosmetics.

Supplies of fresh water are getting smaller because of increasing demands for water by industry. So scientists are trying to improve the process of **desalination**, removal of salt from seawater. If this process can be done efficiently, the sea can become an excellent source of fresh water.

desalination (dee sal uh NAY shuhn)

Figure 14-19 *Oysters are grown in a number of "farms" such as this one in Japan.*

Figure 14-20 *Off-shore drilling rigs, like this one off the coast of Yucatan, Mexico, are a common sight in many places.*

The crust beneath the ocean has proven to be an excellent source of oil and natural gas. These fuels are in great demand. Supplies of oil and gas have been found on the continental shelves of many continents. Off-shore drilling rigs are a common sight in the Gulf of Mexico, off the coast of California, at Cook Inlet (Alaska), and in many other places. (See Figure 14-20.)

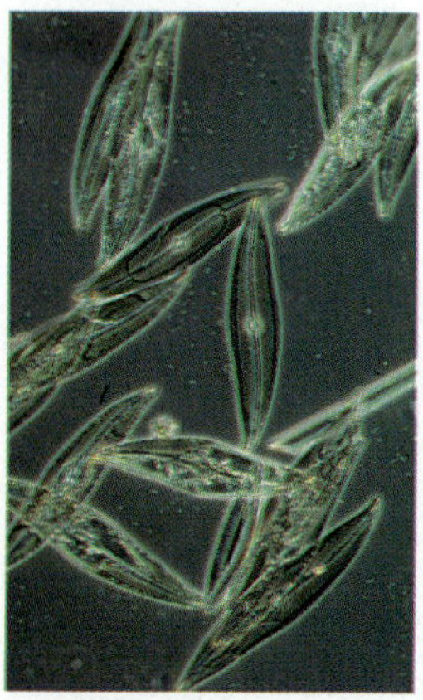

Figure 14-21 *This photograph of diatoms was taken through a microscope.*

Diatoms (DY uh tahmz), microscopic algae in the ocean, have transparent shells of silica. When diatoms die, these shells form thick deposits called *diatomaceous* (dy uh tuh MAY shus) *earth*. (See Figure 14-21.) Because of its strength and porous quality, diatomaceous earth makes excellent filter material for swimming pools. Drug firms also use it to strain out bacteria in the production of medicines. It is used to strengthen concrete, to produce a flat or semigloss finish in paint, and also as an abrasive in toothpaste. The thickest deposits of diatomaceous earth are in California.

A unique feature of many ocean resources is that they are being replenished by deposition of materials from the land and by processes within the ocean. When resources are taken from the land, they cannot be replaced so quickly.

TRY THIS

Dissolve 3.5 g of salt in 100 ml of water in an open flat container. Let this stand at room temperature until all the water has evaporated. What do you find in the dish? Relate this process to the formation of salt flats in Utah, or to salt mines in Russia.

Checkpoint

1. List some of the food sources in the sea.
2. On what part of the ocean floor are supplies of oil and gas found?
3. Name some uses of diatomaceous earth.
4. How can the ocean be used as a source of salt and fresh water?

Ocean Exploration

Modern technology has allowed people to explore the oceans with more ease and to collect more data in the past 30 years than was ever possible before. In order to study the oceans, several problems had to be overcome. For example, the water at extreme depths exerts great pressure. So submarine vessels had to have strong walls. Divers had to wear heavy suits to protect themselves against the cold water if they wanted to leave the vessel. The suits, like the JIM suit used by U.S. Navy deep-sea divers, also had to be designed to maintain normal pressure at extreme ocean depths. Special breathing apparatus was needed. In addition, there is very little light in deep water.

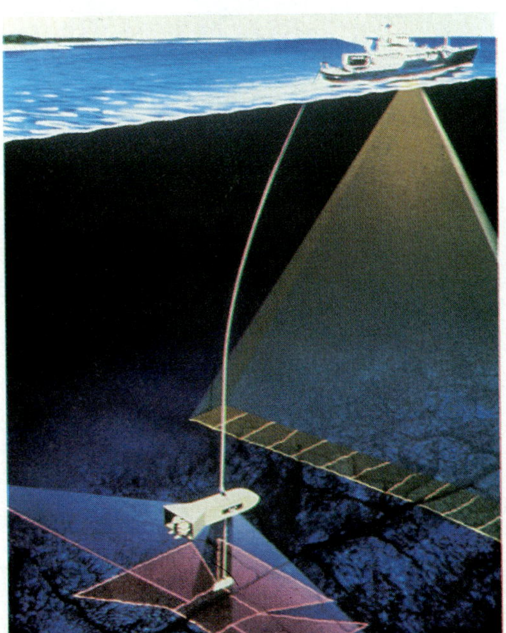

Figure 1

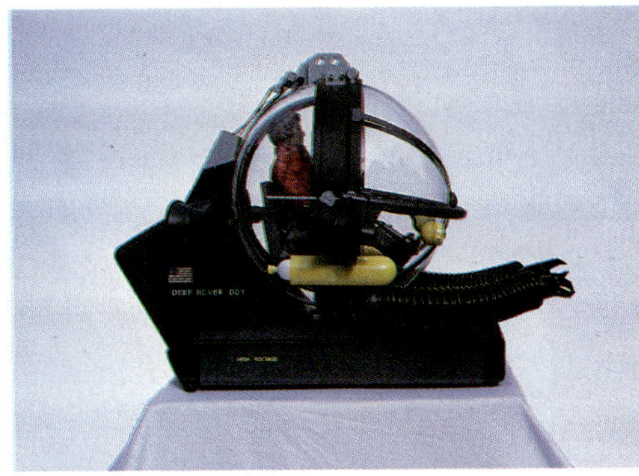

Figure 2

Small, submarinelike vessels called *submersibles* were specially designed to help overcome the difficulties of ocean research. Examples of these early laboratories are the U.S. *Tektite*, the *Trieste*, and the small submersible *Alvin*. But now a new generation of ocean explorers is sending robots to the bottom of the ocean. Electronic devices do the sensing and recording. Messages are sent back to surface ships by fine fiber optic cables. One such system is called *Argo-Jason*. (See Figure 1.) From *Argo* a smaller, self-propelled vehicle, *Jason*, can be sent to get a closer look and gather specimens. Television images can be transmitted by satellite all over the world.

Another new vehicle is a helicopter-like, one-person submersible. *Deep Rover* has mechanical arms, a pressurized cabin, enough oxygen for 100 hours, and safety devices designed so that if anything goes wrong, it will pop back up to the surface like a cork. (See Figure 2.)

Understanding The Chapter 14

Main Ideas

1. Rivers carry dissolved salts from the land to the ocean.
2. The most abundant salt in seawater is sodium chloride.
3. The energy of a wave moves forward, but the water does not.
4. Most waves get their energy from wind.
5. Tides are caused by the gravitational pull of the sun and moon.
6. There are usually two low tides and two high tides per day.
7. Ocean currents are caused by winds, Earth's rotation, and density differences.
8. Ocean sediments fall into two categories: organic and inorganic.
9. Sediments are thicker in the Atlantic Ocean than in the Pacific.
10. By studying core samples, scientists can learn about the history of the ocean.
11. About 60% of the ocean floor is a plain.
12. All oceans do not have the same pattern of features. Deep sea trenches are found in the Pacific Ocean, but not in the Atlantic, which has a mid-ocean ridge.
13. Many islands are chains of volcanic mountains that formed on the ocean floor.
14. Coral reefs are a common feature of warm, shallow water.
15. The sea has a multitude of natural resources.

Vocabulary Review

From the following list, choose the term that best completes each of the statements. Write your answers on a separate piece of paper.

abyssal plain	Nansen bottle
continental rise	neap tides
continental shelf	nodules
continental slope	ocean currents
crest	organic sediments
desalination	salinity
inland seas	seas
inorganic sediments	sonar
mariculture	spring tides
mid-ocean ridge	trough
	tsunamis

1. The highest tides are __?__.
2. __?__ is used to find the depth of the ocean.
3. Part of the continent that is under water is the __?__.
4. __?__ found on the ocean floor are a good source of manganese and other minerals.
5. The percentage of salt in seawater is called __?__.
6. The process of removing salt from seawater is called __?__.
7. Some __?__ are also called oozes.
8. The __?__ covers 60% of the ocean floor.
9. The __?__ is the longest feature on Earth.
10. The farming of the sea by people is called __?__.
11. The highest point in a wave is the __?__.

Chapter Review

Write your answers on a separate piece of paper.

Know The Facts

1. Most of the dissolved salts in seawater come from the ___?___. (continents, atmosphere, sea-dwelling organisms)
2. A swimmer floats higher in seawater than in fresh water because ___?___. (the density of seawater is less, the density of seawater is greater, seawater is a bigger body of water)
3. Which of the following affects the salinity of surface seawater at a given location? (depth, density, evaporation)
4. Most of the gases in seawater come from the ___?___. (continents, atmosphere, life in the sea)
5. What happens to an object floating at the surface when a wave passes? (It moves horizontally, but not up and down; it moves rapidly in the direction of wave travel; it moves in a circle, but ends up in its original position)
6. What is the principal cause of waves in the ocean? (wind, temperature, density differences)
7. What is the primary cause of vertical currents in the ocean? (wind, density differences, shape of the ocean floor)
8. Low tides occur on Earth in places that are ___?___ the moon. (opposite, 90° out of line with, 180° out of line with)
9. Oceanographers use ___?___ to map the ocean floor. (Nansen bottles, core samplers, sonar)
10. A device used to find the temperature of ocean water at various levels is a ___?___. (current meter, core sampler, Nansen bottle)
11. Samples of sediment layers at the bottom of the ocean are brought up to the surface using a ___?___. (Nansen bottle, core sampler, barrel)
12. Tsunamis are caused by ___?___. (wind, earthquakes, gravity)
13. The most abundant salt in seawater is ___?___. (magnesium chloride, calcium chloride, sodium chloride)
14. The moon affects the tides more than the sun because ___?___. (the sun is too big, the moon is closer, the moon is bigger)
15. Trade winds north of the equator cause the ___?___. (Kuroshio current, south equatorial current, north equatorial current)
16. There are more sediments in the Atlantic Ocean than in the Pacific because ___?___. (the Atlantic is bigger, the Atlantic is deeper, the Atlantic has more rivers emptying into it)

Understand The Concepts

17. You know what time a high tide occurs in one place. Explain how you would know approximately when the next low tide will be.
18. Why are there no atolls in the North Atlantic?
19. What happens to the salinity of ocean water when evaporation takes place?
20. Why is the water along the California coast colder than the water along the east coast at the same latitude?

21. Where would you expect to find the densest water in the ocean?
22. What causes the difference in temperature between currents on the eastern and western sides of an ocean basin?
23. What factor has the greatest influence on ocean surface current patterns?
24. Give an example of a chain of volcanic islands.
25. Explain how coral reefs are found in deep water where temperatures are too cold for coral to live.
26. How do currents differ from tides?
27. Where would you expect to find the warmest ocean water? Where would you expect to find the coldest unfrozen ocean water?
28. Benjamin Franklin, when a postmaster, noticed that sailing ships going from North America to Europe took less time than those coming from Europe. What is one good reason for this?

Challenge Your Understanding

29. How can past climates be determined from core samples of ocean sediments? Give an example.
30. Why are ports in Norway near the Arctic Circle open to ships all year long?
31. If you were on the ocean in a ship, why would you not be aware of a tsunami, even if one passed beneath you?
32. Which kind of sediment—organic or inorganic—would be more common on the abyssal plain? Explain your reasoning.

Projects

1. Research the new technology used for getting data about the oceans and the seas of the world.
2. Report on the different kinds of life that have been found on the ocean floor near the mid-ocean ridge.
3. Build a model of the ocean floor showing features such as a continental shelf, a continental slope, a continental rise, an abyssal plain, a mid-ocean ridge, and a trench.
4. Walk through a supermarket and list all of the items that had their source in the ocean.
5. Trace the path of a nearby river to where it enters an ocean. What kind of materials would be carried from your region to the ocean?
6. Read *Kon-Tiki* by Thor Heyerdahl. Report to your class.
7. Interview a newspaper reporter, an environmental lobbyist, or government official about laws governing the use of the oceans for fishing, mining, and oil exploration.
8. Make a model water wheel. List and describe the different machines that could use the energy supplied by the water wheel.
9. Make a drift bottle out of an empty bottle or fruit jar. Place a card in it with your name and address, and instructions for the finder. Put the bottle in a nearby river or stream and wait for an answer.
10. Research the latest ideas for harnessing the energy of tides, currents, or waves.
11. Make a scrapbook of articles about the ocean from magazines and newspapers.

Chapter 15

Astronomy

This radio telescope in the Mojave Desert gathers radio waves from space.

15-1

Planet Earth

Goals

1. To describe Earth's shape.
2. To describe the rotation and revolution of Earth.

universe

The **universe** is everything that exists. It includes all of the stars that you can see on a clear night, for example. It has been said that there are as many stars as there are grains of sand on all the beaches in the world!

The sun is one of the billions of stars in the universe. Moving around the sun are nine large objects called *planets*, one of which is Earth. The study of everything in the universe that can be observed from Earth or space is called **astronomy**.

astronomy

Pictures taken from space show that Earth is "round." It has a spherical shape, like a baseball or basketball. (See Figure 15-1.) However, Earth is not a perfect sphere. The distance around Earth, its *circumference*, is somewhat greater at the equator than around the poles. The average circumference of Earth's "flattened" sphere is 40,008 km. (If you drove that distance nonstop at a speed of 80 km/h, it would take you about 21 days.)

Figure 15-1 *Try to locate Africa, the Arabian Peninsula, and the South Pole.*

Figure 15-2 *Earth's rotation, as recorded by a camera pointed at the North Star.*

Earth is not a stationary object. It is constantly in motion. But its motion is difficult to detect on Earth. When you are riding in a car, you know you are in motion because you see trees or other fixed objects "whizzing" by. But there are no obvious fixed *reference objects* by which to detect Earth's motion. As a result, people thought for centuries that Earth was stationary.

We now know that Earth moves in two basic ways. One is **rotation**, the spinning or turning about a straight line, or *axis*. (A moving wheel also rotates about an axis.) Earth's axis is an imaginary straight line that runs through Earth from the North Pole to the South Pole. Earth completes one rotation every 24 hours. **rotation**

Earth's rotation produces changes that you can see. The sun appears to rise each day in the east and set in the west. The sun's *apparent* motion results from Earth's rotation. The sun does not move relative to Earth. But a different part of Earth's surface faces the sun at any given time, due to Earth's rotation.

Similarly, Earth's rotation accounts for changes that you can see in the night sky. Early in the evening, find a group of stars that you can easily recognize in the eastern sky. An hour or two later, look for the same stars. You will find that they appear to have moved toward the west. Pictures have been taken focusing the camera on the North Star and using a time exposure of several minutes. The circular streaks of light are the star trails made as Earth rotates underneath the stars. (See Figure 15-2.)

The other basic motion of Earth is its **revolution**, or its movement around the sun. Earth completes one *orbit*, or revolution, in one year. Other planets take a longer or a shorter time to revolve around the sun. **revolution**

You can tell that Earth revolves around the sun by looking at the sky at night. We see stars at night, when we are on the side of Earth that faces away from the sun. The stars can serve as "fixed" reference points to help us observe Earth's motion.

People have known for many years that certain patterns of stars, or **constellations**, are visible only at certain times of the year. Imagine sitting at a window table in a slowly-revolving rooftop restaurant. From the restaurant, the view of the city lights below is constantly changing. Similarly, from **constellations**

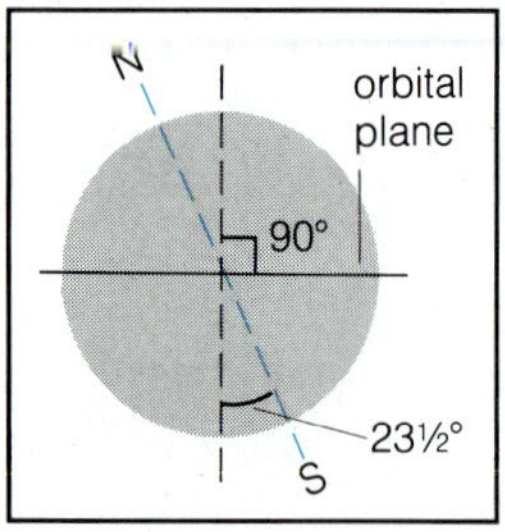

Figure 15-3 *Earth's axis is not at right angles to its orbital plane.*

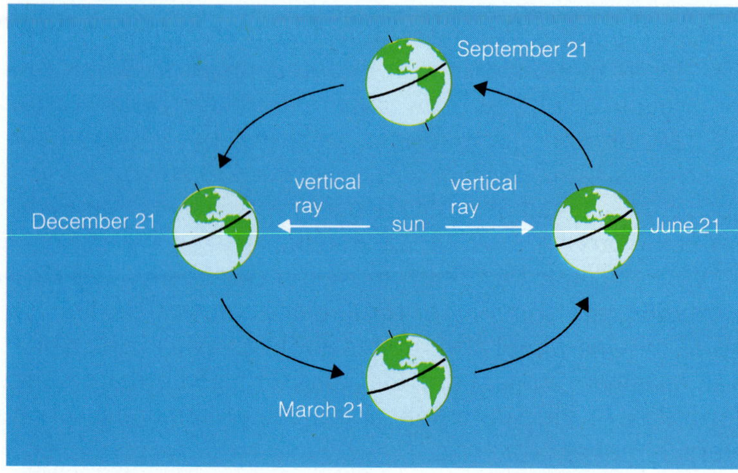

Figure 15-4 *As Earth revolves around the sun, the Northern Hemisphere receives more vertical rays during summer than winter.*

the revolving Earth, the view of the sky changes with time. Each season we see a different view. As a result, we speak of winter, spring, summer, and fall constellations.

The imaginary flat surface, or plane, in which Earth revolves around the sun is called Earth's *orbital plane*. Earth's axis is not at right angles to its orbital plane. Instead, it is tilted 23½° from the right-angle (90°) position. (See Figure 15-3.)

If Earth were not tilted, vertical rays of sunlight would be received only at its equator. Areas receiving vertical rays are warmer. Because of Earth's tilt, vertical rays are shifted north of the equator part of the year and south of it at a different part of the year. (See Figure 15-4.) The 23½° tilt determines how far north and south of the equator the vertical rays of the sun reach. As the location of the vertical rays changes, the amount of energy received at each latitude changes. Thus, the seasons change. Summer begins in June in the Northern Hemisphere (Earth's northern half) and in December in the Southern Hemisphere.

TRY THIS

In a darkened room, shine a bright light onto a globe. The light source should be 1–3 m from the globe. Rotate the globe slowly. Compare the amount of light that falls on each hemisphere as it is (a) tilted toward, (b) tilted away from the source.

Checkpoint

1. Describe Earth's shape.
2. What motion of Earth causes day and night?
3. Explain why we have seasons.

15-2

Earth's Moon

Goals

1. To identify major surface features of the moon.
2. To describe some effects of the moon's motions.

satellite

A **satellite** is an object that revolves around a planet. Earth's natural satellite is the moon. It revolves around Earth, as Earth revolves around the sun. The planet Mercury has no moon. Saturn has at least 17! As moons go, Earth's is large. The circumference of our moon is more than one fourth that of Earth.

When you look at the full moon you can see contrasting light and dark areas. These areas make a pattern, often called the face of the "man in the moon." The light areas are mountains, and the dark areas are flat parts of the moon. (See Figure 15-5.) The moon was first viewed through a telescope by Galileo, in 1609. He thought the dark areas were seas. So he called them *maria* (MAH ree uh), which is the plural of *mare* (MAH ray). *Mare* means "sea" in Latin.

The moon is dotted with many steep-walled circular depressions called *craters*. Most of the craters are located in the mountain regions. Some are more than 100 km across. Some craters have *central peaks*, which are small mountains on the crater floors. (See Figure 15-6.) Most of the craters on

Figure 15-5 *The light areas on the moon are mountains and craters.*

Figure 15-6 *Craters on the moon. Note the central peaks, and smaller craters.*

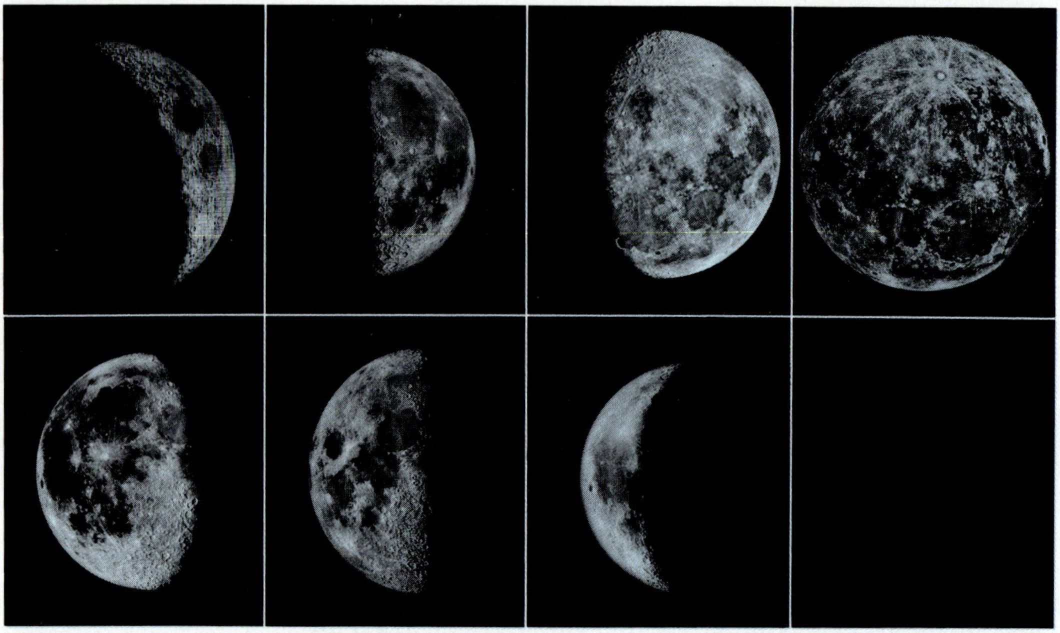

Figure 15-7 *The phases of the moon. Compare these photographs with Figure 15-8.*

the moon were caused by objects hitting the moon. Other craters are of volcanic origin.

A moon day, the time it takes for the moon to rotate once about its axis, is 27⅓ Earth days. This is the same time it takes the moon to *revolve* around Earth. As a result, the same side of the moon always faces Earth. The far side of the moon was a complete mystery until pictures of it were taken from space in 1959.

The moon's long days and nights help to explain the great temperature extremes on its surface. Temperatures range from about −173°C at night to more than 100°C in the daytime. Since the moon has no atmosphere, there is nothing to filter the powerful rays of the sun. Nor is there an insulating blanket of atmosphere to slow the escape of heat at night.

The impact of objects hitting the moon, strong radiation from the sun, and extreme temperature changes combine to break up rock material over time. As a result, the moon's surface is covered with a layer of fine dust and rock fragments.

The moon does not produce its own light. The moon is visible because it reflects sunlight. We see the moon in different parts of the sky during a month because of its motion

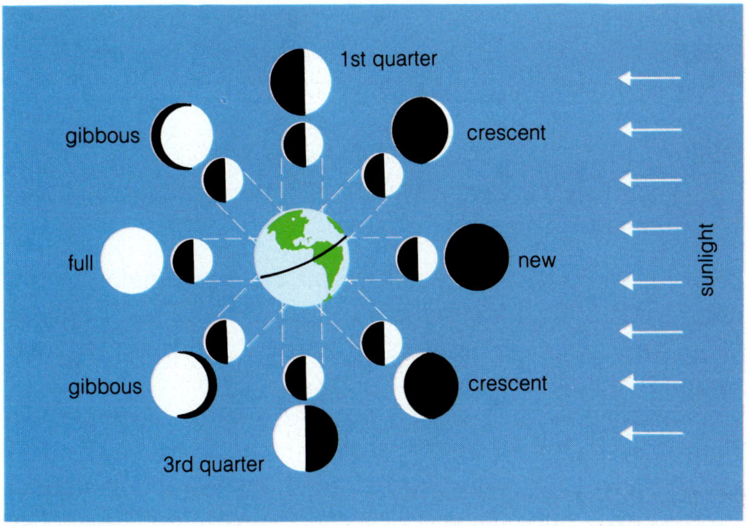

Figure 15-8 *As the moon revolves around Earth, the sun shines on the same side of the moon. But we see the lighted part from different angles.*

around Earth. The apparent shapes of the moon that we see as it completes one revolution are called **phases**. It takes about 29½ days, a lunar month, for the moon to go through a full set of phases. (See Figure 15-7.) The relative positions of sun, moon, and Earth determine what phase we see at any given time. (See Figure 15-8.)

phases

Like all objects in space, Earth and the moon attract each other. So do Earth and the sun. The most noticeable effect of this attraction is seen in the tides. The moon has a greater effect on the tides because it is closer to Earth.

Checkpoint

1. What is a satellite?
2. Describe the main features of the moon's surface.
3. How long does it take the moon to go through a full set of phases?
4. Why does the moon have a greater effect on the tides than does the sun?

TRY THIS

Use a globe or large ball to represent Earth and a smaller ball to represent the moon. Make some markings on one side of the "moon" and move it around "Earth" *without* rotating it. Do you see only one side of the "moon" from "Earth"? Now rotate the "moon" so that the same side is always toward "Earth." What can you say about the speed of the "moon's" rotation?

15-3

The Solar System

Goal

To identify the general features of the planets and other members of the solar system.

solar system

The sun and all the objects in orbit around it make up the **solar system**. Besides Earth and its moon, the solar system includes eight other planets, at least 47 other moons, and numerous smaller objects.

The planets are divided into two groups, according to their characteristics. One group is the rocky planets: Mercury, Venus, Earth, Mars, and Pluto. The second is the gaseous group: Jupiter, Saturn, Uranus, and Neptune. The rocky planets are more dense and smaller than the giant gaseous planets. (See Figure 15-9.) The gaseous planets have thicker atmospheres and more moons than the rocky planets.

Figure 15-9 *Facts about the planets. The minutes, hours, days, and years are Earth time units. The relative size of the sun is shown in yellow.*

Planet	Size	Radius	Time to rotate	Length of year
Mercury		0.38	59 d	88 d
Venus		0.95	243 d	225 d
Earth		1.00	24 h	365.25 d
Mars		0.53	24.7 h	687 d
Jupiter		11.18	9 h 50.5 min	11.9 y
Saturn		9.42	10 h 14 min	29.5 y
Uranus		3.84	10 h 49 min	84 y
Neptune		3.93	15 h	165 y
Pluto		0.50	6.39 d	248 y

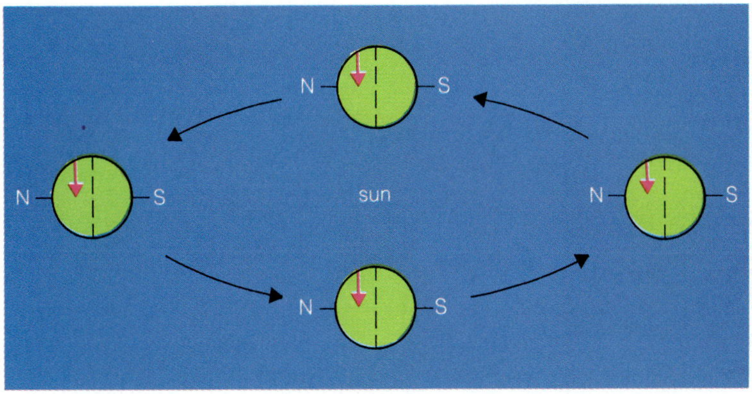

Figure 15-10 *Uranus seems to be lying on its side as it moves around the sun.*

All planets revolve around the sun in about the same plane. The ones closer to the sun move faster than those that are farther away. The length of time it takes a planet to revolve is called its *year*. If you want to look for a planet in the sky, always look on the same path that the sun took during the day.

Each planet rotates around a tilted axis. The tilt of most planets is about the same as Earth's, except for Uranus. The tilt of Uranus' axis is so great that the axis lies nearly in the planet's orbital plane. It looks like a spinning top that has fallen over on its side. (See Figure 15-10.) Uranus and Venus rotate in a direction opposite to Earth's rotation.

TRY THIS

With binoculars or a small telescope, watch the planet Jupiter for several nights. Use a sky chart to help you locate the planet. Draw the planet and the positions of the four largest moons as you see them each night.

It is difficult to have up-to-date data about the planets in any book. Sometimes it seems as though new finds are reported almost daily. For years, astronomers thought Saturn was the only planet with rings. Recently they have discovered rings around Uranus and Jupiter, too. *Rings* are frozen bits of matter that revolve around some planets. Matter in the rings is frozen because the planets the rings encircle are so far from the sun.

asteroids

The **asteroids**, or minor planets, are found mainly between the orbits of Mars and Jupiter. Asteroids are rocky and range in size from less than 1 km to 1000 km across. There are about 100,000 of them. They may be material that never combined to form a planet.

comets

Far out near the edge of the solar system is a nearly spherical cloud of objects. These objects, called **comets**, average only a few kilometers across. They consist chiefly of fine dust and frozen water. Occasionally, one of the billions of comets travels toward the sun.

Figure 15-11 *Halley's comet, in 1910.*

Figure 15-12 *This meteorite fell in Arizona.*

When a comet gets as near to the sun as Jupiter's orbit, it starts to warm up. As it gets closer to the sun, the original comet, or *nucleus*, begins to boil off gases and dust. These stream out behind the nucleus in a *tail* that may be millions of kilometers long. (See Figure 15-11.) The nucleus becomes surrounded by a glowing cloud called the *coma*.

Since comets lose matter each time they near the sun, they may not last through many orbits. Some, however, have returned as bright objects in the sky many times. Halley's Comet is famous because it was the first comet known to appear at regular intervals. It returns every 76 years.

meteoroid

meteors

meteorite

Any small piece of rock or mineral matter traveling through space is known as a **meteoroid**. Meteoroids may be small, stray asteroids or debris from a comet. If meteoroids enter Earth's atmosphere, they usually burn up due to friction. The fiery streaks that are produced in this way are called **meteors**. You may see meteors in the sky on a clear night. They are often called "falling stars." A meteoroid that does not burn up in the atmosphere and hits Earth is called a **meteorite**. (See Figure 15-12.) If large enough, meteorites may make craters in Earth's surface. The last large meteorite fell in Siberia in 1908.

Checkpoint

1. How do rocky planets differ from gaseous planets?
2. Which planet's axis tilts the most?
3. Name some members of our solar system other than planets and moons.
4. How do a meteor and a meteorite differ?

15-4

The Sun

Goals

1. To describe important properties of the sun.
2. To compare a solar eclipse to a lunar eclipse.

Most stars in the universe are found in groups of billions of stars called **galaxies**. Our sun is an average star, one of the many stars in the Milky Way Galaxy.

galaxies

The sun is a huge, gaseous ball. It gives off tremendous amounts of energy. The source of the sun's energy is **nuclear fusion**, the *fusing* of hydrogen nuclei to form helium nuclei. Nuclear fusion occurs only at very high temperatures (millions of degrees Celsius). Such temperatures are produced in the dense interiors of stars.

nuclear fusion (NOO klee ur FYOO zhun)

The sun accounts for nearly 99.9% of the total mass of the solar system. Most of the sun's mass is hydrogen (about 75%) and helium (about 24%). Other elements make up no more than 1–2% of the sun's mass. The sun's volume is about one million times that of Earth. It is estimated that there is enough hydrogen in the sun to continue the fusion process for another 4–5 billion years.

In Figure 15-13, hotter areas of the sun show up as lighter colors. The darker areas are cooler. The leaping *prominence*, or flamelike body, shows that the sun's surface is very active.

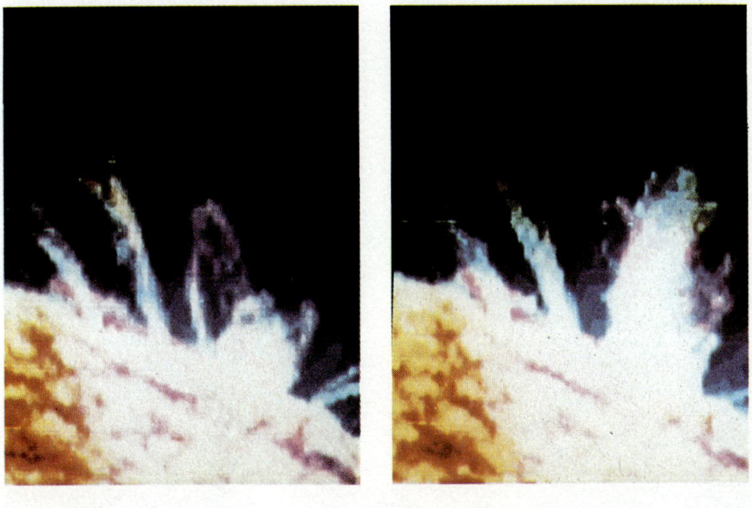

Figure 15-13 *Great explosions of hot gases make the sun's surface "boil." These two photographs, taken a short time apart, show an erupting prominence.*

sunspots

The sun has dark spots on its surface called **sunspots**. Sunspots have a lower temperature than the rest of the sun's surface. Astronomers have found that sunspot activity (the number of sunspots) increases and decreases over an average period of 11 years. Recent periods of high sunspot activity were 1968–69 and 1979–80.

During periods of high sunspot activity, the solar wind increases. The *solar wind* is a continuous flow from the sun of ionized (charged) gases, chiefly hydrogen and helium. The solar wind extends far out beyond Earth. When the solar wind increases, it causes noticeable effects on Earth. Bright-colored areas in the sky, the *northern and southern lights*, are seen more often. (See Figure 15-14.) Interference with radio and television is more frequent. You may have heard the announcement, "Audio (or video) difficulties are due to atmospheric conditions." This interference could be due to the charged particles from the sun entering Earth's upper atmosphere.

solar eclipse

Scientists have found that the sun rotates on its axis once every 27 Earth days. Its equator rotates faster than its polar regions, showing that it is not a solid. The sun also revolves around the center of the Milky Way at a speed of about 250 km/s. The fact that it takes 200 million years for one revolution at that rate shows just how big our galaxy is.

A **solar eclipse** occurs when the moon passes directly between Earth and the sun. The moon can either partially or totally block out the sun's light. (See area *a* in Figure 15-15.)

Figure 15-14 *The northern lights. The light comes from charged solar gases streaming through the atmosphere toward the North Pole.*

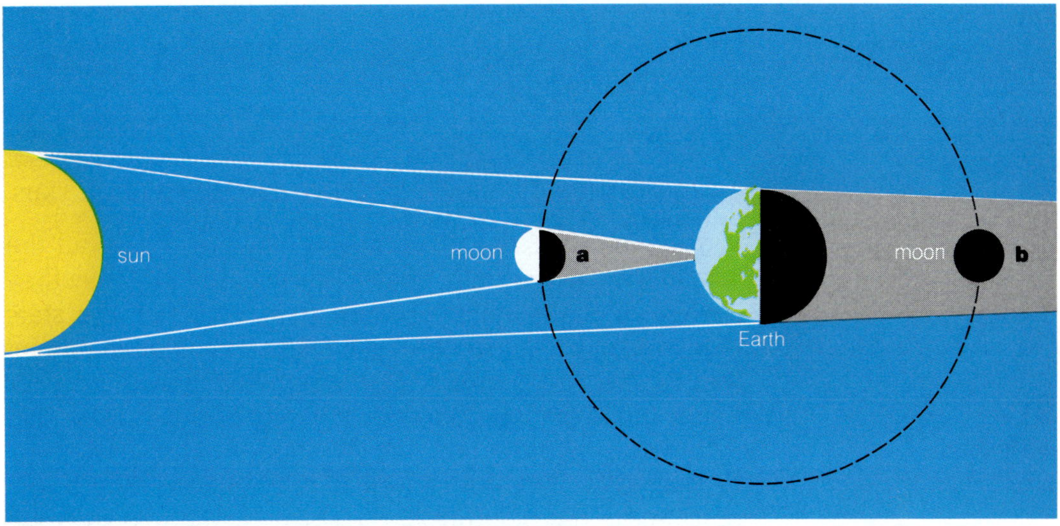

Figure 15-15 *When the moon passes between Earth and the sun, there is a solar eclipse (a). When the moon is in Earth's shadow, the moon is eclipsed (b).*

If you ever have the opportunity to see a solar eclipse, never look at the sun directly. In fact, *at no time* should you view the sun directly. Its powerful rays can permanently damage your eyes.

During a total solar eclipse, daylight gradually disappears, and stars become visible. A solar eclipse must have been very frightening before people understood what was happening. A solar eclipse lasts only a few minutes, because the moon and Earth are both moving very rapidly. A total solar eclipse can be seen from only a small portion of Earth's surface, because the moon's shadow on Earth is quite small.

lunar eclipse (LOO nur)

A **lunar eclipse** occurs when Earth passes directly between the sun and the moon. Earth's shadow is then cast on the moon, as shown in area *b* in Figure 15-15. Lunar eclipses are easier to see than solar eclipses. Everyone on the night side of Earth can see them. Lunar eclipses last longer, too.

TRY THIS

Set up a model to show how eclipses occur. Hold a coin so that it blocks all the light from a light bulb. What represents the sun in this model? the moon? Earth? Is this a solar or lunar eclipse? Show how the other kind of eclipse occurs.

Checkpoint

1. What process is the source of the sun's energy?
2. How does high sunspot activity affect Earth?
3. What is the difference between a solar eclipse and a lunar eclipse?

15-5

The Universe

Goal

To describe different properties and life cycles of stars.

As you look at a constellation, you may notice that all the stars in it are not equally bright. This might be because they are not all at the same distance from Earth. Or, the stars might not all be the same size. It might be a combination of both factors. For example, a large star may look no brighter than a small, dim star because it is much farther away. The observed brightness of a star is called its **magnitude**.

magnitude

Stars differ in color as well as in brightness. The colors indicate the temperature of the stars. The hottest stars are blue, and the coolest stars are red. In between are yellow, orange, and white stars. The sun is a yellow star.

Stars change over long periods of time. They form in huge clouds of gas and dust called **nebulas**. (See Figure 15-16.) A nebula may be the remains of a former star that exploded. Some of the gas and dust may be drawn together into a clump by gravitational attraction, just as objects on Earth are drawn toward its center. As the cloud of gas and dust contracts, it heats up. When the temperature gets high enough nuclear fusion begins, and a new star is born.

nebulas
(NEB yuh luhz)

Figure 15-16 *The Crab Nebula is a cloud of dust and gas among the stars.*

Figure 15-17 *Relative sizes of different kinds of stars, from the largest to the smallest. Notice how they compare with the sun. Black holes are extremely small and dense.*

As long as nuclear fusion takes place at a steady rate, a star is said to be stable. But the hydrogen fuel cannot last forever. When the rate of fusion slows down, the star changes. The sun, for instance, will begin to expand in a few billion years. It will probably become a **red giant**, a star hundreds of times bigger and cooler than the sun is now. (See Figure 15-17.) A red giant eventually becomes smaller as the force of gravity pulls its parts closer together.

red giant

After the red-giant stage, which lasts for millions of years, the sun may go through a variable-star stage. In this stage it would flicker like a light bulb that is burning out. In any case, the sun will probably become a **white dwarf**—a very small, dense, hot star. Even this stage cannot last forever. Eventually, a white dwarf becomes a cold, dark mass in space. This stage is sometimes called a *black dwarf*.

white dwarf

The length of a star's life depends on how big it is to begin with. The more massive a star, the sooner it will die, and the more likely it will die in a very spectacular fashion. It could end up as a **supernova**, a star that explodes, producing a nebula. Within this nebula a very small, very dense star, called a **neutron star**, remains. In a neutron star the atoms are squeezed together so tightly that all protons and electrons are combined, forming neutrons.

supernova

neutron star

If a star is massive enough, it may completely collapse on itself, becoming a **black hole**. A black hole is extremely dense, with such a strong gravitational pull that even light cannot escape from it. Black holes can be detected by the way they affect matter around them.

black hole

The distance between stars is so great that it is measured in special units. A **light-year** is the distance light travels in one year. Light travels at about 300,000 km/s. (At this rate you could make nearly eight trips around Earth in one second!) The nearest star that we can see from the Northern Hemisphere is Sirius (SIHR ee us), which is 8.7 light-years away. This means that the light we see when we look at Sirius left that star 8.7 years ago. Distances within the universe can range up to millions of light-years.

light-year

Checkpoint

1. What two factors determine the brightness of a star as seen from Earth?
2. What is a supernova? a black hole?
3. What stages will our sun probably go through before it dies out?
4. What is meant by a distance of 10 light-years?

TRY THIS

Choose a clear night. Using a cardboard tube (such as one from inside a roll of paper towels) for a viewing tube, look at one part of the night sky. Count the number of stars you see through the tube. Call this "Trial 1" and write down the number of stars you see. Do this nine more times, looking at a different part of the sky each time. Record the number of stars seen each time. Calculate the average number you saw from the ten trials. Multiply this number by 700. This is the average number of stars in our galaxy visible to the unaided eye, from one position on Earth.

15-6

Tools of Astronomy

ENRICHMENT

Goals

1. To describe various kinds of telescopes.
2. To explain how a spectrograph is used to analyze starlight.

Everything we know about the stars has been learned from the energy that they give off. Most instruments astronomers use to study stars are light-gathering devices. The greater the amount of light captured, the better an object in space can be seen. Binoculars and telescopes are examples of such instruments.

visual telescopes

Visual telescopes gather visible light, so that objects can be directly seen or photographed. There are two major kinds of visual telescopes. *Refracting* telescopes contain lenses that gather light. *Reflecting* telescopes use curved mirrors for this purpose. (See Figure 15-18.) Reflecting telescopes are easier to build. They can be made larger than refracting telescopes, also.

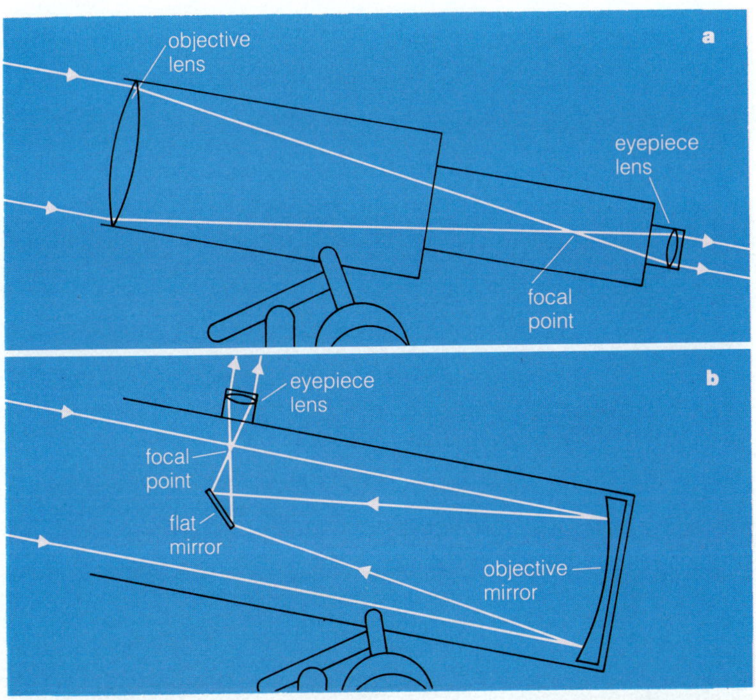

Figure 15-18 *The two kinds of visual telescopes. (a) A refracting type, that uses only lenses. (b) A reflecting telescope, with a concave mirror.*

radio telescopes

Radio telescopes gather radio waves from space. Radio waves can pass more readily than light waves through dust clouds in space and through clouds in Earth's atmosphere. Radio waves can be received 24 hours a day, while visual telescopes can be used only when it is dark. Radio astronomy has come into use only in the last 50 years.

Since radio waves are longer than light waves, the antennas, or *dishes*, of radio telescopes must be very large. One of the more spectacular radio telescopes is the 305-m dish at Arecibo (ah ray SEE boh), Puerto Rico. The dish is in the crater of an extinct volcano. As Earth rotates, the dish picks up radio waves from space. (See Figure 15-19.) Another way of making a large radio telescope is to have groups of dishes spread over the ground. There are many such groups around the world.

spectroscope

The chemical elements present in a star can be identified by analyzing starlight. If a narrow beam of light from a glowing gas is passed through a prism, a colorful *spectrum* is produced. (See Figure 15-20.) A **spectroscope** is used to produce a spectrum and measure the wavelengths of light in it. If the gas is an element, the spectrum is a series of narrow lines. Every element has its own *bright-line spectrum*.

Figure 15-19 *The radio telescope at Arecibo, Puerto Rico.*

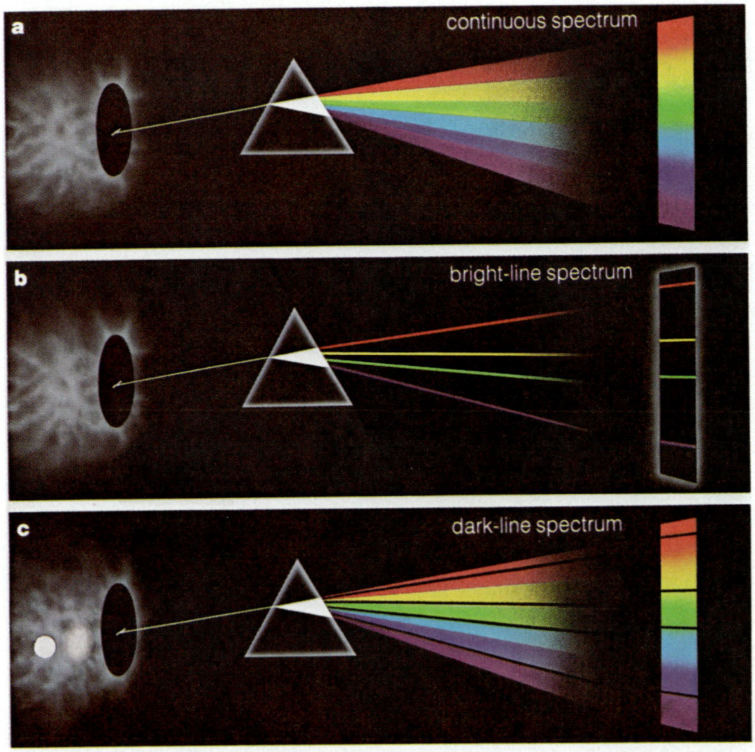

Figure 15-20 *Three kinds of spectra. (a) Light from glowing gas gives a continuous spectrum. (b) A gas of a single element gives a bright-line spectrum. (c) Light passing through cooler gas gives an absorption spectrum of dark lines.*

Its line spectrum can serve as a "fingerprint" to identify the presence of the element.

Most stars don't produce bright-line spectra. The hot gases in the interior of the star radiate a continuous spectrum. As the radiations pass through the star's atmosphere, each gaseous element absorbs the wavelengths of light that appear in its bright-line spectrum. So, a certain number of black lines appear in the spectrum. The *absorption spectrum* of an element can also serve as its "fingerprint."

Checkpoint

1. What two types of visual telescopes are in use?
2. How does a visual telescope differ from a radio telescope?
3. What information can a spectroscope provide?
4. What can astronomers learn about stars by using a spectroscope?

TRY THIS

On a clear night, go outside and find the Big Dipper in the northern sky. Look closely at the second star from the end of the "handle." What do you see? Now take a pair of binoculars and look again at the same area. What do you see now?

Telescopes in Space

The study of astronomy will take a giant step forward when a telescope is launched into space. It is possible that a space telescope will help us to learn more about the universe in a few years than has been learned in all of past history! The greatest advantage of a space telescope is that it will be outside Earth's atmosphere. Earth's thick atmosphere limits the distance telescopes on Earth can "see" into space. The atmosphere also absorbs and refracts much of the light from the stars.

The space telescope that will be put into orbit by the Space Shuttle will be 5 m wide and 15 m long, with a mass greater than 10,000 kg. Its mirror will be 30 cm thick and 240 cm in diameter. Along with the telescope there will be two cameras, two spectrographs, and two instruments to measure distances between stars and changes in their brightness. All of this equipment will be in orbit 500 km above Earth. This is far above the thickest part of Earth's atmosphere. The largest telescopes on Earth can detect objects two billion light-years away. The space telescope will be able to detect light from galaxies 14 billion light-years away. This will extend our view of the universe by 350 times.

Understanding The Chapter 15

Main Ideas

1. Earth is spherical, but it is not a perfect sphere.
2. Earth rotates on its axis and revolves around the sun.
3. Earth's seasons depend on the tilt of its axis.
4. The moon's major surface features are maria, craters, and mountains.
5. The moon's phases are caused by the positions of the moon, the sun, and Earth relative to each other.
6. Planets, moons, comets, meteoroids, and asteroids are members of the solar system.
7. Earth is a part of the solar system. The solar system is in the Milky Way Galaxy. The Milky Way Galaxy is one of many galaxies in the universe. The universe includes all the matter that exists.
8. The sun is an average star. It produces its energy by nuclear fusion, as do all stars.
9. Solar and lunar eclipses may be observed by people on Earth when the sun, Earth, and moon are lined up with each other.
10. Each star has a characteristic size, temperature, brightness, and color.
11. Stars go through a series of changes in their lifetimes. The stages they go through depend on their original mass.
12. Starlight can be studied using visual telescopes, radio telescopes, and spectroscopes.

Vocabulary Review

From the following list, choose the term that best completes each of the statements. Write your answers on a separate piece of paper.

asteroid	orbit
astronomy	phase
black hole	radio telescope
comet	red giant
constellation	revolution
galaxy	rotation
light-year	satellite
lunar eclipse	solar eclipse
magnitude	solar system
meteor	spectroscope
meteorite	sunspot
meteoroid	supernova
nebula	universe
neutron star	visual telescope
nuclear fusion	white dwarf

1. A(n) __?__ is a huge cloud of gas and dust.
2. When the moon moves between the sun and Earth, a(n) __?__ occurs
3. A(n) __?__ is sometimes called a "falling star."
4. The apparent shape of the moon as we see it from Earth is its __?__.
5. High __?__ activity causes the northern lights to increase.
6. A group of billions of stars is called a(n) __?__.
7. The moon is a natural __?__ of Earth.
8. The observed brightness of a star is called its __?__.
9. A(n) __?__ is made of dust and ice and has a long tail.

10. A(n) __?__ is an object that orbits the sun between Mars and Jupiter.
11. The unit used to measure distances between stars is called a(n) __?__.
12. A(n) __?__ uses a lens or a mirror to gather light from a star.
13. A "dish" is the antenna for a(n) __?__.
14. A small piece of rock or mineral matter traveling through space is a(n) __?__.

Chapter Review

Write your answers on a separate piece of paper.

Know The Facts

1. Earth's shape is most like a(n) __?__. (football, basketball, egg)
2. Because of Earth's rotation, the stars appear to move to the __?__ during the evening. (east, west, north)
3. The fact that Earth has seasons is due primarily to __?__. (its revolution around the sun, the tilt of its axis, its distance from the sun)
4. We always see the same side of the moon because __?__. (Earth is moving around the moon, the moon rotates as it moves around Earth, the moon does not rotate as it moves around Earth)
5. The rocky planets are __?__ than the gaseous planets. (larger, smaller, less dense)
6. A star that explodes, producing a nebula, is called a __?__. (nova, supernova, black hole)
7. Comets, planets, and asteroids are similar because they all __?__. (contain ice, move around the sun, have atmospheres)
8. The sun differs from other members of the solar system because it __?__. (is made up of gases, is rotating slowly, is producing energy by nuclear fusion)
9. An instrument that helps us see stars because it gathers light is a __?__. (microscope, spectroscope, telescope)
10. The sun is a __?__. (red giant, white dwarf, yellow star)
11. During periods of high solar activity, Earth experiences __?__. (strong winds, northern lights, meteor showers)
12. A light-year is a unit of __?__. (time, distance, mass)
13. The length of time a star will "live" depends on its __?__. (original mass, temperature, motion)
14. A group of stars that forms a recognizable pattern in the sky and is visible to the unaided eye is called a __?__. (galaxy, solar system, constellation)
15. A __?__ can receive radiation from space on a cloudy day. (radio telescope, refracting telescope, reflecting telescope)

Understand The Concepts

16. The most recent period of high sunspot activity was in 1979–80. What part of the sunspot cycle is the sun in now?
17. How long are periods of daylight on the moon?
18. What evidence is there that the moon rotates on its axis?
19. Draw a diagram to show how images are produced in a refracting telescope.

20. During part of every year the sun never sets at the poles. Is it hot in those areas at that time? Why or why not?
21. Why doesn't it rain on the moon?
22. What information can be learned about a star by studying its absorption (black-line) spectrum?
23. How long is it between a full moon and a third-quarter moon?
24. Describe the life cycle of a star such as our sun.
25. How do Jupiter, Saturn, Uranus, and Neptune differ from other planets in our solar system?

Challenge Your Understanding

26. In order to find the planets in the night sky, why do we look along the same path followed by the sun during the day?
27. If you took a 20-minute time exposure picture of the sky, focusing on the North Star, through how many degrees would the stars move in that time?
28. Name two characteristics of a planet that would allow it to have an atmosphere.
29. The planet Mercury rotates about its axis once every 59 Earth days. It revolves in the same direction around the sun in 88 Earth days. How many Mercury days are there in a Mercury year?
30. Explain why a lunar eclipse lasts longer than a solar eclipse. Use drawings to show what is happening with each.
31. Explain, using a diagram, why the North Star is always visible in the Northern Hemisphere.

Projects

1. Find out what astronomers mean by the term "light pollution" and what they are doing to cope with this problem.
2. Make a series of photos of sunrise or sunset positions. Take two a week for 12 weeks. Can you predict where the sun will set a month beyond your last observation? What causes the changes in position you have observed?
3. Observe the motion of Venus in relation to the sun for a period of three months. Make a large drawing showing what motion Venus is making in relation to Earth and the sun during this time.
4. Build a sundial that keeps good time. Books in the library will give helpful hints.
5. With a constellation chart for the current season, plot the paths of meteors you observe during a meteor shower. Your teacher can give you the dates for yearly meteor showers. The best viewing time is after midnight. You might see up to 25 per hour.
6. In the library, find a photograph of the moon that has major landscape features labeled. Using materials such as clay or plaster of Paris, make a model of a portion of the lunar landscape.
7. Research the topic of celestial navigation used by ships, planes, submarines, and space vehicles.
8. Read A *Connecticut Yankee in King Arthur's Court* by Mark Twain. Plan a class skit based on the "eclipse" scene.

Mining Engineer

Coal mining company is in search of engineers to plan and improve mining operations. Of special concern are mine safety, pollution control, and environmental restoration. Some time will be spent searching for new coal deposits. Experience with systems to bring air and water into underground mines is required. Most of the work involves creative problem-solving.

The position requires an engineering degree with courses in earth science, physics, chemistry, and mathematics. Those hired will begin by learning about coal production and doing routine work under close supervision. Most of the work will be done as part of a team working outside or in mines. Work can be hazardous. Opportunities exist for advancement.

Geologist

Federal government has several positions open for geologists to assist in resource exploration, construction site analysis, and geologic research. Projects include recording plate movements, improving methods of earthquake prediction, monitoring volcanic activity, analyzing ground stability, and searching for oil and minerals. Applicants must be able to travel worldwide and draw maps of their findings. The positions involve working indoors in a laboratory and working outdoors in the field.

A college degree in geology is required with course work in earth science, physics, chemistry, and mathematics. Physical stamina is needed for some of the field work. Applicants must be able to work well with others as part of a team.

Help Wanted

Agricultural Technician

Farm agency seeks technicians to help research scientists who are studying ways to improve farming methods. Research involves how to increase the resistance of farm organisms to disease, how to best breed and feed farm animals, and how to conserve and improve soil. Technicians will talk with farmers and help them apply research findings. The job involves a lot of travel and work outdoors. It also requires the ability to work well with many different people.

Candidates with college or technical school training are preferred. Courses should include earth science, botany, biology, chemistry, and agricultural sciences. On-the-job training will be provided. As experience is gained, there will be opportunities for advancement to a supervisory level.

Aeronautical Technician

Aerospace company has openings for aeronautical technicians to assist engineers and scientists. Technicians perform tests, gather data, write reports, and make sketches that relate to the design of missiles, rockets, and airplanes. Technicians must be able to use highly technical instruments and take readings accurately.

Candidates must have a high school diploma, and preferably some college or technical school training as well. Courses in earth and space science, physics, mathematics, computers, and drafting would be helpful.

Some on-the-job training will be provided. Technicians should be able to work well with others in small teams. When deadlines are near, overtime can be expected.

Understanding The Unit 3

Main Review

Write your answers on a separate piece of paper.

Know The Facts

1. Rocks that contain imprints of past life forms are called __?__. (fossils, igneous rocks, intrusions)
2. Maps that show elevations of mountains are __?__ maps. (geologic, geographic, topographic)
3. Wells for drinking water must be drilled far enough down to reach __?__. (cold water, the water table, pure water)
4. The planet Earth is unique because on Earth water __?__. (exists in all three states, covers 90% of the surface, is blue)
5. An example of physical weathering of rocks is __?__. (acid rain falling on rocks, rocks breaking up by expansion of ice, the reddish color of rocks)
6. The most important factor in the development of soil in any area is the kind of __?__. (bedrock, plant life, climate)
7. The most important agent of erosion is __?__. (ice, water, wind)
8. Glacial deposits are easily detected because they are __?__. (unsorted by size, sorted by size, always found near rivers)
9. Plate tectonics explains the __?__. (Ice Age, pattern of winds, ages of rocks on the ocean floor)
10. Most earthquakes on Earth occur __?__. (at plate boundaries, in the Southern Hemisphere, around the edge of the Atlantic Ocean)
11. Volcanoes can affect many people because __?__. (they occur all over the world, their gases are always poisonous, the fine dust from volcanoes affects the atmosphere)
12. The layer of the atmosphere that contains all the weather is called the __?__. (troposphere, stratosphere, mesosphere)
13. Thunderstorms occur from __?__ clouds. (stratus, cumulonimbus, cirrus)
14. The average salinity of seawater by mass is __?__ of salt per kilogram of water. (.35 g, 3.5 g, 35 g)
15. Waves on water are caused by the __?__. (density of water, moon, wind)
16. Spring tides occur during the __?__ phase. (full-moon, quarter-moon, crescent-moon)
17. Deep-ocean currents are caused by __?__. (ocean trenches, density of water, high temperature)
18. Ocean floor sediments called oozes come from __?__. (volcanic dust, meteoritic dust, organic matter)
19. The best description of Earth's shape is __?__. (oval, pear-shaped, flattened sphere)
20. The motion of Earth that gives us the seasons is __?__. (rotation, revolution, spinning)

21. The length of time the moon takes to go through all its phases is one __?__. (year, day, month)
22. Falling stars are __?__. (meteoroids, meteors, meteorites)

Understand The Concepts

23. What is the difference between rocks and minerals?
24. What determines the classification of rocks?
25. List four properties that help identify a mineral.
26. In what class of rock are most fossils found?
27. Tell the relationship between weathering of rocks and erosion.
28. Describe how sedimentary rocks are formed.
29. How does the theory of plate tectonics explain the location of most earthquakes and volcanoes?
30. How can you tell wind deposits from glacial desposits?
31. Why is air pressure greatest at the surface of the Earth?
32. What conditions determine the kind of precipitation that will fall?
33. Describe the weather that accompanies (a) a typical cold front, (b) a typical warm front.
34. What atmospheric conditions bring about the formation of tornadoes and where do tornadoes most frequently occur?
35. What conditions increase the salinity of sea water?
36. In what part of the ocean would waves tend to be the highest?
37. Why are the sediments on the ocean floor thicker in the Atlantic than in the Pacific Ocean?
38. Explain the cause of the greenhouse effect.
39. What causes (a) the eclipse of the sun, (b) the eclipse of the moon?
40. Besides Earth's revolution around the sun, what other factor helps determine the seasons?
41. What two main elements make up the sun and what is the source of the sun's energy?
42. Why are the polar constellations visible all year?
43. What is the relationship between the temperature of a star and the color of a star?
44. How can astronomers identify the elements present in a star?

Challenge Your Understanding

45. Explain why more people on Earth can see a lunar eclipse than a solar eclipse.
46. Why do giant planets have thicker atmospheres?
47. How can fossils provide evidence of past climates?
48. Describe one fact that supports the idea that the sea floor is spreading.
49. Why are most severe storms related to low-pressure systems?
50. Describe the effect that a nonrotating Earth would have on wind patterns.
51. Why are hurricanes easier to predict than tornadoes?
52. What would happen to tides if Earth and the moon stood still?
53. Why can it be said that the sun is responsible for the water-carved shapes of rocks on seacoasts?
54. Why do some scientists predict that sea levels will rise in the future?

Unit 4 · Physics

The glass fibers used in fiberoptics are so fine that 660 of them can fit through a needle's eye.

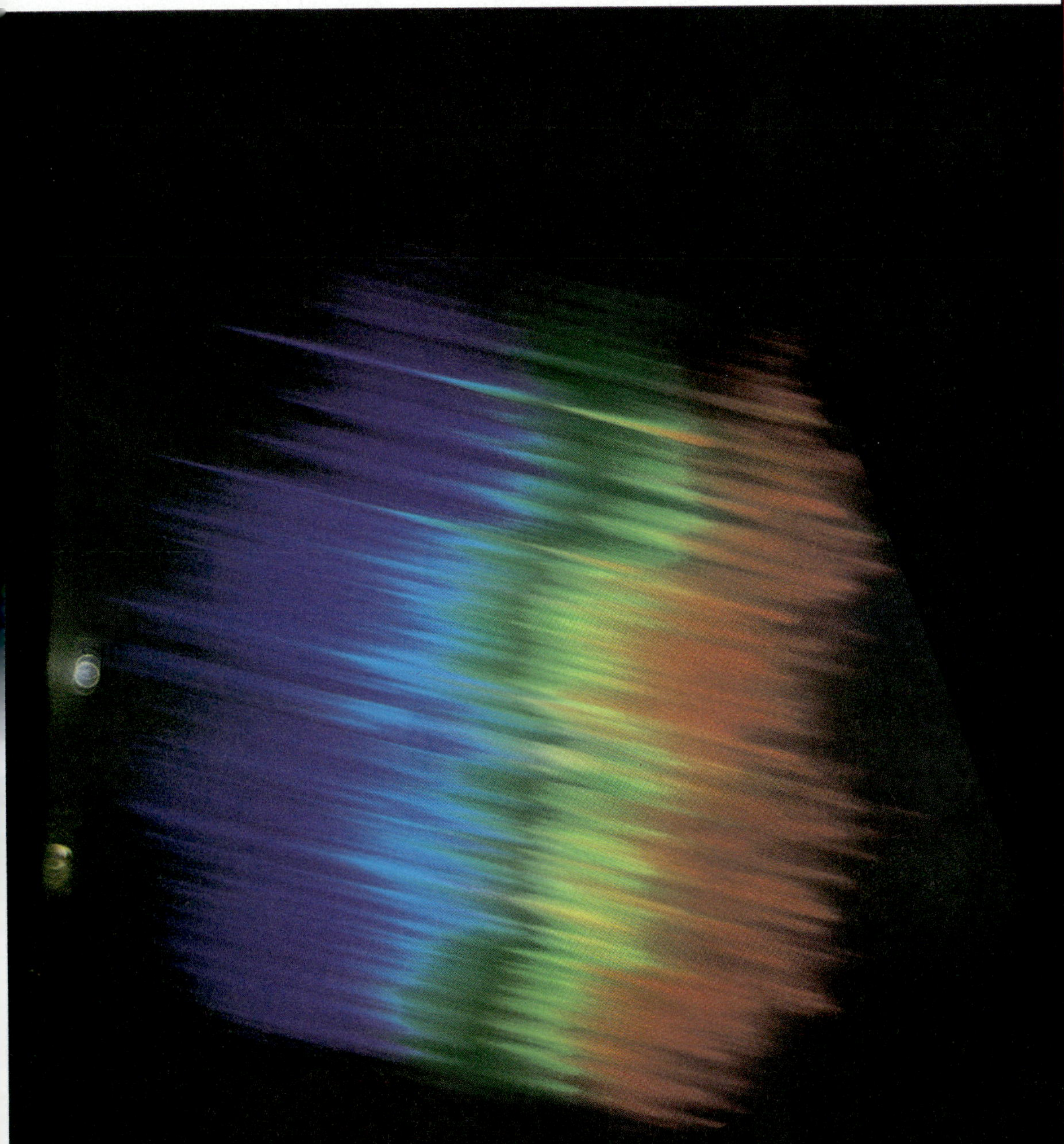

Chapter

16

Motion and Force

Capturing the motion of a speeding car in a photograph is a challenge.

16-1

Motion

Goals

1. To find average speed from the distance traveled in a given period of time.
2. To distinguish between speed and velocity.
3. To define acceleration and give some examples.

It seems that people have always been fascinated by motion. For centuries scientists have followed the paths of the planets across the sky. Every year many people run in marathons, races first held in ancient Greece. Every day people gather somewhere to watch cars or horses race around tracks.

Galloping horses and sleek cars show the beauty and grace of motion. However, in races, beauty and grace are not as important as speed. **Speed** is the distance traveled in a unit of time. You can find average speed by dividing the distance traveled by the time taken.

speed

$$\text{average speed} = \frac{\text{distance}}{\text{time}}, \text{ or } s = \frac{d}{t}$$

EXAMPLE

Find the average speed of an automobile that travels 150 kilometers in two hours.

$$d = 150 \text{ km} \qquad s = \frac{d}{t}$$

$$t = 2 \text{ h} \qquad s = \frac{150 \text{ km}}{2 \text{ h}}$$

$$s = 75 \text{ km/h}$$

Velocity is the speed an object has in a particular direction. An airplane flying due north at a speed of 400 km/h has a velocity of 400 km/h north. Velocities can be represented by arrows. (See Figure 16-1.) The arrowhead shows the direction and the length of the arrow represents speed. Because you must travel in the right direction to reach a chosen destination, it is more useful to know your velocity than your speed.

velocity
(vuh LAHS uh tee)

Figure 16-1 *Velocity arrows show direction and speed. The longer the arrow, the greater the speed.*

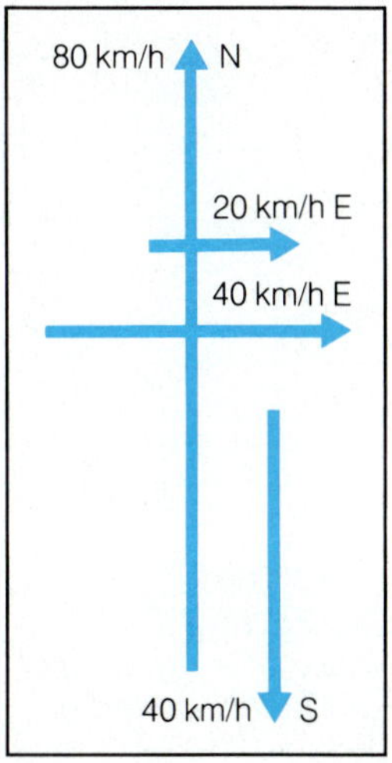

Driving over long distances, you seldom move at the same velocity for long. You may brake the car to avoid an accident. The direction of the highway may change. You may speed up to pass a truck. A car *accelerates* when its velocity changes in speed or direction, or both. **Acceleration** is the change in velocity in a unit of time. To find average acceleration when velocity changes in speed only, divide the change in velocity by the time it takes the velocity to change. Change in velocity is the difference between beginning velocity and final velocity.

acceleration (ak sel uh RAY shun)

The driver of a racing car on an oval track is nearly always accelerating the car. The driver accelerates as the car leaves the starting line and reaches top speed. The driver accelerates (changing speed and direction) while slowing the racing car as it goes into the first turn. (People commonly call a negative acceleration *deceleration*. But scientists use the term *acceleration* for any change in speed.) The driver accelerates while increasing the car's speed out of the first turn. The driver accelerates going around the second turn and once again when gaining speed down the backstretch. Racing cars can go from rest to 100 km/h in about six seconds. This is an acceleration of about 17 km/h/s. This means the car is going 17 km/h faster each second.

Figure 16-2 *The shape of a snow drift can indicate wind direction.*

Your family's automobile doesn't accelerate as quickly as a racing car. Its acceleration is probably about 5 km/h/s. It is more important for your family's automobile to have a short braking distance. This is the distance a car travels after braking and before coming to a stop. The braking distances for an average automobile traveling at different speeds are shown in Table 16-1. The *time* it takes to stop doubles when the speed doubles. However, the *distance* needed to stop is four times as great. This is one reason why speeding is so dangerous.

Table 16-1: Stopping Times and Distances

Initial speed (km/h)	*Time to stop (s)*	*Distance traveled (m)*
10	1	1.4
20	2	5.6
30	3	12.6
40	4	22.4
50	5	35.0
60	6	50.4

Checkpoint

1. Define (a) speed, (b) acceleration.
2. Suppose you are on a train moving southwest. It travels 100 km in two hours. (a) What is its speed? (b) What is its velocity?
3. How does an increase in the speed of an automobile affect the braking distance?

TRY THIS

Place two markers 100 m apart on a playground. Time different people riding a bicycle between the markers. Have a friend signal to you when the bicycle passes the first marker. Start a stopwatch at that time. Have another friend signal when the bicycle passes the second marker. Stop the watch at that point. Determine the bicycle's average speed in meters per second. What is the average speed in kilometers per hour?

16-2

Newton's First Law of Motion

Goals

1. To explain what forces are.
2. To recognize the role of friction in slowing motion.
3. To state Newton's first law of motion and give examples that illustrate the law.

Isaac Newton was an English scientist who did outstanding work in many areas of science. In 1664, at the age of 22, Newton gave an explanation of motion that was true for all objects. By the time of his death, in 1727, his theories about motion were accepted as laws.

force

Newton's *first law of motion* is about objects moving along a straight line. It involves the idea of **force**, a push or a pull on an object. You apply a force when you hit a tennis ball with a racquet, or a baseball with a bat, or move a boat by pulling on an oar.

Like velocities, forces can be represented by arrows. The length of the arrow indicates the strength of the force. In Figure 16-3 the arrows show that the tractor pulls on the rock with a larger force than the horse.

Often, high-flying aircraft are able to take advantage of the forces available in the jet streams — winds far above

Figure 16-3 *A horse and a tractor can pull the object, but the tractor can exert a much greater force.*

Figure 16-4 *If this rider pedals into a strong wind, he will have to work harder to maintain his speed.*

Earth's surface that move at very high speeds. These winds push on the aircraft, adding to the force applied by the plane's engines. Of course, a plane that tries to fly into a jet stream *against* the wind will be slowed down. It would be much like a bike being pedaled into a very strong wind. (See Figure 16-4.)

first law of motion

Newton's **first law of motion** states that *an object in motion at constant velocity along a straight-line path will maintain that motion unless acted on by an outside force. Also, an object at rest will remain at rest unless acted on by an outside force.*

Safety engineers, knowing about Newton's first law of motion, design automobile seat belts to protect you. If the automobile you are riding in suddenly stops, you continue forward. You could bump into or even go through the windshield. Seat belts, fastened around you and attached to the automobile's frame, hold you firmly in place and keep you from continuing forward. An object tends to keep moving forward because of its **inertia**, its resistance to change in its motion. When an automobile stops, you keep moving because you are not part of the automobile.

inertia
(in UR shuh)

Because of its inertia, an object at rest will remain at rest unless acted on by a force. This is the secret of the magician's trick of pulling a tablecloth from under dishes. With a very smooth tablecloth, there is little force on the dishes

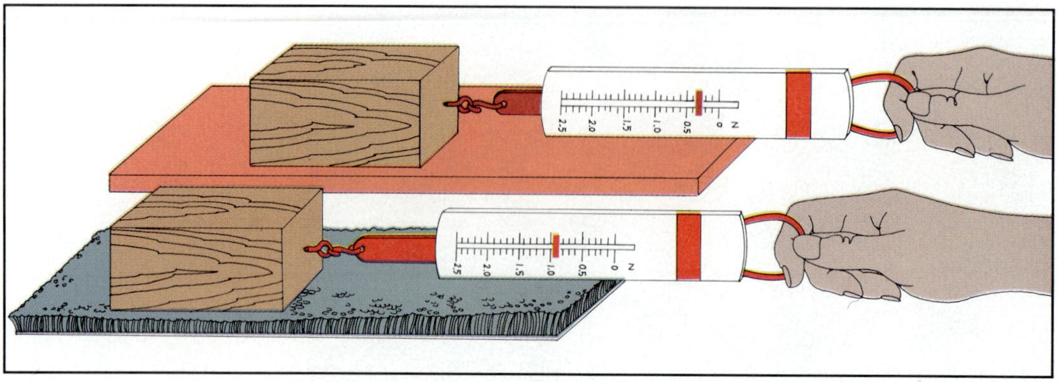

Figure 16-5 *With a spring scale, you can measure how many newtons of force it takes to move an object. Less force is needed to move it over the smoother surface.*

when the cloth moves. Because of their inertia, the dishes remain at rest.

Whenever one object moves over another the motion is opposed by a force. This force, called **friction**, acts against the motion of one surface over another. When you move a box resting on the floor by giving it a push, it soon slows down and stops. Frictional force between the bottom of the box and the floor is what stops the box.

friction

Without friction, an object in motion along a horizontal surface would move at constant velocity forever. There would be no force to oppose its motion. Friction between your shoes and concrete enables you to walk along a sidewalk. An ice-covered path provides very little friction, so it is difficult to walk on.

To reduce friction in machinery and promote freer motion, grease or oil is placed between moving parts. Without these *lubricants* the heat produced by friction could cause the moving parts to wear or even to melt.

The force of friction can be measured with a spring scale. (See Figure 16-5.) A spring scale includes a spring that stretches evenly. The greater the force pulling on the spring, the more it stretches. The scale shows how much force is pulling on the spring. Force is measured in units known as *newtons* (N).

Checkpoint

1. What is a force? Use an arrow to represent (a) a large force to the left, (b) a small force upward.
2. What is friction? How does it affect motion?
3. Give an example to illustrate Newton's first law.

16-3

Newton's Second Law of Motion

Goals

1. To state Newton's second law and give examples to illustrate it.
2. To explain how force, mass, and acceleration are related in the motion of objects.

According to Newton's first law, an object in motion keeps moving at the same velocity unless it is acted on by an outside force. Frictional force, however, is always present. It slows down and stops any moving object, unless a force equal to or greater than friction opposes it.

You can probably think of situations in which frictional force is quite small. (See, for example, Figure 16-6.) By using extremely smooth objects and surfaces, we can produce such a situation. When this is done, a simple relationship between the force acting on an object and its resulting acceleration can be demonstrated. A force applied to an object at rest causes it to accelerate in the direction of the applied force. If the force remains constant, the object will move with a constant acceleration. That is, its velocity will

Figure 16-6 *A hockey puck will slide a long distance over the ice because there is little friction between the ice and the puck.*

increase at a constant rate. For example, if the velocity at the end of one second were 1 cm/s, it would be 2 cm/s at the end of two seconds, 3 cm/s at the end of three seconds, and so on.

Suppose the force applied to the object is doubled. Then the acceleration will double, too. The velocity will be 2 cm/s at the end of one second, 4 cm/s at the end of two seconds, and so on.

Experiments of this kind show that the acceleration of an object increases when the force acting on it increases. There is another factor to consider, however. Suppose the inertia of the object were twice as great. You would have to double the applied force to produce the same acceleration as before.

Newton used the relationship between force and acceleration to measure inertia. He called the measure of inertia *mass*. He defined an object's **mass** as the ratio of an applied force to the resulting acceleration of the object.

mass

$$\text{mass} = \frac{\text{force}}{\text{acceleration}}, \text{ or } m = \frac{F}{a}$$

Figure 16-7 *When the force on the car is doubled (b), the acceleration is doubled. If the mass is then doubled (c), the acceleration is halved.*

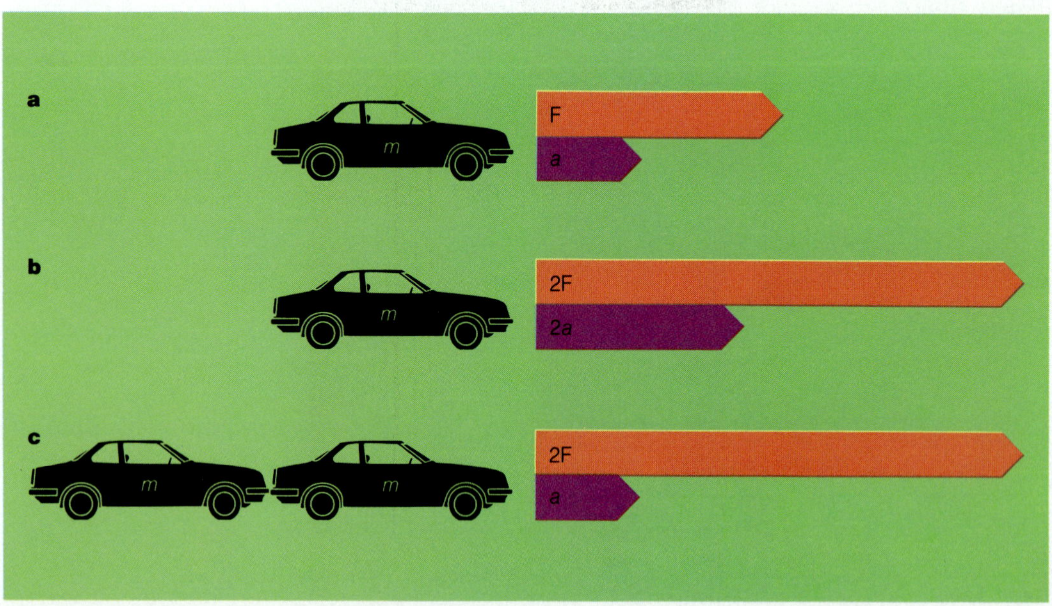

Two objects are of equal mass if the same force applied to each object produces the same acceleration.

The effects of mass and force on acceleration may be summarized as follows:

1. For a given mass, acceleration is greater when the force acting on it is greater.
2. For a given force, acceleration is less when the mass on which the force acts is greater.

Figure 16-7 illustrates the effects of force and mass on acceleration.

second law of motion

Newton summarized his results in the **second law of motion**. *When a force is applied to an object, the object accelerates in the direction of the applied force. The acceleration is greater when the force is greater. The acceleration is less when the mass of the object is greater.*

Newton's first two laws can help you understand the interaction of force and motion that takes place when you ride in a car. A car at rest will not move until a force at least equal to the force of friction is applied. If the applied force is just equal to friction, the car moves at a constant velocity. For a car to accelerate, the force applied must exceed friction. It is the *net* force, the force in excess of friction, that causes acceleration. Once a car is in motion, a force greater than friction must continue to be applied to maintain its acceleration. As the car's velocity increases, the retarding force of air resistance increases, too. When the forces tending to move the car forward and those tending to resist its motion are equal, the car moves at a constant velocity.

Checkpoint

1. What is Newton's second law of motion? Give an example to illustrate this law.
2. To make an object accelerate, what must be true of the force applied to the mass?
3. A force that is 5 N greater than the force of friction is applied to an object. The object accelerates. A force 10 N greater than friction is applied to an object of equal mass. How will the accelerations of the objects compare?

16-4

Newton's Third Law of Motion

Goal

To state Newton's third law of motion and give examples that illustrate it.

When you push on a friend while both of you are ice-skating, your friend slides away from you. But you also move. (See Figure 16-8.) When you push on your friend, you get a push in the opposite direction. A similar thing happens when burning fuel is directed out of the engines of a rocket. The burning fuel pushes on the rocket and the rocket pushes back with equal force. If you've ever watched a rocket blast off, you know that the burning fuel and rocket move in opposite directions.

Figure 16-8 *When one skater pushes on another, the first skater moves, too—backward.*

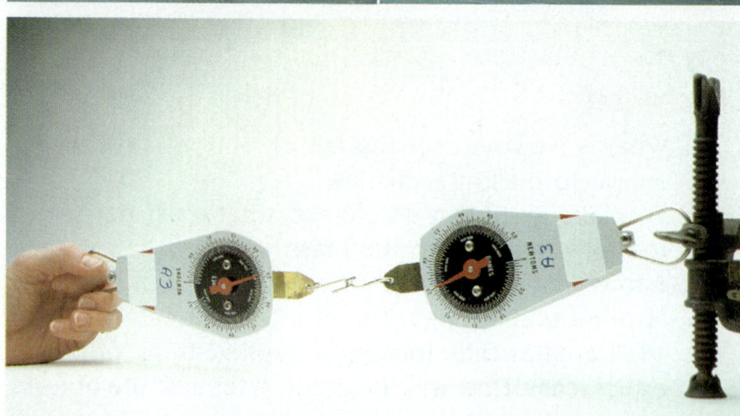

Figure 16-9 *If a person pulls on one scale, and the other is attached to a stationary object, the scales show that the person and the object exert the same force.*

Figure 16-10 *This swimmer is moving fast because the water pushes on her as she pushes on it.*

third law of motion

Newton's third law of motion explains what happens when one object applies a force to another object. The **third law of motion** states that *for every action force there is an equal but opposite reaction force*. This means that if you push or pull on an object, the object will push or pull back with an equal force. (See Figure 16-9.) When an automobile leaves the road and hits a tree, the tree pushes back with the same amount of force. Every time you take a step you have to push against Earth. Earth pushes back with an equal force in the opposite direction, and you move forward. Earth moves also, but its mass is so huge that its acceleration cannot be measured.

Newton's third law explains why it is unwise to step from a small boat that is not tied to the dock. As you step from the boat, it will move away from the dock and your first step could be into the water. Then you might have to swim to shore. Swimming, too, involves Newton's third law. When you take a stroke, pushing against the water, the water pushes back with equal force. With repeated strokes you move forward. (See Figure 16-10.)

TRY THIS

Build a model of a rocket by attaching a balloon to a soda straw on a string as shown in Figure 16-11. As the balloon forces air to the right, the air pushes back on the balloon, forcing it to the left.

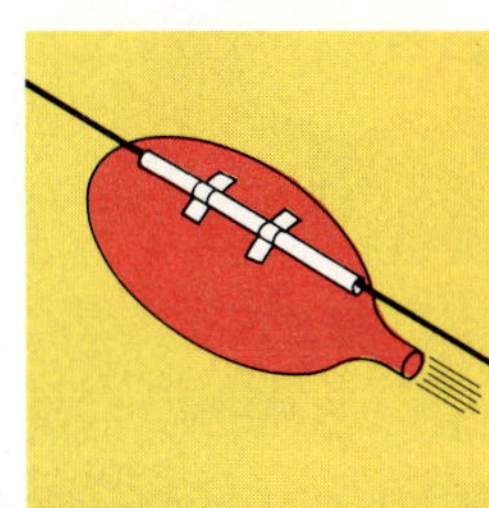

Figure 16-11 *Your balloon "rocket" should look like this.*

Checkpoint

1. State Newton's third law.
2. Explain the act of walking in terms of the third law.
3. Explain how the launching of objects into space illustrates the third law.

16-5

Gravity

Goals

1. To give evidence that Earth attracts objects at or near its surface.
2. To recognize that all objects attract each other with a force that depends on the mass of each object and the distance between them.

When you drop something, it falls to the ground. This is true of all objects near the surface of Earth. A ball, a skydiver, raindrops, and snowflakes all fall toward Earth. Perhaps Newton's most important contribution was explaining how objects fall toward Earth.

Newton was born in the same year that Galileo Galilei, an Italian scientist, died. This was not the only way in which their lives crossed, however. Galileo found that all objects

Figure 16-12 *If the mass of one object is doubled (b), the force of gravity between the objects doubles. If the distance is doubled instead (c), the force becomes one quarter as great.*

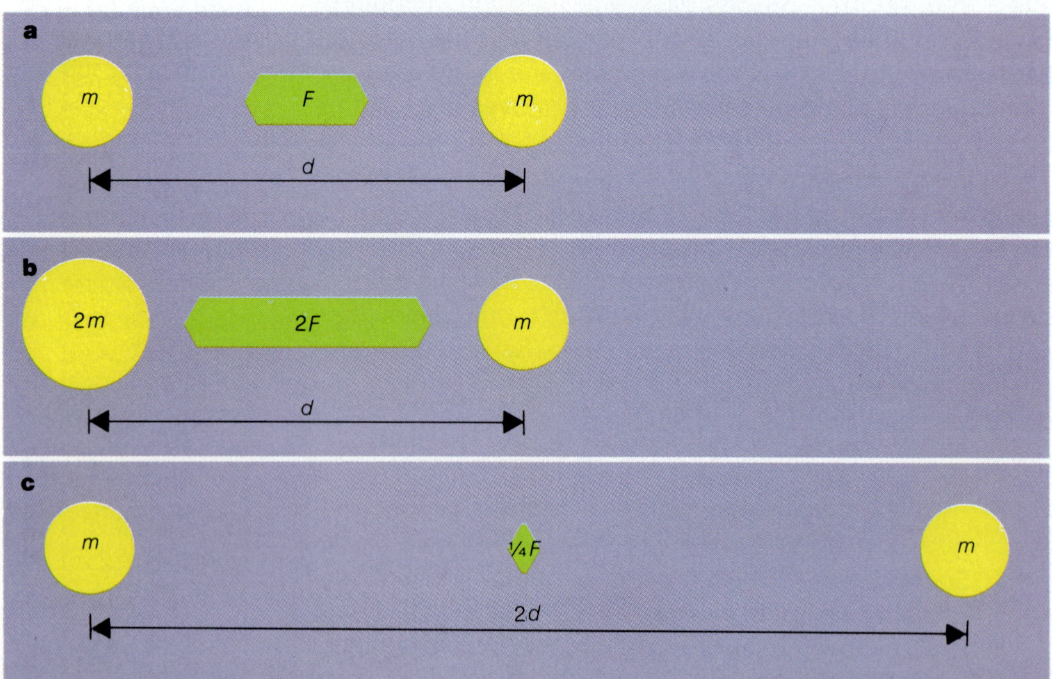

accelerate at the same rate when they fall toward Earth. Galileo tested falling objects when he was a mathematics professor at the University of Pisa, in about 1590. He dropped objects made of heavy metals (gold, lead, and copper) along with light objects, all from the same height. All of them hit the ground at the same time. (You may have heard or read that Galileo dropped objects from the Leaning Tower of Pisa. Historians have not found any proof that he did.)

Using Galileo's experiments with falling objects, Newton suggested that the acceleration must be due to a force. The force, he believed, was the attraction Earth has for other objects.

According to Newton's third law, the force pulling an object toward Earth is equal to the force pulling Earth toward the object. Newton suggested that all objects attract each other in the same way. He was able to explain the motions of Earth (and other planets) around the sun. The explanation was based on the mutual attraction between each planet and the sun. This force of attraction between objects is called the **force of gravity**. The greater the mass of either object, the greater the force of attraction. The greater the distance between the objects, the smaller the attraction. (See Figure 16-12.)

force of gravity

Compared to objects near its surface, Earth is very large. The distances from its center to all objects near its surface are very nearly the same. Since the size of Earth and its mass remain the same, the force of gravity acting on any object on or near Earth depends only on the mass of the object.

Checkpoint

1. Why do objects fall toward Earth?
2. How did Newton know there was a force pulling falling objects?
3. Bill's father's mass is twice Bill's. How will the gravitational force acting on Bill compare with that acting on his father?
4. If Earth's orbit were closer to the sun, would the force of gravity between the sun and Earth be greater or less than it is now?

16-6

Weight and Mass

Goals

1. To explain the difference between weight and mass.
2. To explain how mass can be measured on a balance.
3. To explain why an object's weight is less on the moon than on Earth.

weight

The gravitational force acting on an object is called the object's **weight**. The greater the object's mass, the greater the gravitational force. If the mass of object A is twice that of object B, its weight will be twice B's also. Earth will pull on A with twice as much force as on B. If the weights of two objects, measured at the same location, are the same, they must have the same mass. If the objects are given the same horizontal push, they will have the same acceleration.

Like other forces, weights are measured in newtons (N). An apple weighs about 1 N. You probably weigh between 400 and 1000 N. A spring scale can be used to measure weight. (See Figure 16-13.)

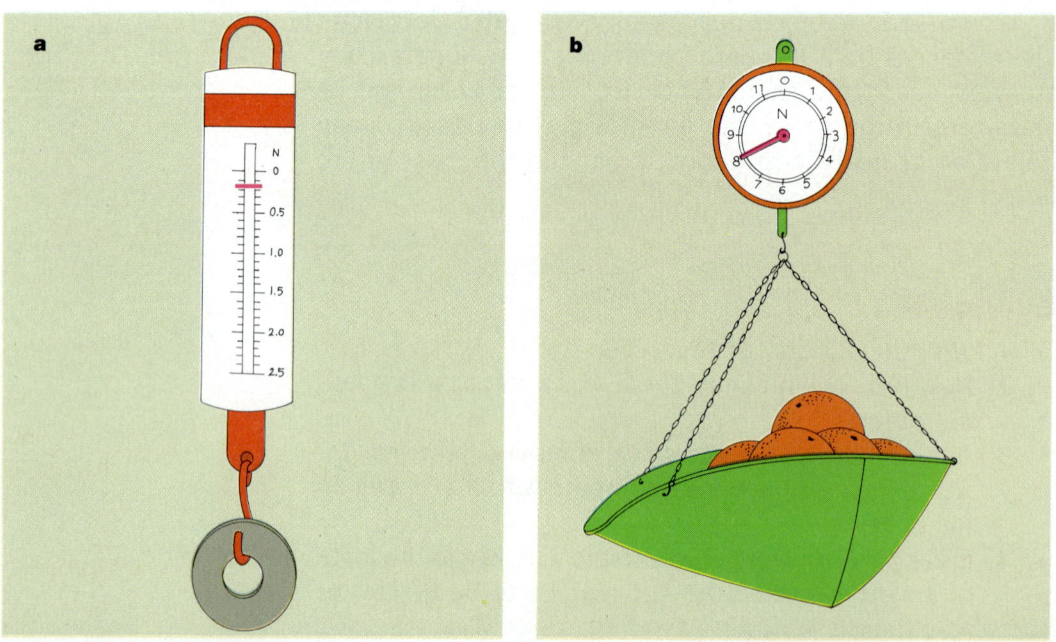

Figure 16-13 *(a) This spring scale measures the force of gravity on a washer. (b) Groceries can be weighed on a similar type of spring scale.*

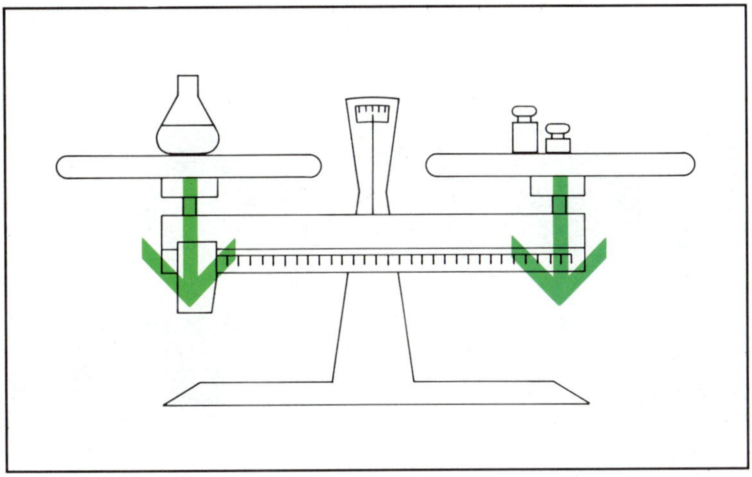

Figure 16-14 *An equal-arm balance. When the mass of the object on the left pan equals the sum of the standard masses on the right pan, the pointer is at zero.*

Mass is measured in *kilograms* (kg). A mass of one kilogram weighs 9.8 N near Earth's surface. Since 9.8 N is almost 10 N, an object that weighs 5 N has a mass of about 0.5 kg, while an object that weighs 20 N has a mass of about 2 kg.

Normally you would find the mass of an object with the use of an equal-arm balance like the one in Figure 16-14. The object of unknown mass is placed on the left pan of the balance. Then standard, or known, masses, such as kilograms or grams, are added to the right pan. When the gravitational force acting on each pan is the same, the balance is level, and the pointer is at the center of the scale. (See Figure 16-14.) This shows that the masses are equal.

Suppose you went to the moon. Suppose, further, that you brought along a balance, standard masses, and a spring scale. With this equipment, you could compare the masses and weights of objects on the moon with the masses and weights of the same objects on Earth.

At your moon station, you would find the mass of any object to be the same as on Earth. But the *weight* of an object on the moon is not the same as on Earth. The spring scale would show that the weight of an object is only about one sixth of its weight on Earth. This is because the moon pulls on an object at or near its surface with only about one sixth as much gravitational force as Earth does. If your mass is 70 kg, you weigh nearly 700 N on Earth. On the moon, you would weigh only about 120 N. Since your muscle strength is the same on the moon as it is on Earth, you would appear to be much stronger on the moon. You could easily lift your

Figure 16-15 *The backpack carried by an astronaut on the moon has a large mass. But, because its weight on the moon is so small, it can be carried easily.*

own mass, for example. The moon's smaller force of gravity allowed astronauts on the moon to move about easily even though they carried backpacks almost equal to their own mass. (See Figure 16-15.)

Checkpoint

1. If object A weighs twice as much as object B, how do their masses compare?
2. What is the weight of a 10-kg mass on Earth?
3. Explain how an equal-arm balance is used to measure mass.
4. How would your mass on the moon compare with your mass on Earth? How about your weight?

TRY THIS

Using a spring scale, find the weights, in newtons, of various known masses. What is the ratio of each object's weight to its mass (divide weight by mass)? Is it the same for all the masses you tried? How would the ratio on Earth compare with the ratio on the moon?

16-7

Falling Objects

Goals

1. To explain why all objects fall with the same acceleration.
2. To describe the effect of air resistance on falling objects.
3. To recognize that the horizontal and vertical motions of falling objects are independent of each other.

Suppose you dropped a marble and a baseball at the same time, from the same height. You would find that they hit the ground at the same time. (See Figure 16-16.) Falling rocks and pebbles, bricks and tennis balls, ice cubes and refrigerators, all cover about the same distance in the same amount of time.

At first, this behavior may seem strange. You might expect the larger of two masses to fall faster, since the gravitational force acting on it is greater. But the larger mass *requires* a larger force to accelerate it. (Remember Newton's second law.)

The same situation applies when different cars are accelerated. Imagine two cars, one having a mass of 500 kg and the other a mass of 1000 kg. Now imagine that a force of 100 N is applied to each of the cars. The more massive car accelerates half as fast as the less massive one. However, if the force acting on the 1000-kg car is doubled, this car accelerates at the same rate as the 500-kg car. Similarly, a 2-kg object that is falling requires twice as much force as a 1-kg object to give it the same acceleration. But the gravitational force (weight) acting on a 2-kg object *is exactly* twice as great as that on a 1-kg object. Thus, the acceleration is the same for both objects.

As an object falls faster, the air resistance to its fall becomes greater. However, air resistance is not the same for all falling objects. A leaf has a larger surface compared to its mass than a stone has. As a result, a falling leaf is slowed down more by air resistance than a falling stone is. For this reason, a leaf and a stone will not fall with the same acceleration. As the stone falls faster and faster the air resistance to its fall becomes greater. Soon air resistance equals the force

Figure 16-16 *The larger ball has the greater mass. But it does not fall faster than the smaller one.*

of gravity. Then the stone falls at constant velocity. A leaf reaches constant velocity much earlier in its fall.

Newton predicted that if a feather and a hammer were to fall in a place where there was no air, both would fall with the same acceleration. Three hundred years later astronauts on the moon tested Newton's prediction and found it to be true. Of course, by then no one doubted that he was correct.

A feather or leaf very quickly reaches constant velocity as it falls through the air. No matter what its mass, an object will reach constant, or *terminal*, velocity if it falls long enough. Skydivers take advantage of this fact to perform stunts as they fall. As a skydiver falls faster and faster, the air resistance increases. After 12 s, the frictional force between skydiver and air becomes equal to the force of gravity. To increase air resistance, the diver may "spread-eagle" to increase the surface exposed to air. (See Figure 16-17.) When the air resistance and gravitational force are equal, the skydiver falls at the terminal velocity. For skydivers, terminal velocity is about 50 m/s. When the skydiver's parachute has unfurled, the large area of the open chute increases the air resistance. The increased upward force on the parachute decelerates the skydiver to a new, smaller, terminal velocity.

Figure 16-18 shows two balls that were released at the same time. One was allowed to fall straight down. The other was pushed horizontally. The horizontal lines in the illustration represent the distance fallen in a given period of time.

Figure 16-17 *Skydivers can increase air resistance and slow their fall by spreading out their arms and legs.*

Figure 16-18 *One ball is pushed horizontally while the other drops straight. Yet they both fall toward Earth at the same rate.*

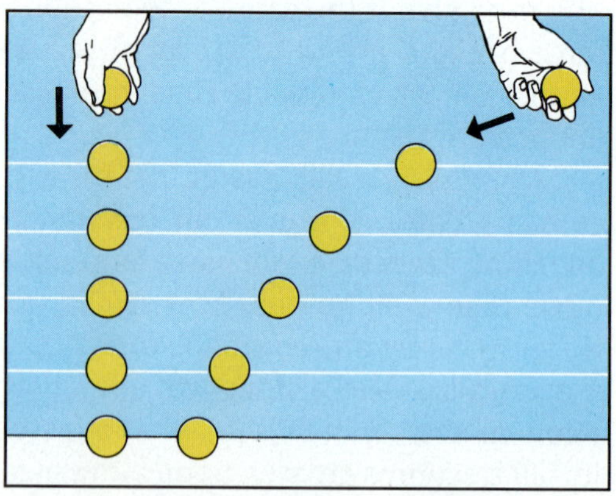

The lines show that both balls fell the same distance during the same time. Horizontal motion had no effect on the vertical motion. The two motions are, as Galileo discovered, independent of one another. Once both balls started to fall, the force of gravity was the only force acting on them. Since gravity gives all falling objects the same acceleration, the balls reached the floor at the same time.

Checkpoint

1. Explain why a marble and a baseball will fall side-by-side.
2. Why do all objects that fall through air eventually reach a terminal velocity?
3. A bullet is fired horizontally over level ground. At the moment the bullet leaves the gun, another bullet is dropped from the same height as the gun barrel. Will both land at the same time? Explain.

TRY THIS

Place two coins 5 cm apart near the edge of a table. Place a flexible ruler as shown in Figure 16-19. Release the ruler, knocking both coins off the table. Listen for the coins to strike the floor. Do both coins reach the floor at the same time? What does this tell you about the *downward* acceleration of the two coins?

Figure 16-19 *For the activity to work, the ruler and coins must be in these positions.*

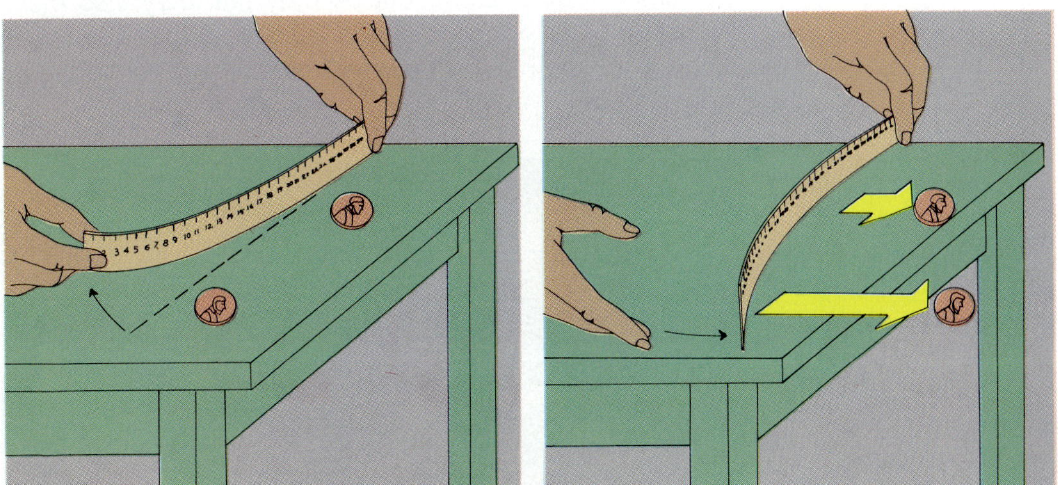

16-8

ENRICHMENT

Circular Motion

Goals

1. To recognize that an inward force is required to make objects move in a circle.
2. To explain weightlessness.

centripetal force (sen TRIP uh tuhl)

The hammer-thrower in Figure 16-20 has to pull inward on the revolving mass to keep it moving in a circle. If he didn't, the hammer would fly off along a straight-line path, as it does when he lets go (Newton's first law). The inward-directed force that keeps an object moving in a circle is the **centripetal force**.

When a car goes around a curve, the tires push outward against the road, and the road pushes inward against the tires (Newton's third law). The inward force keeps the car moving along its curved path. Should the car move onto ice along the curve, the friction between tires and road that provides the centripetal force might become very small. The car would then follow a straight-line path off the road.

Earth *satellites*, including the moon, are in orbit because of Earth's gravity. Newton explained how Earth's gravity provides the centripetal force that keeps the moon in orbit. If gravity could be "shut off," satellites would fly off into space along straight-line paths just as a hammer does when it is released by a thrower. The moon and other Earth satellites are falling toward Earth. But as they fall, they also move sideways. Their sideways speed is so great that when combined with their fall, the satellites follow a curved path around Earth.

Figure 16-20 *The hammer-thrower pulls inward to keep the hammer moving in a circle. When he lets go, the hammer travels in a straight line.*

Figure 16-21 *This is a NASA design for a space station. It will rotate to create artificial gravity.*

Astronauts on an orbiting space shuttle are said to be "weightless." They do not feel the effects of gravity that are felt on Earth. They float freely in the ship's cabin because gravity causes the astronauts and the ship to fall toward Earth at the same rate.

Long periods of weightlessness might weaken the muscles of astronauts. This could occur because they do not use their muscles to support their weight. For this reason, some plans for large space stations where people will live, work, and play for long periods of time call for the space station to rotate. (See Figure 16-21.) The outside of the rotating station will be accelerated inward, creating an artificial gravity. People walking along the outer rim will feel the ship pushing inward against their feet. Thus, "down" in this artificial gravity will be "out" toward the outer edge of the spinning space station. "Up" will be "in" toward the ship's center.

Checkpoint

1. What is a centripetal force?
2. Why do astronauts in orbit feel "weightless"?
3. How might artificial gravity be created in a space station?

Space Shuttle

Its powerful engines blazing, the Space Shuttle lifts off on another flight to orbit. The Space Shuttle Orbiter is the first spaceship to make round trips from Earth to space to Earth and back to space. Its large cargo bay and robot arm allow the Shuttle to carry payloads as large as 29,500 kg and place them into orbit. Satellites can now be carried into space or recovered and returned to Earth to be repaired or modified. Satellites and spaceships can now make more than one voyage into space.

At liftoff, the two rocket boosters and the Orbiter's main engines fire together. The force accelerates the ship to an altitude of 43 km and a speed of 5100 km/h in slightly more than two minutes. The rocket boosters are then released. They parachute into the ocean, where they are recovered.

The Orbiter's engines, fueled by the giant external fuel tank (ET), continue to push the ship higher and faster. At a speed of 28,000 km/h and an altitude of 106 km, the empty fuel tank is released. Fuel needs beyond this point are relatively small. The 44 rockets in the ship's orbital maneuvering system (OMS) and two small rear engines provide the forces needed to bring the ship to a stable orbit.

To return the Orbiter to Earth, one or more of the OMS rockets are fired in the direction of the ship's motion. This produces an equal and oppositely directed force against the Orbiter (in accordance with Newton's third law), causing it to slow down. It re-enters the air that blankets Earth. Here, friction further slows down the ship. Heat produced by the friction would turn the Orbiter into a flaming meteor were it not for the insulating tiles that cover its surface. After its hot flight through the atmosphere, the spacecraft glides to a landing under the control of the astronauts. The Shuttle is then prepared for another voyage into space.

Understanding The Chapter

16

Main Ideas

1. Speed is the distance traveled in a unit of time. Velocity is speed in a particular direction. Acceleration is the change in velocity in a unit of time.
2. An object at rest will remain at rest, and an object moving in a straight line will maintain its motion unless acted on by an outside force.
3. Friction acts against the motion of one surface over another.
4. If enough force is applied to an object, the object accelerates in the direction of the applied force.
5. For every action force there is an equal but opposite reaction force.
6. The force of gravity depends on the mass of each attracting object and the distance between them.
7. Weight is the gravitational force acting on an object.
8. An object has the same mass on the moon as it does on Earth, but it weighs less on the moon.
9. In the absence of air resistance, all objects fall to Earth with the same acceleration.
10. Objects that are falling in air reach a terminal velocity if they fall far enough.
11. The horizontal and vertical motions of falling objects are independent of one another.
12. Inward-directed forces, called centripetal forces, keep objects moving in circular paths.

Vocabulary Review

From the following list, choose the term that best completes each of the statements. Write your answers on a separate piece of paper.

acceleration
centripetal force
force
force of gravity
friction
inertia
mass
speed
velocity
weight
first law of motion
second law of motion
third law of motion

1. A(n) __?__ is a push or a pull.
2. The force required to move an object in a circle is called __?__.
3. When one surface slides over another there is a force called __?__ which slows the movement.
4. According to the __?__ there is an equal and opposite reaction force for every action force.
5. An object's resistance to a change in its motion is its __?__.
6. A 1-kg ball will fall with the same __?__ as a 2-kg ball.
7. The __?__ between two objects depends on each object's mass and the distance between them.
8. The distance an object travels in a unit of time is the object's __?__.
9. A 1-kg __?__ weighs 9.8 N near Earth's surface.
10. An object traveling south at 70 km/h has a(n) __?__ of 70 km/h south.
11. __?__ is the distance traveled in a unit of time.

Chapter Review

Write your answers on a separate piece of paper.

Know The Facts

1. An object's __?__ is greater on Earth than it is on the moon. (weight, mass, velocity)
2. Newton's __?__ law of motion states that an object in motion will maintain that motion unless acted on by an outside force. (first, second, third)
3. Mass is the ratio of the applied __?__ to the resulting acceleration. (weight, speed, force)
4. You can walk because of the force of __?__ between your shoes and the ground. (gravity, inertia, friction)
5. The acceleration that results when you and your friends push a stalled car with a given force can be predicted by Newton's __?__ law of motion. (first, second, third)
6. A pushes on B with a force of 30 N. B pushes back on A with a force of __?__. (20 N, 30 N, 10 N)
7. There is a 10-N force acting on an object. The frictional force on the object is 3 N. The net force on the object is __?__. (10 N, 5 N, 7 N)
8. You drop a baseball at the same time a baseball leaves the pitcher's hand. The force of gravity acting on your baseball is __?__ the force of gravity acting on the pitched ball. (less than, more than, equal to)
9. The terminal velocity of a falling oak leaf is __?__ the terminal velocity of a falling acorn. (less than, equal to, more than)
10. A 1-kg rock falls with an acceleration of 9.8 m/s/s. A 2-kg rock will fall with an acceleration of __?__ m/s/s. (4.9, 19.6, 9.8)

Understand The Concepts

11. If a car travels 15 km in 15 min, what is its average speed in km/h?
12. An object weighs 98 N. What is its mass?
13. The force of friction acting on a car is 100 N. You apply a horizontal force of 150 N. What is the net horizontal force acting on the car?
14. Does the third law of motion apply to the motion of jet planes? Explain.
15. A person who weighs 800 N stands on a floor. What force does the floor exert on the person?
16. Explain why the moon moves in a nearly circular path around Earth.
17. Why does the terminal velocity of a piece of paper increase when it is squeezed into a ball?
18. Why could astronauts jump higher on the moon than on Earth?
19. Why is it more difficult to bicycle uphill than downhill?
20. Explain why it takes longer to bring an automobile to a stop on a wet or icy road.
21. Why are astronauts orbiting in a space vehicle "weightless"?
22. What force do skiers utilize in skiing downhill? What force can skiers decrease by waxing their skis?
23. Explain how the third law of motion applies to the rowing of a boat or the paddling of a canoe.

Challenge Your Understanding

24. An object is accelerating at a rate of 10 km/h/s. What would its acceleration be (a) if its mass were doubled, (b) if the force acting on it were doubled, (c) if both mass and force were doubled at the same time?
25. Explain how an equal-arm balance works.
26. A car traveling at 50 km/h is given an acceleration of 2 km/h/s. How fast will the car be going 10 s after it starts to accelerate?
27. An object weighs 49 N on Earth. What is the object's mass? About how much will it weigh on the moon if the moon's force of gravity is one sixth that of Earth?
28. Two boxes are at rest on the same floor. One box weighs twice as much as the other one. If each of the boxes is pushed across the floor, for which one will the force of friction be greater? How do you explain this?
29. Describe several swimming strokes, such as the backstroke and breaststroke. Explain how each stroke causes the swimmer to move.
30. The sun and the moon both pull on Earth's oceans, causing tides. Explain why the moon's pull has the greater effect.

Projects

1. Try to pull a sheet of plastic or smooth cloth out from under some unbreakable objects, such as plastic dishes or silverware, without disturbing the objects.
2. Find what the posted speed limits are in your neighborhood or along a highway near your home. What do you think are the reasons for the different speed limits?
3. While skating with a friend, stand facing each other. Ask your friend to give you a push. In which direction does your friend move? In which direction do you move? Who moves faster? Try this experiment with someone much bigger or smaller than you. What effect does your partner's mass have on the result?
4. Prepare a library report on the life and work of Galileo or Newton.
5. Interview some people at your local highway department. Find out how they decide on the location of highway markers such as those that advise motorists to slow down or that tell them where to get on or off a highway. Report your findings to the class.
6. Punch a hole in the side of a styrofoam cup near the bottom of the cup. Fill the cup with water and hold it above a sink. Notice how a stream of water emerges from the side of the cup. What happens to the stream when you let go of the cup and it falls to the sink? Try to explain what you observe.
7. Use a library to find the minimum speed a satellite must have to orbit Earth. How do such speeds affect astronauts?
8. Check government regulations on the crash resistance of automobile bumpers. Why do you think these regulations exist? Do you think they help save lives?

Chapter
17
Work and Energy

Machines are used to do work.

17-1

Work

Goal

To define and give examples of work.

Every day you do things that take time and effort. You sweep the floor, mow the lawn, complete your school lesson, and plan tomorrow's activities. But not all of these things fit the scientific definition of work. To do work in the scientific sense, you must apply a force that causes an object to move through a distance. You do work when you push a mower across a lawn or lift your school books. You do not do much work in the scientific sense when you solve your math problems or read your history lesson. (Of course, you do push your pencil and turn a few pages.)

Figure 17-1 *This person is trying hard to lift the window. Because it does not move, no work is done.*

If the force acting on an object is less than the friction opposing its motion, no work is done. This is because the object does not move. The distance it travels is zero. If you push or pull on an object, and it does not move, no work is done. (See Figure 17-1.)

If you know the force acting on an object and the distance the object moves, you can find out how much work is done. **Work** is the force acting on an object multiplied by the distance the object moves.

work

$$\text{work} = \text{force} \times \text{distance, or } W = F \times d$$

EXAMPLE

You apply a horizontal force of 40 N to push a mower a distance of 30 m across a lawn. Find the work done.

$$W = F \times d$$
$$W = 40 \text{ N} \times 30 \text{ m}$$
$$W = 1200 \text{ N}\cdot\text{m}$$

Notice that the unit for work is the newton-meter (N • m). This unit has a special name. The work done when a force of 1 N acts over a distance of 1 m is 1 N • m, or one **joule** (J). The unit was named in honor of James Joule, a 19th-century English scientist.

joule (jool)

A person who walks up a flight of stairs must do work to

overcome gravitational force. Remember that the amount of this gravitational force equals the person's weight. Since gravitational force is directed vertically downward, the force that the person exerts must be directed vertically upward. Thus, the work that the person does is the product of the force (the person's weight) and the vertical distance through which the person "is lifted" (the vertical height of the stairs).

The amount of work done in climbing straight up a ladder is the same as that done in walking up a very gentle incline. This is so because the vertical distance covered is the same in each case. Of course, with a gentle incline, the work is "spread out" over a longer period of time. So walking up the incline is easier than climbing the ladder.

EXAMPLE

You use a vertical force of 10 N to lift an object to a height (distance) of 1 m. Find the work done, in joules.

$$W = F \times d$$
$$W = 10 \text{ N} \times 1 \text{ m}$$
$$W = 10 \text{ N} \bullet \text{m} = 10 \text{ J}$$

EXAMPLE

A person who weighs 420 N walks up a flight of stairs that has a vertical height of 5 m. What work, in joules, does the person do?

$$W = F \times d$$
$$W = 420 \text{ N} \times 5 \text{ m}$$
$$W = 2100 \text{ N} \bullet \text{m} = 2100 \text{ J}$$

TRY THIS

Use your mass in kilograms to find your weight in newtons. (Remember, a 1-kg mass weighs 9.8 N.) Figure out how much work you do when you walk up a certain flight of stairs.

Checkpoint

1. What is the scientific definition of work?
2. While rearranging furniture, you apply a force of 70 N to move a couch 6 m. How much work is done in moving the couch? (Remember, $W = F \times d$.)
3. A person who weighs 500 N walks up a flight of stairs that has a vertical height of 2 m. How much work is done in climbing the stairs?

17-2

Machines

Goals

1. To define a machine and give some reasons why people use machines.
2. To explain how a lever works.

machine

A **machine** changes the direction or strength of a force. Machines can save time and effort. Using machines, such as a bicycle or roller skates, you can get where you are going more quickly than if you walk. Lifting heavy loads can be made easier with machines like a jack or a crane. (See Figure 17-2.) With a small jack, you can lift an automobile.

Figure 17-2 *This crane is a machine that can easily lift heavy freight.*

A seesaw is a familiar machine. It can change both the direction and the strength of a force. By making one side of the seesaw longer than the other, a smaller person's weight on one end of the plank can balance or lift a larger person on the other end. (See Figure 17-3.) The *downward* force acting on the woman can produce an upward force on someone at the opposite end of the seesaw. The distance the woman moves downward is more than the distance the man is lifted. The work (force × distance) done *by* the woman is very nearly the same as the work done *on* the man.

The seesaw is an example of a basic type of machine called the *lever*. You can use another form of the lever to move a heavy rock that you cannot budge without using

Figure 17-3 *This seesaw is balanced because the smaller person is farther from the middle.*

fulcrum

some kind of machine. (See Figure 17-4.) Levers turn, or pivot, about a point called the **fulcrum**. The fulcrum divides the lever into two parts or *arms*. The *effort arm* is the distance from the fulcrum to the *effort force*, the force applied to the machine. The *resistance arm* is the distance from the fulcrum to the *resistance force*, or *load*, the force resisting the machine's effort.

A small effort force applied to a lever can match a large resistance force. This happens when the fulcrum is placed so that the effort arm is much longer than the resistance arm. Levers can be used to increase the effect of an effort force or to make things more convenient. Many persons with handicaps depend on machines based on the lever to make their lives easier. (See Figure 17-5.)

The human body contains a number of levers. When you lift an object, your lower arm acts as a lever. The biceps muscle, which is attached to the shoulder at one end and to

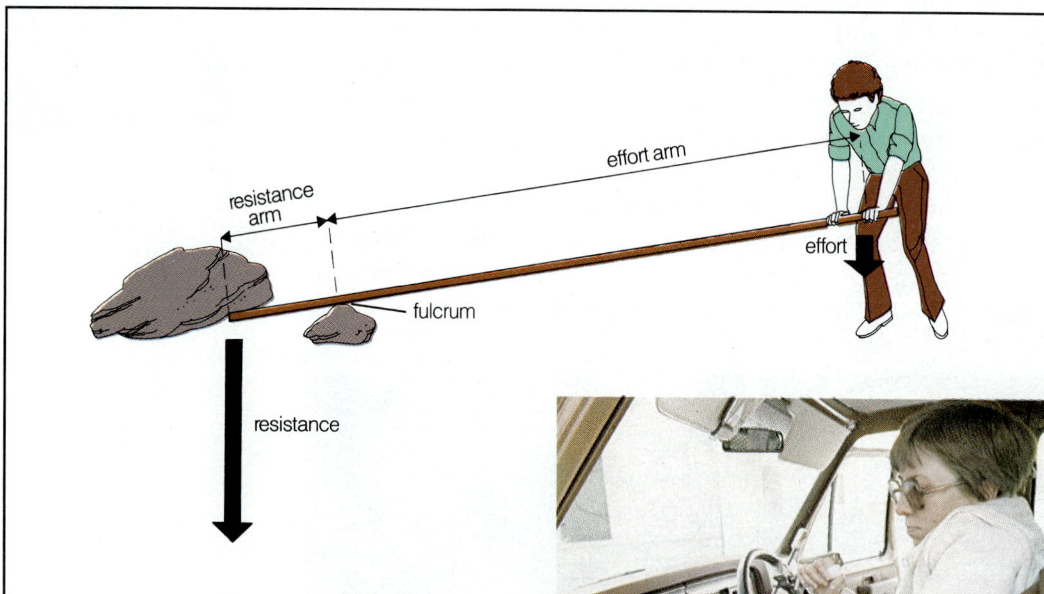

Figure 17-4 *A small force at the end of a long effort arm can lift a large weight at the end of a short resistance arm.*

Figure 17-5 *A paraplegic person can drive this van, using a hand-controlled lever. Moving it back and forth controls speed. Moving it from side to side steers the van.*

Figure 17-6 A human forearm acts as a lever when it lifts an object.

one of the bones of the lower arm at the other end, contracts. This causes a force to be applied to the lower arm. (See Figure 17-6.) The work done in picking up the object is the product of the object's weight and the height through which it is raised. The lower-arm lever does not increase the effect of the force applied by the biceps muscle. Instead it increases distance. The muscle contracts a relatively short distance compared to the distance the load moves.

Checkpoint

1. What is a machine? Why do people use machines?
2. Use a diagram to show where you would place the fulcrum of a lever to move a heavy rock. Identify the effort arm and the resistance arm.
3. How do the effort arm and resistance arm compare when a child is balancing an adult on a seesaw? Explain.

TRY THIS

Place a large book at the edge of a table. Slide a wooden ruler or pencil several centimeters under the book. Let the rest of the pencil or ruler extend beyond the table surface. Push down on the ruler at a point several centimeters out from the table's edge. (The edge of the table will serve as a fulcrum.) Note how hard you have to push to lift the book with the ruler. Now push on the ruler at a point farther out from the fulcrum. What happened to the force needed to lift the book as you moved the force farther from the fulcrum?

17-3

Simple Machines

Goals

1. To explain the difference between simple and complex machines
2. To give examples of simple machines.

simple machines

complex machine

The **simple machines** are devices from which all of the other machines are made. The lever is one of six simple machines. The others are the pulley, the wheel and axle, the inclined plane, the wedge, and the screw. A **complex machine** is made up of two or more simple machines.

You can change the direction of a force with a *fixed pulley*. You can lift an object by pulling downward as shown in Figure 17-7a. This is easier than bending to pick up the object in your arms. With a *movable pulley* the effort can be increased. A force less than the weight of the load can be used to lift it. (See Figure 17-7b.) Fixed and movable pulleys can be used together in a *block and tackle*, as shown in Figure 17-7c. Block and tackle machines are very important on a sailboat. Using them the crew can quickly change sails to take advantage of a shift in the wind. (See Figure 17-8.)

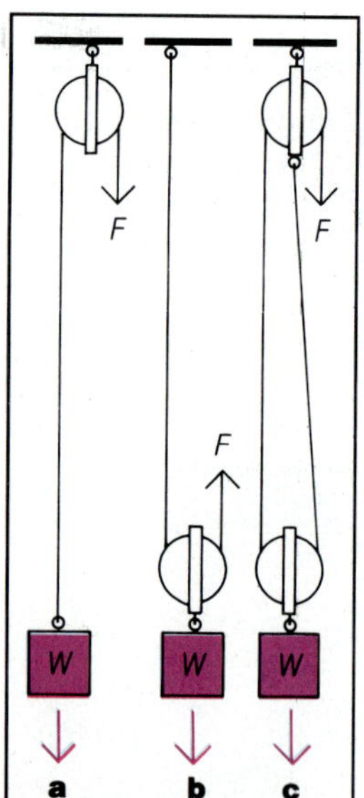

Figure 17-7 *A pulley is a simple machine. (a) Fixed pulley. (b) Movable pulley. (c) Block and tackle.*

Figure 17-8 *Sails can be lifted by a block and tackle. The fixed pulley is on the left and the moving one on the right.*

A *wheel and axle* machine consists of a wheel or crank attached to an axle, or shaft. A doorknob is a familiar wheel and axle. When you rotate the knob (wheel), the shaft (axle) to which it is attached turns, also. The wheel and axle shown in Figure 17-9a can be used to lift a weight. The effort force applied to the crank is increased by the machine to match the resistance force (the weight). A pencil sharpener is a wheel and axle machine. (See Figure 17-9b.) The gears in a car or bicycle are more complicated forms of the wheel and axle. The pedals on a bicycle connect to a gear that turns smaller gears through a connecting chain. (See Figure 17-9c.) This produces the rotation of the rear wheel about its axle.

An *inclined plane* makes it easier to raise an object through a height. A ramp is an inclined plane. (See Figure 17-10.) Less force is needed to lift an object using an inclined plane than that needed to lift the object straight up. However, the distance through which the force must be applied is greater. Either way, the same amount of work is done on the object.

If you put two inclined planes together, as shown in Figure 17-11, you will have a *wedge*. Chisels, axes, knives, and nails are wedges. An inclined plane wound around a central

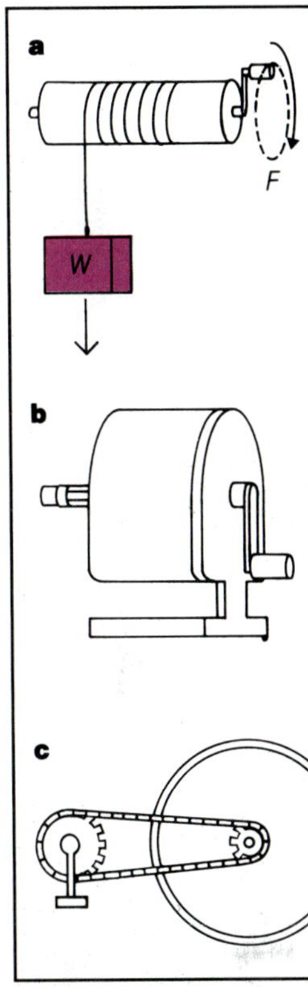

Figure 17-9 *(a) Simple wheel and axle. (b) Pencil sharpener. (c) Bicycle drive-system.*

Figure 17-10 *A ramp helps this man get his load into the van. He will use less force than he would need to lift the load straight up.*

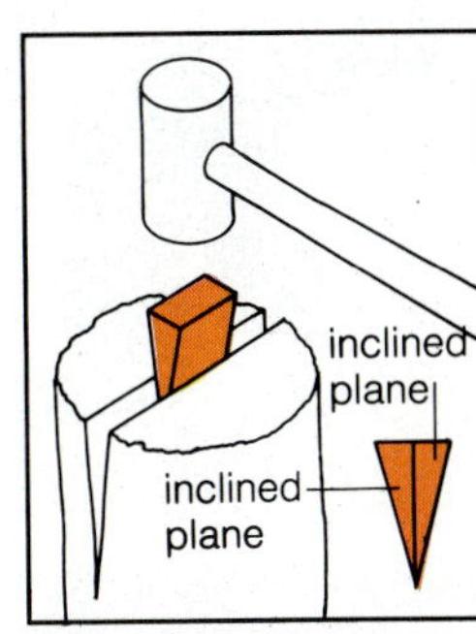

Figure 17-11 *A wedge is like two inclined planes put together.*

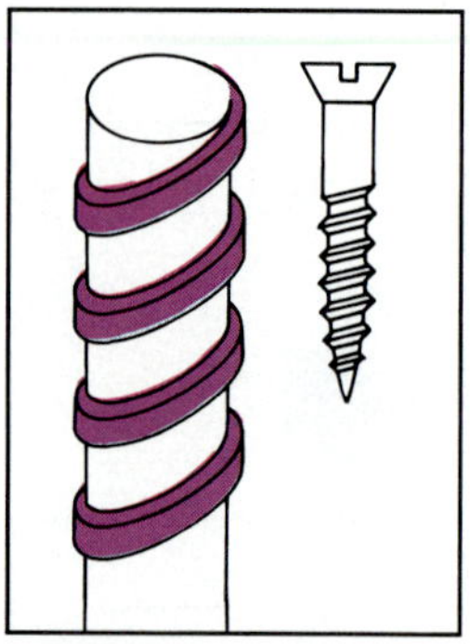

Figure 17-12 A wood screw is made by wrapping an inclined plane around a cylinder.

cylinder forms a *screw*. (See Figure 17-12.) The smaller the distance between the threads of a screw, the easier it is to turn the screw. Just as with an inclined plane, you need less force if you apply the force through a greater distance.

Most machines are complex machines. Automobiles, airplanes, and bicycles are complex machines. If you look at them closely, you will find that they are made up of many simple machines.

Checkpoint

1. What is a simple machine? a complex machine?
2. Name the six simple machines. Give an example of each one.
3. Explain why it is easier to raise an object using an inclined plane than to lift it to the same height.

TRY THIS

Ask two friends to hold onto two sticks, as in Figure 17-13. With your "block and tackle" you will be able to move the sticks together no matter how hard they pull. What happens as you add more turns (pulleys) to the sticks?

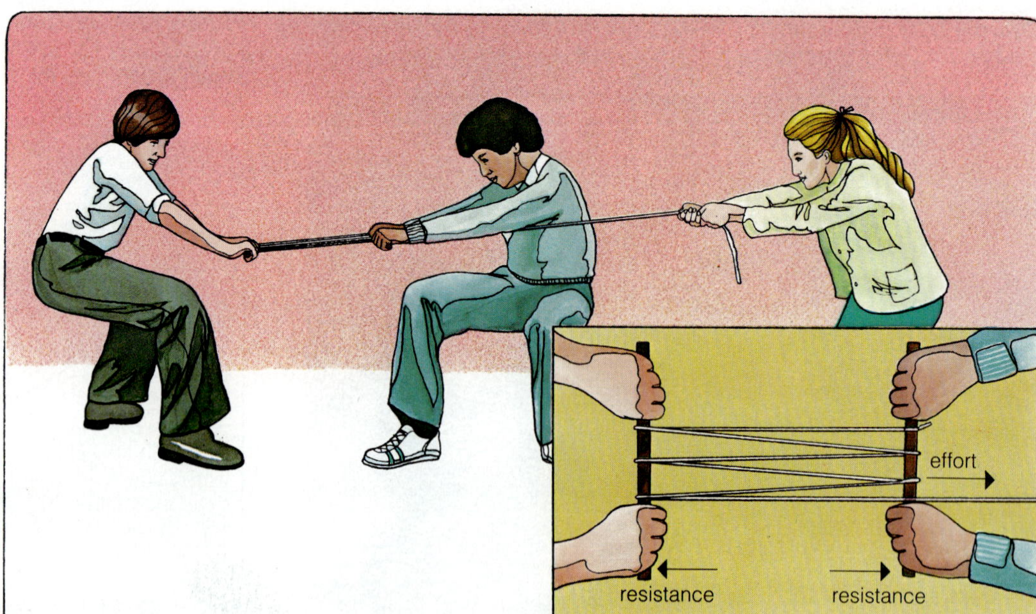

Figure 17-13 This is the way to test your "block and tackle."

17-4

Efficiency of Machines

Goals

1. To define the mechanical advantage of a machine.
2. To define the efficiency of a machine.
3. To explain why complex machines are less efficient than simple machines.
4. To recognize that efficiency is not the only measure of a machine's usefulness.

Machines are designed for different purposes. Some are designed so that a force can move a larger load. Other machines are designed to decrease the time needed to do the work. For example, a fishing rod is designed to give you great speed of movement at its end. (See Figure 17-14.)

A fishing rod is a third-class lever. With a third-class lever, a small movement of the effort arm can result in a large movement of the resistance arm. There are also first- and second-class levers. The class of a lever depends on the relative positions of the effort, resistance, and fulcrum.

Figure 17-14 *When a fishing rod is used to cast a lure, its far end moves much faster than its near end. The great speed flings the lure far out into the water.*

Figure 17-15 *A crowbar is a first-class lever, a wheelbarrow is a second-class lever, and a fishing rod is a third-class lever.*

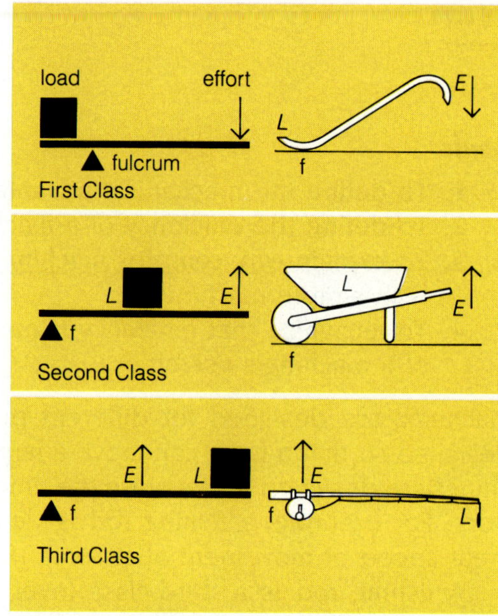

Figure 17-16 *With the lever drawn here, the effort force is one-fifth the resistance force. The lever has a mechanical advantage of 5.*

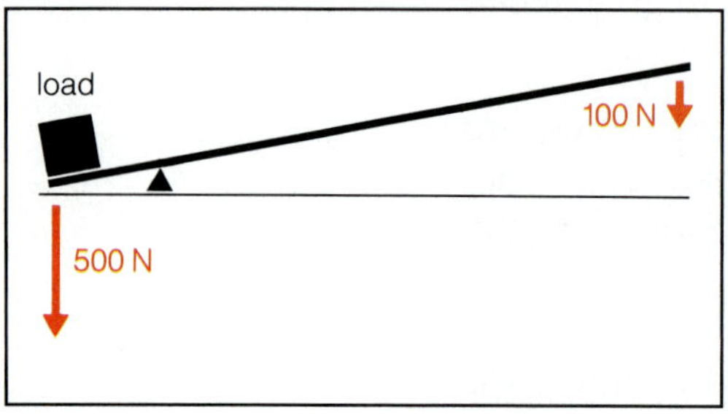

Figure 17-15 shows first-, second-, and third-class levers, and some common applications of them.

It is often important to know how much force a machine will produce when a certain amount of force is applied. The **mechanical advantage** (MA) of a machine is the ratio of the resistance force to the effort force. The *ideal* mechanical advantage (IMA) is the ratio that would be obtained in the absence of friction. Under ideal conditions, the lever shown in Figure 17-16 exerts a resistance force of 500 N when an effort force of 100 N is applied.

mechanical advantage

$$\text{IMA} = \frac{\text{resistance force}}{\text{effort force}} = \frac{500 \text{ N}}{100 \text{ N}} = 5$$

A simple first-class lever can come very close to its ideal mechanical advantage. The friction involved in its use is very small. But for most machines, the amount of effort needed to overcome friction must be taken into account. Suppose the force of friction for the lever shown in Figure 17-16 were 25 N. The effort force would then become 100 N + 25 N, or 125 N. The mechanical advantage (MA) of the lever would be 4.

$$\text{MA} = \frac{\text{resistance force}}{\text{effort force}} = \frac{500 \text{ N}}{125 \text{ N}} = 4$$

You might think that the greater the mechanical advantage, the more desirable the machine. But this is not always the case. A fishing rod has a mechanical advantage of less than 1. Its real advantage is the speed of motion at its end. All third-class levers have a mechanical advantage of less than one.

It is often more helpful to compare the input and output of a machine in terms of work rather than force. The work input for a lever is the effort force multiplied by the distance through which it moves ($W = Fd$). The work output is the product of the resistance force and the distance through which the load moves.

efficiency

The **efficiency** of a machine is the ratio of work output to work input. In the ideal case, the work output and input are equal, and the machine's efficiency is 100%. But because there is always some friction to overcome, the efficiency of any machine is less than 100%.

In general, complex machines are far less efficient than simple machines. They are made up of one or more simple machines, each of which contributes some friction. Even with the use of lubricants to reduce friction, complex machines are less efficient than most simple machines. But the efficiency of a machine may be less important to people than other advantages it may provide, such as convenience or safety.

Hand-operated can openers come in two types, one simple and the other more complex. One type is a second-class lever with a sharp edge on the resistance arm. (See Figure 17-17a.) The second type contains gears, a wheel and axle, a lever, and a sharp wheel. (See Figure 17-17b.)

Figure 17-17(a) A simple can opener is a second-class lever. (b) This opener is a combination of simple machines.

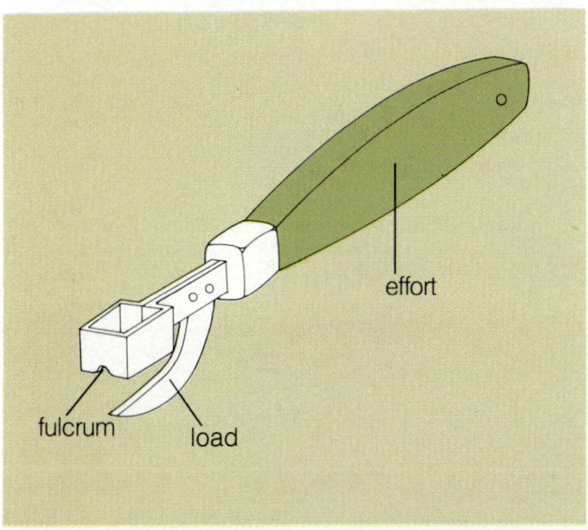

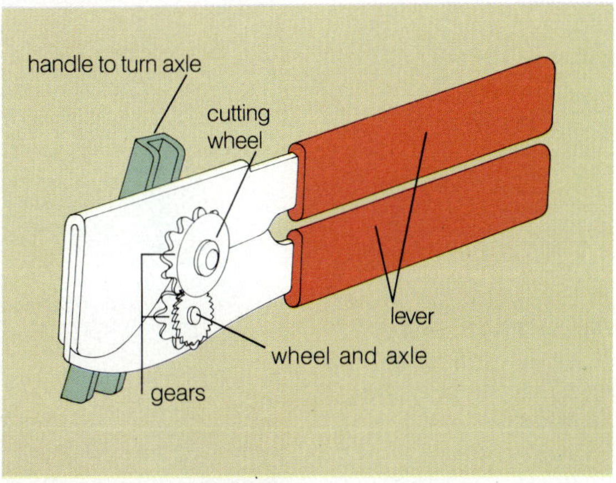

The first type of opener is crude, but quite efficient. Although there is some friction between the resistance arm and the can being opened, most of the effort goes into opening the can. But this type of opener is difficult and, at times, dangerous to use. The second type of opener is more convenient and safer to use. But when this machine is used, some of the effort goes into overcoming friction in the gears and at the axle. Less of the effort goes into the useful work of opening the can.

Checkpoint

1. What is mechanical advantage? ideal mechanical advantage?
2. Give an example of a machine that illustrates (a) a first-class lever, (b) a second-class lever, (c) a third-class lever.
3. What is the efficiency of a machine? Why do machines have efficiencies of less than 100%?
4. Why are complex machines less efficient than simple ones? Why do people use complex machines?

TRY THIS

Build a simple pulley like the one shown in Figure 17-18. First find the weight of the object being lifted. Then measure the force (F) that you apply to the string to lift the object. Also measure the height (h) to which the object is lifted and the distance (d) through which you apply force F. Find the efficiency of this simple machine.

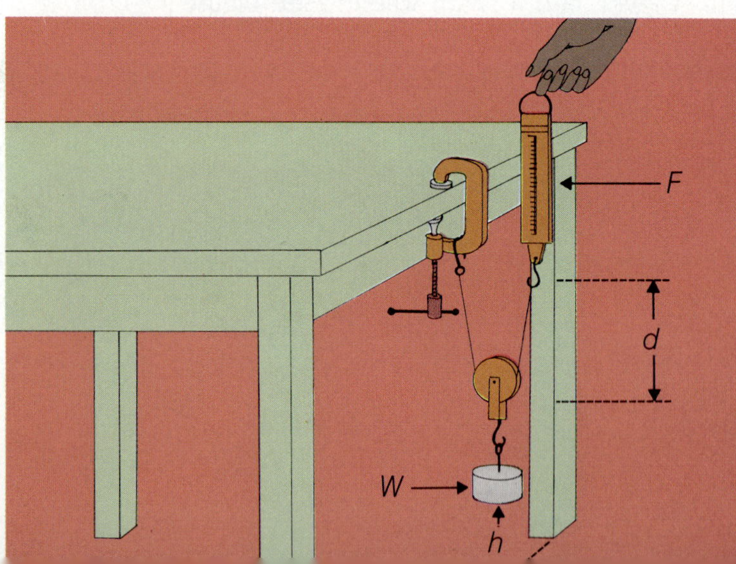

Figure 17-18 *This is the setup for finding the efficiency of a simple pulley.*

17-5

Work and Energy

Goals

1. To define energy and distinguish between potential and kinetic energy.
2. To give examples of objects that possess kinetic or potential energy.

energy

Energy can be defined as the ability, or capacity, to do work. Anything that can do work has energy. Wind can turn the blades of a windmill. Falling water can turn a water wheel or the turbines in a hydroelectric power plant. Windmills, water wheels, and turbines are all wheel and axle machines. (See Figure 17-19.)

kinetic energy

You can give energy to an object by "doing work on it." For example, you can push a billiard ball and cause it to move. A moving ball has energy. It can cause another ball, which is not moving, to move when the moving ball hits it. The energy that a moving object has due to its motion is called **kinetic energy**.

If you push a billiard ball harder, it will move faster. A faster-moving ball has more energy than one that moves more slowly. Any ball that it hits will go farther, as a result. If a marble and a billiard ball were traveling at the same speed, however, any ball hit by the billiard ball would move

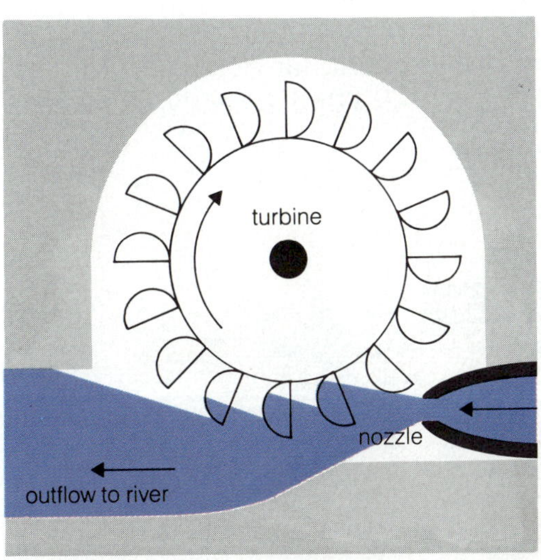

Figure 17-19 *The drawing shows a turbine in the base of a dam, as if you were looking down on it. The photograph shows turbines in a hydroelectric power station.*

faster than one hit by the marble. A great variety of experiments has shown that the energy of a moving object depends on both its mass and its speed. Furthermore, its speed, or velocity, has the greater effect on a moving object's energy.

$$\text{kinetic energy} = \tfrac{1}{2} \times \text{mass} \times \text{velocity} \times \text{velocity}$$
$$\text{k.e.} = \tfrac{1}{2}mv^2$$

A truck traveling at the same velocity as a car has more kinetic energy than the car because it has much more mass. But a fast-moving car may have as much kinetic energy as a slow-moving truck. When the velocity of a car doubles, its kinetic energy becomes four times as great. A car going 100 km/h can do four times as much "work," or damage, as the same car going 50 km/h.

Falling objects, like falling water, can do work. The amount of work that they can do depends on their height above the ground. Suppose an acrobat climbs to a high platform. The work done to reach the platform is the product of the platform's height and the acrobat's weight. In jumping from the platform, the acrobat converts this work into kinetic energy. The kinetic energy can be converted into work if the acrobat lands on one end of a seesaw. (See Figure 17-20.) The acrobat standing on a platform has the *potential* to do work. The acrobat "stored" energy by climbing to the platform. Stored energy, which an object may possess due to its position, is called **potential energy**.

potential energy

There are a number of ways of storing potential energy. It can be stored in a diving board, for example, or in the string of a bow. (See Figure 17-21.) Until recently, all watches

Figure 17-20 *The acrobat stores potential energy by climbing, converts it to kinetic energy in jumping, and converts that into work on the seesaw.*

Figure 17-21 *Energy stored in the bowstring can be used to shoot the arrow.*

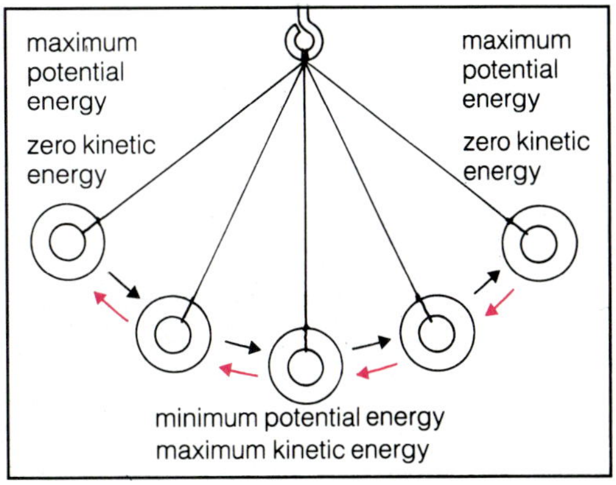

Figure 17-22 *A pendulum bob stores energy as it swings up.*

were run by springs. A wound watch spring contains stored energy. As the watch spring unwinds, a portion of its energy is transformed into the kinetic energy of the moving hands.

A pendulum (PEN ju lum) stores energy in its bob every time it swings to the top of its path. (See Figure 17-22.) When it swings back it converts potential energy to kinetic energy. Then as it moves again in its upward path it gains potential energy. In an ideal pendulum (no friction), the kinetic energy the bob acquires on its downward swing equals the potential energy it loses. The potential energy the bob has at its highest position equals the work done moving the bob up to that point. In most pendulums, frictional forces are very small, so the pendulum can keep swinging for a long time.

TRY THIS

To build a simple pendulum, screw a cup hook into an old golf ball. Tie one end of a thread to the hook. Tie the other end to a paper clip and attach the paper clip to the top of a door frame. Does the ball return to its original position after making a swing?

Checkpoint

1. What two factors determine the kinetic energy of an object?
2. Give two examples of things that have (a) kinetic energy and (b) potential energy.
3. A force of 10 N is exerted in lifting a certain object to a height of 2 m above the ground. How much work was done on the object? (Remember, $W = F \times d$.) How much potential energy is stored in the object at that height?

Pole Vaulting

Pole vaulting involves a number of energy transformations. A pole vaulter begins by sprinting down a track to develop kinetic energy. When he places the pole firmly against the ground, he heads skyward. Some of the pole vaulter's kinetic energy is stored in his body as potential energy due to his position above the ground. Some is stored in the bent, springy pole, as shown in Figure 1. As the pole straightens, potential energy is converted into kinetic energy that helps to propel the vaulter over the cross bar. At this point, (see Figure 2), the vaulter pushes against the pole to get the maximum possible leverage.

When the pole vaulter reaches the highest point in his path, his potential energy is at its maximum. At this point, the force of gravity causes the vaulter to begin to accelerate downward. As he falls, his kinetic energy increases and his potential energy decreases. When the vaulter hits the cushioned pads on the ground, all of the kinetic energy that he gained in his fall is lost. Most of it is transformed into heat.

Figure 1

Figure 2

17-6

Perpetual Motion Machines

ENRICHMENT

Goal

To describe one type of perpetual motion machine and explain why such machines cannot work.

Many people think a perpetual motion machine is something that will keep moving forever. But that is not the kind of perpetual motion machine that countless people have spent years trying to build. The machine these dreamers dream about is one that could provide a work output greater than the work input.

One machine that has been proposed is shown in Figure 17-23. The machine is a series of levers on a wheel. All of the levers are the same length and have the same "hammerhead" mass on one end. The red levers on the right apply *effort forces* that are larger than the *resistance forces* exerted by the blue levers on the left. Because of the greater force of the right-hand levers, the wheel, once started, would continue to turn, according to its designer. It would even gain speed as it rotated.

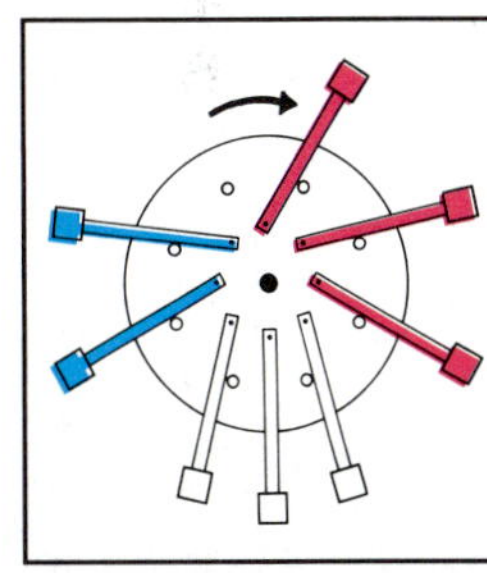

Figure 17-23 *This perpetual motion machine is based on a design from the 13th century.*

Before you rush off to build this machine and beat the original designer to the patent office, take another look. All work done by the falling hammers on the right is balanced by the work required to lift the hammers on the left. Inventors have devised a number of similar machines, none of which has ever produced "perpetual motion."

No machine has ever been designed that will put out more work than that put in. Indeed, a moving object cannot even retain the kinetic energy that it has. A rolling marble will eventually stop because of friction. Even satellites close to Earth eventually fall back to Earth. (The moon is far enough away to be in nearly empty space, where friction is extremely small.)

Checkpoint

1. What was the advantage of the perpetual motion machine described in this section *supposed* to be?
2. Why didn't the machine work as planned?
3. Why is any machine less than 100% efficient?

17-7

Power

Goal

To define power and explain its importance.

It is often useful to know not only how much work is done, but how quickly it is done. For example, a farmer can move a bale of hay from the ground to a hayloft more quickly with a block and tackle than by carrying it up. **Power** is the amount of work done in a certain length of time.

power

$$\text{power} = \frac{\text{work}}{\text{time}}, \text{ or } P = \frac{W}{t}$$

EXAMPLE

A motor does 10,000 J of work in 10 s. What is its power?

$$P = \frac{W}{t}$$

$$P = \frac{10{,}000\ \text{J}}{10\ \text{s}} = 1000\ \text{J/s}$$

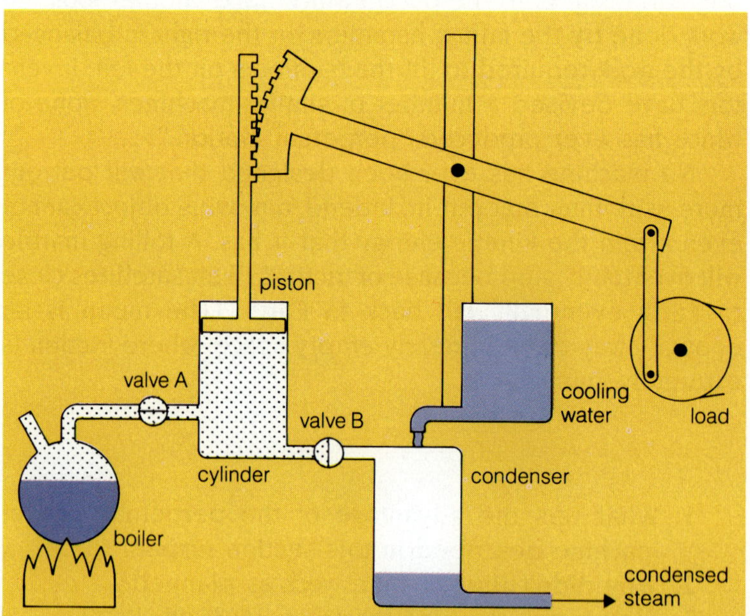

Figure 17-24 *A simplified diagram of James Watt's steam engine.*

Figure 17-25 Light bulbs and appliances, such as this electric blanket, are marked to show their power ratings.

The joule per second has a special name. It is called a **watt** (W). The watt is named after James Watt, an 18th-century Scottish engineer who invented a practical steam engine. (See Figure 17-24.) You have probably noted that light bulbs are rated in watts. A 60-W light bulb converts 60 J of electric energy to light and heat every second. The power of most appliances is printed somewhere on them. (See Figure 17-25.) The power produced in large electrical machines is often expressed in *kilowatts* (kW). One kilowatt equals 1000 watts.

watt

The power of trucks, cars, airplanes, and motors is often given in *horsepower* (hp). This unit was defined in earlier days when horses, rather than machines, did much of the work. One horsepower is equal to 746 W.

TRY THIS

Examine a number of light bulbs and appliances around your home. Find the power required to operate each one. You can usually find the power rating printed somewhere on the machine or appliance.

Checkpoint

1. What is power?
2. What important information does *power* give that *work* does not give?
3. You do the same amount of work against gravitational force whether you walk or run up a flight of stairs. In which case do you develop more power?

Understanding The Chapter

17

Main Ideas

1. Work is the force acting on an object multiplied by the distance the object moves.
2. A machine changes the direction or strength of a force.
3. Complex machines are made up of two or more simple machines.
4. The six simple machines are the pulley, wheel and axle, lever, inclined plane, wedge, and screw.
5. There are three classes of levers. The class of a lever depends on the positions of the effort, resistance, and fulcrum.
6. The mechanical advantage of a machine is the ratio of the resistance force to the effort force.
7. The efficiency of a machine is the ratio of its work output to its work input.
8. Despite their inefficiency, complex machines save time and effort.
9. Anything that has the capacity to do work has energy.
10. An object's stored energy is called its potential energy.
11. An object's energy due to its motion is called kinetic energy.
12. Potential energy can be transformed to kinetic energy and vice versa.
13. Machines cannot produce output work greater than input work.
14. Power is the rate at which work is done.

Vocabulary Review

From the following list, choose the term that best completes each of the statements. Write your answers on a separate piece of paper.

complex machine	mechanical advantage
efficiency	potential energy
energy	power
fulcrum	simple machines
joule	watt
kinetic energy	work
machine	

1. __?__ is the net force acting on an object multiplied by the distance the object moves in the direction of the force.
2. __?__ is the work done divided by the time taken.
3. Anything that has the capicity to do work has __?__.
4. An object gains __?__ as it falls.
5. A(n) __?__ is a unit of power that is equal to 1 J/s.
6. __?__ are devices from which all other machines are made.
7. A lever turns about its __?__.
8. A machine's __?__ is the ratio of its work output to its work input.
9. A wound watch spring has __?__.
10. A(n) __?__ changes the direction or strength of a force.
11. The __?__ of a machine is the ratio of the resistance force to the effort force.
12. __?__ are made up of two or more simple machines.

Chapter Review

Write your answers on a separate piece of paper.

Know The Facts

1. The _?_ is a simple machine. (pulley, automobile, bicycle)
2. The effort force applied to a lever is used to overcome the _?_. (resistance, power, fulcrum)
3. The kilowatt is a unit used to measure _?_. (energy, friction, power)
4. Most machines have an efficiency _?_ 100%. (equal to, less than, more than)
5. A 10-N weight at rest 1.0 m above the floor has 10 J of _?_ with respect to the floor. (kinetic energy, potential energy, power)
6. An object's _?_ can be found using the mathematical expression $\frac{1}{2}mv^2$. (potential energy, power, kinetic energy)
7. Perpetual motion machines _?_. (create energy, have more work output than work input, are impossible)
8. If an input force of 100 N lifts a load of 400 N, the machine has a mechanical advantage of _?_. (1, 4, 8, 16)
9. One watt is _?_ (equal to, less than, greater than) one joule per second.
10. A machine's work input of 500 J produces a work output of 400 J. The machine's efficiency is _?_. (40%, 100%, 80%, 125%)
11. An example of a wheel and axle machine is a _?_. (block and tackle, windmill, chisel)
12. At the top of its swing, a pendulum has its maximum _?_. (kinetic energy, potential energy, power)
13. The distance from the fulcrum of a lever to the load that the lever is moving is the lever's _?_. (effort arm, resistance arm, efficiency)

Understand The Concepts

14. You push on a wall with a force of 100 N. The wall does not move. How much work did you do on the wall?
15. An effort force of 100 N moves through a distance of 0.5 m. Find the work input.
16. Describe some objects having (a) potential energy, (b) kinetic energy.
17. Why does a ball rolling on a horizontal surface come to rest?
18. Explain the energy transformations that occur when a rubber ball falls from your hand to the floor and bounces once.
19. Explain how a screw resembles an inclined plane.
20. What advantages might a complex machine have that would compensate for its low efficiency?
21. Explain how one machine could have both a lower efficiency and a higher power rating than another.
22. A power rating of 120 W is how many J/s? kW? hp?
23. Explain why the perpetual motion machine described in this chapter does not work.
24. Describe how you would adjust the fulcrum of a seesaw so that a small person can balance a larger one. Where would each person sit?

25. What is the advantage of using a fixed pulley to raise an object rather than lifting it yourself? What is the advantage of using a movable pulley rather than a fixed pulley?
26. Explain why complex machines are usually much less efficient than simple machines.
27. Which light bulb requires more energy to burn for one hour, a 60-W bulb or a 100-W bulb? Explain your answer.
33. You lift a 10-N weight to a height of 5 m. How much work do you do on the weight? How much potential energy does the weight have after you raise it? How much kinetic energy will the weight have just before it hits the floor?
34. Describe the changes in potential and kinetic energy experienced by a skier. Start with the skier about to go up a slope. End when the skier has arrived at the bottom of the slope again, after skiing down.

Challenge Your Understanding

28. An effort force of 100 N is applied through a distance of 20 m. What is the maximum distance that the effort force can move a load of 500 N?
29. Explain why it is easier to roll a heavy barrel up an inclined plane than to lift it straight up.
30. Explain how the kinetic energy of a moving object changes if (a) the mass is doubled but the velocity remains the same, (b) the velocity is halved but the mass remains the same, (c) both the mass and the velocity are doubled.
31. In 5s, a machine does 500 J of work. Find the power developed by this machine.
32. The effort force on a lever is 75 N. If the effort force moves a resistance of 150 N, what is the ideal mechanical advantage of the lever?

Projects

1. Use the library to investigate the origin of "horsepower" as a unit of power.
2. Build a lever, a pulley, and an inclined plane. Test their mechanical advantages and efficiencies.
3. Find as many simple machines as you can in your home and school.
4. Go to the library and find out what kinds of machines were used to build the pyramids in Egypt.
5. Ask your school's driving instructor what lubrication is recommended for automobiles. Why do you think the engine oil should be changed periodically?
6. Find the mechanical advantage for a seesaw in your local playground.
7. Take apart a complex machine, such as an old clock. How many simple machines can you find?

Chapter 18

Heat

This solar furnace in France produces a great deal of heat.

18-1

Heat and Temperature

Goals

1. To define heat and temperature.
2. To describe a common thermometer.
3. To compare the Celsius and Kelvin temperature scales.

Have you ever rubbed your hands together briskly to warm them up? If so, you were using friction to produce heat. The work done to overcome the friction between the surfaces of your hands appeared as heat. In this case, you wanted to produce heat. Often, however, the heat produced by friction can be a nuisance. When a car traveling at high speed stops suddenly, the brakes must provide a tremendous amount of stopping force. The friction involved produces a great deal of heat. For this reason, the brakes on a car must include heat-resistant materials. Disc brakes transfer heat to the air more rapidly than do other types of brakes. (See Figure 18-1.)

heat

Heat is the energy transferred *to* an object when it warms or *from* an object when it cools. An object can gain heat when work is done on it or when heat is transferred from a hotter object. Rubbing your hands together warms them. Putting your hands in hot water warms them, also.

Figure 18-1 *The metal disc in a disc brake is attached to a wheel and rotates with it. Braking occurs as two friction pads are squeezed against the spinning disc.*

All matter is made of extremely small particles. These particles are in constant motion. When heat is transferred to an object, the motion of these particles increases. The faster a particle moves, the greater its kinetic energy. To measure the average kinetic energy of the particles in a substance is to measure how hot, or cold, the substance is. The **temperature** of a substance is a measure of the average kinetic energy of the particles in it.

temperature

If water in a small glass is at the same temperature as water in a large pitcher, the average kinetic energy per water particle, or molecule, is the same in each container. But there is more water in the pitcher. Thus there are more molecules in that water than in the glass of water. It follows that the total energy of the water in the pitcher is greater than that of the water in the glass. Temperature is not a measure of total energy. It is the measure of the average kinetic energy *per particle*.

Temperature is usually measured with a thermometer. Many thermometers contain a liquid. The liquid expands when it gains heat and contracts, or shrinks, when it loses heat. The liquid is usually mercury or alcohol. Mercury has a silver color. Alcohol is colorless, so a red dye is normally added to it, to make it show up. The liquid is sealed into a narrow glass tube. The tube is narrow so that a small change in the space taken up by the liquid in the bulb causes a noticeable change in the height of the liquid in the tube.

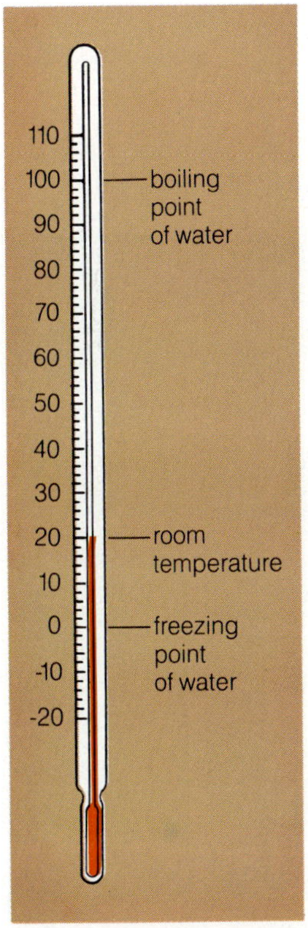

Figure 18-2 *A typical laboratory thermometer with a Celsius scale.*

A common thermometer is shown in Figure 18-2. Its scale is marked off in *Celsius* (SEL see us) degrees (°C). The Celsius scale is named after Anders Celsius, a Swedish scientist who invented the scale in the early 1700's. On the Celsius scale the temperature at which pure water freezes is 0°C. The temperature at which water boils is 100°C. Pure water is readily obtainable and it is relatively easy to freeze or boil. Thus it is a good substance on which to base a thermometer scale.

To make a Celsius thermometer, a liquid-containing tube is put in a mixture of ice and water. The 0°C mark is made on the tube, at the top of the liquid column. Next, the tube is placed in boiling water. The liquid in the tube rises and a mark is made for 100°C. Then the distance between the 0°C mark and the 100°C mark is divided into equal parts. The Celsius degree is one one-hundredth of the distance between the 0°C and 100°C marks.

With the size of the Celsius degree established, the scale can be extended above 100°C and below 0°C, as shown in Figure 18-2. There is apparently no limit to how hot a substance can become. But there is a limit to how cold anything can be. At −273°C, the particles in a substance have their lowest possible energy. No experiments have produced a temperature as low as −273°C, although scientists have come within 0.001°C of this temperature. Since nothing can be colder than −273°C, this temperature is called **absolute zero**.

absolute zero

For scientists doing calculations, a temperature scale without negative numbers is useful. The Kelvin scale has absolute zero as its lowest point. The divisions of the Kelvin scale, called *kelvins* (K), are the same size as those of the Celsius scale. That is, one kelvin is equal to one Celsius degree. (See Figure 18-3.) The Kelvin scale is named after Lord Kelvin (William Thomson), a pioneer in the science of heat.

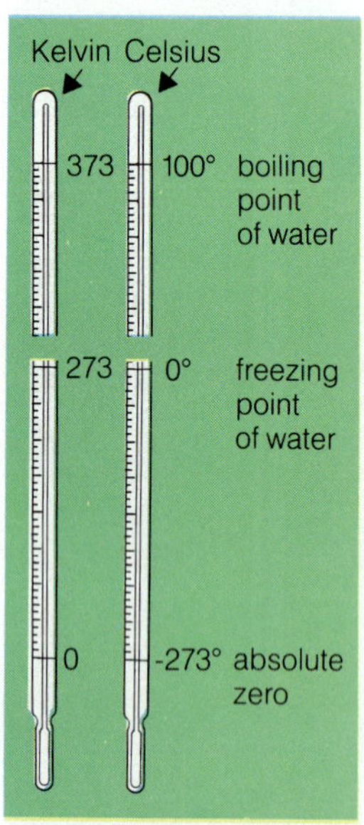

***Figure 18-3** The Kelvin scale compared to the Celsius scale.*

Checkpoint

1. What is heat?
2. What is temperature?
3. Describe a common thermometer and explain how it works.
4. Compare the Celsius and Kelvin temperature scales.
5. What is absolute zero?

TRY THIS

1. Using a thermometer, look for warm and cold places in and around your home or school. How do the temperatures near a ceiling and floor compare? How about the north and south sides of a building on a sunny day?

2. Put one hand in a pan of ice water. Put the other hand in very warm water. Leave both hands in for about one minute. Then plunge both hands into a pan of lukewarm water. How warm does each hand feel?

18-2

Temperature and Pressure

Goal

To define pressure and explain the effect of temperature on gas pressure.

A person standing on snow sinks into it less when wearing skis (or snowshoes). (See Figure 18-4.) The force that the person exerts on the snow is the same with or without skis. But, with skis, the force is spread over a greater area, and the force on any given area is less. **Pressure** is the ratio of a force to the area over which it acts.

pressure

$$\text{pressure} = \frac{\text{force}}{\text{area}}, \text{ or } P = \frac{F}{A}$$

Most substances expand when they are heated. In a gas, such as air, there is more space between particles than in liquids or solids. So gas particles move more independently, and gases expand more noticeably than liquids or solids when heated.

Figure 18-4a *The weight of a person puts great pressure on the snow directly below each foot.*

Figure 18-4b *When a person wears snowshoes, the force is spread out. The pressure on any one spot is much less.*

Figure 18-5 *When air inside a balloon warms, the air particles move faster and with more force. The pressure on the inside surface increases and the balloon expands.*

A balloon may burst if the air inside heats up enough. The tiny particles in the air inside the balloon are constantly colliding with the balloon's inside surface. (See Figure 18-5.) As the air in the balloon gains heat, the particles move faster. They hit the inside surface of the balloon more often and with more force. As a result, the pressure on the inside surface of the balloon increases. There is a limit to how much the balloon itself can expand, however. Eventually, the balloon bursts.

Similarly, the pressure exerted by the air inside an automobile tire will rise when the air gets warmer. The air in the tire warms up as the car is driven. Thus its pressure increases. That is why tire air pressure should be measured "cold."

Checkpoint

1. What is pressure?
2. What happens to its volume when gas is heated?
3. How are gas pressure and temperature related?

TRY THIS

Find the air pressure in the tires of a car that has not been driven for an hour or more. Check the pressure again, immediately after the car has been driven for an hour or so. What is the change in the air pressure?

18-3

Specific Heat

ENRICHMENT

Goals

1. To define specific heat.
2. To describe how the specific heat of certain substances can affect local climates.

People who live near the ocean in temperate climates enjoy winter and summer temperatures that are relatively moderate. Temperatures along a coastline generally do not vary as much day to day or season to season as temperatures inland. The coast is warmer in the winter and cooler in the summer than inland locations at the same latitude. The differences between day and night temperatures are usually greater inland at any season. People who live near large lakes experience a similar effect. This is because a lot of heat is needed to warm a large body of water.

The amount of heat needed to increase the temperature of a substance depends on the *amount* of the substance. The more particles the substance contains, the greater the amount of heat needed to increase their average kinetic energy. It takes more heat to increase the temperature of a pot of water than to increase the temperature of a cupful by the same amount.

The amount of heat needed to increase the temperature of a substance depends on the *nature* of the substance, also. The type of particles the substance consists of and their arrangement are important factors. It takes more heat to raise the temperature of 1 g of water 1°C than to raise the temperature of 1 g of iron 1°C. The particles in water are different from the particles that make up iron. It takes more heat to raise the temperature of 1 g of liquid water 1°C than to raise the temperature of 1 g of ice 1°C. The arrangement of the particles is different in water and ice.

The amount of heat needed to raise the temperature of 1 g of any substance 1°C is called the **specific heat** of the substance. It takes 4.2 J of heat to raise the temperature of 1 g of water 1°C. Thus the specific heat of water is 4.2 J/g/°C. (Read this as 4.2 joules per gram per degree Celsius.) Table 18-1 shows the specific heats of various substances.

specific heat

Table 18-1: Specific Heats of Various Substances

Substance	Specific heat (J/g/°C)	Substance	Specific heat (J/g/°C)
aluminum	0.92	iron	0.46
brass	0.38	lead	0.13
copper	0.39	olive oil	2.0
glass	0.84	silver	0.24
glycol	2.6	water	4.2
gold	0.13	zinc	0.38
ice (0°C to −20°C)	2.1		

Of all the substances in Table 18-1, water has the largest specific heat, by far. That is why large bodies of water have moderating effects on temperature changes. A large lake or ocean is slow to warm up in spring and slow to cool off in fall. Air moving over the lake or ocean is warmed or cooled by the water. This air in turn keeps the land near the water warmer or cooler than land away from the water.

The specific heat of sandy soil is small. As a result, desert temperatures may quickly rise to very high levels (40°C or more) during the day. But at night temperatures quickly fall to 0°C or below.

Checkpoint

1. It takes 0.24 J to warm 1 g of a metal 1°C. What is the specific heat of the metal? What metal could it be?
2. The temperatures of equal masses of aluminum, copper, gold, and zinc were all raised from 10°C to 30°C. To which metal was the most heat transferred?
3. Why do the temperatures of coastal cities generally not vary as much as those of inland cities?

TRY THIS

Place a stack of pennies on a flat piece of ice. Place an equal mass of metal washers on a similar piece of ice. When the washers stop sinking into the ice, remove them. Which metal melted more ice? Which metal has the larger specific heat?

18-4

Transferring Heat

Goal

To describe three ways in which heat is transferred.

Heat always "flows" from a warmer to a cooler body. If someone placed one end of a metal spoon in a flame, that person would soon let go of the other end. It would get too hot to hold. Heat from the flame caused the kinetic energy of the particles to increase in the part of the spoon in the flame. This energy was then transferred to nearby particles along the length of the spoon. The "flow" of energy quickly reached the person's hand, raising its temperature.

conduction

conductor

insulator

The transfer of heat from particle to particle is called **conduction** of heat. A material through which heat passes readily is called a **conductor**. Metals are good conductors of heat. We use metal pots and pans for cooking, because heat passes quickly from the heat source to the food being cooked.

A material through which heat passes very slowly is called an **insulator**. We pour hot drinks into cups made of ceramic or plastic materials. Because these materials are good insulators, the cups are easier to hold than metal cups. In addition, such cups keep their contents warm longer. Air is a good insulator. In fact, most good insulators are filled with tiny air spaces. But air can transfer heat in a way that is different from conduction.

Figure 18-6 *When the burner is going, the air inside a hot air balloon is warmed. The air expands, becomes less dense, and the balloon rises.*

When a mass of air is heated it expands. As a result it occupies a larger space, or a larger *volume*. The expanded gas, however, is less *dense* because there are fewer particles in a given volume of the gas. A less dense substance will rise above a denser substance. For example, if you pour salad oil into water, the oil will soon float to the top because it is less dense than water. Heated air will rise above cooler air. A hot air balloon rises because the hot air in the balloon is less dense than the air through which the balloon is rising. (See Figure 18-6.)

Your hand does not get hot when you hold it near the *side* of a flame. But if you hold it quite far *above* a flame, you will feel warm air. The air that is warmed directly by the flame expands, becomes less dense, and rises. This creates a movement of warm air upward from the flame. The movement of gases or liquids due to differences in density is called **convection**. Convection currents in a liquid are shown in Figure 18-7. Heat is transferred within a room by convection currents. Figure 18-8 shows how warm air near a room heater rises, cools, settles downward, and returns to the heater.

convection
(kahn VEK shuhn)

The sun's heat cannot reach Earth by conduction because there is nothing in the vast emptiness of space to conduct heat. The heat cannot be transferred by convection because there are no liquids or gases that link the sun to Earth. Heat moves from the sun to Earth in the same way that light does, but the heat is invisible. This type of heat transfer is known as radiation. In general, **radiation** is a transfer of energy that does not require matter to transmit it. It can occur in empty space.

radiation

The heat given off by hot objects, such as the sun, is often called *thermal* radiation. A fire in a fireplace gives off thermal radiation. We cannot see thermal radiation, but we can detect it. It affects some types of film in the same way that light affects ordinary film. Film sensitive to thermal radiation can be used to detect heat leaking away from a building. (See Figure 18-9.)

Figure 18-7 *The dye in this container of hot water shows that convection currents form in liquids as well as in air.*

Figure 18-8 *Warm air next to a radiator or from a warm air duct will start a convection current in a room.*

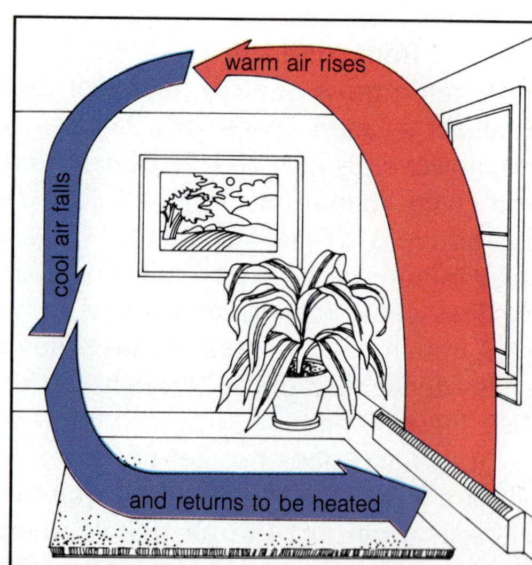

Figure 18-9 *This photograph was taken with special film to detect heat leaking from the house. Different colors on the film indicate different degrees of heat loss.*

Checkpoint

1. Describe three ways in which heat is transferred.
2. What are conductors?
3. What are insulators? Give two examples.
4. Explain why heated air rises.
5. Describe the circulation of air in a typical heated room.
6. How can thermal radiation be detected?

TRY THIS

Add a few drops of food coloring to a small amount of hot water. Using a medicine dropper, slowly add a few drops of the hot, colored water to the bottom of a transparent container of cold water. Watch the convection currents form as the warm water rises.

18-5

Transformation of Energy

Goals

1. To describe some energy transformations.
2. To state the law of conservation of energy.
3. To discuss the nature of heat engines.

Count Rumford (Benjamin Thompson), an American-born scientist, was one of the first to recognize that work can be converted into heat. In the 1780's he was supervising the boring of cannons in Munich (in what is now West Germany). He noticed that when the cannons were bored great amounts of heat were released. By boring a cannon in water, Rumford showed that heat was produced as long as work was done. In fact, with enough work he could even make the water boil.

Working about 50 years after Rumford, James Joule made careful measurements of the heat produced when work was done. One of Joule's experiments showed that the potential energy of a slowly falling mass could be transformed into heat. (See Figure 18-10.) The falling mass caused a stirrer to turn in a well-insulated container of water. The thermometer

Figure 18-10 *In this setup, the falling mass makes the stirrer turn, which heats the water. Potential energy is converted into heat.*

Figure 18-11 *This cutaway view of a typical modern water heater clearly shows its well-insulated wall.*

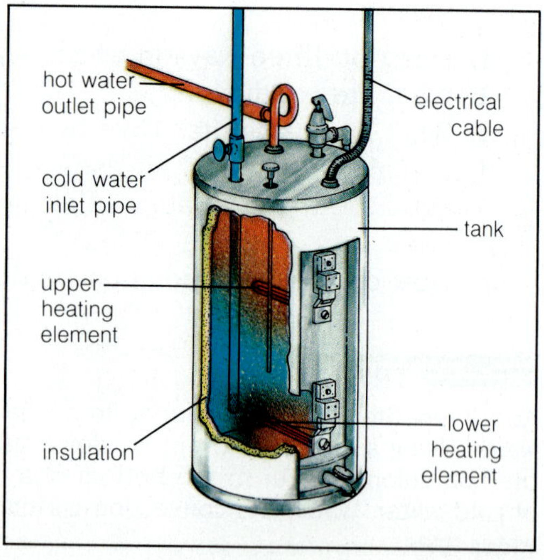

Figure 18-12 *An electric motor can be used to transform electric energy into potential energy stored in a raised object.*

measured the temperature of the water. The amount of energy gained by the water could be calculated from the temperature increase. The energy lost by the falling mass could be calculated, too. Joule showed that the energy lost by the mass was *proportional* to the energy gained by the water. In other words, the more work done, the more heat produced, and the greater the temperature increase.

Joule also used electricity to heat water. He found that the water's temperature increase depended on the amount of electric energy used. In many homes, water is heated by the use of electricity. Figure 18-11 shows a typical electric water heater. Note the insulation around the tank to help keep heat from escaping.

Electric energy can be used to produce work. An electric motor can be used to lift a heavy object, as shown in Figure 18-12. Some of the electric energy is transformed into potential energy as the object is raised. Some is transformed into heat, which raises the temperature of the motor and its surroundings.

Many experiments have shown that whenever one form of energy disappears, an equal amount of another form, or forms, appears. Energy can be transformed, but the total amount of energy does not change. This is known as the **law of conservation of energy**.

law of conservation of energy

Figure 18-13 A steam engine provides energy to move the locomotive. A steam engine is a heat engine.

heat engine

Work can be transformed completely into heat, as Joule's experiments showed. But there are limits to the amount of heat that can be transformed into work. Despite these limits, heat engines are widely used. A **heat engine** is a machine that changes heat into mechanical energy. Gasoline engines and steam engines are heat engines. Table 18-2 shows the efficiencies of some heat engines.

Table 18-2: Efficiencies of Some Heat Engines

Heat engine	*Efficiency*
airplane engine	10%
steam locomotive	10%
automobile engine (gasoline)	22%
steam turbine	30%
diesel engine	35%

TRY THIS

Hold a paper clip to your lips. How warm does it feel? Now bend the paper clip back and forth rapidly 10 times. Again hold it to your lips. How warm does it feel now? What was the effect of the work you did on the paper clip?

Note that even the diesel engine is only 35% efficient at converting the energy in fuel to useful work. The efficiencies of heat engines are necessarily low, no matter how well they are designed and maintained. The human body uses fuel much more efficiently. But when you want to get somewhere quickly, speed can be more important than efficiency.

Checkpoint

1. Name two of the energy transformations that Joule studied.
2. State the law of conservation of energy.
3. What is a heat engine?

18-6

Solar Energy

Goal

To define solar energy and give examples of how it can be used to decrease the use of fossil fuels.

Have you ever had to shovel the walk after a heavy snowstorm? If so, you might envy a ski-lodge owner in Colorado. He owns a solar "snow plow." Sunlight reflecting off a series of curved mirrors melts any snow that collects on the sidewalk or steps. (See Figure 18-14.) This is just one way energy from the sun, or **solar energy**, can be used.

solar energy

Buildings today often have many south-facing windows to capture winter sunlight. Thick concrete floors or walls, or large containers of water, are used to absorb and store the sun's heat. This stored energy is then available when there is no sunlight.

Solar energy can be used to generate electricity. *Photovoltaic cells* produce electricity when they absorb solar energy. They are now used in remote places, including outer space. This means of producing electricity is too expensive for more general use today. But the cost of photovoltaic cells has been decreasing dramatically.

Figure 18-14 *The solar "snow plow" in Colorado. (a) The house, with the curved mirror over the front door. (b) A close-up view of the mirror.*

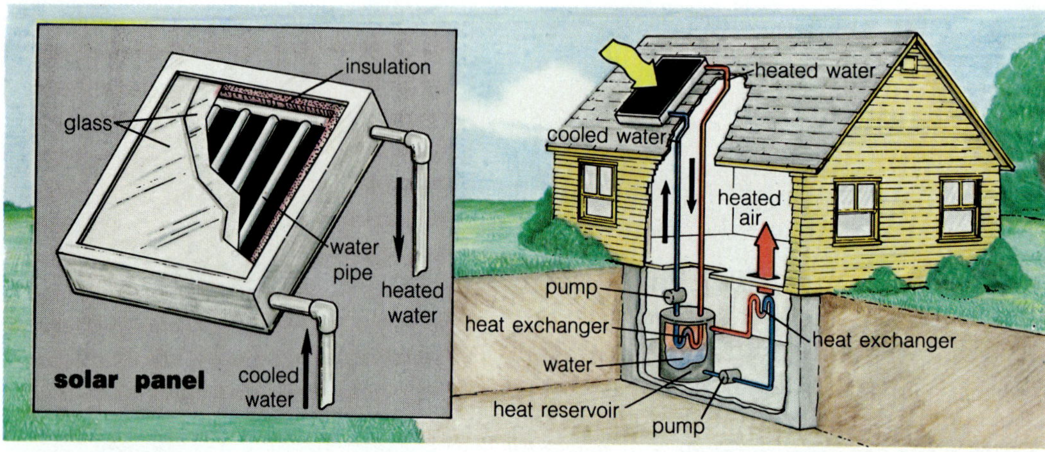

Figure 18-15 *The parts of a solar heating system. This one uses water to transfer the heat.*

Most of the energy now used to heat homes and water, provide transportation, generate electricity, and turn the wheels of industry comes from *fossil fuels*. They are mainly coal, oil, and natural gas. Fossil fuels are the remains of organisms that lived millions of years ago. While living, plants transformed energy that they received from the sun into stored energy called *chemical energy*. When fossil fuels burn they release energy from the sun that was captured by living organisms millions of years ago.

renewable energy source

To reduce the use of expensive fossil fuels, people are turning directly to solar energy to heat their homes and water. A solar heating system uses black collector plates in solar panels mounted on a south-facing roof to absorb heat from the sun. The heat is transferred to cool air or water passing over the plates. The warmed air or water is then used to heat the building. (See Figure 18-15.)

Once fossil fuels have been burned, they are gone forever. But solar energy keeps coming to Earth in the same amounts no matter how much is used by humans. Solar energy will last as long as the sun shines, probably another 10 billion years. An energy source that is produced about as fast as it is used is called a **renewable energy source**. Solar energy is renewable. Fossil fuels are not.

TRY THIS

Fill a clear glass or plastic bottle with water and cap it. Fill an identical bottle with water that has been darkened with black ink. Place both bottles in the same sunny place. After an hour or so find the temperature of the water in each bottle. Try to explain your results.

Checkpoint

1. What are fossil fuels?
2. Describe two ways in which solar energy is used.
3. Explain what a renewable energy source is.

18-7

Other Energy Sources

Goal

To describe alternatives to fossil fuels other than solar energy.

Figure 18-16 *A modern power-generating windmill at Clayton, New Mexico.*

Approximately two percent of the solar energy that reaches Earth is transformed into the kinetic energy of winds. (Winds are created by the unequal heating of Earth.) Winds can turn windmills, which can generate electricity and do other work. (See Figure 18-16.) Sweden has plans to build 3300 wind generators. They will produce a substantial portion of that country's electricity.

hydropower

Solar energy contributes to **hydropower**, the use of falling water to do work. Water on Earth absorbs solar energy and evaporates. When it condenses some water falls back to Earth at higher levels as rain or snow. Eventually the water flows downhill in streams or rivers. Water flowing through tunnels in a dam across a river can turn turbines and generate electricity. (See Figure 18-17.)

In some regions sunlight warms the upper layer of the ocean to temperatures well above those of the cold depths below. These temperature differences could be used to produce *ocean thermal energy*. Heat from the warm layer could

Figure 18-17 *Roosevelt Dam near Phoenix, Arizona. Such dams help produce hydropower.*

evaporate a liquid such as ammonia. The gas produced could be used to turn a turbine. Electricity produced by the turbine could be sent ashore via cables or used to decompose water into oxygen and hydrogen. The hydrogen can be used as a fuel, also.

Plants are an energy resource with great potential. Wood stoves are becoming a familiar sight again in many areas where wood is plentiful. Wood can be a renewable energy source if forests are carefully managed and harvested. Other plants, such as sugar cane and sugar beets, are being grown for fuel. Sugar can be fermented to produce alcohol, which can be burned in engines instead of fossil fuels. In Brazil, alcohol has already begun to replace gasoline as a fuel for motor vehicles.

In some places solid waste is a serious problem. Many cities have used up their burial sites for trash and garbage. Foods and paper products make up much of this waste. Munich, West Germany obtains one-eighth of its electricity by burning garbage and trash. Baltimore, Maryland has a similar recovery program.

The tides that rise and fall twice each day can be a source of energy. The energy that causes the ocean to bulge

Figure 18-18 *The tidal power plant at Saint-Malo, France. As the tide rises and falls, water moves first into the river then out into the ocean through openings in the dam. The flow of water turns the turbines.*

Figure 18-19 *Geothermal energy is readily available in some regions of Earth. This plant in northern California uses geothermal energy to generate electricity.*

outward comes from the gravitational pull of the moon (and, to a lesser extent, the sun). By harnessing the ebb and flow of tidal water to turbines, it is possible to generate electricity. The French have built a plant that uses tidal energy on the coast of Brittany. (See Figure 18-18.) For fifty years there have been proposals to build a similar energy plant in North America in the Bay of Fundy.

Hot springs, geysers (GY zurz), and the molten rocks from volcanoes reveal the huge storehouse of energy within Earth. This is called **geothermal energy**. It can be used to heat buildings, generate electricity, and power industry. Iceland uses geothermal energy to heat two-thirds of its homes. Nearly all the greenhouses, which produce most of that country's fresh vegetables, are heated this way, too.

geothermal energy (jee oh THURM ul)

The potential for tapping geothermal energy is limited to those regions where Earth's crust is thin. Nevertheless, many such sites exist. Canada is one of a number of nations planning to build industrial plants or electric power plants on these sites. Canadians are also considering using geothermal heat as a source of space heating for remote villages in the Northwest Territories.

Checkpoint

1. Name the three renewable sources of energy that are "powered" by solar energy.
2. What is the source of the energy that produces Earth's tides?
3. Give two practical uses of geothermal energy.

18-8

Heating and Cooling Systems

Goal

To describe various heating and cooling systems.

Colonial homes are noted for large fireplaces that were used to heat them. Few homes today use the fireplace as a primary heat source. In fact, in many homes the warm air drawn up the chimney when a fireplace is used actually increases the rate at which heat is lost from the house.

The wood stove, which was North America's primary source of heat for homes through most of the 19th century, is again finding wide use. A centrally located modern stove is about 50% efficient in transferring the energy stored in the wood to heat in the home. The stove radiates heat, conducts heat to air in contact with it, and sets up convection currents as the heated air rises and cooled air falls.

Most homes and businesses are heated by circulating hot air or water. The air or water gains heat from the burning of fuel oil or natural gas by conduction. Then convection currents carry heat to the rooms. During this process some heat is transferred by radiation, also. (See Figure 18-20.) A good oil or gas furnace transfers about 80% of the energy in the fuel to the building. The rest of the energy goes up the chimney.

Figure 18-20 *Diagrams of typical hot air and hot water heating systems. Radiation, conduction, and convection are involved.*

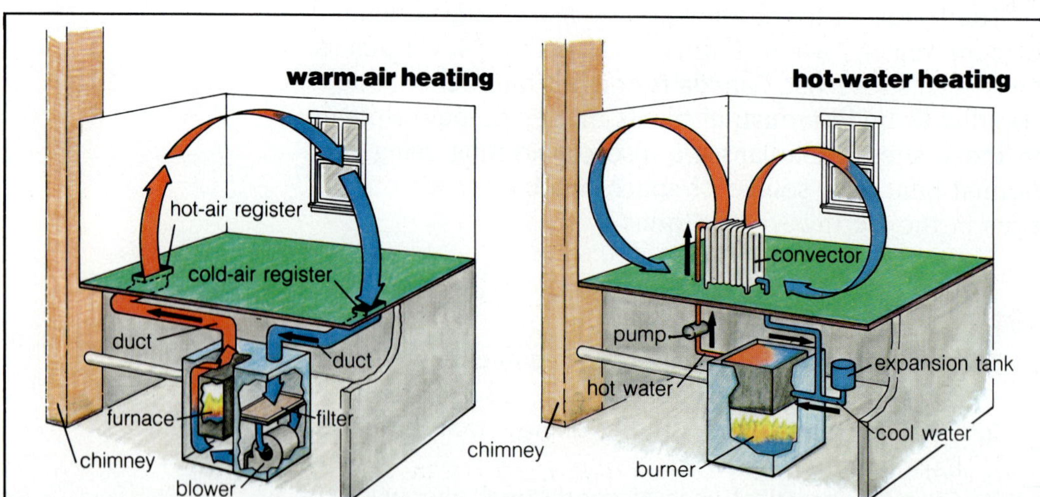

Some homes are heated by electric coils in baseboards, floors, or ceilings. Since all the electric energy is transformed into heat, the efficiency is 100%. However, the generation of electricity at the power plant is probably only about 30% efficient. This is why electric heat is generally more expensive than other forms.

In some parts of North America, cooling is a more important concern than heating. To cool buildings, energy is extracted from the air inside and transferred to the outside. This requires work, since heat does not flow naturally from a cool area to a warm one. Electricity is commonly used as the energy source for cooling buildings. To reduce the costs of air conditioning, many homeowners install fans to pull warm air out of the house and set up cooling convection currents.

Several office buildings are being cooled by circulating ice water. During winter months a snowmaking machine produces a huge mound of snow in a lined pit that is then covered with insulation. In the summer the melted snow is circulated through nearby buildings to absorb excess heat.

In moderate climates, *heat pumps* are increasingly being used to both warm and cool buildings. A refrigerator is a type of heat pump. If you put your hand behind a refrigerator that is running, you can feel warm air. A refrigerator extracts energy from the air inside it and transfers the energy to the air outside it.

In the summer, a heat pump works like a refrigerator. It transfers heat from a building to the outside. In the winter, a heat pump transfers heat from the outside air, water, or ground to the building. Even "cold" matter has energy that can be extracted as heat. But work must be done to extract it. (See Figure 18-21.)

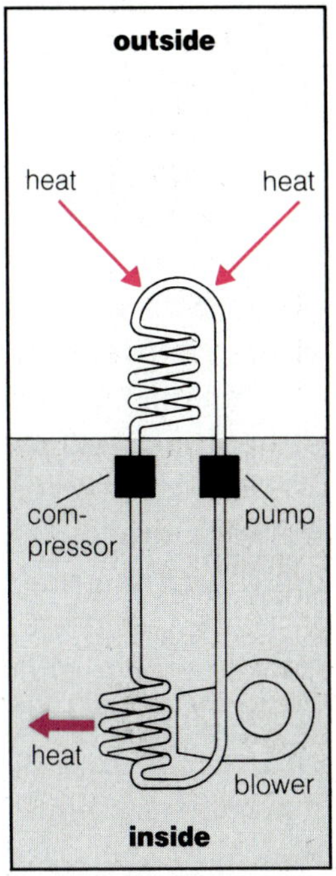

Figure 18-21 *A diagram of a heat pump. A liquid is pumped outside, where it expands and absorbs heat. Then it is moved inside, where it is compressed and gives off heat.*

Checkpoint

1. Name some energy sources that can be used to heat buildings.
2. Describe some common heating systems.
3. Why is electric heating generally more expensive than heating with oil or natural gas?
4. Which energy source is commonly used to cool buildings?

Insulation

Soaring energy costs make it necessary to reduce the amount of energy used to heat or cool buildings. Insulation is one of the best ways to reduce the transfer of heat between the inside and outside of a building. *Insulation* slows the transfer of heat into or out of a building.

In most wooden buildings the space between the inside and outside walls is filled with mineral wool batts or cellulose (tiny pieces of paper). In brick buildings large sheets of rigid insulation are usually placed between the bricks and the inner wall. All such insulating materials are filled with tiny air spaces that slow the conduction and convection of heat. Because warm air rises, the most important area to insulate is the ceiling. Most new homes have at least 22 cm of insulation in the attic over the ceilings. The illustration shows insulation being installed in an attic.

The insulating quality, or resistance to heat flow, of various materials is expressed as their *R-values*. The larger the R-value, the better the insulating quality.

During the winter the warm air inside a house holds much more moisture than the cold air outside. If moisture in the air flows from the living area of a home into the cooler walls, it may condense. When wet, insulation conducts heat more quickly. This reduces its effectiveness. To avoid this, the inner side of the insulation is covered with a thin plastic sheet or thick paper. This cover, called a *vapor barrier*, prevents water vapor from reaching the insulation.

To provide more insulation, many modern homes have double outer walls. The windows often are made of two or more sheets of glass with air space between them. To further reduce heat transfer through glass windows, many homes have insulated shades or heavy drapes that can be drawn over the glass.

Many new homes are being built at least partially underground. Because the ground itself is a good insulator, underground temperatures remain constant throughout the year. Underground houses are therefore less expensive to heat or cool.

Understanding The Chapter 18

Main Ideas

1. Heat is the energy transferred to an object when it warms or from an object when it cools.
2. Matter is made of tiny particles that are in constant motion. When an object gains heat the kinetic energy of these tiny particles increases.
3. Temperature is a measure of the average kinetic energy of the particles in a substance.
4. The scales on thermometers are established by fixed points, such as the melting and boiling points of water.
5. Pressure is the ratio of a force to the area over which it acts.
6. The heat needed to change the temperature of a fixed mass of a substance depends on the nature of the substance.
7. Heat can be transferred by conduction, convection, and radiation.
8. Energy can be transformed from one form into another, but some of the energy transferred always appears as heat.
9. Energy can be transformed, but the total amount does not change.
10. A heat engine changes heat into mechanical energy.
11. An energy source that is produced about as fast as it is used is a renewable energy source.
12. Heating and cooling systems for buildings still depend primarily on energy from fossil fuels.

Vocabulary Review

From the following list, choose the term that best completes each of the statements. Write your answers on a separate piece of paper.

absolute zero	law of conservation of energy
conduction	pressure
conductor	radiation
convection	renewable energy source
geothermal energy	solar energy
heat	specific heat
heat engine	temperature
hydropower	
insulator	

1. Energy that is transferred from a warm body to a cooler one is called __?__.
2. The lowest possible temperature is __?__.
3. __?__ is the ratio of a force to the area over which it acts.
4. Heat may be transferred from place to place by __?__, __?__, and __?__.
5. A thermometer is used to measure __?__.
6. The amount of heat required to raise the temperature of one gram of a substance one degree Celsius is the __?__ of that substance.
7. __?__ is energy from within Earth.
8. The tides can be thought of as a __?__.
9. A material through which heat passes very slowly is a(n) __?__.
10. A __?__ is a machine that changes heat into mechanical energy.
11. A metal is a good __?__ of heat.

Chapter Review

Write your answers on a separate piece of paper.

Know The Facts

1. ___?___ is a measure of the average kinetic energy of the particles that make up an object. (heat, temperature, pressure)
2. ___?___ is a renewable energy source. (natural gas, coal, hydropower)
3. The lowest point on the Kelvin scale is ___?___. (−100°C, the freezing point of water, absolute zero)
4. Air is a good ___?___. (conductor, insulator, fuel)
5. The pressure exerted by an enclosed gas ___?___ when the temperature of the gas increases. (increases, decreases, stays the same)
6. ___?___ is a form of heat transfer involving the movement of gases or liquids due to density differences. (conduction, convection, radiation)
7. ___?___ is energy transferred from one object to another. (heat, pressure, temperature)
8. The specific heat of water is ___?___ J/g/°C. (0, 1, 4.2)
9. Energy is transferred from the sun to Earth by ___?___. (conduction, convection, radiation)
10. Ocean thermal energy arises from ___?___. (solar energy, fossil fuels, tidal power)
11. A(n) ___?___ is the most efficient heating system of the three listed. (oil furnace, wood stove, fireplace)

Understand The Concepts

12. How could you use an eggbeater to warm water?
13. Explain why solar energy is the ultimate source of the energy stored in gasoline.
14. The temperature of an object is 0°C. What is its temperature on the Kelvin scale?
15. What is the main advantage of a heat pump compared to other heating systems?
16. You are given equal masses of silver, gold, copper, and lead. Which metal will require the most heat to raise its temperature 10°C? (See Table 18-1.)
17. Think of two appliances or machines that you use every day. Name them. What energy transformations occur in each one?
18. Why can't the sun's heat be transferred from the sun to Earth by convection?
19. You pedal a bicycle up a hill. The work you do increases the potential energy of both you and the bicycle. What other form of energy can you be sure has been produced, too?
20. List three renewable sources of energy and three nonrenewable sources.
21. Describe a home heating system that is based on solar energy.
22. Explain how wind and hydropower depend on solar energy.
23. How can plants such as sugar cane reduce the use of fossil fuels?

Challenge Your Understanding

24. The particles in the air inside a tire strike the walls of the tire, exerting pressure. Since there are particles in the air outside of the tire, also, why does the tire stay inflated?

25. How do hot air balloonists make their balloons ascend and descend?

26. Why is an increase in the price of fossil fuels often accompanied by increases in electric bills?

27. If 420 J of work are transformed completely into heat that is absorbed by 100 g of water, what temperature change will occur in the water?

28. Equal masses of water at 20°C and 30°C are mixed. What will be the final temperature of the mixture?

29. Why are there gaps between the ends of joining sections of a metal bridge?

30. A one-gram sample of anthracite coal, when burned, produces 30,000 J of heat. If the coal is used to generate electricity, each gram burned produces only 10,000 J of electricity. What is the efficiency of this energy transformation?

Projects

1. Find an unmarked thermometer or cover the scale of a thermometer with masking tape. Then invent your own temperature scale. How can you convert readings on your scale to temperatures in degrees Celsius?
2. Try to obtain permission to visit some solar homes in your vicinity. What fraction of the heating needs of each of the homes you visit is obtained from solar energy?
3. Visit a store that sells insulation and examine the various kinds of insulation. How is the thickness of a particular type of insulation related to its R-value?
4. See if you can find out what factors are responsible for heat losses from a building. Then see if you can estimate the amount of heat that is lost from your home or school in one year. Try to find ways to reduce these heat losses.
5. Use your library to write a brief biography of Count Rumford or Sir James Joule.
6. Have a contest with your classmates. See who can design the best ice-cube keeper. You will want to develop some rules first. Obviously, the keeper cannot be placed in a cold place, and you will want to establish a standard-size cube. Then go to work and see who can build a device that will keep an ice cube from melting for the longest time.
7. Try to find out how the R-values of insulating materials are determined experimentally.

Chapter 19

Electricity and Magnetism

This photo of the effects of electric charge was taken at the Ontario Science Center in Canada.

19-1

Electric Charge

Goals

1. To explain how electric charges are produced.
2. To describe the behavior of charged objects.

You reach for a door knob after walking across a rug and a spark "jumps" between your hand and the knob. When you take off your sweater in a dark room, you see small flashes of light and hear crackling sounds. During the winter, you notice that your hair tends to "fly away." If you brush it, you may see sparks snapping between your hair and the brush. All of these events result from the buildup of *electric charge*. For example, your shoes moving across a rug cause electric charge to build up on your body. If the charge is large enough, a spark will jump between your hand and a metal object, such as a door knob.

People were aware of the effects of electric charge long before they understood it. The ancient Greeks knew that amber (shown in Figure 19-1) could be rubbed vigorously to produce a charge. In fact, our word *electricity* is derived from the Greek word *elektron*, which means amber.

Figure 19-1 *A piece of amber. Rubbed amber was known to attract light objects (because it had a charge) as long ago as about 300 B.C. Amber is the hardened sap of extinct pine trees.*

Suppose two objects are charged in the same way, say by rubbing them with a woolen cloth. They will then move away from, or *repel*, each other. (See Figure 19-2a.) Figure 19-2b shows two objects that were charged in different ways. These two objects move toward, or *attract*, each other.

Charged objects always either attract or repel each other. This is because there are only two kinds of charge.

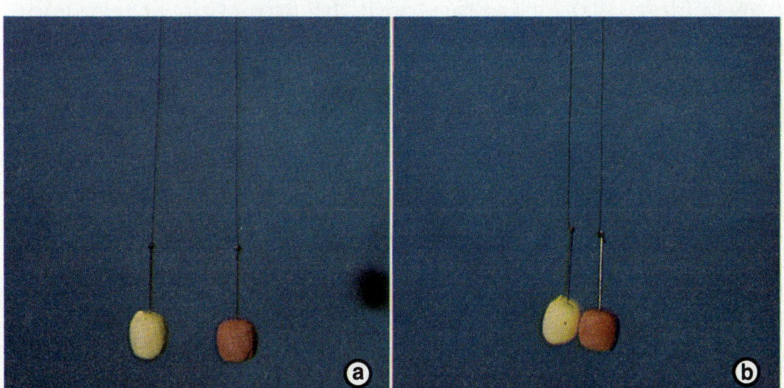

Figure 19-2 *(a) When two pith balls have the same charge, they repel each other. (b) Oppositely charged pith balls attract each other.*

Figure 19-3 *Benjamin Franklin proved lightning is electricity. With a kite in a storm cloud, he detected electricity at a key hanging from the end of the kite string. He was very lucky—he might have been killed.*

Benjamin Franklin (Figure 19-3), well known for his early studies of electricity, called the two types of charge *positive* and *negative*. Two positively charged objects repel each other. Two objects that are negatively charged also repel each other. But suppose one object is positively charged and the other is negatively charged. Then the objects attract each other. The important thing to remember is that like charges repel each other and unlike charges attract.

Two factors determine how much two objects attract or repel each other. The first factor is the amount of charge. If the charge is large on one or both objects, the attraction or repulsion too is large. If the charge is small, the attraction or repulsion is small. The second factor is the distance between the objects. If they are close, the attraction or repulsion is large. If they are far apart, the attraction or repulsion is small.

Checkpoint

1. Give at least one way that charge can be built up on an object.
2. How many kinds of charge are there? What are they called?
3. If two objects have the same type of charge, will they repel or attract each other? If the charges are different, how will the two objects react?

TRY THIS

Rub a balloon on your clothing. "Stick" it to a wall or ceiling. What holds it in place? How long does it stay attached?

19-2

The Electrical Nature of Matter

Goal

To explain how the atomic model of matter is related to electric charges.

All matter is made up of tiny particles called *atoms*. Atoms are too small to be seen, even with the best microscopes available. Atoms themselves are made up of even smaller particles. These are called *protons*, *neutrons*, and *electrons*.

Protons and neutrons are found in the central core, or *nucleus*, of an atom. Neutrons carry no charge. Thus, they are *neutral*. Protons carry a positive charge. The particles called electrons are in constant motion around the nucleus. Each electron carries a negative charge equal to the positive charge of a proton. The force of attraction between positive protons and negative electrons gives the atom its structure. A diagram of a model atom is shown in Figure 19-4. In a real atom, electrons are much farther from the nucleus than in the model.

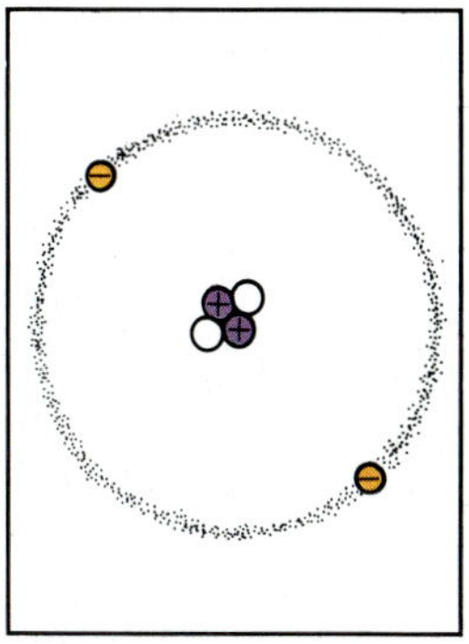

Figure 19-4 *A diagram of a model helium atom. It has two protons and two neutrons in the nucleus and two electrons outside. The + and − show the charges of the particles.*

Different kinds of atoms have different numbers of protons, neutrons, and electrons. Normally, whatever the number of particles is, an atom contains equal numbers of protons and electrons. The total positive charge balances the total negative charge, so the atom itself is neutral.

Atoms can gain or lose electrons. For example, if you rub a glass rod with a piece of silk, the silk will gain electrons and become negatively charged. The glass rod loses electrons. Because its atoms now have more protons than electrons, the glass is positively charged. A **negative charge** results from an excess of electrons. A **positive charge** results from too few electrons to balance the protons in the atoms of an object.

negative charge

positive charge

Checkpoint

1. Name the three main particles that make up an atom. What is the charge on each particle?
2. Why is an atom normally neutral?
3. How can an object become negatively charged? positively charged?

19-3

Static Electricity

Goals

1. To use the atomic model to explain static electricity.
2. To define and illustrate induction and grounding.
3. To describe some practical uses of static electricity.

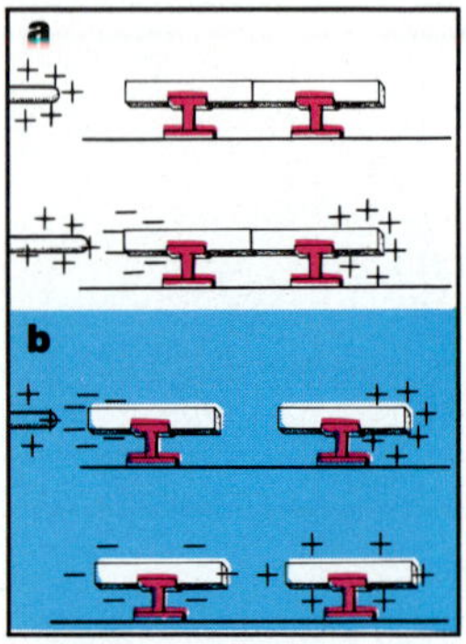

Figure 19-5 *(a) A positively charged rod close to a pair of neutral metal bars that are in contact attracts negatively charged particles. (b) When the bars are separated, they are oppositely charged.*

Electrons are more easily removed from some atoms than from others. If you rub a plastic comb with cloth, the comb will become charged. If you rub a metal rod with a cloth, however, it does not become charged. This is because electrons move easily along a metal rod. Any charge that builds up flows away immediately. If the metal rod is in your hand, the extra electrons flow into your body. Electrons move easily along some materials. Such materials are called **conductors** of electricity. Metals are good conductors.

conductors

Materials that keep electrons from moving are electrical **insulators**. Examples of electrical insulators are rubber, glass, wool, and many plastics. Because insulators do not conduct electricity, any charge they gain stays in one place. Anything that stays in one place is said to be *static*. Thus, charges that stay on an object are called **static electricity**.

insulators
(IN suh layt uhrz)

static electricity

When electricity "flows," it is because electrons move. They tend to move from areas that have an excess of electrons to areas that have too few. Electrons can also move to areas that are neutral, that is, to areas neither positively nor negatively charged.

Figure 19-5 shows a positively charged rod being held near two neutral metal bars on insulating holders. Note that the bars are in contact. The electrons in the metal bars are attracted to the positive rod. Because dry air is a good insulator, however, the electrons cannot leave the bars and move to the rod. Instead, the electrons collect on the metal bar nearest the positive rod. This bar becomes negatively charged. Because electrons have moved away, the other bar becomes positively charged.

If the two bars are now separated, one carries a positive charge, the other a negative charge. When a neutral object is charged by being placed near a charged object, the process is called charging by **induction**. These metal bars were charged by induction.

induction
(in DUHK shuhn)

Figure 19-6 *A bolt of lightning striking the Empire State Building in New York City. It will not harm the building because it is striking a lightning rod at the top of the roof-top tower.*

If a negatively charged conductor touches an uncharged conductor, charge will move from the charged body to the neutral one. Movement stops when the charges are spread over both conductors. This leaves the original charged object with a fraction of its original charge. Suppose the neutral object is Earth. Earth, of course, is very large compared to ordinary objects. Thus, once the charge is distributed, the small object is left with practically no charge at all. This process of a charge flowing into Earth is called **grounding**.

grounding

The most spectacular example of grounding is the action of a lightning rod during a storm, as shown in Figure 19-6. A lightning rod is a pointed metal conductor placed at the top of a building. Lightning rods are connected by heavy wires to metal pipes buried in the ground. When lightning strikes, the charges are grounded when they travel along the heavy wires to the metal pipe in the ground. This grounding protects the building from damage.

Figure 19-7 *These smokestacks contain Cottrell precipitators to catch smoke particles. At the left, the precipitators are turned off. At the right, they are turned on.*

Charge tends to collect on the outside surface of materials, especially metals. During a thunderstorm, you are safest if you are inside a building or car. Stay away from high ground, tall trees, power lines, and open boats. Never swim during a thunderstorm.

Static electricity can be bothersome and even dangerous, but it can be useful too. For example, static electricity is used in painting automobiles. The surface of the car is given a charge. The paint is then given the opposite charge. When the auto body is sprayed or dipped, the paint droplets are attracted to the body's surface. The result is a strong bond between paint and metal.

Cottrell precipitators use static electricity to control industrial air pollution. (See Figure 19-7.) They are placed in the chimneys of plants that burn coal. Each precipitator is made up of pairs of screens bearing static charges. The charged screens attract oppositely charged soot and smoke particles that would otherwise escape.

TRY THIS

Bring a plastic comb or ruler near a pile of tiny pieces of paper. Be careful not to touch the paper. What happens? Now rub the comb or ruler through your hair or on some cloth to put a charge on the plastic. Again, bring the comb or ruler near the paper. What happens this time?

Checkpoint

1. What is static electricity?
2. Name one conductor of electricity.
3. What is an electrical insulator? Give two examples.
4. Explain how lightning rods protect buildings.
5. Describe one way in which static electricity can be useful.

19-4

Current Electricity

Goals

1. To describe the parts of a simple electric circuit.
2. To explain Ohm's law.

Electrons flowing in wires can be used to make bulbs light, motors run, heaters heat, and television sets form images. A stream of flowing electrons is called an **electric current**. It takes more than wire to make a current flow. Electrons must move along a path from a region that has an excess of electrons to an area where there are too few. This path is called an **electric circuit**.

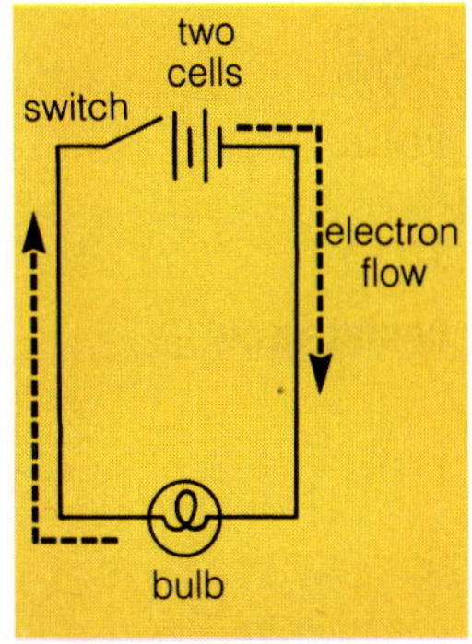

Figure 19-8 *A diagram of a simple electric circuit.*

Figure 19-8 is a diagram of a simple electric circuit. The *switch* controls the circuit. When the switch is closed, the circuit is complete and electrons can flow freely. They flow from the negative pole of one *cell* (an area of excess electrons) to the positive pole of the other. (What most people call a flashlight "battery," scientists call a *cell*.) Along the way, the electrons flow through the bulb. In the bulb, some of the energy of the electrons is converted to light and heat.

Figure 19-9 shows symbols used in circuit diagrams. An *ammeter* is a device that measures current. In effect, it counts electrons flowing by one point in a circuit. The ammeter measures current in units called **amperes**.

electric current

electric circuit

amperes
(AM pihrz)

Figure 19-9 *There is a symbol for each part of an electric circuit.*

symbol	meaning	symbol	meaning
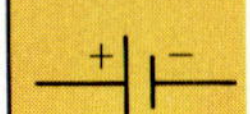	One cell. The long line is the positive (+) side of the cell. It can also be drawn with the short line on the left and the long one on the right.		Resistor. This is a wire that gets hot when electrons travel in it. There is no special number of zigzags.
		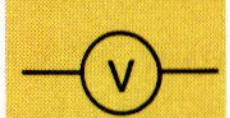	Voltmeter.
	Three cells connected together. The positive (+) side of each cell is connected to the negative (−) side of the next cell.		Ammeter.
		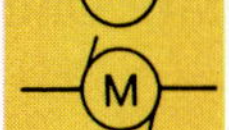	Electric motor.
	Switch, always shown open in diagrams.		Alternating current electric energy source. This is what you plug into when you use electricity in your house.
	Connecting wire.		
	Lamp.		Fuse.

Another kind of meter, the *voltmeter*, is used to measure the energy of the electrons that flow in the circuit. This energy is measured in units called **volts**. The number of volts measured by the meter is called the *voltage*.

volts

Some types of wire allow a large current to flow. Others offer more **resistance** to electron flow. Wires that allow a large current to flow are said to have a low resistance. Wires that allow only a small current to flow have a high resistance. In most cases, a wire's resistance increases if the wire is made thinner or longer.

resistance

The filaments in most light bulbs are thin tungsten wires with a high resistance. As a result of this resistance, electric energy is converted to light and heat inside a bulb. A toaster works the same way. Inside the toaster, electric energy is converted to heat, which browns the bread. Electric heating systems in buildings also depend on resistance.

Georg Ohm, a German physicist, studied the relationship between voltage, current, and resistance. He found that the current in any circuit is equal to the voltage divided by the resistance. This relationship is called Ohm's law. It means that for a constant voltage, current is less when resistance is greater. Or, for a constant voltage, current is greater when resistance is less. Units of resistance are called **ohms**, after Georg Ohm.

ohms

$$\text{current (amperes)} = \frac{\text{voltage (volts)}}{\text{resistance (ohms)}}$$

Checkpoint

1. What is an electric current?
2. How is electric current measured?
3. What does a voltmeter measure?
4. Give an example of a household appliance that must have a large electric resistance.
5. State Ohm's law.

TRY THIS

Using one D-cell, a flashlight bulb, and a 15-cm length of copper wire, see if you can light the bulb. How many different ways can you find to make the bulb light? Make a circuit diagram for each way that works.

19-5

Series and Parallel Circuits

Goals

1. To compare series and parallel circuits.
2. To explain how fuses work.

A basic circuit is made up of three parts: a power supply, an energy-consuming device, and a control, such as a switch.

series circuit

In a **series circuit** there is only one path the electrons can follow from the power supply and back. The power supply and the energy consumers are placed one after the other, in a series. (See Figure 19-10a.) When two or more cells are connected this way, they form a **battery**. A battery has more voltage than a single cell. When cells are connected in series, the negative pole of one is connected to the positive pole of the next.

battery

parallel circuit

In a **parallel circuit**, electrons have more than one path to follow (Figure 19-10b). When two cells are connected in parallel, their negative poles and their positive poles are connected.

Look again at the circuit in Figure 19-10a. If you open either switch A or B, both lights go out. In this circuit the bulbs are in series. Any current that flows through A must flow through B.

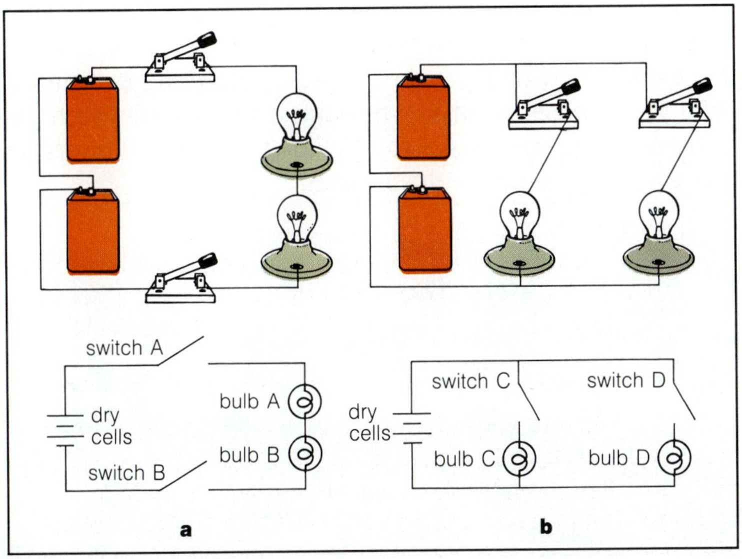

Figure 19-10 *(a) A series circuit and the corresponding circuit diagram. (b) A parallel circuit with its circuit diagram.*

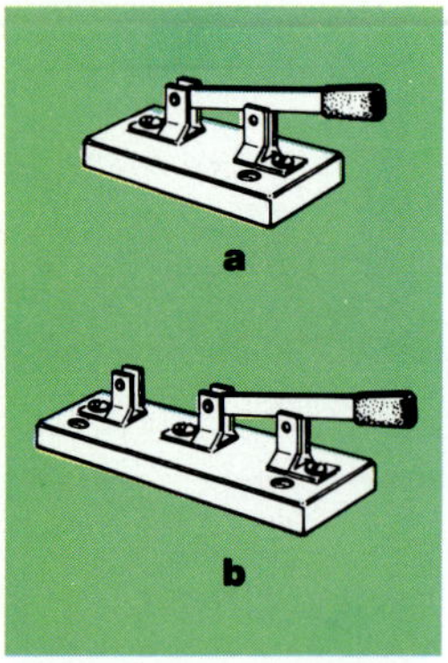

Figure 19-11 *(a) A toggle switch with one handle, or pole. The pole can be "thrown" only one way. (b) A double-throw switch. The single pole can move two ways. This is a three-way switch in that it can be closed one way, opened, and closed the other way.*

Now look at Figure 19-10b. If you open switch C with D closed, only bulb C goes out. If you open switch D with C closed, only bulb D goes out. In this circuit the bulbs are wired in parallel. Current from the power supply divides. Some electrons flow through bulb C at the same time that other electrons flow through bulb D. In a home, circuits are wired in parallel. In series, all the lights and other appliances would have to be either on or off at the same time.

Several kinds of switches are used in home-wiring circuits. (See Figure 19-11.) The simplest is the *toggle switch*, or single-pole, single-throw switch. The single-pole, double-throw switch is more complicated. It is a three-way switch. It might be used in a circuit with a light at the top of a stairway. With two single-pole, double-throw switches, the light can be controlled at the top and bottom of the stairs.

Suppose you are upstairs and the light is on, as shown in Figure 19-12a. Current is flowing through the right-hand wire between the switches. Then you walk downstairs and flip the lower switch. (See Figure 19-12b.) The circuit is broken, and the light goes out. Later on, you want to go back upstairs. So you flip the lower switch, which closes the circuit again. The light goes on. (See Figure 19-12c.) Then you walk up the stairs and flip the upper switch. The circuit is broken, and the light goes off. (See Figure 19-12d.)

Electricity from a power plant goes to homes in parallel circuits. The electricity travels from the power plant at very high voltage. This voltage is much higher than the voltage in your house. In fact, if this voltage were used with your appliances, it would damage them. Thus, before the current enters your house, it passes through a *transformer*. The transformer lowers the voltage to 110 V or 220 V. Transformers can be used to increase voltage, also.

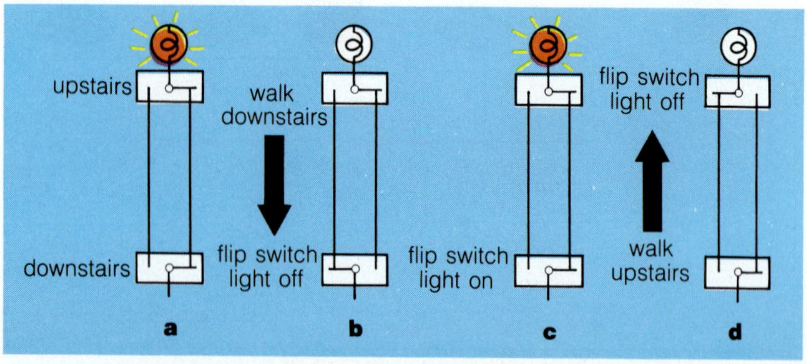

Figure 19-12 *Double-throw switches make it possible to control the light above a stairway from the top or bottom of the stairs.*

Household electric circuits have fuses or circuit breakers in them in order to protect the circuits from carrying too much current. In a *fuse*, a metal connector will melt if too large a current flows through it. Once the connector has melted, the circuit is broken. A *circuit breaker* is like a switch. It opens when too much current flows through it.

Figure 19-13 *A three-pronged plug. The round middle prong is for the ground connection.*

In a home, too many electric appliances are sometimes plugged into one circuit and used at the same time. Then a large current flows in the circuit. The wires overheat, and a fuse burns out or a circuit breaker trips. The fuse or circuit breaker opens the circuit. This prevents the possibility of a fire being produced by overheated wires. Fuses are designed for your safety. Never replace a fuse with a penny. A penny can carry a current large enough to cause a fire.

Your body is a reasonably good conductor, especially when it is wet. Suppose you touch an uninsulated wire or switch when you are connected to the water system. You then become part of a circuit. You might be standing in a bathtub or touching a faucet. A current large enough to kill you can go through your body to Earth. Current can pass from the wire through you to the faucet or tub and then to the ground.

Many modern appliances have plugs with three prongs, as shown in Figure 19-13. The round prong carries the ground wire. If the appliance develops a defect, grounding provides an easy path for the current to pass rapidly to Earth.

Never use an appliance if the cord appears to be worn. Wires in the cord might make contact with grounded objects, injuring you or starting a fire. If you receive a shock from an electrical appliance, shut it off. Do not use it again until it has been checked by an electrician and repaired.

Checkpoint

1. Make sketches showing a light bulb and a motor wired in series and in parallel.
2. How do you know that the electric lamps in your home are wired in parallel?
3. Why are fuses used in circuits?
4. List some safety procedures you should follow to avoid electrical shocks or fires.

Figure 19-14 *Magnetic lines of force around a bar magnet (above) and a horseshoe magnet (below).*

19-6

Magnetism

Goal

To describe magnetic fields, including Earth's magnetic field.

magnets

magnetic field

You have probably seen **magnets**, objects that attract anything containing iron. If you put a bar-shaped magnet into a box of paper clips, some clips stick to it. Most of the clips stick to one end or the other of the magnet. These regions where the magnetic attraction is strongest are called *poles*.

Magnets can attract or repel objects without touching them. In other words, the magnet can exert a force (a push or pull) on an object. If you place a paper clip on a piece of paper and then move a magnet under the paper, the paper clip will move. A clearer way to show this is to pour iron filings on a paper covering a magnet. The iron filings line up, making a picture of the invisible force around the magnet. The region in which a magnet can affect an object is called its **magnetic field**. Figure 19-14 shows the magnetic *lines of force* that make up a magnetic field.

Earth itself has a magnetic field. It has a north and a south magnetic pole, also. Any magnet, freely suspended and far from other magnetic materials, lines up along a north-south direction. Knowing this, you can use a magnetic compass to determine direction. The needlelike indicator is a small magnet. (See Figure 19-15.) The end of the magnet that points north is called a *north-seeking pole* (N). The oppo-

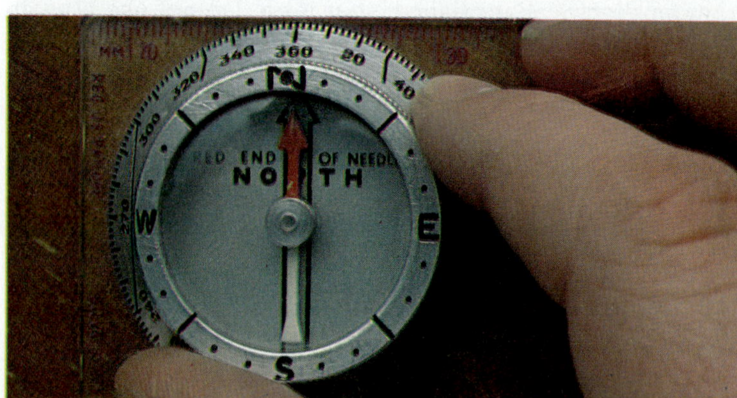

Figure 19-15 *A person using a magnetic compass.*

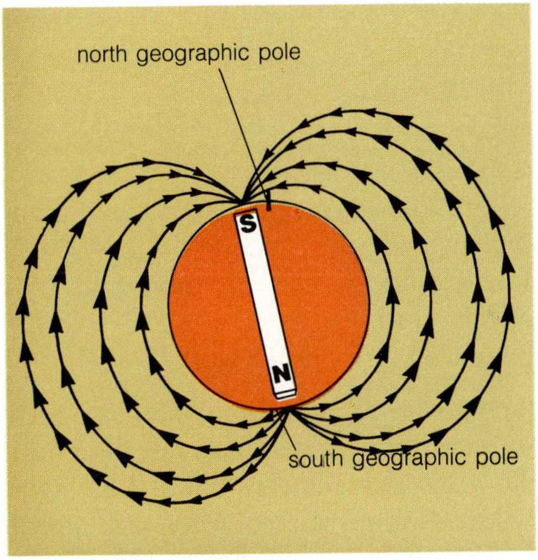

Figure 19-16 *Magnetic lines of force around Earth. They are drawn as they would be if there were no streams of particles from the sun to bend them.*

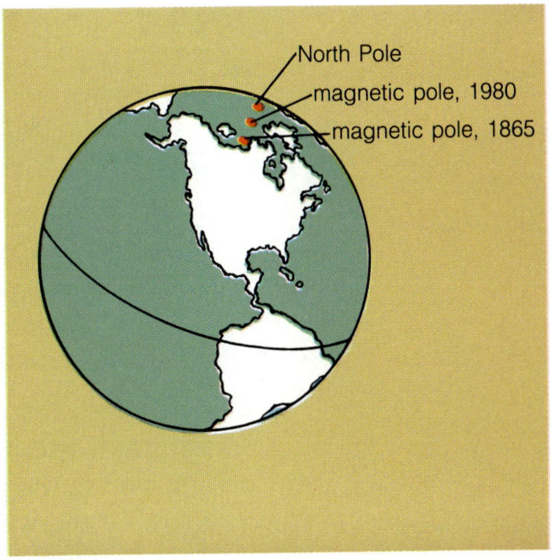

Figure 19-17 *The positions of Earth's magnetic poles have changed. This shows how far the north (south-seeking) magnetic pole has moved since 1865.*

site end of the magnet is called a *south-seeking pole* (S). A compass needle lines up in a north-south direction. The north-seeking pole of one magnet attracts the south-seeking pole of another magnet. Like magnetic poles repel each other.

Maps of Earth's magnetic field lines have the general shape shown in Figure 19-16. Earth can be thought of as a sphere with a giant bar-shaped magnet inside. The south-seeking pole of this very large magnet is in northern Canada. The poles of Earth's magnetic field do not coincide with its geographic poles. Therefore, compass needles in most locations do not point to true (geographic) north. Earth's magnetic field changes slightly every year. (See Figure 19-17.)

Checkpoint

1. What can be learned from a picture of a magnetic field?
2. What happens when the north-seeking poles of two bar magnets are brought close to one another?
3. How do scientists know that Earth's magnetic pole in northern Canada is a south-seeking pole?

TRY THIS

Stroke a paper clip with a bar magnet. Do the same to several other paper clips. How do these paper clips now behave?

19-7

Electricity from Magnetism

Goals

1. To show that there is a magnetic field around a current-carrying wire.
2. To explain how electricity can be produced by a changing magnetic field.
3. To explain how an electric motor works.

Current flowing through a wire will cause a compass needle to turn. If you place a wire above a compass needle and in line with it, the needle turns when current flows. (See Figure 19-18.) If the wire is placed below the compass, the needle turns in the opposite direction. If you reverse the connections to the battery poles, the direction that the needle turns is also reversed.

electromagnet (ih lek troh MAG nit)

These experiments show that a current-carrying wire is surrounded by a magnetic field. By winding the wire into a coil, you can produce a magnetic field similar to that around a bar magnet. Placing a soft iron core within the coil makes the field even stronger. A wire coil wound around an iron core is an **electromagnet**. It behaves like a magnet only when electricity flows through the coil.

After people learned that electricity can produce a magnetic field, they tried to turn the process around. They tried to produce electricity from magnets. At first, attempts to

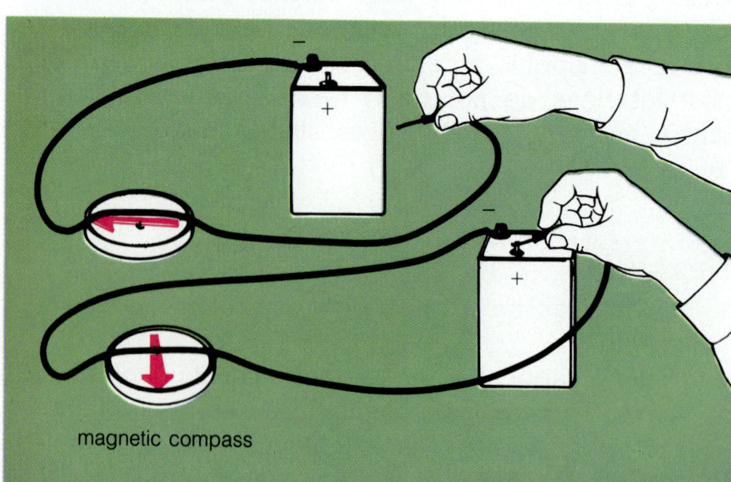

Figure 19-18 *When current flows, a compass needle turns as shown.*

produce electricity from strong magnets were not successful. But in 1831, Michael Faraday, a self-taught English scientist, uncovered the secret. For electricity to be produced in a coil of wire, the magnetic field through the coil must be changing.

Look at Figure 19-14 (on page 484) again. Notice that the number of lines in the field is much greater at the poles than at the center of the magnet. The number of lines in an area indicates how strong the magnetic force is—more lines, stronger force.

If a bar magnet is pushed into and out of a coil of wire, the number of lines of force affecting the coil is constantly changing. When this happens, a current is induced in the coil. The current can be detected with a sensitive ammeter. (See Figure 19-19.) A current is also induced when the magnet is stationary and the coil is moved.

Most of the electricity used in homes and industry is produced at power plants. Power plants make use of electric *generators*. These consist of large magnets turning inside giant wire coils. Current flows as the magnetic field through the wire coils changes.

Figure 19-19 *A coil of wire can be attached to an ammeter. If a magnet is moved back and forth through it, the ammeter will show that a current has been induced. A current also may be induced by moving the coil back and forth around a stationary magnet.*

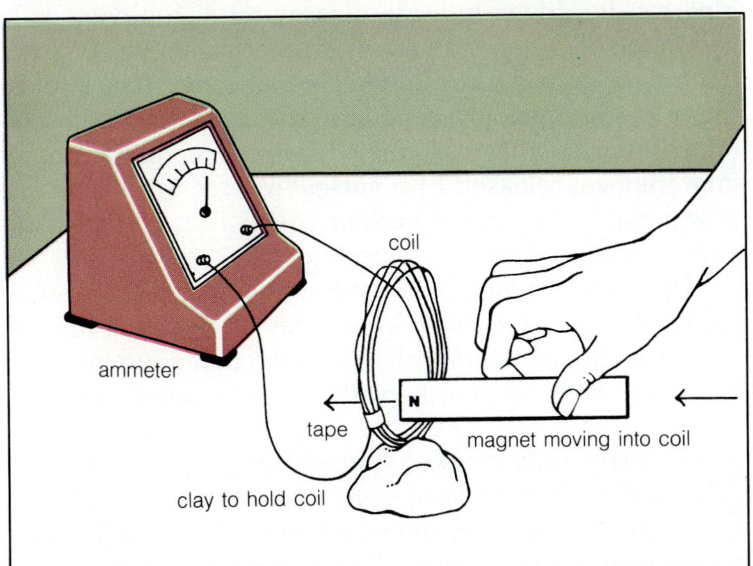

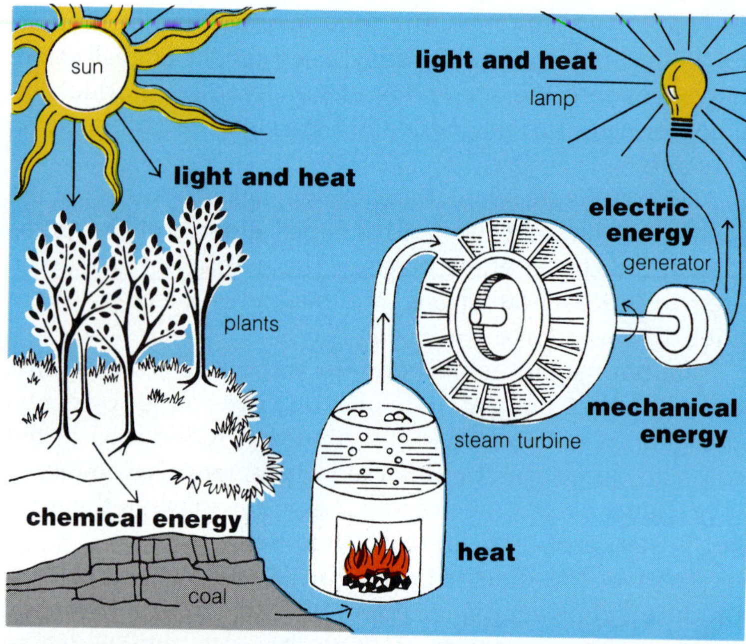

Figure 19-20 *Modern light and heat come from ancient light and heat. Energy from the sun was stored by ancient plants, which then became coal. Today, burning coal gives off heat that is transformed into the energy used by a lamp.*

The generator converts kinetic energy to electrical energy. The kinetic energy comes from the spinning magnet, which is powered by a turbine. (See Figure 19-20.) The turbine can be driven by water flowing through a dam, as in hydroelectric power stations. The potential energy of water behind the dam is converted to kinetic energy, then to electricity. More commonly, generators are driven by steam produced from burning fossil fuels (coal, oil, or natural gas) or from the heat released by a nuclear reactor. So, in a typical power plant, chemical or nuclear energy is converted to kinetic energy and then to electricity.

The current produced in these generators changes direction as the magnetic field changes its direction through the coils. The back-and-forth movement of the current in the coils and wires leading from the coils is called *alternating current* (AC). This current differs from the current produced by flashlight cells or car batteries. Such current is called *direct current* (DC) because it flows in only one direction.

Kinetic energy from steam or flowing water can be used to produce electric energy. Electric energy, in turn, can be used to produce kinetic energy. *Electric motors* transform electric energy to kinetic energy. The operation of an electric motor is just the reverse of the way a generator works. In fact, electric motors can be used as generators and generators can be made to run as motors. To make an electric motor, a magnet is suspended inside a coil. Then an alternating current is sent through the coil. This produces a magnetic field around the coil. This magnetic field is constantly changing in strength and its poles are constantly switching.

If the switching is timed correctly, the magnet is alternately attracted and repelled. This makes it spin. The spinning movement is kinetic energy that can be used to do work, such as turning a fan.

The motor just described is a very primitive one. Most of the motors you use do not contain permanent magnets. In most cases both the coil and the spinning part are electromagnets.

Checkpoint

1. What happens if you place a magnetic compass over a current-carrying wire?
2. Describe an electromagnet.
3. What did Faraday discover about producing electricity from magnets?
4. What does a generator do? What does an electric motor do?
5. What is the difference between alternating and direct current?

TRY THIS

Wrap 2–3 m of enameled copper wire around an iron (or steel) nail. (See Figure 19-21.) Be sure to wrap the wire in one direction only. Sandpaper the ends of the wire to remove the enamel insulation. Connect the ends of the wire to the poles of a battery and you will have an electromagnet. (**CAUTION**: Don't leave the wires connected for very long or you will wear out the battery quickly.)

How many paper clips can your electromagnet lift? How can you make the electromagnet stronger? weaker?

How strong is your magnet when you wind half the wire turns one way and the other half in the opposite direction? How can you explain this?

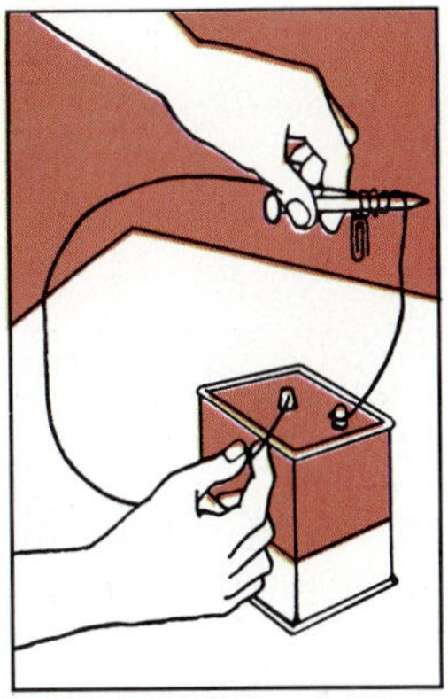

Figure 19-21 *This illustration shows how to make an electromagnet.*

Figure 19-22 Electric meters like this register the kilowatt-hours of electricity used in a home.

19-8

Using Electricity at Home

Goal

To explain how the amount of electrical energy used is measured.

The *electric meter* in Figure 19-22 is a familiar sight. Every customer serviced by an electrical utility company has a meter. As the meter's disc turns, it indicates the rate at which electricity is being used. Thus, an electric meter is similar to a speedometer in an automobile.

The dials on an electric meter resemble an automobile's odometer. Suppose you want to find how far a car has traveled in a period of time. You simply subtract an earlier odometer reading from the present reading. Similarly, your utility company can determine how much electrical energy you have used by reading the dials on your meter. Utility companies measure and charge for electrical energy in *kilowatt-hours* (kW · h). A kilowatt is a unit of power equal to 1000 watts.

Power is energy used per unit of time. If power is multiplied by time, the product will be energy.

$$\text{power} \times \text{time} = \text{energy}$$

or

$$\text{kW} \times \text{h} = \text{kW} \cdot \text{h}$$

For example, to find the electrical energy transformed into heat by a toaster, multiply the toaster's power by the time it operates. A 1.2 kW toaster uses 1.2 kW · h in 1 hour, 2.4 kW · h in 2 hours, 3.6 kW · h in 3 hours, and so on. If the toaster is used for 15 minutes each day, the energy required for the toaster each day is:

$$1.2\ \text{kW} \times 0.25\ \text{h} = 0.3\ \text{kW} \cdot \text{h}$$

The total electric bill each month is calculated by multiplying the total kilowatt-hours of energy used by the cost of electricity per kilowatt-hour.

Table 19-1: Energy Use of Common Home Appliances

Electric appliance	*Average wattage*	*Average hours used per year*	*Kilowatt-hours per year*
Kitchen			
broiler	1400	70	98
coffee maker	900	120	108
dishwasher	1200	300	360
range	12,200	100	1220
microwave oven	1450	130	189
toaster	1200	35	42
refrigerator (340 L)	240	3000	720
frostless refrigerator (340 L)	320	3800	1216
frostless freezer (425 L)	440	4000	1760
Laundry			
clothes dryer	4800	200	960
iron	1000	140	140
washing machine	500	200	100
water heater	2500	1600	4000
quick-recovery water heater	4500	1000	4500
Lighting			
light bulbs	660	1500	990
Comfort			
air conditioner	900	1000	900
electric blanket	180	830	149
dehumidifier	250	1500	375
fan (window)	200	850	170
humidifier	180	900	162
Health/Beauty			
hair dryer	750	50	38
shaver	14	80	1
sun lamp	280	60	17
Entertainment			
radio	70	1200	84
B/W television			
tube	160	2200	352
solid-state	55	2200	121
color television			
tube	300	2200	660
solid-state	200	2200	440
Housewares			
clock	2	8760	18
vacuum cleaner	630	75	47
sewing machine	75	140	11

Table 19-1 lists the average power, hours of use, and energy used for a number of common appliances throughout a one-year period. To find the average annual operating cost of any appliance, multiply the annual energy required in kilowatt-hours by the cost per kilowatt-hour.

Checkpoint

1. What does an electric meter measure?
2. How could you calculate the energy used by an electric appliance?

19-9

ENRICHMENT

Saving Electricity

Goal

To list ways people can reduce their use of electricity.

The rising cost of fossil fuels is causing the price of electricity to soar. To keep costs down, people can use certain *energy conservation* measures. These reduce the use of electricity, and thus its cost.

If you use an electric stove, be sure to cover pans when you cook food in boiling water. To make the best use of the available heat, select pans that cover the entire heating element. Those are two ways to get as much heat as possible to the food. Turn off heating elements, or the oven, several minutes before food is thoroughly cooked. The food will continue to cook while the elements are cooling. Do not open the oven to check on food while it is cooking. Use heating elements on the top of the stove, rather than using the oven, whenever possible. The heat is transferred more directly to the food.

If you have an automatic dishwasher, do not run it unless you have a full load. After the rinse cycle, let the dishes air dry instead of using the heating element.

To keep a refrigerator and freezer efficient, the inside of the refrigerator should be no colder than 4°C. The doors of the refrigerator should close tightly. If the rubber strips that seal the door are in good condition, they should grip a strip of paper firmly. A freezer that is full of food is more efficient than one that is half empty. The freezer or refrigerator should be well insulated so that heat from the surroundings does not easily enter the cold space. Because cold air tends to sink, a chest type freezer is more efficient than an upright model. A frostless model of either of these appliances needs more energy to operate.

Figure 19-23 *Drying clothes on a clothesline saves the energy needed to operate a dryer.*

To wash clothes, use cold or warm rather than hot water whenever possible. Use the washer and dryer only when you have enough clothes for a full load. Dry large volumes of clothing in consecutive loads. That way the dryer stays hot from one load to the next. Dry clothes on a clothesline whenever possible. You can do this either outside or inside. Wear wash-and-wear clothing that does not require ironing.

Figure 19-24 *The old TV set with tubes uses much more energy than a modern solid-state set.*

There are other important ways in which people can conserve electrical energy:

1. Insulate hot-water heaters and pipes.
2. Turn off the electric water heater when going away for a day or more.
3. Lower the thermostat on the water heater to the temperature at which the water is normally used.
4. Take quick showers instead of baths.
5. Turn off lights and appliances that are not being used.
6. Use a single high-wattage light bulb in multi-bulb fixtures, rather than many low-wattage bulbs.
7. Buy a solid-state television set (Figure 19-24) with as small a screen as can be reasonably used in the selected location.

TRY THIS

List the appliances used in your home throughout one day. After studying the ways in which electricity is used in your home, think of ways to reduce your family's use of electricity. Try your ideas for one month. Is there a difference in the electric bill?

Checkpoint

1. Why is it more energy efficient to use the elements on the top of the stove instead of the oven?
2. What is the best operating temperature for a refrigerator?
3. List three ways to make a hot-water heater more efficient.

Magnetic Organisms

Scientists have assumed for many years that pigeons can sense Earth's magnetic field, and use that sense to help them fly in the right direction. In 1978, a scientist discovered that pigeons have small grains of magnetite in their heads. Magnetite is the kind of iron mineral that is a magnet itself. The grains are in a pocket at the base of the beak.

Scientists believe magnetite helps pigeons tell the difference between north and south. Perhaps the grains are affected by Earth's magnetic field, and nerve cells sense the changes.

In the last few years, magnetite has been found in other animals, including dolphins, whales, tuna, mice, and monkeys. It is easy to see why birds, fish, and dolphins would find a magnetic sense of direction useful in their travels. But can you imagine magnetic bacteria?

Magnetic bacteria were first found in 1975, in marshes along the east coast of North America. (See Figure 1.) They seem always to move toward the North Pole. Scientists tested some bacteria by putting them in a reversed magnetic field. Sure enough, they moved in a reverse direction.

Other magnetic bacteria have been found in the Southern Hemisphere, in New Zealand and Tasmania. They always move toward the South Pole.

Each magnetic bacterium contains a string of magnetic particles. The particles are very big and heavy compared to the rest of the organism. When the particles react to a magnetic field, the rest of the bacterium has to go along.

Earth's lines of force are parallel to Earth's surface in the middle latitudes. But at higher latitudes, the lines of force angle downward. Bacteria that follow the lines swim downward into the bottom mud from the ocean above. The bottom mud is where they find the best living conditions.

One of the latest discoveries is that Monarch butterflies have magnetite in their bodies. (See Figure 2.) Perhaps it helps them on their yearly migration from Canada to Mexico.

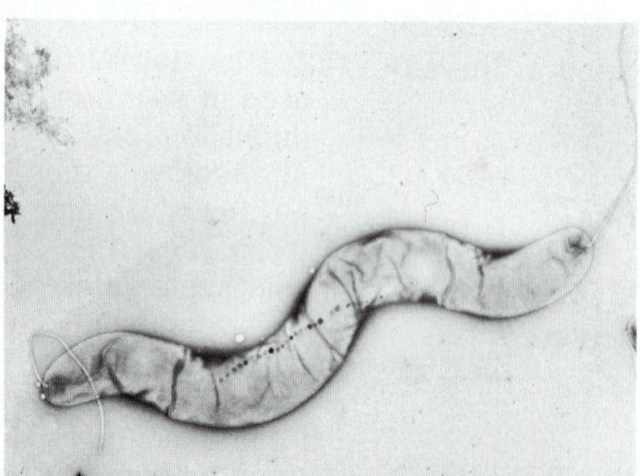

Figure 1

Figure 2

Understanding The Chapter 19

Main Ideas

1. Friction between objects can create static electricity.
2. Like charges repel one another. Unlike charges attract.
3. Atoms can gain or lose electrons.
4. Electrons flow along materials that are good conductors of electricity.
5. A neutral object held near a charged object can be charged by induction.
6. Grounding can protect buildings and people.
7. An electric circuit provides a complete path for electrons. The flow of electrons along a circuit is an electric current.
8. Circuits may be wired in series or parallel. Household circuits are always wired in parallel.
9. A magnet is surrounded by a magnetic field that can be represented by lines of force.
10. There is a magnetic field around a current flowing in a wire.
11. When the magnetic field through a coil of wire changes, an electric current is produced.
12. A generator turns kinetic energy into electricity; a motor reverses the process.
13. Use of electric energy is measured in kilowatt-hours.
14. Wise use of home appliances can reduce the amount of electric energy used.

Vocabulary Review

From the following list, choose the term that best completes each of the statements. Write your answers on a separate piece of paper.

amperes	magnetic field
battery	negative charge
conductor	ohms
electric circuit	parallel circuit
electric current	positive charge
electromagnet	resistance
grounding	series circuit
induction	static electricity
insulator	voltmeter
magnets	volts

1. A(n) __?__ is a stream of flowing electrons.
2. Resistance in an electric circuit is measured in units called __?__.
3. A poor conductor of electric charge is a(n) __?__.
4. A(n) __?__ is a circuit with only one path for electrons.
5. Like poles of __?__ repel.
6. Charges that don't move are __?__.
7. __?__ is one way of generating a static electric charge.
8. A(n) __?__ is the kind of circuit used in homes.
9. The rate of electron flow in a circuit is measured in units called __?__.
10. Wires that allow a large current to flow have a low __?__.
11. Your body is a reasonably good __?__ of electricity.

12. A compass needle lines up with a(n) ___?___.
13. A positive charge attracts a(n) ___?___.
14. A wire coil wound around an iron core is a(n) ___?___.
15. A lightning rod used for conducting and ___?___ electric charge.

Chapter Review

Write your answers on a separate sheet of paper.

Know The Facts

1. To obtain electric charge, rub a(n) ___?___ rod with silk. (glass, copper, aluminum)
2. A charged object will attract opposite charges in a neutral object by means of ___?___. (induction, insulation, magnetism)
3. The particles in an atom's nucleus are ___?___. (electrons and protons, protons and neutrons, neutrons and electrons)
4. Electric current can be measured with a(n) ___?___. (voltmeter, ammeter, generator)
5. A(n) ___?___ conducts charge in a circuit. (open switch, wire, insulator)
6. Electrons move from areas that have ___?___ electrons. (an excess of, too few, an equal amount of)
7. A(n) ___?___ can be used to determine direction. (ammeter, voltmeter, magnetic compass)
8. A magnetic compass points to the ___?___. (geographic north, magnetic north, south)
9. A transformer decreases or increases ___?___. (electric charge, resistance, voltage)
10. To produce electricity in a coil of wire, the magnetic field through the coil must be ___?___. (constant, changing, weakened)
11. A small compass is placed near a bar magnet. The direction of the north-seeking pole of the compass needle gives the direction of ___?___. (the magnet's magnetic field, Earth's magnetic pole, north)
12. The energy, in kilowatt-hours, required to operate a 120 watt light bulb for 10 hours is ___?___. (1.2, 12, 1200)
13. Resistance in a wire is measured in ___?___. (volts, ohms, kilowatt-hours)

Understand The Concepts

14. Use Table 19-1 to find how much energy the average family would save in one year if it bought a regular refrigerator instead of a frostless model. At 10¢ per kilowatt-hour, how much money would be saved?
15. Explain why socks may stick together when you take them out of a clothes dryer.
16. Why is it important that lightning rods be well grounded?
17. Explain why insulators develop static electricity but conductors don't.
18. Why are you safe in an automobile during a thunderstorm, even though you are surrounded by metal?
19. You should never replace a fuse with a penny. What is the difference between a fuse and a penny that makes the fuse safe?
20. How do scientists know there is a magnetic field around an electric current?
21. How can magnets be used to produce electricity?

22. A meter reader from a utility company reads an electric meter once and reads it again one month later. The reading is 35678 the first month and 36678 for the second. If the company charges 10¢ per kW·h, what would the bill be?
23. Name five ways someone who heats water with electricity could save energy.
24. If two objects that are rubbed together become charged, will the charges be the same or different? Will the objects repel or attract?

Challenge Your Understanding

25. Draw a diagram showing an electric circuit with a lamp and a motor wired in parallel. On your diagram show where you would place an ammeter to measure the current through (a) the lamp and (b) the motor.
26. A fuse is needed to protect several appliances wired into the same circuit. Should the fuse be wired in series or in parallel with the rest of the circuit? Why?
27. Explain what is happening to electrons when you rub a balloon on your clothing and then stick the balloon to a wall.
28. Think of a way to clean the particle collectors of a Cottrell precipitator. Describe your cleaner.
29. Explain how using an appliance with a worn cord could cause a fire.
30. Two small charged spheres each carry a positive charge. They repel each other with a force F. How will F change if: (a) the charge on one sphere is doubled, (b) the distance between the charges is doubled?

Projects

1. Report on Faraday's work in electricity and magnetism.
2. Do some library research to answer these questions:
   (a) How does a transformer work?
   (b) What is a thermocouple? How can it be used to generate electricity?
   (c) What are ferromagnetic and paramagnetic materials?
   (d) What is lodestone? How is it related to magnetism? Where was it first used?
3. Visit an electric power plant and see electricity being produced on a large scale.
4. Find a magnetic compass, a battery, and a piece of wire. What happens when you place the compass needle over and parallel to the wire when the wire is connected to the battery? What happens if you put the compass under the wire?
5. Using the materials shown in Figure 19-19, demonstrate Faraday's discovery.
6. Consult reference books to learn how to build a simple electric motor. Then build the motor.
7. Find the power ratings of several appliances in your home. Keep a record of the time they are in use. Then estimate the energy required to run them for one year. Estimate the cost to operate them for one year.
8. Contact your local electric company. Ask them for information about ways to save electricity in your home. Most electric companies have such information.

Chapter

20

Light and Sound

Still water reflects light the way a mirror does, forming images.

20-1

Introduction to Light

Goals

1. To describe some sources of light and the way light travels from a source.
2. To describe what may happen to light when it strikes an object.

Light is a form of energy. A source of light must produce or transform energy. Powerful nuclear reactions in the sun provide the energy that causes it to give off light. Energy released when substances burn causes light to be given off by candles, kerosene lamps, and wood fires. Electric energy from power plants, batteries, and solar cells is transformed into light by light bulbs.

Light from a source, such as the sun or a bare light bulb, generally spreads out in all directions, in straight lines. Light given off by a bulb can be concentrated into a *beam* by a flashlight or spotlight. (See Figure 20-1.)

Figure 20-1 *Performing musicians are often highlighted by beams from spotlights.*

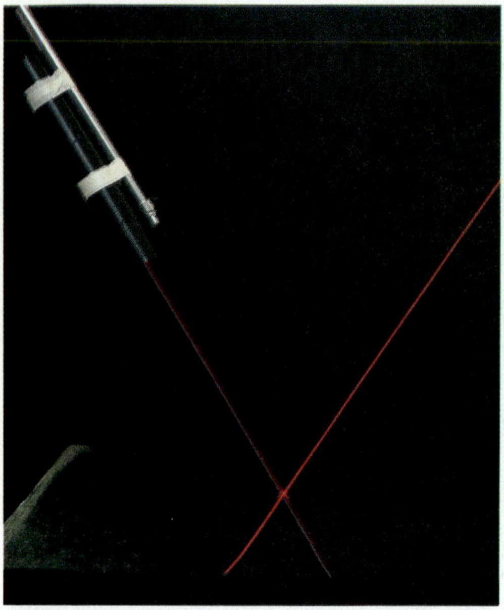

Figure 20-2 *A beam of light that strikes a mirror bounces off the reflective surface.*

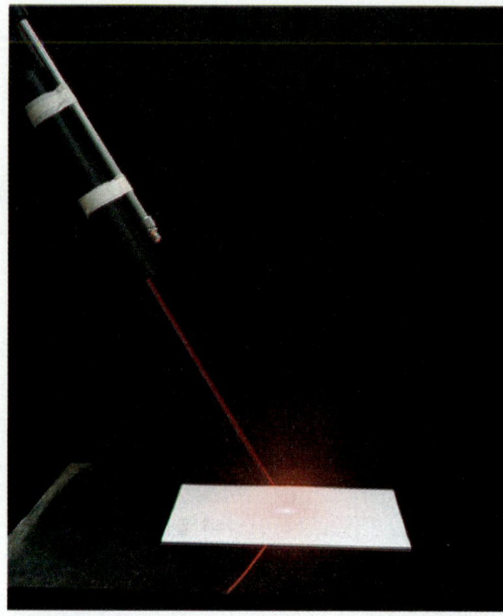

Figure 20-3 *Light is scattered by a translucent object*

When a beam of light strikes a mirror or some other smooth surface, most of it bounces off. (See Figure 20-2.)

reflection

The bouncing of light off an object is called **reflection**. Most objects reflect at least some light.

Reflection changes the direction in which light is moving. You see objects that reflect light toward you. You see the moon because it reflects sunlight. When you shine a flashlight in the dark, you see only those objects that reflect its light toward you.

transparent

When light strikes a **transparent** object, most of the light passes through. Clear glass and plastic are typical transparent materials. Windows provide light inside a building in the daytime, since window glass is transparent.

translucent
(tranz LOO sunt)

Most of the light that strikes a **translucent** object is *scattered* (reflected in many different directions). (See Figure 20-3.) You cannot see things clearly through a translucent object. Frosted glass, waxed paper, and many lampshades are translucent.

opaque
(oh PAYK)

Most of the light that strikes an **opaque** object is absorbed. You cannot see through opaque materials. They are usually dark in color and often rough in texture. Wood, brick, and soil are opaque materials. The light absorbed by

Figure 20-4 *Notice how the black solar collector on this house contrasts with the lighter roof.*

opaque materials is transformed into heat. The parts of solar collectors that absorb sunlight are usually black. (See Figure 20-4.)

Most objects are not totally reflective, transparent, or opaque. For example, transparent materials such as glass and water reflect some light. You have probably seen your reflection in still water or window glass.

Checkpoint

1. Give some examples of light sources. Identify the form of energy that enables them to give off light.
2. Why can people see some objects but not others?
3. Give an example of a material that is (a) transparent, (b) translucent, (c) opaque.

TRY THIS

Make a list of transparent, translucent, opaque, and reflecting objects that you see during a day.

20-2

Mirrors

Goals

1. To explain the formation of images by plane and concave mirrors.
2. To state some practical uses for plane, convex, and concave mirrors.

Mirrors are everywhere: in cars, purses, bathrooms, bedrooms, and lobbies. You find them at amusement parks and in telescopes. You find them in unexpected places too—on a calm lake, the hood of a car, or the surface of a spoon.

Ancient Egyptians learned to make mirrors by polishing metals. Coated glass mirrors were first made in Venice in the 1500's. Mercury was placed on tinfoil and covered with paper, making a "sandwich." A sheet of glass was placed on the paper. When the paper was carefully withdrawn, the mercury and tin formed a reflecting layer that stuck to the glass. In 1835, von Liebig (vahn LEE bik) discovered that a solution containing silver nitrate, when heated, would deposit a thin layer of silver on glass. This process is similar to the method used today to make ordinary mirrors.

When a beam of light strikes a mirror, the angle at which it is reflected is equal to the angle at which it strikes the surface. (See Figure 20-5.) This behavior makes it possible to explain how mirrors form images.

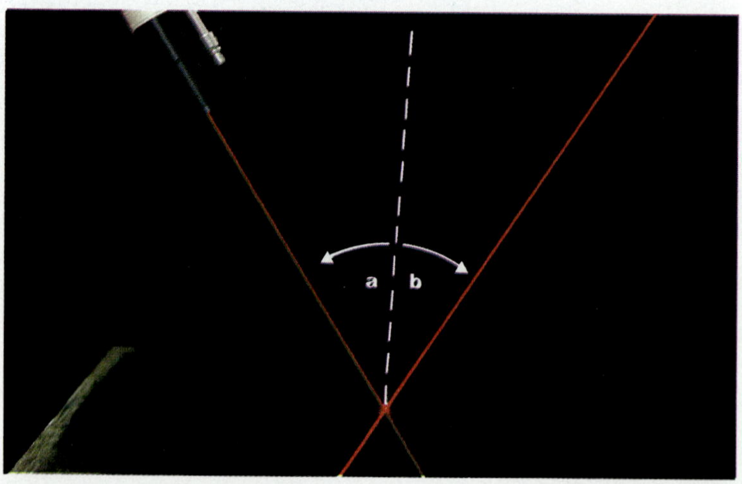

Figure 20-5 *The angle at which light strikes the surface of a mirror (a) is the same as the angle at which it is reflected (b).*

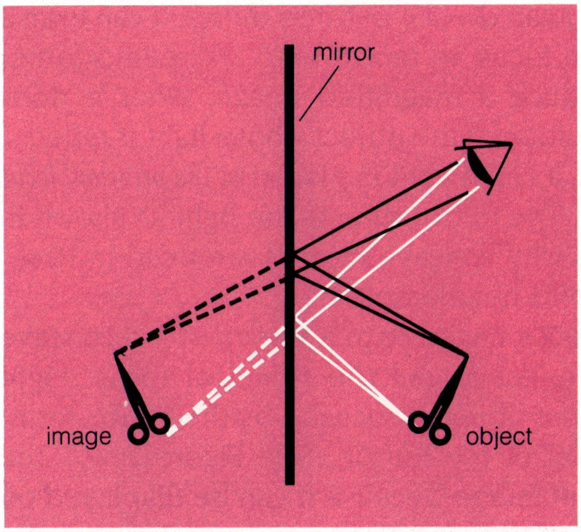

Figure 20-6 *The observer sees the light reflected by a plane mirror as if it had traveled in straight lines from behind the mirror.*

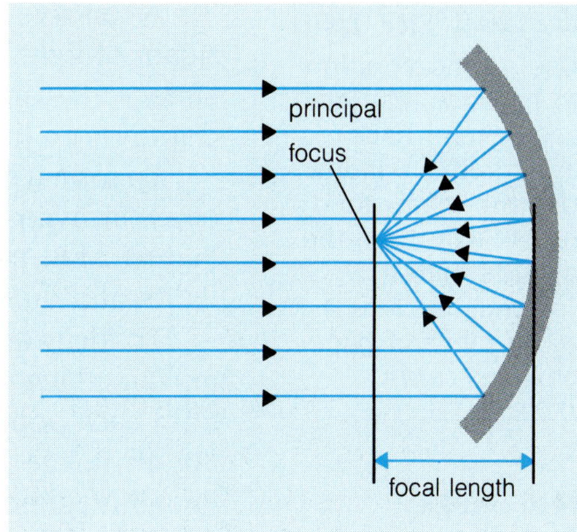

Figure 20-7 *A concave mirror focuses parallel light rays at a point called the principal focus.*

An ordinary flat mirror is called a **plane mirror**. Figure 20-6 shows how images are seen in plane mirrors. The light reflected by the object and the mirror is represented by imaginary lines called *rays*. The rays represent only a few of the countless "paths" that the reflected light takes. The light reflected by the object strikes the mirror at an angle and is reflected by the mirror at an equal angle. The viewer sees the reflected light as if it had traveled in straight lines from behind the mirror.

plane mirror

Your image, as seen in a plane mirror, looks just like you (except that left and right are reversed). But if you look at your image in an amusement park mirror, you may find it stretched, inverted, thin, fat, even unrecognizable. This happens because these mirrors are curved instead of flat.

Both sides of a shiny metal soup spoon can act as a mirror. A **concave mirror** has a curved reflecting surface similar to the soup-holding side of a spoon. Figure 20-7 shows light rays striking a concave mirror. The rays shown striking the mirror are *parallel*. (They remain the same distance apart and never touch or cross.) A concave mirror will *focus* (bring together) parallel rays. The point where these reflected rays meet is called the **principal focus**. The distance between a mirror and its principal focus is called the **focal length** of the mirror. The more curved the mirror, the shorter the focal length.

concave mirror (kahn KAYV)

principal focus

focal length

TRY THIS

Use a concave mirror to produce real images. Stand back away from a window. Use the mirror to reflect light from the outside onto a "screen," such as a white piece of paper or index card.

A concave mirror can do one of three things. It can form a beam of light. It can focus an image onto a screen or other surface. It can produce a magnified image. What it does depends on the position of the object whose light it reflects.

To form a beam, a small light is placed *at the principal focus* of a concave mirror. The mirror directs the light as shown in Figure 20-8. That is why concave mirrors are used in flashlights and automobile headlamps.

real image

To focus an object's image onto a screen with a concave mirror, place the object *beyond* the principal focus. Light from each point on the object is focused by the mirror, forming an image in front of the mirror. (See Figure 20-9.) The image is called a **real image** because it can be displayed on a screen. The size of the image depends on how far the object is from the mirror.

Light from any distant object is very nearly parallel. When light from the sun strikes a concave mirror, a small, hot, real image of the sun forms at the principal focus of the mirror. The temperature of an object can be increased by placing it at the mirror's principal focus. A solar cooker works this way. (See Figure 20-10.)

virtual image

A concave mirror produces a magnified image of an object placed *between* the mirror and its principal focus. Make-up and shaving mirrors are concave mirrors. The image seen in the mirror is called a **virtual image** because it cannot be displayed on a screen. The images formed by plane mirrors are virtual images, also.

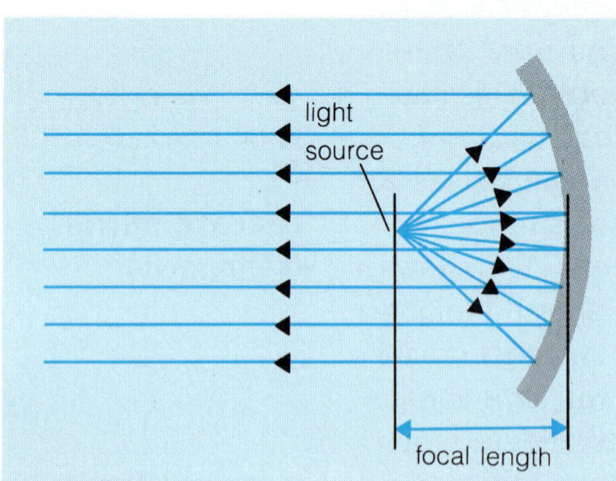

Figure 20-8 *If you place a light source at the principal focus of a concave mirror, the light is reflected as a beam.*

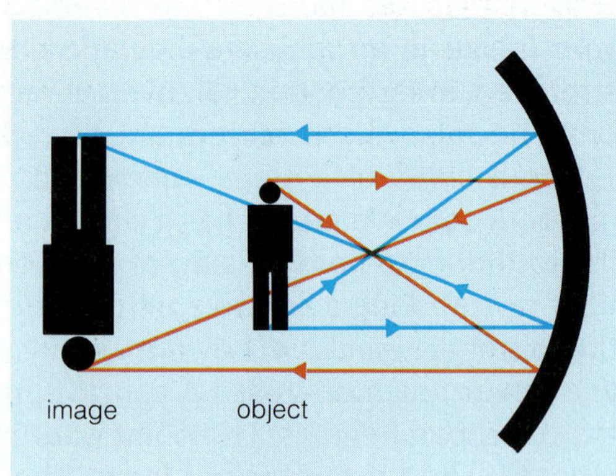

Figure 20-9 *A real, upside down image is formed by a concave mirror if you place an object beyond the principal focus.*

Figure 20-10 *You can use sunlight to cook if you place the food at the principal focus of a concave mirror.*

Figure 20-11 *(a) A convex mirror reflects a larger view with distorted images. (b) In a plane mirror, images are not distorted.*

A curved reflecting surface shaped like the side of a spoon that does not hold soup is called a **convex mirror**. Images formed by a convex mirror are always virtual, right side up, and smaller than the objects they reflect. Convex mirrors are used as rearview mirrors on trucks or cars. They are also used in stores to help detect shoplifters. Convex mirrors reflect a larger view than comparable plane mirrors, but the images they form are distorted. (See Figure 20-11.)

convex mirror (KAHN veks)

Checkpoint

1. Use a light-ray diagram to show (a) how a plane mirror forms a virtual image, (b) how a concave mirror forms a real image.
2. What is the difference between real images and virtual images?
3. State one use each for plane, concave, and convex mirrors.
4. Describe how a solar cooker works.

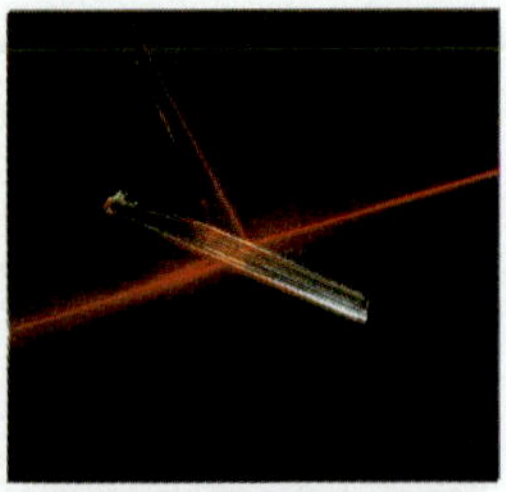

Figure 20-12 A beam refracted by glass. The beams that go up are reflections.

20-3

Lenses

Goals

1. To define refraction of light.
2. To explain how concave lenses and convex lenses form images.
3. To state some practical uses of lenses.

When light passes from one transparent material into another, it changes direction. Figure 20-12 shows that its path bends as light passes from air into glass and from glass into air. The bending of its path as light passes from one transparent material into another is called **refraction**.

refraction

Refraction produces some rather strange effects. For example, if you look at a ruler that is partly under water, it looks bent. (See Figure 20-13.) This happens because the light reflected from the part of the ruler that is under water is refracted when it passes into the air. Since the light reflected from the part of the ruler that is above the water line is not refracted, the ruler appears to bend at the water line.

Refraction of light is put to practical use in a number of devices, including cameras, telescopes, and movie projec-

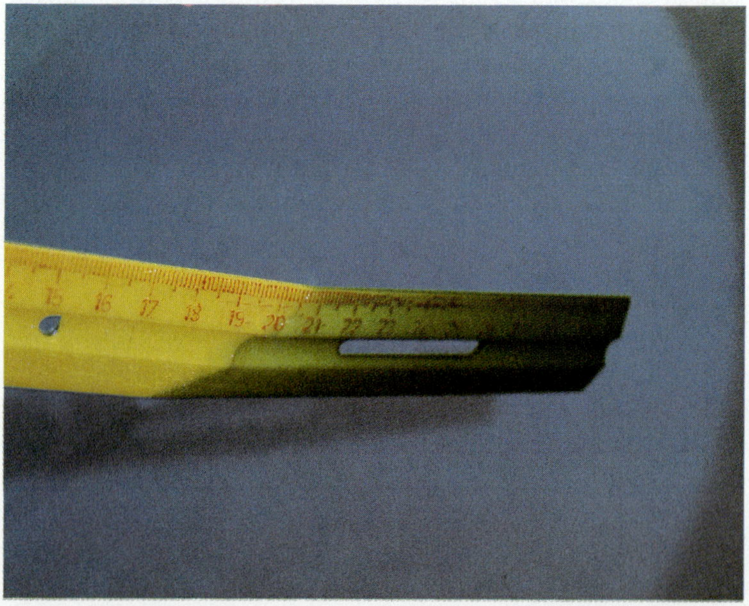

Figure 20-13 When you put something partly into water, it appears to bend. The lower end of this ruler is not where it seems to be.

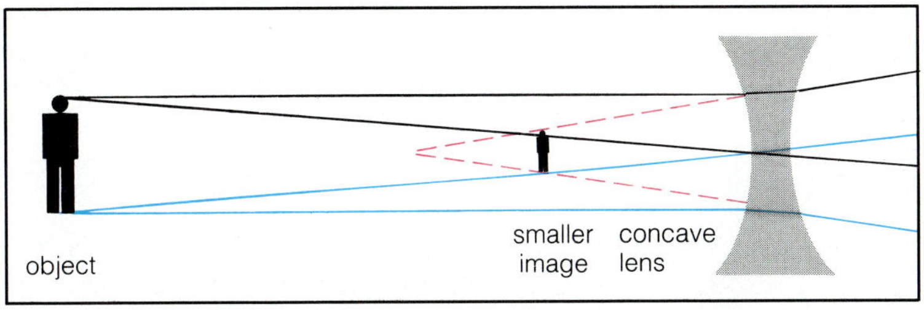

Figure 20-14a *The light rays spreading from a concave lens make the image look as if it were where the dashed lines go on the diagram.*

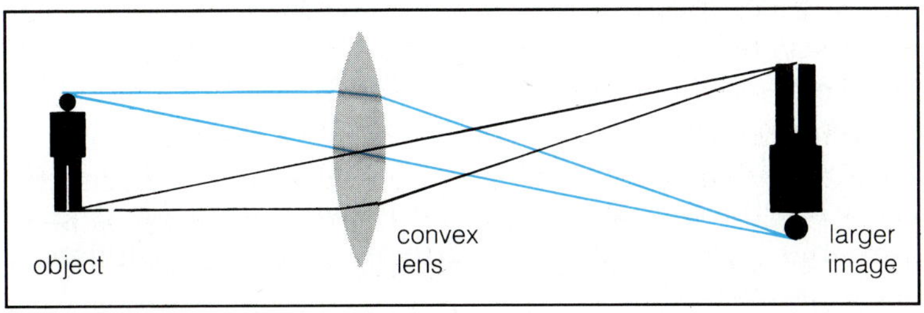

Figure 20-14b *A convex lens forms an image that is upside down and larger than the object.*

lenses

concave lens

convex lens

tors. These devices use **lenses**, transparent objects with at least one curved surface, to control light. Lenses are usually made of clear glass or plastic. Light is refracted both when it enters a lens and when it leaves the lens. The path of light as it leaves a lens depends on the shape of the lens.

A **concave lens** is thicker at the ends than in the middle. It spreads light, producing images that are right side up and smaller than the objects. (See Figure 20-14a.) A **convex lens** is thicker in the middle than at the ends and brings light rays together. (See Figure 20-14b.)

A single convex lens is often used as a magnifier. The earliest microscopes were nothing but powerful magnifying glasses. Two convex lenses can be used together to make a more powerful microscope. Figure 20-15 shows how a two-lens microscope works.

The two-lens (or *compound*) microscope can magnify objects much more than a single-lens microscope can. But using two lenses together has a drawback. If you have used a microscope, you know that the image is reversed and upside down. Also, when you move the object to the right and toward you, the image moves left and away.

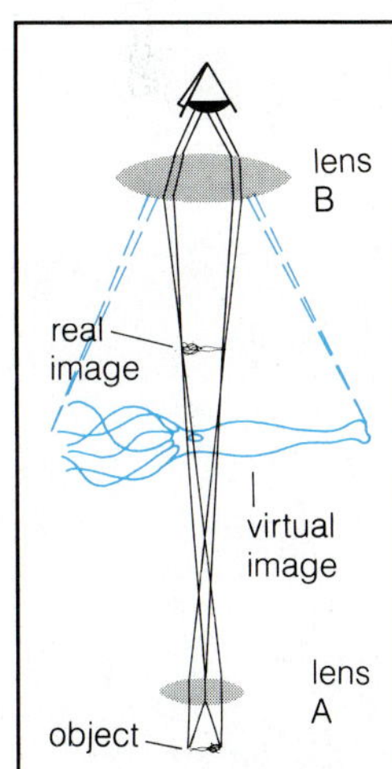

Figure 20-15 *The eyepiece lens, B, magnifies the image from the objective lens, A.*

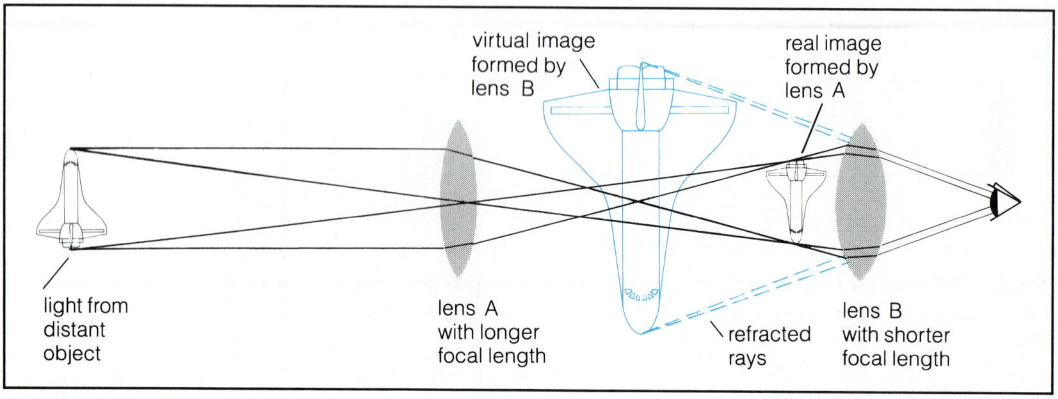

Figure 20-16 *A telescope's objective lens has a long focal length to make a large image.*

Two convex lenses can be put together to make a telescope, also. (See Figure 20-16.) Because the object to be magnified is far away, the objective lens (lens A) has a long focal length. This is to make the image it forms as large as possible. Lens A should be wide, to collect enough light to make a bright image.

The human eye contains a convex lens. This lens forms real images on the *retina* (RET n uh), the inner lining of the eye. (See Figure 20-17.) These images stimulate nerve endings in the retina. Nerves carry the resulting impulses to the brain, where the images are perceived. The image on the retina is upside down. In the brain, the image is turned right side up.

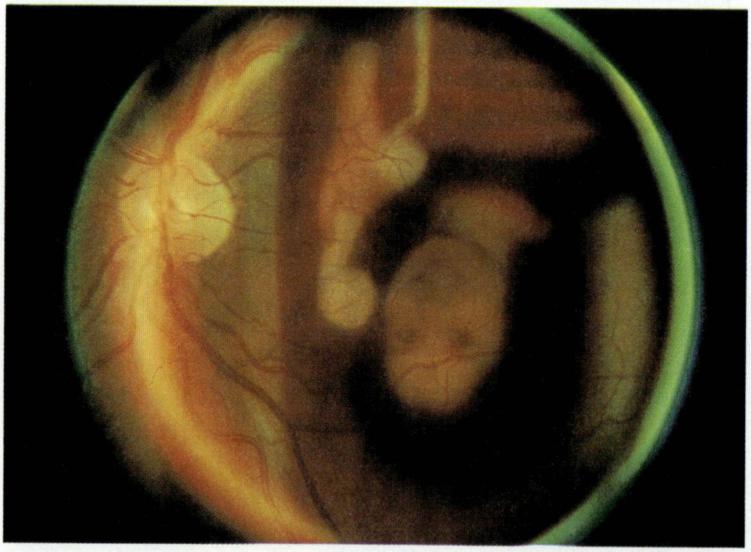

Figure 20-17 *This photograph shows the inside of a human eye. Notice the upside down image on the retina.*

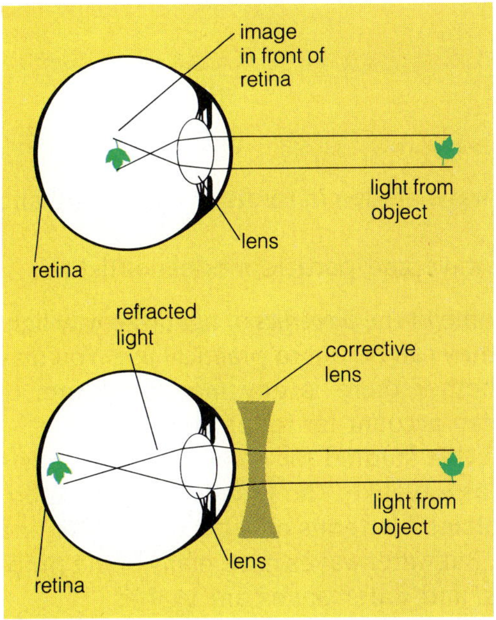

Figure 20-18 *Above: The eye of a nearsighted person. The image forms in front of the retina, and the person's vision is not clear. Below: A corrective lens causes the image to form on the retina.*

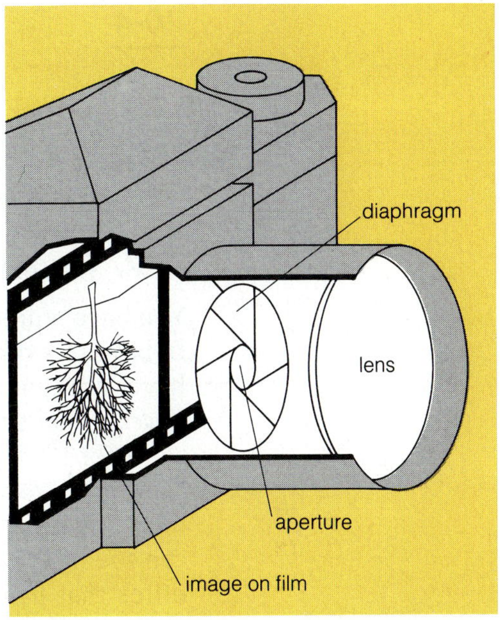

Figure 20-19 *A camera is similar to a person's eye. Light goes through the lens, which focuses the image on the film. The diaphragm is like the pupil of your eye, controlling the amount of light.*

If the lens does not focus light into clear images on the retina, corrective lenses can be used. These may be contact lenses or in eyeglasses. Concave lenses are used to correct the vision of nearsighted people. (See Figure 20-18.) Convex lenses can correct farsightedness.

A camera works very much like the human eye. In place of a retina there is sensitive film to make a permanent record of images. (See Figure 20-19.)

Checkpoint

1. Explain what is meant by refraction. Use an example to illustrate your explanation.
2. How can you tell convex from concave lenses?
3. Using diagrams, explain how images are formed by convex and concave lenses.
4. List some uses for convex lenses.

TRY THIS

Place a tiny drop of water on a letter of a word found on a shiny page of a magazine. Does the drop act as a lens? How can you tell? Try to explain what you observe.

20-4

The Nature of Light

Goals

1. To describe wave motion in terms of wavelength and frequency.
2. To compare wave and particle models of light.

You have studied some of the *properties* of light (the way light behaves) and how they can be put to practical use. You may have wondered whether there is any mental picture, or *model*, of light that can account for its properties.

After light had been studied for many years, scientists came to think of it as wavelike. The results of their experiments could be explained in terms of "light waves." No one can see light waves. But water waves have many of the properties that light has, and water waves are visible.

If you drop an object into still water, waves spread out, or radiate, in all directions from the point of impact. (See Figure 20-20.) Some of the object's energy is transformed into wave energy. A cork floating on water bobs up and down as waves pass by. The waves provide the force needed to move the cork upward. Similarly, light energy radiates in all directions from a source, such as a light bulb or the sun.

Figure 20-21 shows a side view of water waves. The highest points on the waves are called *crests*, and the lowest

Figure 20-20 *Waves radiate out from the point at which a drop of water from the paddle hits the surface of still water.*

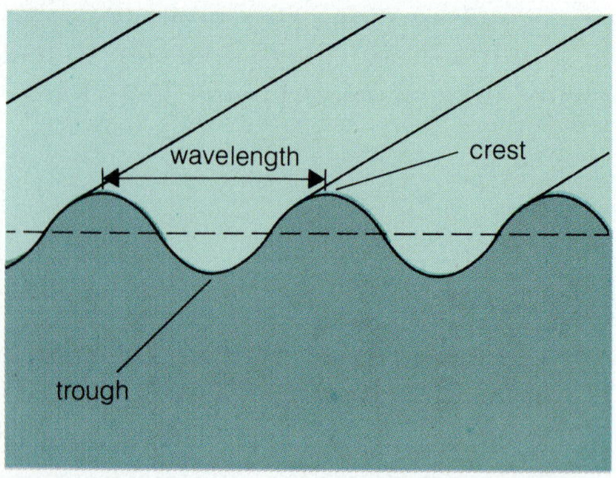

Figure 20-21 *This cutaway view of water waves indicates how the wavelength is determined.*

points are called *troughs* (trawfs). The distance from the crest of one wave to the next is the **wavelength**. The number of wavelengths, or waves, that pass a given point in one second is the **frequency**. Frequency is measured in units called *hertz* (Hz). A frequency of 1 Hz is 1 wave per second, 2 Hz is 2 waves per second, and so on.

wavelength

frequency

A wave can be identified by either its wavelength or its frequency. As shown in Figure 20-22, the frequency of a wave decreases as its wavelength increases and increases as its wavelength decreases.

Water waves can be produced and photographed. Figure 20-23 shows water waves striking a smooth surface. The angle at which the waves are reflected is the same as the angle at which they strike the surface. This is what happens when light is reflected from a smooth surface.

As waves pass from deep to shallow water, their path bends, or refracts. (See Figure 20-24.) Their path is similar to that of light moving from air to glass.

Light and water waves have a number of similar properties. In general, a wave model can explain many of the properties of light. But light has some properties that do not fit the wave model.

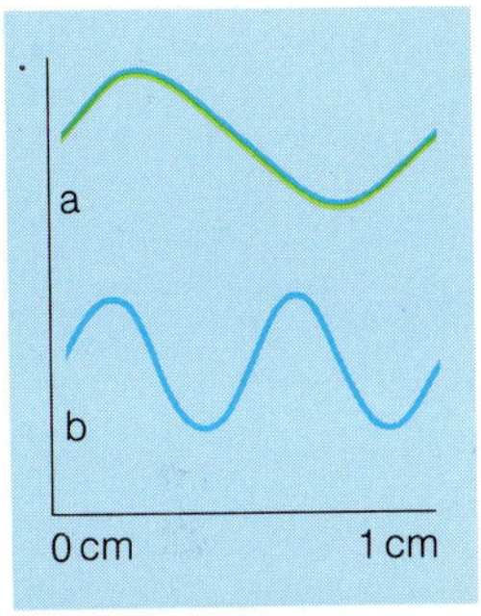

Figure 20-22 The wavelength of wave b is half as long as that of wave a. Thus its frequency is twice that of wave a.

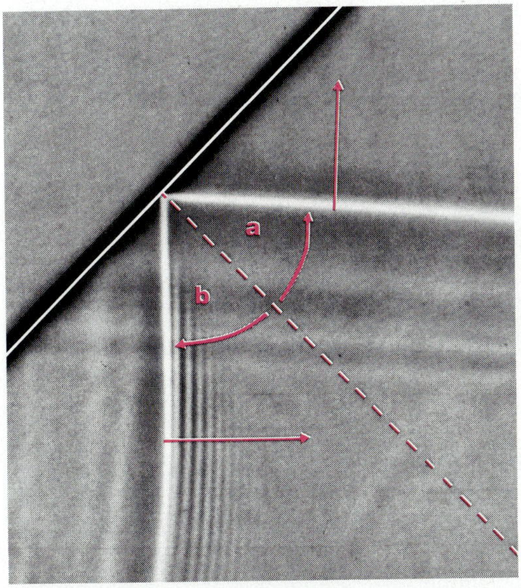

Figure 20-23 In this photograph, the angle of reflection (a) is the same as the angle at which the waves strike the surface (b).

Figure 20-24 Water waves are refracted when they strike shallow water. The waves bend where the depth changes.

Figure 20-25 The array in the photograph contains photovoltaic cells, which convert sunlight into electricity.

In some cases, light energy absorbed by a material can cause electrons in the material to move as if they had been struck by tiny bullets. Sometimes the electrons are knocked right out of the material. In order to explain this property of light, it is necessary to think of light as a stream of tiny bullet-like particles. These particles are called *photons*. The transfer of light energy to electrons is the basis for devices that convert sunlight into electricity. (See Figure 20-25.)

Both a wave model *and* a particle model are needed to explain the way light behaves. Neither model can account for all its properties.

Checkpoint

1. Draw a side view of a series of water waves. Label crests, troughs, and wavelength.
2. What is the frequency of a wave? Explain what is meant by a frequency of 2000 Hz.
3. Name two different models used in describing light.

TRY THIS Put some water in a shallow pan or a sink. One end of the pan can represent a "mirror." Use a medicine dropper to let a drop of water fall a few centimeters in front of the "mirror." Watch the waves spread out from this point source of "light." Watch the circular waves that bounce off the mirror. Find the center or origin of these reflected waves. Where is the "image" of the "object" in front of the "mirror"?

20-5

The Electromagnetic Spectrum

Goals

1. To describe the visible spectrum and how it can be produced.
2. To identify various components of the electromagnetic spectrum.

When sunlight passes through droplets of rain in the air, a rainbow may result. Similarly, "white" light given off by a bulb can be spread out into a rainbowlike spectrum of colors. One way to do this is to pass the light through a wedge-shaped, transparent object called a *prism* (PRIZ um). (See Figure 20-26.) If the wavelengths, or frequencies, of the separate colors are measured, they are found to be different. For example, red light has a lower frequency than violet light. The colors in white light can be spread out by refraction because each frequency, or wavelength, is refracted to a different extent.

The band of colors produced when white light passes through a prism is called the **visible spectrum**. It is a spectrum of colors, and also of wavelengths or frequencies. Since each frequency corresponds to a different energy, it is an energy spectrum, as well.

visible spectrum

Visible light given off by hot, glowing gases in the sun spreads out, or radiates, in all directions. It consists of constantly varying electromagnetic waves, which can travel through empty space at the highest possible speed (300,000 km/s).

Figure 20-26 *When white light passes through a prism, it is separated into the colors of a rainbow.*

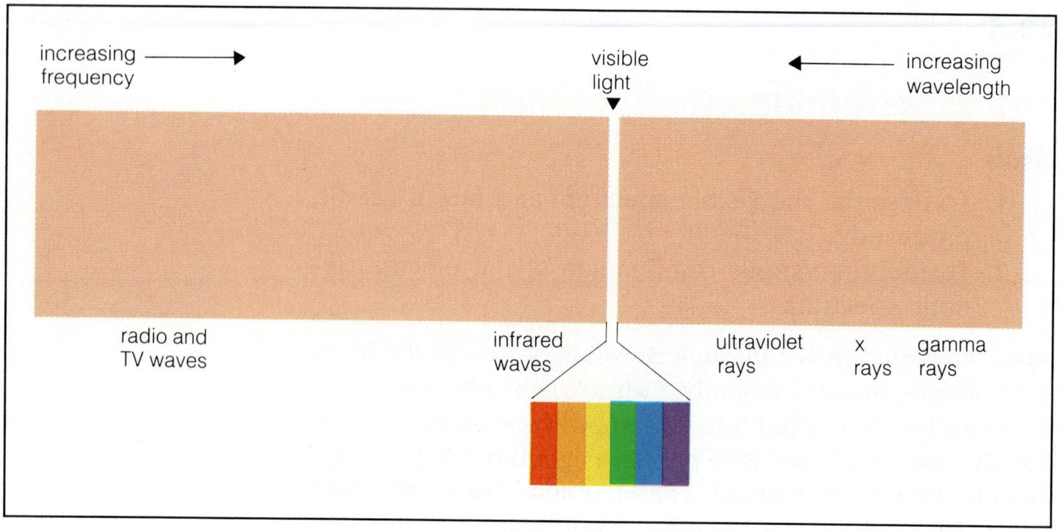

Figure 20-27 *Visible light is only a small portion of the entire electromagnetic spectrum.*

electromagnetic spectrum (il lek troh mag NET ik)

Visible light is only a tiny portion of a broad spectrum of electromagnetic radiation called the **electromagnetic spectrum**. (See Figure 20-27.) Different types of electromagnetic radiation have different wavelengths, or frequencies, and thus different properties. The radiation ranges from very low frequency radio waves to very high frequency gamma rays.

infrared radiation (in fruh RED)

ultraviolet radiation (uhl truh VY uh lit)

Sunlight contains invisible radiation as well as visible light. Radiation with somewhat lower frequencies than those of red light is called **infrared radiation**. Heat radiation consists of infrared rays. Radiation with somewhat higher frequencies than those of violet light is called **ultraviolet radiation**. The ultraviolet rays in sunlight are the rays that can give you a sunburn. Sun screen products can help prevent a serious burn.

Checkpoint

1. What is the visible spectrum? How can it be produced?
2. Compare the wavelength and frequency of visible, infrared, and ultraviolet radiation.
3. Which type of electromagnetic radiation has longer wavelengths, radio waves or x rays?

20-6

Color

ENRICHMENT

Goals

1. To explain how people see color.
2. To describe how lights or pigments can be combined to produce colors.

You see an object because the light reflected from that object reaches your eyes. The colors that you see depend on what wavelengths of light are reflected by the object.

When white light strikes an object, the light may be totally reflected. If so, the color that you see is white. If white light is completely absorbed by an object, the object looks black. No light is reflected from the object.

White light is a mixture of different wavelengths, or colors, of light. When white light strikes an object, some of these colors may be absorbed and others reflected. For example, if all the colors except red are absorbed by an object, only red light is reflected. As a result, the object looks red.

One way to produce red light is to pass white light through a transparent red object, such as red cellophane. All colors except red are absorbed by the object. An object of this kind is called a *filter*. It "filters out" the unwanted colors and allows only the desired color to pass through.

Figure 20-28 shows what happens when equal amounts of red, blue, and green light shine on a white object. A white

Figure 20-28 *When the primary colors (red, blue, and green) are combined in equal amounts, they produce white light. Combining the primary colors in varying amounts produces other colors.*

color is seen where all three colors overlap. Red and blue combine to produce a color called *magenta* (muh JEN tuh). Blue and green combine to produce a color called *cyan* (SY an). Red and green combine to produce yellow. By varying the amounts and combinations of red, green, and blue light, any color can be produced. Therefore, red, green, and blue are called the *primary colors of light*.

A *pigment* is a substance that is added to other substances to produce a desired color. Yellow, cyan, and magenta are called the *primary colors of pigments* because they can be mixed to produce any color.

When white light shines on a mixture of magenta and cyan pigments, red and green light are absorbed and blue light is reflected. (See Figure 20-29a.) Similarly, a mixture of yellow and cyan pigments absorbs red and blue light and reflects green light when white light shines on the mixture. (See Figure 20-29b.) When white light shines on a mixture of yellow and magenta pigments, red light is reflected because blue and green light are absorbed.

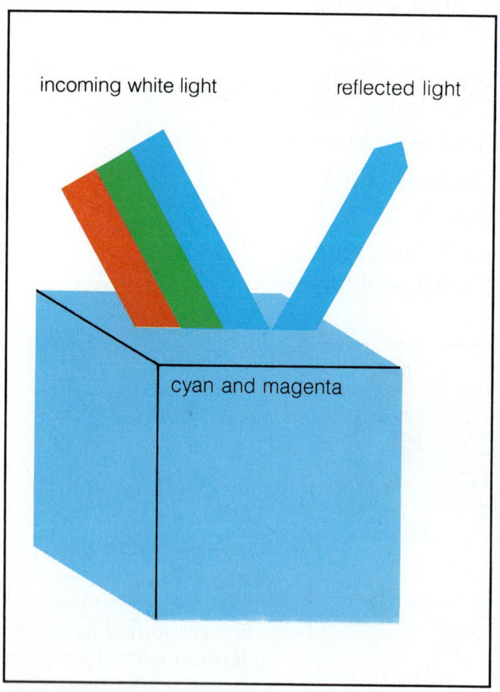

Figure 20-29a *A mixture of magenta and cyan pigments absorbs red and green light and reflects blue light.*

Figure 20-29b *A mixture of yellow and cyan pigments absorbs red and blue light and reflects green light.*

Figure 20-30 *This is a test for red-green colorblindness. Can you see the number?*

The light that reaches your eyes is not the only factor that determines what colors you see. The way your eye responds to the light is also a factor. People who do not see the same colors that most other people see are said to be *colorblind*. The ability to perceive colors accurately is important to many people. Electricians must identify color-coded wires and photographers must control the color in their pictures. You can take the test in Figure 20-30 to see if you have red-green colorblindness.

TRY THIS

Draw a picture using crayons, paints, or pens. Look at the drawing under different colors of light in an otherwise dark room. (Make color filters, if necessary.) Try to explain why the colors in the drawing seem to have changed.

Checkpoint

1. Explain how a person with normal color vision sees an object as "red."
2. What two colors of light would you shine onto a white piece of paper to produce a yellow color?
3. What primary pigments would you mix to produce a green color?
4. What colors would you expect the pigments in a black sweater to absorb?

Fiber Optics

Optics, the science that deals with light, could change our world in many ways. Scientists working in *fiber optics* have developed a way to send light through thin strands of glass no thicker than the width of a hair.

Each flexible glass strand, or fiber, consists of a glass *core* coated with a thin layer of glass called the *cladding*, as shown in Figure 1. The glass in the cladding is different from the glass in the core. As a result, when light strikes the core-cladding boundary at a great enough angle, it is reflected back into the core. This behavior is shown by the arrows in Figure 1. This *internal reflection* allows light to travel from one end of an optical fiber to another, as shown in Figure 2. Light will even travel around curves and corners in this manner. Doctors and dentists are using this property of optical fibers to bring light to hard-to-see places within the body. (See Figure 3.)

Fiber optics may be the answer to the growing need for communications pathways. The major advantage to using optical fibers is that they carry many more messages at one time than the metal wires now in use. Optical fibers can be used in telephone cables, to transmit television pictures, and to link computers. Someday soon, optical fibers may even link a computer in your home to such things as weather reports, your bank account, or even the latest baseball standings.

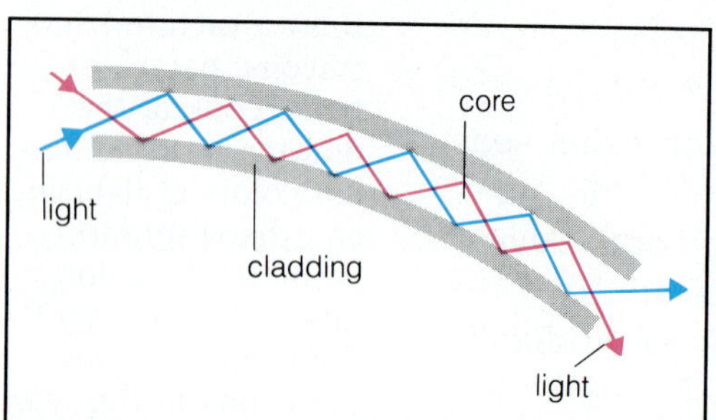

Figure 1

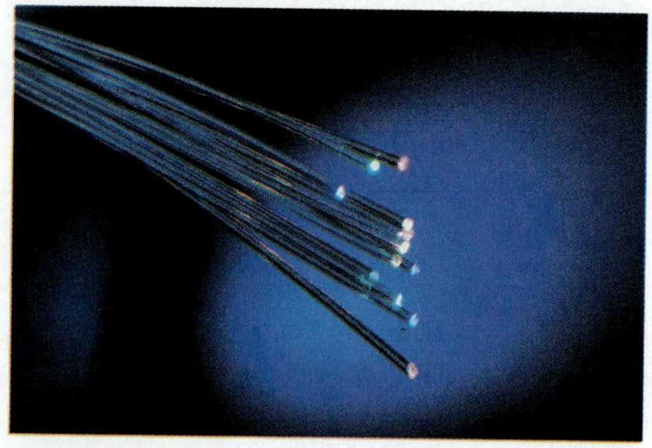

Figure 2

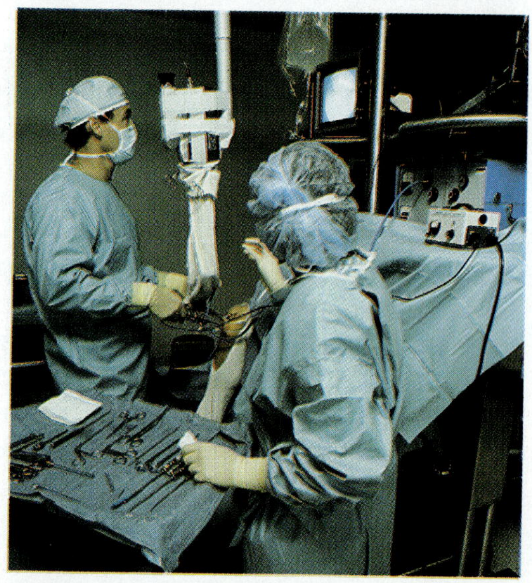

Figure 3

20-7

Sound

Goals

1. To describe how sound is produced.
2. To discuss some properties of sound that depend on its frequency or loudness.

Sound can help us to understand our surroundings. A doctor listens to the sound of a heartbeat through a stethoscope. A mechanic listens to the sounds of an engine. In each case, sound tells them about something they cannot see.

You commonly use sound to communicate. Within your larynx (LA ringks), or Adam's apple, are vocal cords that *vibrate* (move back and forth) when air from your lungs moves over them. If you put your fingers over your larynx as you talk you can feel the vibrations.

Vibration is the source of most sound. The number of vibrations per second, or frequency, determines the *pitch* of the sound that is produced. (The sounds that a canary makes have a higher pitch than those made by a bullfrog.) The greater the frequency, the higher the pitch. The vibration frequency of vocal cords depends on their length, thickness, and tautness, or tension. This is true for similar sources of sound, such as guitar strings.

Ideally, humans can hear sounds with frequencies of 20 to 20,000 vibrations per second (20–20,000 Hz). Some animals can hear higher frequencies. "Silent" dog whistles produce a pitch so high that it cannot be heard by humans, but it can be heard by dogs. Frequencies greater than 20,000 Hz are called *ultrahigh frequencies*.

Bats can produce and hear ultrahigh frequency sounds. By responding to the echoes of these sounds, bats can locate objects that they cannot see. Some ships use a batlike device called *sonar* (*so*und *n*avigation *a*nd *r*anging) to detect and locate underwater objects. Another use of ultrahigh frequencies is to produce "pictures" of objects inside the human body. For example, doctors can check the development of a fetus within its mother's body. The appropriate use of ultrahigh frequencies for this purpose can be extremely valuable. However, there is some concern that use on a routine basis may be unwise.

Sound can cause problems as well as pleasure. Loud noises can be disturbing and even frightening. They can damage your ears, also. Loudness of sound is measured in units called *decibels* (db). The loudness of ordinary conversation is about 60 db. A sound of 120 db, which is not twice but a million times as intense as 60 db, can produce permanent damage to your ears. How much damage is done depends on the sound's loudness, frequency, and duration. Even at 80 db, long or repeated exposure can cause hearing loss. High-pitched sounds are more harmful than low-pitched sounds. The noise levels of some common sounds are listed below in Table 20-1.

Table 20-1: Noise Levels

Source of sound	*Noise level (db)*
faintest sound detectable	0
rustling leaves	20
whisper	25
conversation	60
street traffic	70
subway train	100
amplified music	115
jet plane 30 m away	140

Checkpoint

1. What is the cause of most sounds?
2. What range of frequencies can humans normally hear?
3. List two uses of ultrahigh frequency sound.
4. What factors can determine whether a sound may damage your ears?

TRY THIS

Press one end of a wooden ruler firmly against a desktop, with the free end extending horizontally. Set the free end of the ruler vibrating by plucking it with your other hand. Repeat the experiment several times. Each time move more of the ruler onto the desk so that less of the ruler is free to vibrate. In what way does the frequency of vibration change as the length of the ruler that is free to vibrate changes?

20-8

Sound and Light

Goals

1. To state similarities and differences between sound and light.
2. To describe how sound waves move through air.

Like light, sound can be reflected. That is why we can hear echoes. Sound shares other properties with light, including refraction. But sound cannot be the same as light. We see light from the sun and stars that has passed through nearly empty space. But we do not *hear* the raging fires of the sun. Sound cannot travel across empty space. The vibrations of a source of sound must be "passed along" through some medium.

The sounds that we hear most commonly travel through air. Air is made up of many tiny invisible particles (molecules) that are constantly moving about at high speeds. On the average, however, a given region of air contains a fairly uniform, or even, distribution of molecules. When a plucked guitar string vibrates, it pushes together, or *compresses*, the air nearby. (See Figure 20-31.) As the guitar string moves in the opposite direction, a region of *rarefaction* (rair uh FAK shun), in which there are fewer molecules than average, results. Some of the molecules from the region of *compression* move back into the region of rarefaction. As the pulses of alternating compression and rarefaction continue in response to the

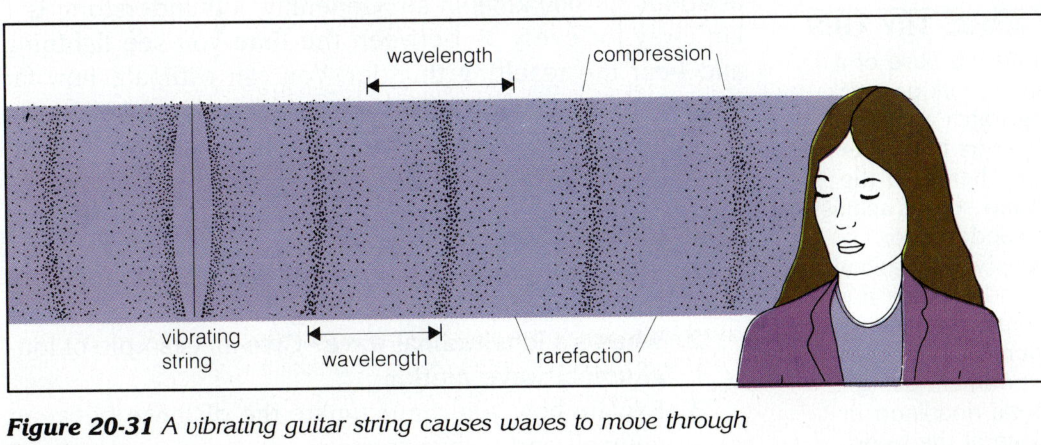

Figure 20-31 *A vibrating guitar string causes waves to move through the air.*

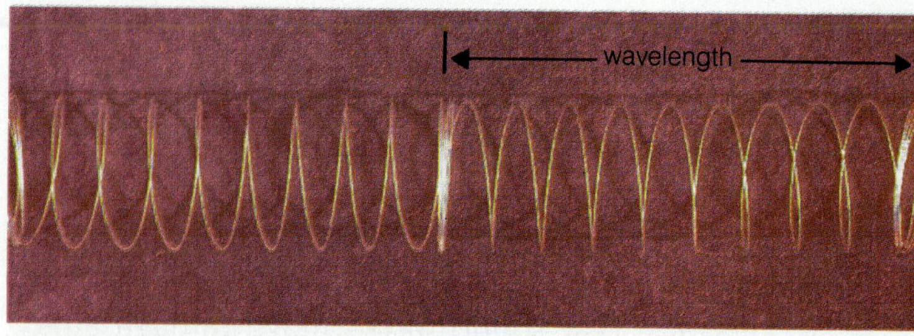

Figure 20-32 *This photograph of a spring shows two complete moving waves.*

motion of the guitar string, they cause pulses in the air next to them. In the same way, these pulses produce other pulses, creating a wave of alternating regions of compression and rarefaction. As a result, waves travel out in all directions from the vibrating string. The molecules move back and forth within a small region of space. It is the wave that travels from a guitar string to your ear, *not* the molecules.

Sound waves are similar to water waves (ripples), since it is the waves that travel, while the medium only vibrates. Water vibrates (up and down) at right angles to the direction of wave motion. Waves of this type are called *transverse waves*. Light waves are transverse waves. When sound waves travel in air, the molecules move back and forth (vibrate) in the direction of wave motion. Waves that travel in the same direction as the vibrations of the medium are called **longitudinal waves**. Longitudinal waves can be shown in a coiled spring. (See Figure 20-32.)

longitudinal waves (lahn juh TOOD un ul)

During a thunderstorm, you see lightning before you hear thunder because light travels much faster than sound. The speed of sound in air is only 331 m/s at 0°C. Light travels at about 300,000 km/s in air. Generally, a thunderstorm is 1 km away for every 3s between the time you see lightning and hear the resulting thunder. You can estimate how far away a storm is by counting (at one-second intervals) from the time you see lightning until you hear thunder.

TRY THIS

Hold the base of a tuning fork or a ticking watch a few centimeters from your ear. Then hold the sound source against a wooden door. Does wood transmit sound? To see if metals conduct sound, repeat the experiment using a metal doorknob in place of the wood.

Checkpoint

1. State two properties of sound that are (a) similar to (b) different from those of light.
2. What is a longitudinal wave? Give an example of longitudinal wave motion.
3. Explain how you can estimate the distance between yourself and a thunderstorm.

Understanding The Chapter 20

Main Ideas

1. Light given off by a source generally radiates outward in all directions, in straight lines.
2. Light may be reflected, transmitted, or absorbed when it strikes an object.
3. Plane mirrors and convex mirrors reflect light, forming images, but convex mirrors can reflect light from a larger area.
4. Concave mirrors can collect and focus light to produce real images or parallel beams.
5. Light is refracted when it passes from one transparent material into another.
6. Concave and convex lenses can be used to form images in a variety of practical devices.
7. A wave model can explain many properties of light. A particle model is needed to explain other properties of light.
8. White light can be separated into a spectrum of colors by a prism.
9. The frequencies and wavelengths of the various types of electromagnetic radiation form a broad spectrum. This spectrum is known as the electromagnetic spectrum.
10. The color of an object is the color of the light it reflects.
11. Most sounds are produced by vibrating objects.
12. Important properties of sound include frequency and loudness.

Vocabulary Review

From the following list, choose the term that best completes each of the statements. Write your answers on a separate piece of paper.

concave (lens, mirror)
convex (lens, mirror)
electromagnetic spectrum
focal length
frequency
infrared radiation
lens
longitudinal wave
opaque
plane mirror
principal focus
real image
reflection
refraction
translucent
transparent
ultraviolet radiation
virtual image
visible spectrum
wavelength

1. The bending of light as it passes from one transparent material into another one is called __?__.
2. A __?__ can be produced by a concave mirror and captured on a screen.
3. The number of wavelengths per second is the __?__.
4. __?__ materials let light pass through them.
5. We can hear echoes due to the __?__ of sound.
6. The __?__ is the distance between a concave mirror and its principal focus.
7. Light is reflected or absorbed completely by __?__ materials.
8. When white light is passed through a prism it spreads out into a band of colors called the __?__.

9. X rays, ultraviolet rays, and radio waves are part of the __?__.
10. A sound wave is a __?__.
11. A __?__ lens is thicker in the middle than at the ends.
12. When you look in the mirror you see a __?__.
13. Waxed paper is an example of a __?__ material.
14. Heat radiation consists of __?__, which has lower frequencies than those of red light.

Chapter Review

Write your answers on a separate piece of paper.

Know The Facts

1. Light that strikes a plane mirror is __?__. (refracted, reflected, absorbed)
2. High-pitched sounds arise from __?__ vibrations. (rapid, slow)
3. __?__ lenses should be used to correct nearsightedness. (concave, convex)
4. A concave mirror will produce a parallel beam of light if a small light bulb is placed __?__. (at the principal focus of the mirror, between the principal focus and the mirror, beyond the principal focus)
5. When red and green light shine on the same part of a white screen at the same time, the color you see on the screen is __?__. (yellow, magenta, cyan)
6. When a ray of light strikes a plane mirror, the angle at which it is reflected is __?__ the angle at which it strikes the mirror. (greater than, less than, equal to)
7. __?__ is *not* transmitted through empty space. (light, sound, infrared radiation)
8. __?__ mirrors are commonly used in stores to detect shoplifters. (plane, concave, convex)
9. The electromagnetic spectrum does *not* include __?__. (sound, light, x rays)
10. __?__ mirrors are used in flashlights to produce a light beam. (plane, convex, concave)
11. The speed of sound is __?__ the speed of light. (less than, greater than, equal to)
12. The __?__ of sound is measured in decibels. (pitch, loudness, speed)
13. The wavelengths that are present in white light are __?__ to different extents when the light passes through a prism. (reflected, refracted, absorbed)
14. If the frequency of a wave traveling at a certain speed increases, the wavelength __?__. (increases, decreases, stays the same)

Understand The Concepts

15. Explain the difference between real and virtual images.
16. Explain how a convex lens can be used to start a campfire.
17. Account for the fact that cats seem to hear sounds that you cannot hear.
18. Draw a diagram to show how a concave mirror can be used to make a spotlight.
19. What primary colors of pigments would you mix to produce a blue color? Explain how the blue color is produced.

20. Which would you expect to absorb more heat from the sun, a black- or white-painted brick? Explain.
21. Compare the human eye with a camera.
22. At a race the starter, who is far from the timekeeper, uses both a flag and a gun to signal the start of the race. How do you explain this?
23. An echo of your shout reflected from a cliff wall returns one second after you shouted. How far away is the cliff?
24. Explain how a solar cooker works.
25. Most of the light that strikes still water passes through it. What else happens to the light that strikes the water?

Challenge Your Understanding

26. Draw a diagram to show how you would position two plane mirrors in order to see the back of your head.
27. Explain why sound travels faster through iron than through air at the same temperature.
28. A large source of light, such as the sun, often produces fuzzy shadows when it strikes small objects. Show, using a diagram, why this should be so.
29. Draw a diagram to show how an image is formed by a convex mirror.
30. Explain why you can be heard farther away if you use a megaphone, or simply cup your hands around your mouth.
31. You can see a rainbow when sunlight passes through the fine spray produced by a garden hose. Explain why.

Projects

1. Set two plane mirrors together at angles of 120°, 90°, 45°, and so on, and find the number of images formed in each case.
2. Find out how many different ways you can produce a visible spectrum. Demonstrate each one.
3. Cameras, slide projectors, and movie projectors all use lenses to produce images. Examine one or more of these devices closely. Try to figure out how they work.
4. Bees are able to communicate without the use of sound. Do some library research to find out how bees communicate.
5. Optical illusions, such as mirages, can be explained in terms of the properties of light. Find out how they are produced.
6. Research and write a report on the history of artificial lighting (the use of light sources other than the sun or moon).
7. Find out what regulations your community has concerning noise pollution.
8. Find out what means of communication are available to people in your community who have impaired sight or hearing.

Industrial Engineer

Positions are available for industrial engineers with a large consulting firm that advises banks, insurance companies, hospitals, and government agencies. Engineers analyze a client's operations and future plans and suggest ways of reducing costs and increasing efficiency, productivity, and safety. Some engineers will assist clients in making relocation plans. Applicants should demonstrate creativity and flexibility and should enjoy working as part of a team. Much time will be spent in travel.

A degree in industrial engineering is required with courses in physics, mathematics, and business. Candidates will begin by assisting experienced engineers. As experience is gained, there will be opportunities for advancement.

Computer Service Technician

Computer company with offices nationwide seeks technicians to install, maintain, and service computers. Maintenance involves adjusting, oiling, and cleaning all parts of the computer. Servicing requires quickly identifying the cause of a breakdown and repairing it. Service technicians must also explain to customers how computers work and respond tactfully to complaints.

Candidates need one to two years of training in electronics after high school. Courses in physics and mathematics also are necessary. Technicians will be trained for one year. After that, technicians should expect to work a forty-hour week with some emergency repairs done during odd hours. As experience is gained, opportunities exist for travel nationwide.

Help Wanted

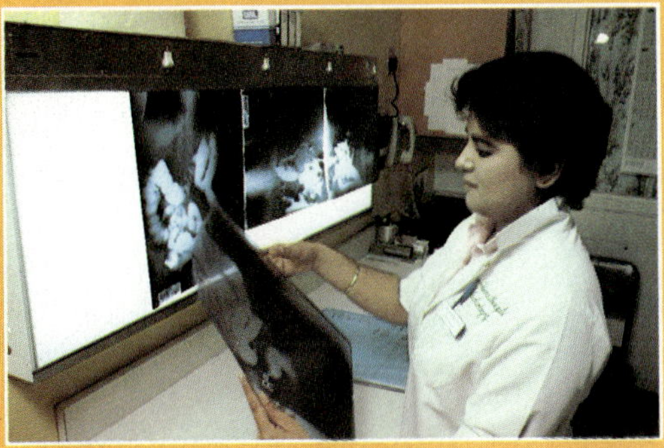

Radiologic Technician

Local hospital has openings for several kinds of radiologic technicians. *X-ray technicians* take x rays of patients. *Radiation therapists* treat cancer patients with prescribed doses of radiation. *Nuclear medicine technologists* prepare radioactive solutions and administer them to patients. The technologists then prepare radiographs, or photographs produced by the radiation.

Technicians must keep accurate records and follow safety procedures. All work is supervised by physicians. The job requires standing for long periods, and often lifting or supporting patients.

A degree from a two-year accredited program in radiology is necessary. A bachelor's or master's degree will improve chances for advancement.

Technical Writer

Technical writers are needed by a large company that makes electronic parts for computers. Writers help produce research reports, user manuals, annual reports, and new product information for customers. Clear, concise writing is essential. Writers must be able to absorb and organize large amounts of technical data. Writers also must be able to explain technical information in terms that can easily be understood by nontechnically-oriented readers.

Applicants should be college graduates with a major in electronics and a minor in writing. Experience with computers is necessary. Applicants should be able to work well as part of a team. Most writers should expect to work a regular forty-hour week, except during rush periods.

Understanding The Unit 4

Main Review

Write your answers on a separate piece of paper.

Know The Facts

1. A body in motion tends to move with a constant __?__ unless acted upon by an outside force. (speed, velocity, force)
2. If the __?__ of an object increases, the force needed to give it a certain acceleration will increase too. (mass, velocity, speed)
3. A 1-kg object and a 2-kg object fall from a rooftop. The acceleration of the larger mass will be __?__ that of the smaller mass. (twice, half, the same as)
4. A satellite orbits Earth. The force that keeps the satellite in orbit is __?__. (centrifugal, centripetal, motion)
5. Work is equal to force times __?__. (speed, distance, acceleration)
6. If the velocity of a car doubles, its kinetic energy __?__. (halves, doubles, quadruples)
7. A watt is one __?__ per second. (newton, kilogram, joule)
8. Power is the __?__. (rate of doing work, amount of work done, product of force times distance)
9. Pressure is force __?__ area. (added to, divided by, multiplied by)
10. The larger the R-value of a material, the better its __?__ quality. (conducting, convecting, insulating)
11. A temperature of 273 K equals __?__. (−273°C, 0°C, 273°C)
12. The electric energy per charge can be measured with a(n) __?__. (ammeter, voltmeter, coulombmeter)
13. Electric energy can be measured in units of __?__. (kW, W, kW-h)
14. You can see the moon because of __?__ light. (reflected, refracted, diffracted)
15. Lenses form images by __?__ light. (reflecting, refracting, diffracting)
16. As the wavelength of light increases, the frequency of the light waves __?__. (decreases, increases, remains the same)
17. To explain all the properties of light, __?__. (a wave model is sufficient, a particle model is sufficient, both models are needed)
18. Moving from red to violet light in the visible spectrum, the wavelength __?__. (decreases, increases, stays the same)
19. A __?__ will transmit light but not sound. (gas, metal, vacuum)
20. Sound travels through air at about __?__ m/s. (331, 33, 3)

Understand The Concepts

21. A cheetah can run at a speed of 100 km/h. How far can the animal can run in 6 minutes?
22. Under a net force of 2 N, a cart's velocity increases by 2 cm/s every second. What net force is needed

to increase the cart's velocity by 4 cm/s every second?

23. Calculate the weight of a 10-kg object (a) on Earth, (b) on the moon.
24. A person who weighs 1000 N climbs a ladder to a height of 10 m. How much work has the person done?
25. A first-class lever has an effort arm of 1.0 m and a resistance arm of 20 cm. What is the ideal mechanical advantage?
26. If an input of 100 J is needed to obtain an output of 15 J, what is the machine's efficiency?
27. Car A is traveling at 50 km/h; car B is going 100 km/h. How do their kinetic energies compare?
28. Explain how you would increase the potential energy of an object.
29. A girl does 1000 J of work in 5 seconds. Calculate her power (a) in watts, (b) in horsepower?
30. Why does the pressure inside a tire increase when the tire gets hotter?
31. Describe the purpose of a vapor barrier.
32. How can you distinguish between an electrical conductor and an insulator? Give examples of each.
33. How do you know that the circuit elements in your house are wired in parallel?
34. How can you show that a magnetic field surrounds an electric current?
35. The average power used in a certain home during the day is 2 kW. How much energy is used in this home in 24 hours?
36. How can dogs hear a whistle that humans cannot hear?
37. What are the primary colors (a) for light and (b) for pigments?

Challenge Your Understanding

38. How does the motion of an accelerating rocket in space demonstrate Newton's third law of motion?
39. Why do astronauts in orbit about Earth feel weightless?
40. For an object falling in a vacuum, what will its speed be after it has fallen for 1 s? 2 s? 5 s? 10 s?
41. Why would anyone want to use a third-class lever which has a mechanical advantage of less than one?
42. Two containers, one with 10 mL of water and one with 1 mL of water, both have a temperature of 50°C.
    (a) How does the average kinetic energy of the molecules in the two volumes compare?
    (b) How does the internal energy in the two volumes compare?
    (c) Which volume of water could transfer more heat?
43. It takes 45 J to raise the temperature of 10 g of a certain metal 10°C. What is the specific heat of the metal?
44. Give evidence that the sun's heat does not reach Earth by convection or conduction.
45. Give examples of renewable and nonrenewable sources of energy.
46. Why will an object that carries no electric charge often be attracted to a body that carries an excess of positive or negative charge?
47. Explain why convex mirrors and concave lenses cannot produce real images.
48. Suppose that a yellow object is viewed in green light. What color will the object appear to have?

Activities

Contents

Working Safely

The activities in this book can be done safely if you follow directions and act responsibly. The following general guidelines can help you to perform the activities safely.

1. Maintain a serious attitude at all times. Fooling around can be dangerous for you and your classmates.
2. Never perform unauthorized procedures. Always check with your teacher before trying anything different from what you are directed to do.
3. Always read the assigned activities carefully before you do them. Try to understand the reasons for the given procedures. If you have any questions about how to proceed, check with your teacher first.
4. Wear safety goggles whenever directed to do so. Follow all other rules for safe procedure that you are given.
5. Never taste or touch unfamiliar substances unless you are directed to do so. Do not directly inhale any fumes given off by a substance.
6. Keep any materials that will burn away from open flames and other heat sources.
7. Clean up any chemical spills at once. If you spill something on yourself, rinse it off immediately with lots of water. In either case, report the spill to your teacher as soon as possible.
8. Always keep your work area clean and well-organized. Make sure that nothing, such as gas or electricity, is left on when not in use.
9. Know the location of safety equipment, such as fire extinguishers, safety showers, and first aid kits.
10. Report any accident or injury to your teacher immediately.

ACTIVITY 1

Using a Microscope

PURPOSE To learn how to make a slide and view it under a microscope.

MATERIALS microscope, dropper, water, slide, cover slip, piece of newspaper with writing

Procedure

1. Hold the slide by the edges so that you do not get fingerprints on it.
2. Put the piece of newspaper in the middle of the slide. Place one or two drops of water on the newspaper. Pick up the cover slip by the edges and lower it carefully on the newspaper as shown in Figure 1. Try not to get air bubbles under the cover slip. Tap on the cover slip very gently with the tip of a pencil to get out any air bubbles you may have trapped. Set the slide aside.
3. Move the low-power objective into position over the diaphragm. You should hear a click when the objective is in place. Open the diaphragm to the widest possible setting. Adjust the mirror so that you can see a bright circle of light when looking through the eyepiece. **CAUTION**: Do not let direct sunlight fall on the mirror.
4. Raise the lens using the coarse adjustment. Place the slide you prepared on the stage. Position the slide so that the object you will be viewing is over the diaphragm. Move the clips so they hold the slide in place.
5. While looking at the microscope from the side—*not* through the eyepiece—use the coarse adjustment to lower the low-power objective toward the stage. (See Figure 2.) Bring the objective to within 0.5 cm of the slide. **CAUTION**: Do not let the objective touch the slide.
6. Look through the eyepiece. (Keep

Figure 1

both eyes open.) Focus the lens by slowly raising the objective using the coarse adjustment. When the image is nearly in focus, use the fine adjustment to bring it clearly into focus. **CAUTION**: Never use the coarse adjustment to lower the objective while looking through the eyepiece. You could break the slide and damage the lens.

If you cannot see the object when you look through the eyepiece, check to see if it is centered above the diaphragm and under

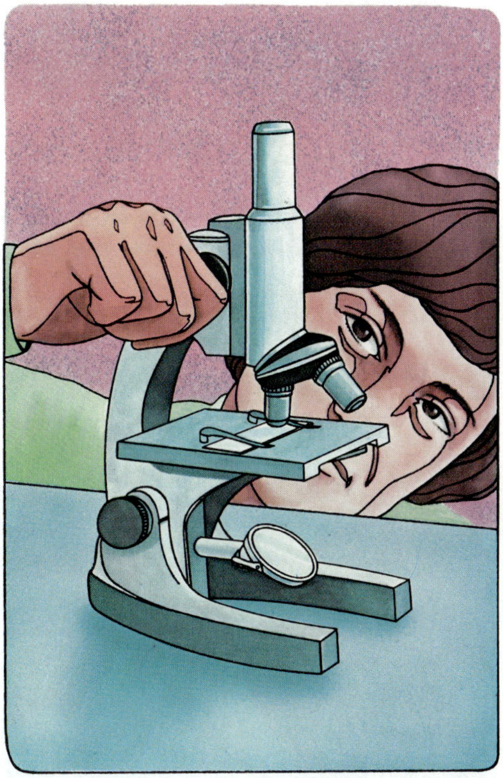

Figure 2

the lens. Lower the lens while watching from the side and start over again.

7. Try keeping both eyes open while looking at the slide. Draw a picture of the object as it appears under low power.
8. Without moving the coarse or fine adjustment, switch the objective to high power. **CAUTION**: Watch from the side to be sure that the lens does not touch the slide. Use the fine adjustment to focus the image. Draw a picture of the object as it appears under high power.
9. Observe what happens when you move the slide slowly right and left on the stage. What happens when you move it towards you and away from you?

Discussion

1. Compare the views of the object under low and high power. How are they alike? How are they different?
2. Which objective lens magnified more? What was the total magnification of the object viewed through that lens?
3. Did you see more of the object under low or high power? Explain.
4. When you moved the slide to your right, which way did the object appear to move when viewed through the microscope?
5. When the object on the slide was right-side up, how did it look to you when you viewed it through the microscope?

ACTIVITY 2

Observing Cells

PURPOSE To observe human cheek cells and onion skin cells.

MATERIALS microscope, slide, cover slip, iodine solution, water, dropper, toothpick, onion section, tweezers

Procedure

1. Place a drop of water in the middle of your slide.
2. Gently scrape the inside of your cheek with the flat part of a toothpick.
3. Spread the cheek scrapings in the drop of water on the slide.
4. Stain the cells to make their structure show up better. Do this by adding a drop of iodine solution to the cheek scrapings.
5. Add a cover slip and examine the slide under low power, then high power.
6. Draw the cells as they appear under the microscope. Label the cell parts on your drawing.
7. Wash and dry the slide and cover slip.
8. Place a drop of water in the middle of the slide.
9. Pull away the thin skin of the onion section with your tweezers, as shown in the illustration. Do not let the onion skin curl or fold over.
10. Place the onion skin so that it is lying flat in the drop of water on the slide. Add a drop of iodine solution and cover the skin with a cover slip.
11. Observe the slide under low power. Draw one onion skin cell as it appears under the microscope. Label the cell parts on your drawing. Switch to high power. Draw and label one cell.
12. Wash and dry the slide and cover slip.

Discussion

1. Compare your drawings of the two kinds of cells. How are the cells alike? How are they different?
2. What is the main function of cheek cells? of onion skin cells?

ACTIVITY 3

Observing One-Celled Organisms

PURPOSE To compare the structure and means of movement of different one-celled organisms.

MATERIALS microscope, slide, cover slip, dropper, paramecium culture, amoeba culture

Procedure

1. Place a drop of the paramecium culture on your slide. Cover the drop carefully with a cover slip.
2. Examine the slide through a microscope, using low power. Move the slide around slowly until you locate one or more of the paramecia. Observe the paramecia for a few minutes. Watch how they move. Paramecia move very quickly. You may have to move the slide to follow their paths.
3. Switch to high power. Observe and draw one paramecium. Label the parts on your drawing.
4. Use the dropper to return the paramecia to the culture. Be sure you put them back in the correct container.
5. Wash the dropper, slide, and cover slip.
6. Obtain a drop from the amoeba culture and repeat the procedure outlined in Steps 1–5 above. As you observe an amoeba, note how its structure and movement differ from those of the paramecium you observed earlier.

Discussion

1. Compare your drawings of a paramecium and an amoeba. How are the organisms alike? How are they different?
2. Compare the way a paramecium and an amoeba move.
3. How does the way an amoeba moves make it well-suited to life at the bottom of a pond?

ACTIVITY 4

Observing Onion Root-Tip Cells

PURPOSE To observe the stages of mitosis in plant cells.

MATERIALS microscope, prepared slides of onion root-tip cells

Procedure

1. Examine each section of the slide under low power and high power. Look for cells containing stained chromosomes. These cells were undergoing mitosis when the slide was made.
2. Locate an area on the slide where several cells are in different stages of mitosis.
3. Prepare drawings of each stage of mitosis from the cells you see through the microscope. Label the chromosomes, cell wall, and spindle fibers on your drawing. Number your drawings in the order in which the stages occur during mitosis.

Discussion

1. Mitosis is going on continually in onion root-tip cells. What does this indicate about that part of the plant?
2. In what other areas of a plant is mitosis probably taking place?
3. Where is mitosis likely to occur in humans?
4. What is the function of mitosis?

ACTIVITY 5

Measuring Lung Volume

PURPOSE To measure the volume of air your lungs can hold.

MATERIALS large jar, water, bucket, cardboard, rubber tube, graduated cylinder, wax marking pencil, ruler

Procedure

1. Make a table for recording normal and deep breath volumes.
2. Put water in the bucket to a height of about 6 cm.
3. Fill the jar with water. While holding the cardboard over the opening of the jar, turn the jar upside-down and place it in the bucket as shown.
4. Remove the cardboard. Slip one end of the rubber tube in the mouth of the jar. Take a normal breath, hold your nose, and breathe out normally into the rubber tube. **CAUTION**: Don't breathe in while the tube is in your mouth.
5. Straighten the jar and mark the top of the water with a wax pencil.

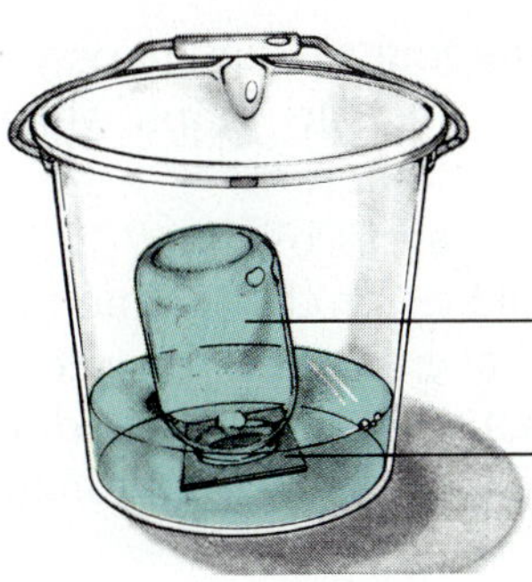

6. Turn the jar right-side up. Empty water from the jar until the water level reaches the pencil line. Measure the volume of the remaining water using a graduated cylinder. (This volume is the same as that of the air that you breathed into the tube.) Record the volume in your table.
7. Now find the volume of air you can force out after a deep breath. Repeat the procedure, only this time take a *deep* breath and exhale as much air as you can into the rubber tube. Record the results.

Discussion

1. Compare the exhalations following your normal and deep breaths. About how many normal exhalations would be equal in volume to one forced exhalation following a deep breath?
2. Compare your lung volumes with those of other members of the class. Which type of exhalation shows greater variation in volume?
3. Under what conditions do you use more lung volume than usual?

ACTIVITY 6

Mapping Taste Areas of the Tongue

PURPOSE To map the location of taste areas on the tongue.

MATERIALS 8 cotton-tipped swabs, dropper, 1 test tube each of sugar solution, salt solution, lemon-juice solution, and quinine solution, test-tube rack

Procedure

1. Make four outline drawings of a tongue. Label one *sweet*, another *salty*, another *sour*, and the last one *bitter*.
2. Use the dropper to place some sugar solution on one cotton-tipped swab. Touch the swab to one spot on your partner's tongue. If your partner could taste the sugar solution, make an X on the tongue drawing labeled *sweet* to show where the taste was detected. If your partner could not taste it, place an O on the tongue map. (See the illustration.)
3. Continue to put small amounts of sugar solution on different areas of your partner's tongue, recording the results after each trial, until you have a "map" of the areas that can taste *sweet*.
4. Now have your partner map the sweet-tasting areas of your tongue. Use a fresh cotton-tipped swab and a new tongue drawing.
5. Rinse the dropper and repeat the procedure, using salt solution, fresh cotton-tipped swabs, and the tongue drawing labeled *salty*.
6. Repeat again using the lemon juice, then the quinine solution. Be sure to rinse the dropper and use fresh swabs for each solution.

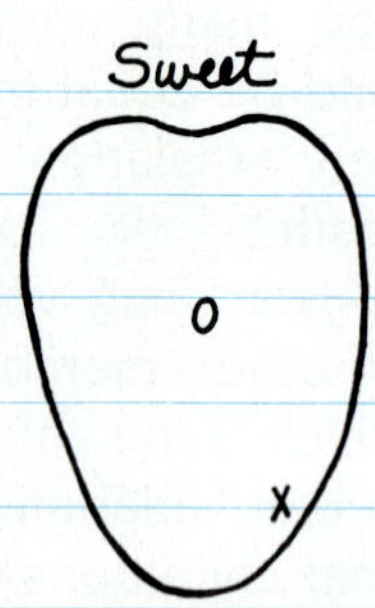

Discussion

1. Compare the four tongue maps you made. How are the maps alike? How are they different?
2. Do any taste areas overlap?
3. Are there places on the tongue that appear to have no taste receptors?
4. What conclusions can you draw as a result of looking at your tongue maps?

ACTIVITY 7

Life in a Square Meter of Ground

PURPOSE To look for living organisms and evidence of life in a square meter of ground.

MATERIALS string, stakes, small garden tools, meterstick, dissecting microscope (optional)

Procedure

1. Get permission to investigate a square meter of ground in your back yard, school yard, or a wooded area. Mark it off with stakes and string, as in the illustration.
2. Write descriptions and make sketches of the plants in the square meter. Do the same for any evidence you might find of animal life, such as a bird feather or snail shell.
3. If permissible, dig out a cube of soil with sides one decimeter long. Look through this soil sample for living animals, such as worms, snails, spiders, and insects. Describe and sketch any that you find.
4. If a dissecting microscope is available, use it to look at some of the soil. The soil should be moist, but not wet. With the microscope you can see smaller animals than those you saw before. Describe and sketch any animals that you may see through the microscope.

Discussion

1. Try to identify each of the living plants and animals that you found. Make a list of all that were in your square meter.
2. List any evidence that animals other than those you saw had been in your square meter.
3. Do your lists include all the living organisms present in your square meter? Explain.
4. How do the plants and animals you found obtain food?
5. What consumer orders are represented by the animals in your square meter? Give examples to support your answer.

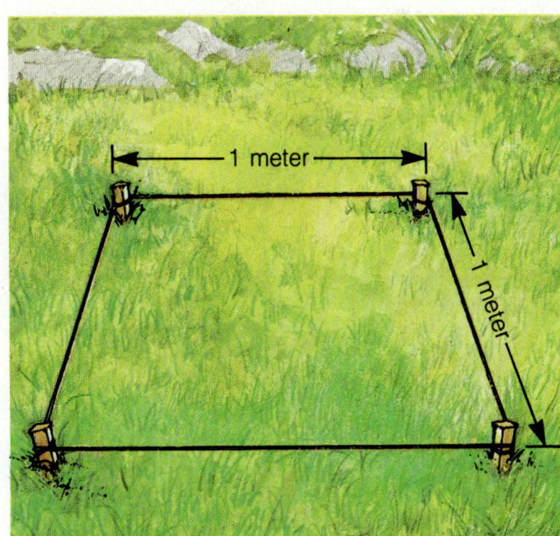

ACTIVITY 8

Bird Beak Adaptations

PURPOSE To match bird beaks with the feeding behaviors for which the beaks are adapted.

MATERIALS pencil, paper

Procedure

Match the beaks shown on this page with the following feeding behaviors.

1. Feeds on insects that live in the trunks of trees.
2. Pries seeds out of the inside of pine cones.
3. Catches slippery fish in its beak.
4. Drinks nectar from flowers.
5. Catches flying insects.
6. Feeds on small animals.
7. Eats seeds that have hard shells.
8. Eats fruit that has tough skin.
9. Scoops insects from the surface of the water in lakes or ponds.
10. Digs up organisms that live in the mud under shallow water.

Discussion

Compare your answers with those of your classmates and discuss the reasons for the choices made.

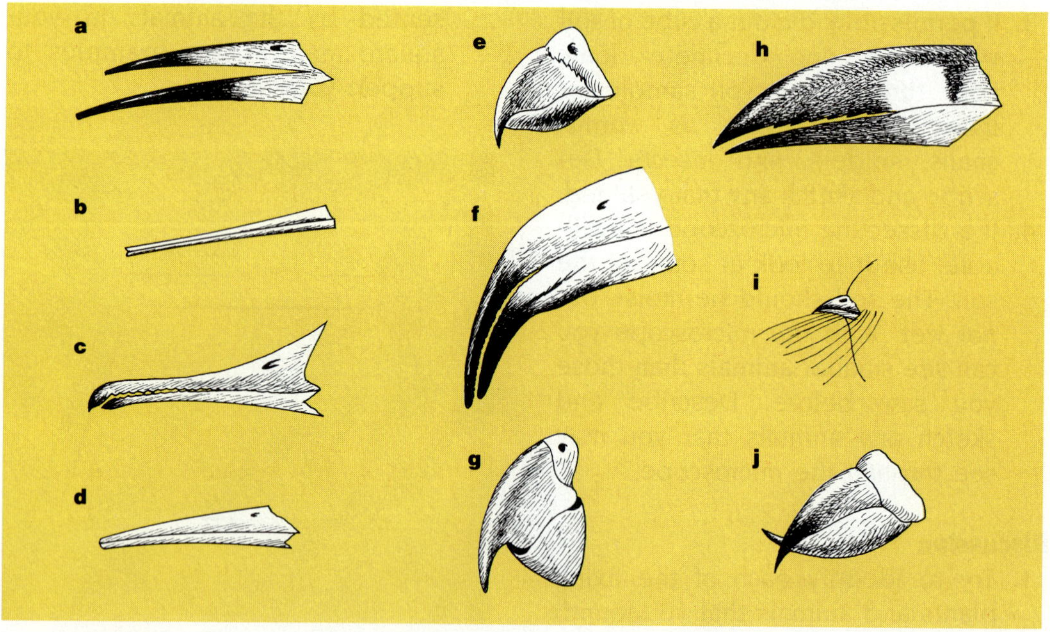

ACTIVITY 9

Measuring Density

PURPOSE To determine the density of some solids and liquids.

MATERIALS laboratory balance with standard masses, thread, centimeter ruler, graduated cylinder, solid and liquid samples

Procedure 1

1. Obtain a solid sample from your teacher. If it is cubic or rectangular, make a table like the one in Figure 1. If it is irregularly shaped, make a table like the one in Figure 2. As you obtain data, record it in the appropriate table.
2. Use a laboratory balance to determine the mass of the sample.
3. If the sample is cubic or rectangular, use a ruler to measure its length (*L*), width (*W*), and height (*H*).
4. If the sample is irregularly shaped, you can find its volume by measuring the amount of water that it displaces. First, half-fill a graduated cylinder with water. Read and

Density of a Cubic or Rectangular Solid

Sample Number	Mass (g)	Dimensions L(cm)	W(cm)	H(cm)	Volume $L \times W \times H$ (cm^3)	Density (g/cm^3)

Figure 1

Density of an Irregularly Shaped Solid

Sample Number	Mass (g)	Volume 1 (mL)	Volume 2 (mL)	Volume of Sample (mL)	Density (g/mL)

Figure 2

record (as Volume 1) the volume of water in the cylinder. Then carefully lower the sample into the cylinder. If possible, tie the sample to the end of some thread and lower the sample into the water as shown in Figure 3. Make sure that the sample is entirely under water. Then read and record (as Volume 2) the volume of the water plus the sample.

Discussion 1

1. Compute the volume of each cubic or rectangular sample you tested by multiplying the length times the width times the height ($L \times W \times H$). Then compute the density by dividing the mass by the volume.
2. Compute the volume of each irregularly shaped object you tested by subtracting Volume 1 from Volume 2. Then compute the density of each object by dividing the mass by the volume.

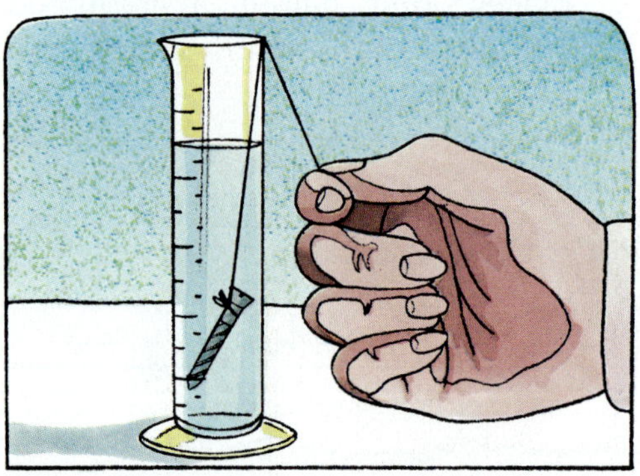

Figure 3

Procedure 2

1. Make a table like the one in Figure 4.
2. Find the mass of a graduated cylinder and record it as Mass 1.
3. Pour a liquid sample into the cylinder. Find and record (as Mass 2) the mass of cylinder plus sample.
4. Read and record the volume of the liquid in the cylinder.
5. Dispose of the liquid as your teacher directs. Wash and dry the cylinder.

Discussion 2

1. Compute the mass of each liquid sample you tested by subtracting Mass 1 from Mass 2.
2. Compute the density of each sample by dividing the mass of the sample by the volume.

Density of a Liquid

Sample Number	Mass 1 (g)	Mass 2 (g)	Mass of Sample (g)	Volume of Sample (mL)	Density g/mL

Figure 4

ACTIVITY 10

Paper Chromatography

PURPOSE To separate mixtures using paper chromatography.

MATERIALS 2–3 large test tubes, test-tube rack, filter paper, solvent, food coloring, 2–3 toothpicks, scissors, ruler, 2–3 small paper clips, aluminum foil

Procedure

1. Cut two strips of filter paper, about 2 cm × 15 cm. Cut one end to form a point, as in the illustration. Fold the strips lengthwise down the middle.
2. Put two dry test tubes in the rack. Add just enough solvent to fill the curved part of each tube.
3. Use the small end of a toothpick to put two small spots of food coloring on one piece of filter paper, as in the illustration.
4. In a paper cup, mix together one drop each of two or three different colors of food coloring, using a toothpick. With the toothpick, put two small spots of the mixture on the second piece of filter paper.
5. After the spots on both strips of filter paper have dried for at least three minutes, place the strips in separate test tubes, as in the illustration. The spots must not be below the top of the solvent. Cover the test tubes with aluminum foil.
6. Observe the paper in the test tubes every 5 min for at least 30 min.

Discussion

1. What do you see in each test tube?
2. Is the food coloring you used by itself a mixture? How can you tell?

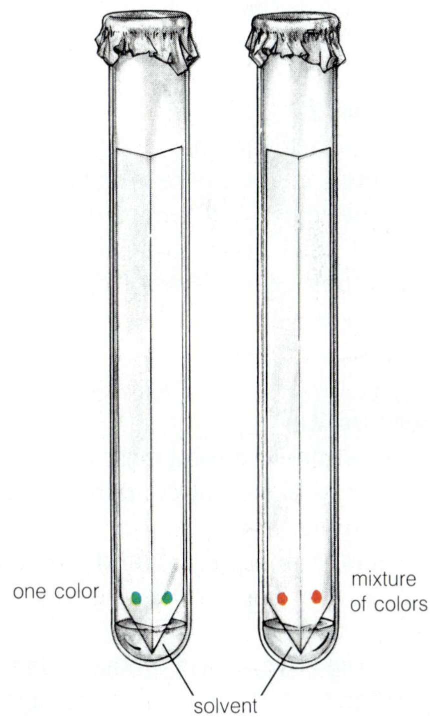

ACTIVITY 11

Conservation of Mass

PURPOSE To demonstrate the law of conservation of mass for both a physical and a chemical change.

MATERIALS laboratory balance and set of standard masses, vial about one-third full of lead nitrate solution, vial about one-third full of potassium iodide solution or sodium iodide solution, vial full of crushed ice, three covers for the vials

Procedure 1

1. Wipe away any condensed liquid on the covered vial of ice as you determine its mass.
2. After the ice has melted, carefully dry the surface of the covered vial and find the mass again.

Discussion 1

1. How does the mass of the vial of melted ice compare with the mass of the vial of ice before it melted?
2. What can you conclude about whether the mass of matter changes when it undergoes a physical change?

Procedure 2

1. Determine the total mass of the two covered vials that contain different solutions.
2. Remove the covers and carefully pour the contents of one vial into the other.
3. Put the covers back on the vials and determine their total mass again.

Discussion 2

1. What evidence of a chemical reaction did you see when you mixed the two solutions?
2. How does the mass of the matter before the reaction compare with the mass after the reaction?

ACTIVITY 12

Some Model Reactions

PURPOSE To demonstrate the law of definite proportions using atomic and molecular models.

MATERIALS paper fasteners and washers (at least six of each), laboratory balance and set of standard masses

Procedure 1

1. Use washers to represent atoms of element "W" and paper fasteners to represent atoms of element "F."
2. Combine as many of your "atoms" as you can to make "molecules" of the compound "WF." (You may have an excess of one of the elements.)
3. Determine the total mass of all the molecules of WF.
4. "Decompose" the molecules of WF by separating the molecules into atoms of elements W and F.
5. Determine the total mass of element W that was in compound WF.
6. Determine the total mass of element F that was in compound WF.

Discussion 1

1. What was the percent of element W in WF? To find out, divide the mass of W by the mass of WF and multiply this result by 100.
2. What was the percent of element F in WF? To find out, divide the mass of F by the mass of WF and multiply this result by 100.
3. Compare your results with those of your classmates. Does the percentage composition of WF depend on the number of molecules in the WF sample?

Procedure 2

1. Use washers and paper fasteners to make molecules of W_2F.
2. Repeat Procedure 1 using compound W_2F instead of WF.

Discussion 2

1. Find the percent of W and of F in W_2F.
2. Divide the percent of W by the percent of F in (a) W_2F, (b) WF. How does the ratio of W to F in W_2F compare with that in WF? How do you account for this ratio?

ACTIVITY 13

Metals and Nonmetals

PURPOSE To investigate the electrical conductivity and melting temperature of some metals and nonmetals.

MATERIALS candle, Bunsen burner, alcohol burner, small pieces of candle wax, sugar cubes, roll sulfur (pea-size pieces), tin (foil or mossy), copper wire, iron wire, aluminum foil, platinum wire, pair of tongs, C or D cell, connecting wires, rubber band, small bulb, bulb holder, matches

Procedure 1

1. Make a table like the one shown in Figure 1.
2. Use a setup like the one shown in Figure 2 to test each assigned sample. Be sure that both wires are firmly in contact with the sample being tested. If the bulb lights, the sample you are testing is a conductor of electricity.

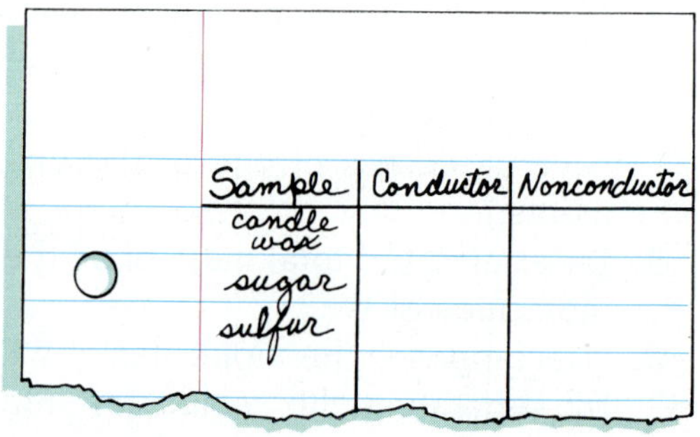

Figure 1

Figure 2

Discussion 1

1. On the basis of your data, which of the substances do you think are metals? nonmetals?
2. Which of the substances do you think have low melting points? high melting points?

Procedure 2

1. Make a table like the one shown in Figure 3.
2. Put on your SAFETY GOGGLES. Then use a pair of tongs to hold each of the assigned samples in a candle flame. (See Figure 4.) Note whether or not each sample melts in the flame. Record the results in your table.
3. Repeat Step 2 using (a) an alcohol burner and (b) a Bunsen burner.

Sample	Candle	Alcohol burner	Bunsen burner
candle wax			
sugar			
sulfur			

Figure 3

Discussion 2

1. Which flame has the lowest temperature? the highest temperature?
2. How do the melting temperatures of the nonmetals you tested compare with the melting temperatures of the metals you tested?
3. Ask your teacher for the melting temperatures of the substances you tested. Compare them with your results. Use this comparison to estimate the range within which the temperature of each flame you used lies.

Figure 4

ACTIVITY 14

Unsuspected Electrochemical Cells

PURPOSE To show that electrochemical cells can be made using common fruits and metal electrodes.

MATERIALS lemons and other available fruits, such as apples, pears, olives, pickles, and oranges; galvanometer or microammeter; two insulated wires, preferably with alligator clips; copper, aluminum, and iron nails; zinc strips; centimeter ruler; paper towels and newspapers

Procedure

1. Make a table like the one shown in Figure 1. Measure the lengths of two electrodes, a zinc strip and a copper nail. Record the lengths (as Length 1) in the table.
2. Insert the electrodes into a lemon, as in Figure 2. Insert both to the same depth. Measure the length of each electrode above the surface of the lemon. Record the lengths (as Length 2) in the table. (The depth is the difference between Lengths 1 and 2.)
3. Measure and record the distance between the electrodes.
4. Connect the electrodes to a microammeter or galvanometer, as shown in Figure 2. In order to get a positive reading on the meter, you must connect the negative electrode to the negative side of the meter, and the positive electrode to the positive side. The connections that give you a positive meter reading, therefore, indicate the type of charge (positive or negative) on each electrode. Record the

Electrode pair	Length 1 (cm)	Length 2 (cm)	Depth (cm)	Distance apart (cm)	Electrode charge (+, −)	Cell current (microamperes)
1. zinc copper						

Figure 1

electrode charges. Read and record the current. Disconnect the electrodes from the meter.

5. Investigate how the electrode depth affects the amount of current produced. Keeping the distance between electrodes the same, change the depth of insertion. Connect the electrodes to the meter. Read and record the current. Repeat the procedure for other electrode depths.
6. Investigate how the distance between electrodes affects the amount of current produced. Keeping the depths the same, change the distance between the electrodes. Connect the electrodes to the meter. Read and record the current. Repeat the procedure for other electrode distances.
7. Replace the zinc electrode with a second copper nail. Connect the electrodes to the meter. Read and record the current.

Figure 2

8. Compare other electrode pairs to the copper-zinc pair. Pairs that you could try include copper and aluminum, copper and iron, and aluminum and zinc. All electrode pairs should be compared at the same depth of insertion and distance apart. For each electrode pair, note and record (a) the charge on each electrode and (b) the amount of current produced. After each trial, clean the electrodes.
9. Make electrochemical cells using fruits other than lemons. Find the amount of current that each produces, trying to keep electrode depth and distance apart the same for each fruit.

Discussion

1. Compare the charges that you found for each of the electrode pairs that you tested. How might this data enable you to predict the charges of the electrodes in pairs that you did not test?
2. What do you conclude about the use in an electrochemical cell of two electrodes that are made of the same metal?
3. How is the current produced by a cell affected by (a) the depth to which the electrodes are inserted, (b) the distance between the electrodes?
4. What does a fruit contain that enables it to act as an electrochemical cell when electrodes are added?

ACTIVITY 15

Acids and Bases

PURPOSE To use indicators to test for acids and bases and to investigate an acid-base reaction.

MATERIALS stirring rod, 4 test tubes, medicine dropper, evaporating dish, graduated cylinder, phenolphthalein solution, red and blue litmus paper, *p*H paper (optional), safety glasses, dilute hydrochloric acid, dilute solutions of sodium, potassium, and calcium hydroxide, heat lamp

CAUTION: Be sure to wear SAFETY GOGGLES throughout this activity.

Procedure 1

1. Pour about 5 mL of each hydroxide solution into a separate test tube. Test each solution. Use a stirring rod to remove a drop of solution from the test tube and place it on litmus paper, as shown in Figure 1. Use both red and blue litmus paper and, if available, *p*H paper. Record your results.
2. Add a drop of phenolphthalein solution to each hydroxide solution. Record your results.
3. Pour about 5 mL of hydrochloric acid into a fourth test tube. Test the acid with litmus paper and, if available, *p*H paper. Then add a drop of phenolphthalein solution to the acid. Record your results.

Discussion 1

1. What can you say about the *p*H of each hydroxide solution? of the acid solution?
2. What is the color of phenolphthalein solution in an acid? in a base?

Figure 1

Procedure 2

1. Put about 10 mL of sodium hydroxide solution in an evaporating dish. Add 2 drops of phenolphthalein. Stir slowly as you carefully add hydrochloric acid, drop by drop, until the solution suddenly changes from pink to colorless. (See Figure 2.) Test the solution with *p*H paper, if available, or with red and blue litmus paper. Record your results.
2. Evaporate the solution to dryness under a heat lamp, as shown in Figure 3. Note the appearance of the substance that remains in the dish after all of the liquid has been evaporated.

Discussion 2

1. Explain why the phenolphthalein changed color when you added hydrochloric acid to the sodium hydroxide solution. Why did you get the results you noted with *p*H or litmus paper?
2. What is the white solid that was left after you evaporated the water from the solution? Write an equation for the reaction that took place between the acid and the base.

Figure 2

Procedure 3

Test some common substances to find out whether they are acidic, basic, or neutral. Try to determine how strongly acidic or basic each acid or base is, as well. Use red and blue litmus paper and, if available, *p*H paper. You might test substances such as tap water, rainwater, lemon juice, grapefruit juice, quinine water, and a stomach "alkalizer" tablet dissolved in water. **CAUTION**: Be sure to check with your teacher before testing any substance.

Discussion 3

1. On the basis of the class results, classify each substance that was tested as acidic, basic, or neutral.
2. Rank the (a) acidic and (b) basic substances in order of strength.

Figure 3

ACTIVITY 16

Half-Life

PURPOSE To observe the "decay rate" of cubes and determine their "half-life."

MATERIALS 100 cubes (sugar or wooden), each with a dot on one face, or 100 dice; container to hold cubes or dice

Procedure

1. Put 100 cubes in the container. Then pour the cubes onto a table. Count all the cubes that have the side with one dot on it face up. Record this number. Each of the cubes with the dot face up represents a decay.
2. Put all the cubes back in the container. Repeat Step 1 four times.
3. Add the total number of cubes with one dot face up in all five trials and divide by five. This represents the average number of "radioactive atoms" that decayed in the first time period. Remove this number of cubes from the container. Record the number of cubes that remain. They represent atoms that have not decayed.
4. Pour the remaining cubes onto a table, as in Step 1. Count the number of cubes with one dot face up. Repeat the procedure four times.
5. Average the five results as before. Remove this number of cubes from the container and record the number that remain in the container.
6. Continue the procedure until 20 or fewer cubes are left in the container.
7. Make a bar graph to display your results. Plot time periods on the horizontal axis and the number of atoms that remain undecayed at the end of a time period on the vertical axis. In what time period does the half-life lie?

Discussion

1. How many cubes will "decay" in the tenth time period?
2. Can you predict *which* cubes will "decay" in the tenth time period?

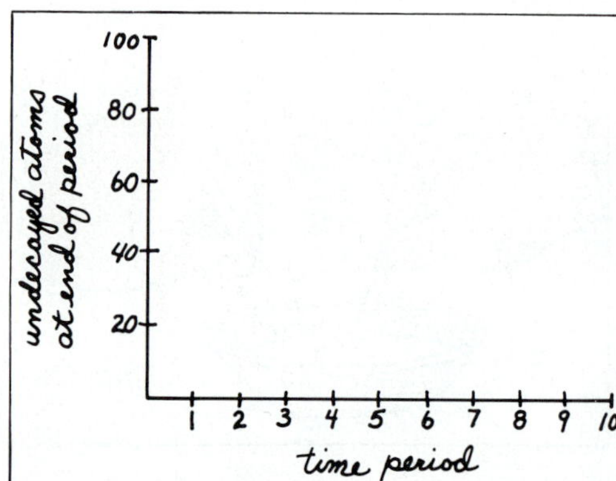

ACTIVITY 17

Testing Mineral Hardness

PURPOSE To compare the hardness of mineral samples.

MATERIALS mineral samples, copper penny, iron nail, steel scissors, glass slide

Procedure

Test each of the mineral samples supplied by your teacher for hardness. Find and record the hardness in terms of the hardness scale given on this page.

Discussion

1. Your teacher will give you the accepted hardness values for the samples you tested. How close were your estimates?
2. How can the hardness test help you in identifying a mineral sample? What are some other properties of minerals that can be used to tell them apart?

Number	*Hardness Test*
1	Easily scratched by fingernail.
2	Barely scratched by fingernail.
3	Barely scratched by copper penny.
4	Scratched by iron nail.
5	Scratched by steel scissors.
6	Easily scratches glass.
7	Easily scratches steel and glass.
8	Scratches all other common minerals.
9	Scratches topaz.
10	Hardest mineral (diamond).

ACTIVITY 18

Moving Land Masses

PURPOSE To use diagrams of land masses to determine how these masses may have fit together in the past.

MATERIALS tracing paper, scissors, transparent tape, world map

Procedure

1. Put a piece of tracing paper over the next page. Trace the diagrams on the page, including all the details. Use the diagrams that you have traced for all of the following steps.
2. Write the name of each land mass on it. (Use a world map, if necessary, to identify the land masses.) Using scissors, cut out each land mass diagram. Cut along the broken lines, where they exist. If there is no broken line, cut along the outermost solid line.
3. Arrange the land masses on a flat surface in their present location, as shown on a map.
4. Move the land masses together, matching up similar areas. Consider the shape of the continental margins, the similarity of mountains and rocks, and directions of glacial markings in the matching process. When you have finished moving the land masses together, use small pieces of tape to hold them in position.

Discussion

1. Between what land masses do the continental margins match closely?
2. Between what continents do the glaciers appear to have moved? In what general direction would you expect the glaciers to have moved? (Hint: In what general direction do such glaciers move today?)
3. Between what land masses do the (a) ancient mountains, (b) rocks of similar ages match up?
4. Did any of the land masses rotate as they moved into their present positions? If so, which ones?
5. List as much evidence as you can to support the theory that these land masses were together at one time.

Key to the Land Mass Diagrams

The broken line around each land mass represents the continental margin. The arrows show the directions in which glaciers moved. The darkly shaded areas represent ancient mountains that were formed at about the same time. The lightly shaded areas represent rocks of about the same age.

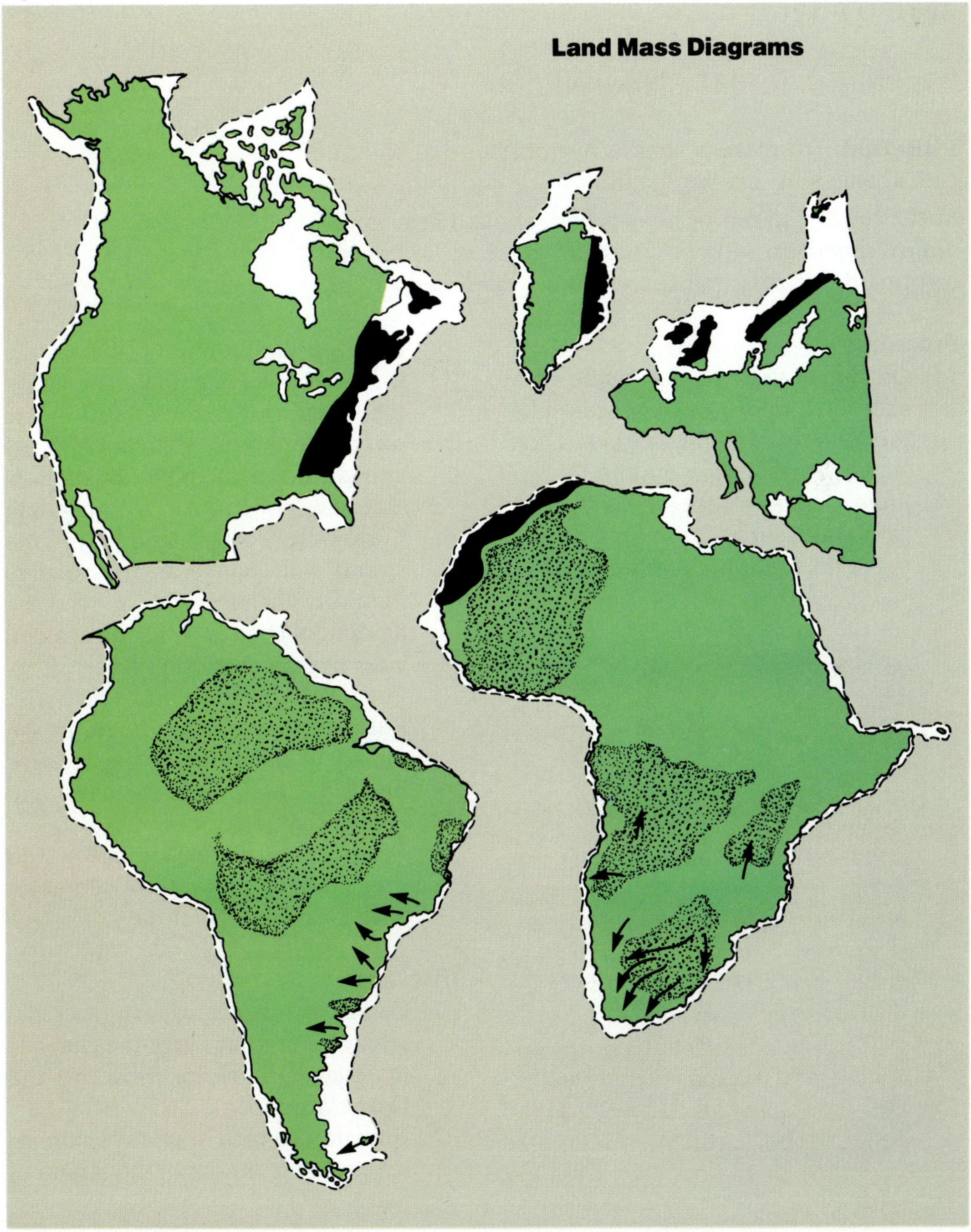
Land Mass Diagrams

ACTIVITY 19

Making a Simple Hygrometer

PURPOSE To make a simple hygrometer that will indicate changes in relative humidity.

MATERIALS strand of human hair, about 15 cm long; empty shoe box, without lid; small piece of lightweight cardboard; scissors; transparent tape; centimeter ruler

Procedure

1. On the bottom of the inside of a shoe box, make a 25-cm scale like the one in the illustration. Then stand the shoe box on end.
2. From a small piece of lightweight cardboard, cut an arrow like the one in the illustration. Tape the arrow to one end of a strand of human hair that is about 15 cm long. Then tape the other end of the strand to the top of the shoe box, so that it hangs straight down, as in the illustration.

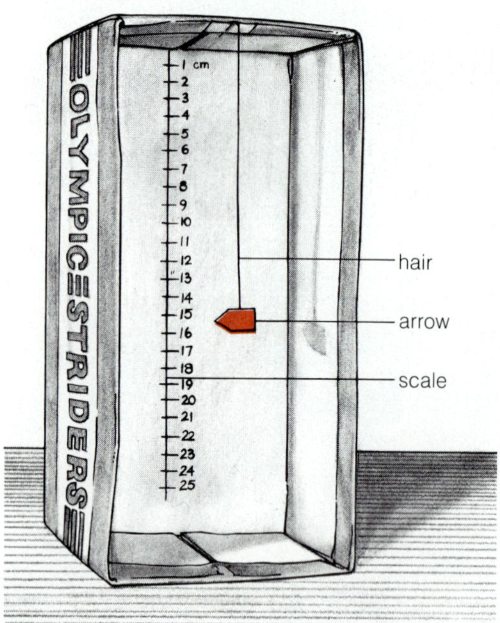

3. You have constructed a human-hair hygrometer. The length of the strand will change as the relative humidity changes. You can read the amount of change using the scale of your hygrometer.
4. Put your hygrometer in a safe place. Read and record the length of the strand of hair. Obtain the relative humidity from your teacher and record that, also.
5. Record the relative humidity and the reading on your hygrometer each day for several days.

Discussion

1. Compare the changes in your hygrometer readings with the changes in relative humidity. How are the two related?
2. How do changes in relative humidity affect the length of the strand of hair?

ACTIVITY 20

Investigating Density Currents

PURPOSE To observe the behavior of solutions of different densities when they are put in less dense water.

MATERIALS food coloring, test tube, artificial seawater (different densities), meter-long plastic column with cap, ring stand and clamp, watch or clock with second hand

Procedure

1. Set up the equipment as shown in the drawing. Fill the plastic column two-thirds full of tap water.
2. Put some artificial seawater in a test tube. Add three or four drops of food coloring. Swirl the solution to mix uniformly.
3. Pour the seawater into the water in the plastic column. Record the time in seconds needed for the leading edge of the colored solution to reach the bottom of the column.
4. Repeat Steps 2 and 3, using fresh tap water in the column and a different salt solution.
5. Repeat Step 4, using the third salt solution.

Discussion

1. Which of the seawater solutions took the longest time to reach the bottom of the tube? Which took the least time?
2. Find out from your teacher which of the artificial seawater solutions was the most dense and which one was the least dense. Use this information to explain your results.

ACTIVITY 21

Seasons of the Year

PURPOSE To investigate the relative positions of the sun and Earth at various times of the year.

MATERIALS Styrofoam ball (8-10 cm diameter), probe or teasing needle, flexible protractor, flexible ruler or tape measure, lamp with 25-watt bulb, several pieces of yarn large enough to encircle the ball, world map or globe

Procedure

1. Use an unmarked Styrofoam ball to represent Earth. Wrap a piece of yarn around the ball to represent the equator. (The yarn should stick to the ball without the aid of tape or glue.) Locate and mark the North and South Poles.
2. Use a map or globe to find the latitudes (in degrees) of the Tropic of Cancer, Tropic of Capricorn, Arctic Circle, and Antarctic Circle.
3. Use a protractor lined up along the equator of your model Earth to locate these latitudes. Use a piece of yarn to represent each of them. Your model Earth should look like the one in Figure 1.
4. Stick a probe into your model Earth to serve as a handle. If you stick the probe in somewhere along the Antarctic Circle, the model will tilt at the proper angle ($23\frac{1}{2}^\circ$) when you hold the probe straight up, as shown in Figure 2.
5. Use a lamp without a shade to represent the sun. Move your model Earth around the lighted lamp to represent Earth's revolution around the sun. Be sure to keep the amount and direction of the tilt constant. (Do not try to rotate Earth about its axis.)
6. As you move Earth around the sun, note the direction of Earth's tilt relative to the sun. On the first day of summer the Northern Hemisphere tilts directly toward the sun. On the first day of winter it tilts directly away from the sun. On the first day of spring and fall it tilts neither toward nor away from the sun.
7. Make a drawing of Earth, as seen from above, for each of the days given in Step 6. In each drawing, indicate the North Pole with a dot. Draw in any portions of the yarn circles that can be seen from above. Show which portions of Earth are lighted by the sun and which ones are in shadow.

Discussion

1. Where is the top of the shadow on

the first day of summer? How much of Earth above the Arctic Circle is lighted? Where is the bottom of the shadow? How much of Earth below the Antarctic Circle is lighted?

2. On the first day of winter, where is the shadow line in each hemisphere? Which polar area is fully lighted?
3. On the first day of spring and fall, where does the shadow line appear in relation to the poles? How does this affect the number of hours of daylight and darkness over the globe?
4. Predict what would happen to the shadow lines if the tilt of the axis were 45° or 0°. If time permits, repeat the activity to test your predictions.

ACTIVITY 22

Measuring Walking Speed

PURPOSE To determine the speed at which someone walks.

MATERIALS meterstick or metric measuring tape, metronome or stopwatches, distance markers

Procedure

1. Choose a level surface, such as a hallway or sidewalk, on which a volunteer will walk. Use a meterstick or tape measure to mark off 10-m intervals along the walkway. Station a volunteer at the end of each interval to record the time at which the walker just passes that checkpoint. (See Figure 1.)
2. As the walker crosses the starting line, the timers start their stopwatches. When the walker passes the checkpoints, the timers stop their watches. (Alternatively, you could start a metronome, set to sound once each second, when the walker crosses the starting line.)
3. When the walker has completed the course, collect the times from the timers. Record them in a table like the one in Figure 2.

Figure 1

Discussion

1. Compute the walker's speed over each interval. Did the walker walk at a fairly constant speed?
2. Compute the walker's average speed over the entire course.
3. (Optional) Plot a graph of distance vs. time from the data. Let the vertical axis represent distance and the horizontal axis represent time.

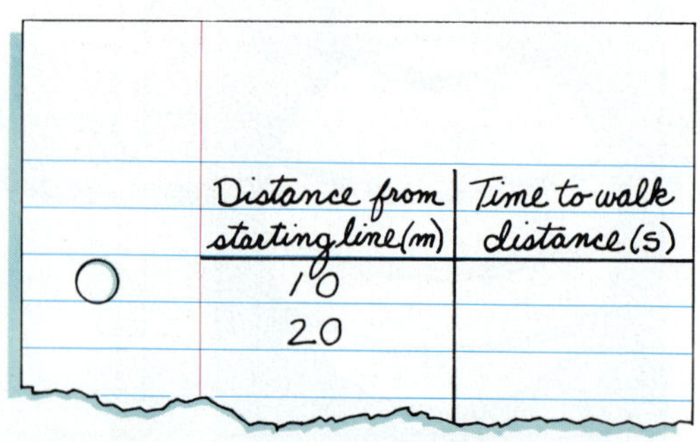

Figure 2

ACTIVITY 23

Measuring Forces of Friction

PURPOSE To measure the force of friction between different surfaces.

MATERIALS smooth, heavy wooden blocks; smooth, level board or other surface; thumbtacks; paper clips; string; sandpaper; aluminum foil; waxed paper; newspaper; wide rubber bands; spring scale

Procedure 1

1. Open a paper clip, so that the two portions are at right angles to each other. Hold one portion flat against one face of a wooden block. Push a thumbtack into the block to hold the paper clip, as in Figure 1.
2. Place the block at one end of a smooth, level board. Attach the hook of a spring scale to the paper clip, as in Figure 2 (on the next page). Pull on the other end of the scale with just enough force to make the block move along the board at a steady rate. Read and record the force shown on the scale.

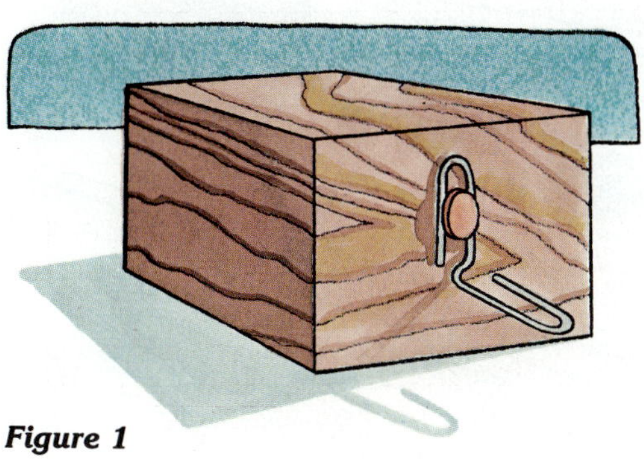

Figure 1

3. Modify the part of the block that is in contact with the board by taping a material such as waxed paper, aluminum foil, or newspaper to it. You can produce a rubber surface by wrapping two or three wide rubber bands around the block. Repeat Step 2 using each of the different surfaces that you prepare.

Discussion 1

1. What does the force that you have measured represent?
2. Rank the forces of friction for the different block surfaces that you tested, from greatest to least.

Procedure 2

1. Find out what happens when you increase the force with which the plain wooden block presses down on the board. You can do this by placing a heavy object on top of the block. You could use a small container half-full of washers, for example. Find and record the force of friction between the block and the

board. Repeat the procedure, this time with the container full of washers. Again read and record the force of friction.

2. Find out what happens when you change the amount of surface area of the block that is in contact with the board. If you have a rectangular block, you can simply stand it on a different face to change the surface area. Find and record the force of friction for at least one surface area that is different from the one you used in the preceding steps.

Discussion 2

1. What happened to the force of friction when you increased the force with which the block pressed down on the board?
2. What happened to the force of friction when you changed the area of block surface that was in contact with the board?
3. What factor(s) must you keep the same when you are comparing the forces of friction that are produced by objects moving over the same surface?

Figure 2

ACTIVITY 24

Measuring Power

PURPOSE To measure the power produced by a person in running up a flight of stairs.

MATERIALS meterstick, stopwatch, stairs

Procedure

1. Use a stopwatch to measure the amount of time it takes for one of your classmates to run up a flight of stairs.
2. Measure the vertical height of the flight of stairs, as indicated in the illustration.
3. Repeat Step 1 for several other classmates.

Discussion

1. Determine the weight, in newtons, of each person who ran up the stairs. Use the weight to compute the work, in joules, done in running up the stairs.
2. Compute the power, in watts, that was produced by each person in running up the stairs. You may want to use this result to compute the power in terms of horsepower, also.
3. Compare the amount of power produced by your classmates who ran up the stairs. What factors were most important in determining how much power was produced? What physical characteristics produce the highest horsepower rating in this activity?

ACTIVITY 25

Testing Insulation

PURPOSE To compare the effect of different insulating materials on the rate of cooling of warm water.

MATERIALS thermometer (0–100°C); one small and one large container made of plastic, glass, or paper; clock or watch; small and large pieces of paper, to cover containers; insulating materials, such as mineral wool, vermiculite, and Styrofoam

Procedure

1. Pour 100 mL of hot water into the small container. Let the water cool to a starting temperature between 35°C and 40°C. Record the starting temperature. Let the water cool for 15 min more. Then again measure and record the temperature of the water.
2. Put an insulating material into the large container to a depth of 3 to 4 cm. Place the small container on the material. Surround it with more material, as in the illustration.

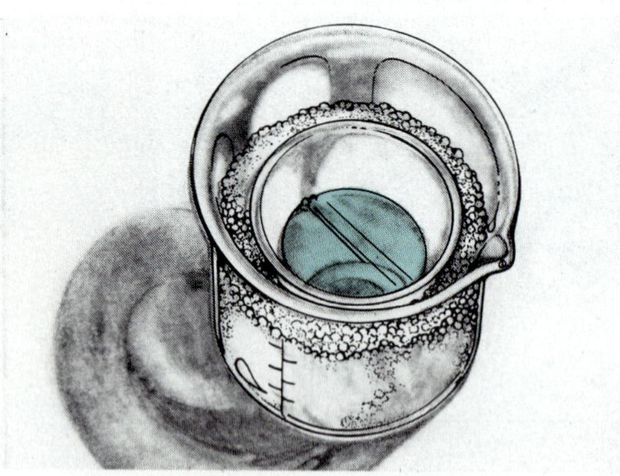

3. Pour 100 mL of hot water into the small container. When the water has cooled to the starting temperature used for Step 1, cover the small container with a piece of paper. Then fill the large container to the top with more insulating material. Place a large piece of paper on top.
4. Let the water cool for 15 min. Then measure and record its temperature.
5. Repeat Steps 2–4 for each of the other insulating materials that you are testing.

Discussion

1. How many degrees cooler did the water in the uninsulated container become in 15 min?
2. How many degrees cooler did the water become when insulated by each of the materials?
3. How effective was each type of insulation in slowing the rate at which the water cooled?

ACTIVITY 26

Conductors of Electricity

PURPOSE To test various solid materials in order to find out how well they conduct electricity.

MATERIALS 4 D cells or 6-V battery, insulated electrical wires, 2 or more alligator clips (optional), light bulb, bulb holder, solid objects (made of wood, metal, plastic, paper, chalk, glass, or some other material)

Procedure

1. Use wires to connect a battery and other objects, as shown in the illustration. This setup is called an *electric circuit*. The wires and the bulb are conductors of electricity. If the object being tested is a conductor, the circuit is said to be *complete*. That is, electrons can flow from the negative terminal of the battery, through the wires and other objects, back to the positive terminal of the battery. They flow in this way due to chemical reactions that take place in the battery. If the circuit is complete, the bulb lights. If not, the bulb does not light.
2. Test various objects. Be sure the bare wire ends or clips are firmly in contact with the object being tested. Note whether or not the bulb lights. If it lights, note how brightly it lights. Record your results.

Discussion

1. What materials that you tested conduct electricity? Do some of these materials conduct electricity better than others? Give evidence for your answers.
2. What materials that you tested do not conduct electricity? Give evidence for your answer.
3. What name is used to describe materials that do not conduct electricity?

ACTIVITY 27

A Look into a Plane Mirror

PURPOSE To find out how to locate an image in a plane mirror.

MATERIALS small plane mirror, small straight pins, white paper, cardboard sheet, large straight pin, ruler, clay

Procedure 1

1. Place a sheet of white paper on top of a piece of cardboard. Then stick a large pin into the center of the paper.
2. Place two small pins 7 to 10 cm apart in a straight line with the large pin. Do this by sighting along the pins as shown in the illustration. Remove the small pins and use a ruler to draw a *sight line* through the pin holes to the large pin. Repeat the procedure to make two more sight lines.

Discussion 1

1. Where do the sight lines that you drew cross?
2. How can you locate an object's position using sight lines?

Procedure 2

1. Draw a straight line about one-quarter of the distance from one end of a sheet of white paper. Place the paper on the cardboard. Use clay to set a small mirror upright on the line, facing the large portion of the paper.
2. Stick the large pin into the paper a few centimeters in front of the mirror. Use two small pins to make a sight line to the large pin's IMAGE. Repeat the process to obtain two more sight lines.
3. Remove the mirror. Extend the sight lines until they meet. Then measure the distance from this point to the mirror line. Measure the distance from the large pin to the mirror line, also.

Discussion 2

How is the distance an image *appears* to be behind a plane mirror related to the distance from the object to the mirror?

APPENDICES

Contents

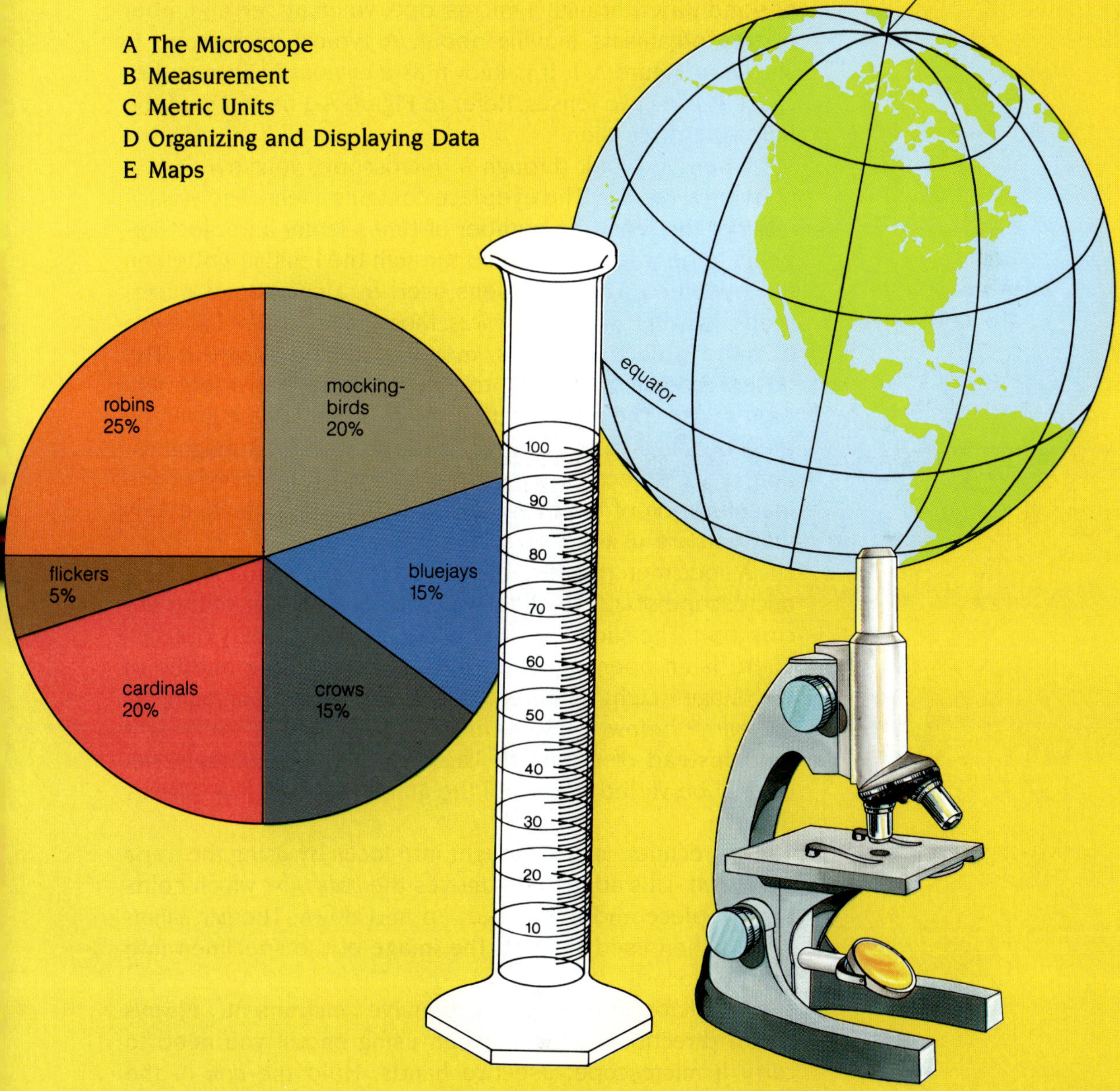

A The Microscope

With the use of a microscope, you can see things that you could not see otherwise. For example, if you look at a drop of pond water through a microscope, you may see a number of tiny organisms moving about. A typical microscope is shown in Figure A-1. It is known as a *compound microscope* because it uses two lenses. Refer to Figure A-1 as you read the following discussion.

When you look through a microscope, your eye is just above the *eyepiece*. The eyepiece contains a lens. The *magnification* of the lens (the number of times larger an object appears to be when it is viewed through the lens) is written on the eyepiece. The other lens used in a compound microscope is called the *objective lens*. Most microscopes have two or more objective lenses, mounted on the *nosepiece*. The nosepiece can be moved to select the objective lens you want to use. The magnifications of these lenses are given on their housings, also. You can compute the total magnification of an object that you are viewing by multiplying the magnification of the eyepiece by that of the objective lens that you are using.

A specimen that is to be viewed is usually mounted on a microscope *slide*. The slide is placed on the *stage* of the microscope. The slide is held in place on the stage by the *clips*. There is an opening called the *diaphragm* in the middle of the stage. Light is directed through the diaphragm by the *mirror* below it. (Some microscopes have a small light bulb instead of a mirror.) The size of the diaphragm can usually be varied to control the amount of light that passes through.

A specimen is first brought into focus by using the *coarse adjustment*. This adjustment moves the *body tube*, which holds the eyepiece and nosepiece, up and down. The *fine adjustment* is then used to bring the image of the specimen into sharper focus.

A microscope is an expensive instrument. Always follow directions carefully when using one. If you need to carry a microscope, use two hands. Hold the *arm* of the microscope, the part that holds the body tube and the adjustment knobs, in one hand. Use your other hand to support the *base*. Always carry a microscope in an upright position.

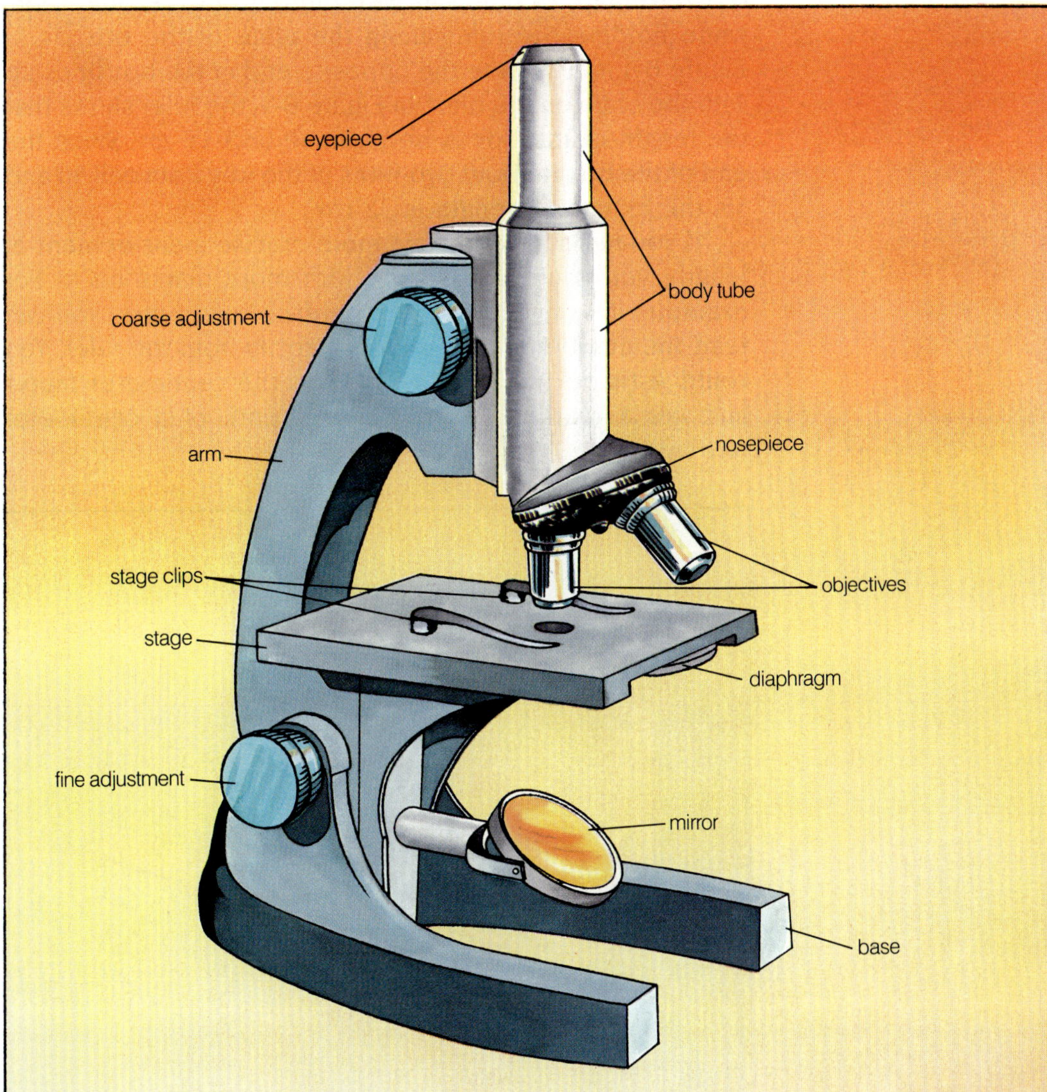

Figure A-1

B Measurement

Length

Length is commonly measured with a ruler. Suppose you want to measure the length of a cereal box to find out whether it can be stored upright in a cupboard. First, line up the zero mark of the ruler with one end of the box. Then, while holding the ruler along the edge of the box, as in Figure B-1, note where the opposite end of the box lines up with the scale on the ruler. In Figure B-1, the ruler shows that the length of the box is between 30 and 35 cm. Since the end of the box is about two-fifths of the way between 30 and 35, the estimated length is 32 cm.

If you needed to make a more precise measurement of length, you could use a ruler with more marks on it. For example, if a ruler has a mark for each centimeter, you can read the number of centimeters directly from the ruler. You could estimate the length between the centimeter marks and get a reading to the nearest tenth of a centimeter (0.1 cm).

Figure B-1

Volume

You can find the volume of a box if you know the length of each of its sides. Suppose you found the lengths of the sides of a cereal box to be 32 cm, 24 cm, and 8 cm. The volume of the box is the product of the lengths of the sides: 32 cm $\times$ 24 cm $\times$ 8 cm = 6144 cm^3, or 6144 cubic centimeters. The volume of any regularly shaped container, such as a box, a cylinder, or a sphere, can be found in a similar way.

What is the volume of a milk jug? It would be very difficult to find out by making measurements of its dimensions. The best way to find its volume is to fill it full of liquid, then pour the liquid into a special container like the one shown in Figure B-2. This container is called a *graduated cylinder*. The scale marked on the graduated cylinder enables you to read the volume of the liquid in the cylinder.

The scale on a graduated cylinder is read in essentially the same way as is the scale on a ruler, but there is an important difference. A liquid in such a container does not form a flat surface. Instead, it is curved, as shown in Figure B-3. The scale is read at the bottom of the curved portion, called the *meniscus*. When you read a volume, your eye should be at the same level as the bottom of the meniscus. Otherwise, the reading will be inaccurate.

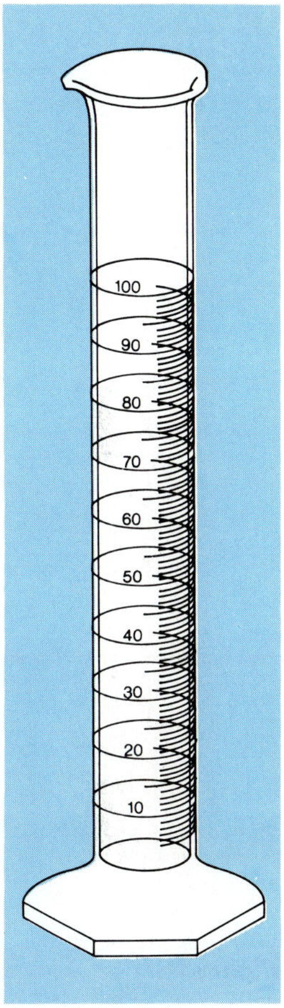

Figure B-2

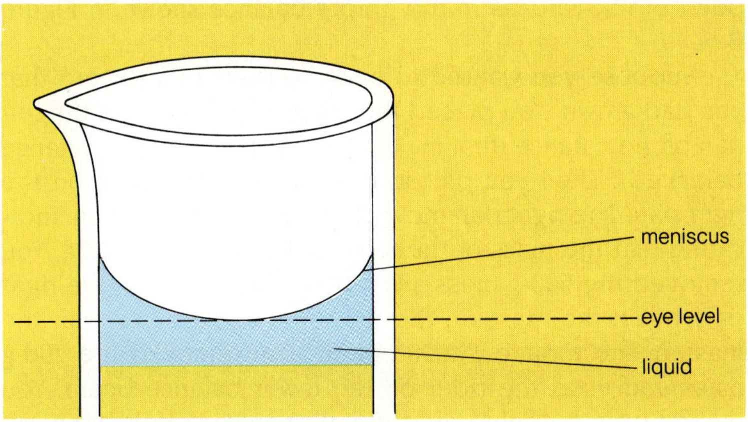

Figure B-3

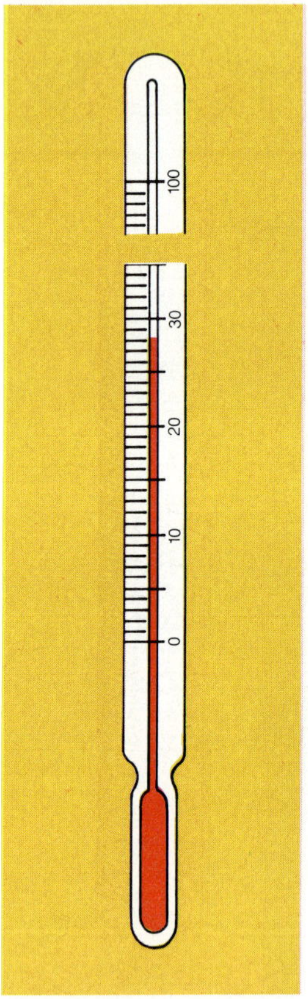

Figure B-4

Temperature

Temperature is usually measured with a thermometer. Common thermometers contain a liquid in a very narrow tube. The liquid expands or contracts as the temperature of its surroundings increases or decreases.

Thermometer scales are usually marked off in units called degrees, such as degrees Celsius (°C). The scales are read in essentially the same way as are those on rulers. Your eye should be level with the top of the liquid column when you read the scale. The temperature shown by the thermometer in Figure B-4 is 28°C.

Mass

The mass of an object is commonly measured on a *balance*, such as the one shown in Figure B-5. The object of unknown mass is placed on the left pan of the balance—the pan on your left as you face the balance. Known masses are added to the right side of the balance until the known and unknown masses just balance. Known mass may be added by placing *standard masses*, objects of known mass, on the right pan or by moving to the right the *riders* on the beams just below the balance pans. When the known and unknown masses balance, the pointer will come to rest at the midpoint of its scale, as in the empty balance shown in Figure B-5.

Suppose you wanted to find the mass of a tomato that you had grown. You placed the clean, dry tomato on the left pan of the balance (first making sure that the empty balance balanced). Then you placed a 500-g standard mass on the right pan. The right pan moved lower than the left pan, indicating that the mass of the tomato was less than 500 g. You removed the 500-g mass and put a 200-g mass on the right pan. Again the mass on the right pan was greater than the mass of the tomato. At this point, you removed the 200-g mass and used the rider on the lower balance beam. You moved it to the right until the right pan sank below the left one, then moved the rider back a notch. Then you moved the rider on the upper beam to the right until the two pans were in balance. The reading on the lower beam was 110 g and the reading on the upper beam was 3.7 g, as shown in Figure B-6. The mass of your tomato was 113.7 g.

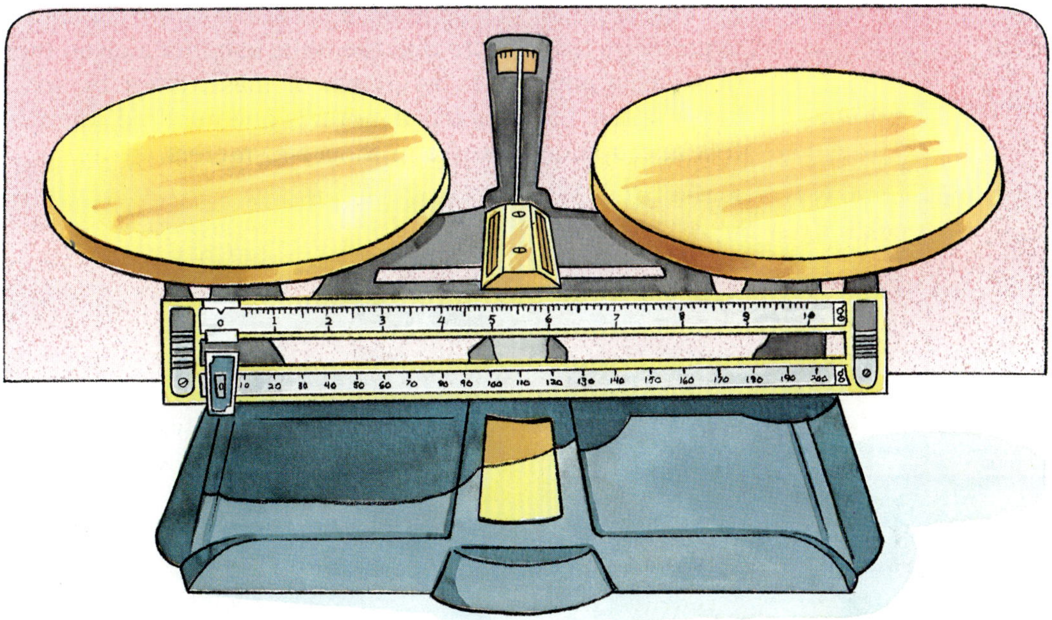

Figure B-5

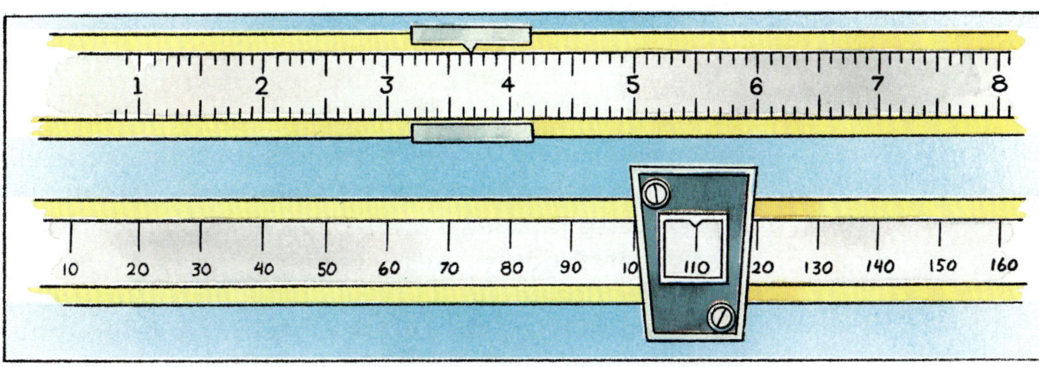

Figure B-6

C Metric Units

The units used in scientific measurements are normally those of the metric system. Scientists in all parts of the world have long used metric units. In recent years, the metric system has been updated. The updated metric system is called the International System of Units—SI, for short.

The table on this page includes some of the commonly used metric units. Note that the units used for a given type of measurement, such as length or mass, differ by some factor of ten. For example, a meter is one hundred centimeters and a kilometer is one thousand meters.

Some Commonly Used Metric Units

	Unit	*Symbol*	*Equivalent*
Length	centimeter	cm	0.01 m
	meter	m	
	kilometer	km	1000 m
Volume	cubic centimeter	cm^3	1 mL
	milliliter	mL	0.001 L
	liter	L	
Mass	gram	g	0.001 kg
	kilogram	kg	
Temperature	degree Celsius	°C	

The United States is the only large country that does not use metric units in everyday life. In terms of units used in the United States, the meter is 3.3 feet and the liter is 1.056 liquid quarts. The weight of a 1-kilogram mass at sea level is 2.2 pounds. A Celsius degree is nearly twice as large as a Fahrenheit degree. On the Celsius scale, pure water freezes at 0°C and boils at 100°C. On the Fahrenheit scale, the corresponding temperatures are 32°F and 212°F. Room temperature is about 21°C, or 70°F.

D Organizing and Displaying Data

People use graphs when they want to compare two sets of numbers. The numbers can be observations or measurements made during an experiment. The graph helps anyone looking at it to understand exactly what was happening during the experiment.

Suppose you wanted to find out how the mass of water changes when its volume is increased. In order to do so, you poured water into a container, 10 mL at a time, finding the mass of the container and its contents after each addition. Before you did this, you made a table in which to record your data. The table with the data that you obtained was as follows:

Volume of water in container (mL)	***Mass of container plus contents (g)***	***Mass of water in container (g)***
0	150	0
10	160	10
20	170	20
30	180	30
40	190	40
50	200	50
60	210	60
70	220	70
80	230	80
90	240	90
100	250	100

The results show that the amount of mass added is equal to the volume of water added. For example, 10 mL of water has a mass of 10 g.

Suppose you decided to display your results in the form of a graph. In order to do this, you first drew the *x*- and *y*-axes for the first quadrant of a coordinate system. (The other quadrants involve negative values, which you did not obtain.) You made each axis just long enough to be able to include all your data.

You chose to plot volume along the vertical (*y*) axis and mass along the horizontal (*x*) axis. You labeled each axis, so that a person looking at the graph can tell what was being measured and the unit in which it was measured. Then you marked each axis with numbers to show the scale.

When you plotted your data, you obtained a graph like the one shown in Figure D-1. The best line through the plotted points, which are circled to make them show up better, is a straight line that passes through the origin.

The graph in Figure D-1 is a line graph. Other kinds of data might be more suitably displayed using a different type of graph. Here is an example: On a certain day, a bird-watcher recorded the following sightings: 3 bluejays, 4 cardinals, 3 crows, 1 flicker, 4 mockingbirds, and 5 robins. These data are displayed in the form of a bar graph in Figure D-2. Another common way of displaying data is to use a circle, or pie, graph like the one shown in Figure D-3. Each "slice" of the pie shows a percentage or fraction of a total.

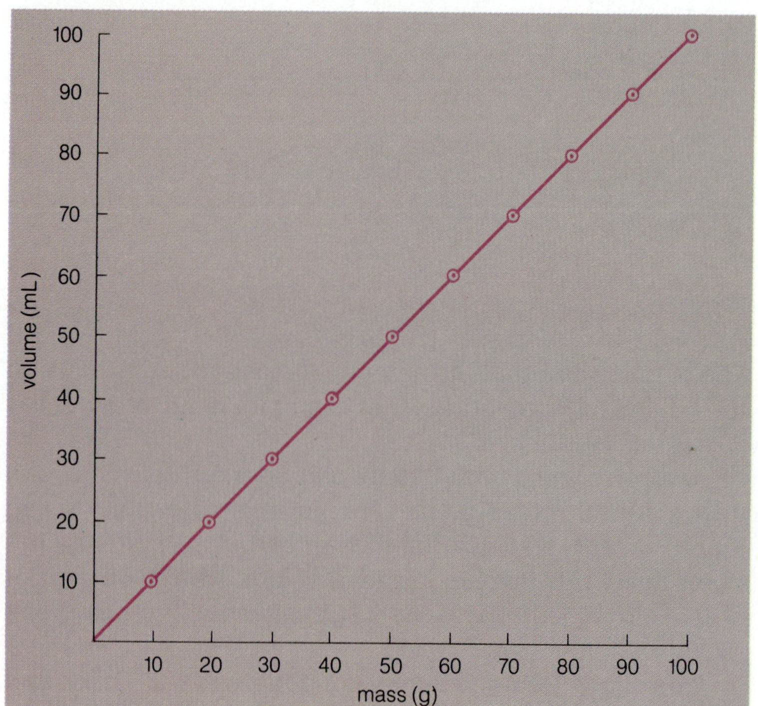

Figure D-1

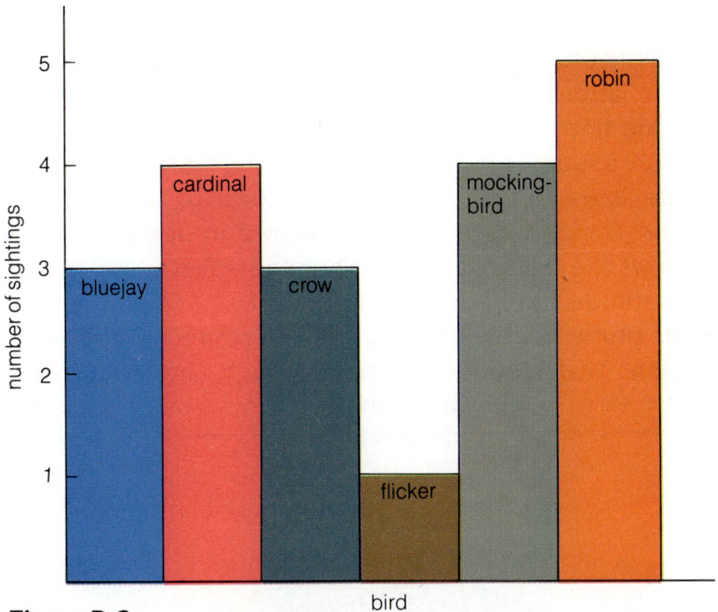

Figure D-2

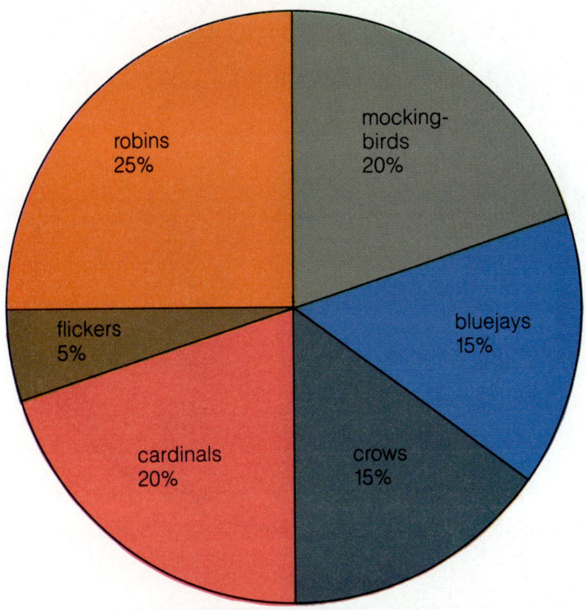

Figure D-3

E Maps

You can locate any point on Earth if you know its *latitude* and *longitude*. Latitude and longitude lines are shown in Figure E-1. The latitude lines run east-west around the globe. The longitude lines run north-south, from pole to pole.

The equator is halfway between the poles. The other latitude lines represent distances north or south of the equator. These distances are measured in degrees. Figure E-2 shows the relationship between the equator and the 60° north latitude line.

Longitude lines are measured with respect to the Prime Meridian. This meridian passes through Greenwich, Eng-

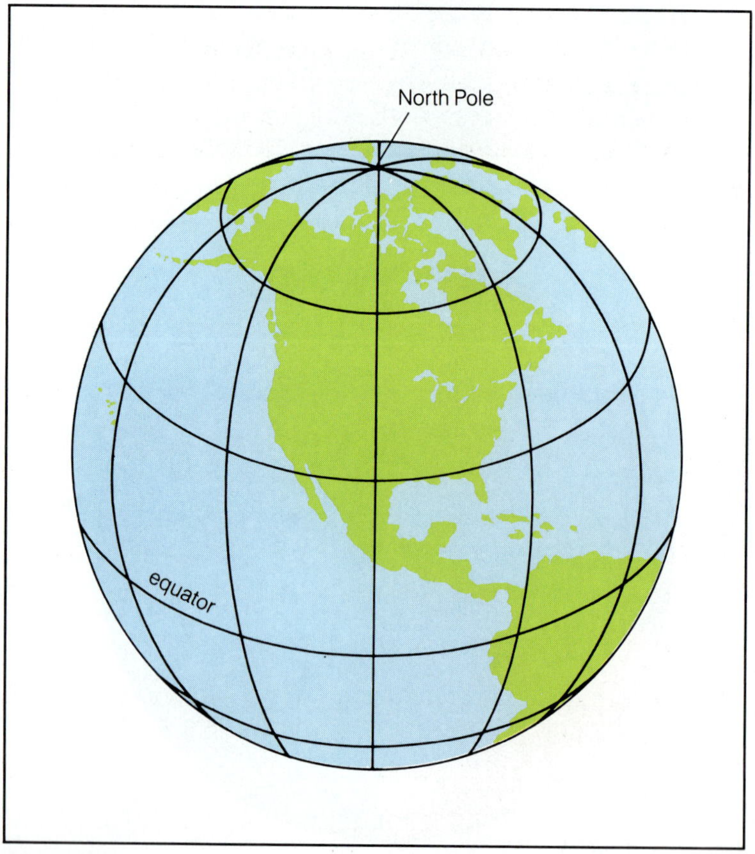

Figure E-1

land—an arbitrary choice. Other meridians, or longitude lines, are drawn to the east and west of the Prime Meridian. Figure E-3 shows the relationship between the Prime Meridian and 60° east longitude. As before, the angle is measured from the center of Earth. The meridian on the side of Earth opposite the Prime Meridian is 180° east *and* west.

Maps of Earth, including lines of latitude and longitude, can be transferred, or projected, onto two-dimensional surfaces, such as a flat piece of paper. Such maps can accurately represent locations and distances on Earth's surface.

Locations may be indicated by the intersection of latitude and longitude lines. For example, New Orleans, Louisiana is located where the 30°N latitude line and the 90°W longitude line cross. Its location can be stated as 30°N,90°W. Similarly, Columbus, Ohio is at 40°N,83°W. Can you find Perth, Australia on a map? Its location is 32°S,116°E.

Most maps include a scale for estimating distances between points. For example, a scale might indicate that a distance of 1 cm on the map represents an actual distance of 100 km.

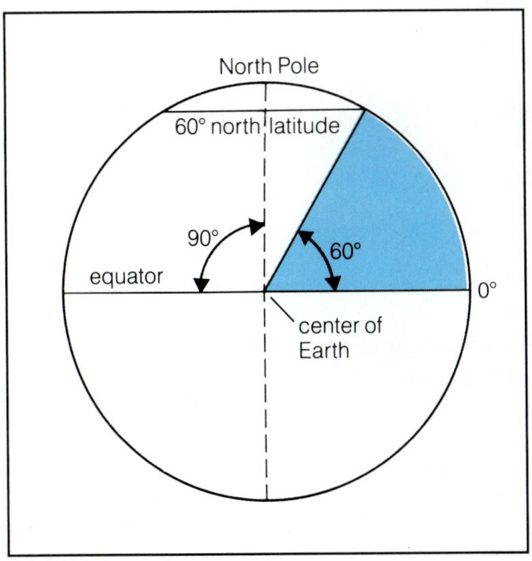

Figure E-2

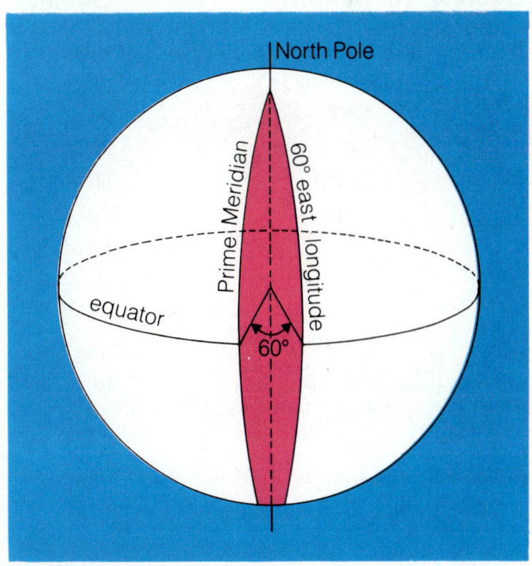

Figure E-3

Pronunciation Guide

Some words in this book may be unfamiliar to you or hard to pronounce. These words are followed by a pronunciation guide in parentheses. The pronunciation guides appear either in the margins or in the text. The respelled words are divided into syllables. The guide syllable printed in small capital letters should be given the greatest emphasis when you say the word. The following list contains some of the symbols used in the pronunciation guides in this book, along with examples of words that contain the sounds represented by those symbols.

You can find an example of a pronunciation guide for the word "mitosis" in the margin of page 44: (my TOH sis). When you say the word, the sound "toe" (TOH) should be given the greatest emphasis. You may wish to sound out the various pronunciation symbols listed below to make sure you understand them.

Symbol	Sound
a	pat
ah	father, pot
air	care
ar	car
aw	paw
ay	pay
e	pet
ee	bee
eer	ear
eh	essay
ehr	errand
eye	iron
i	pit
ih	divide
ihr	irritate
oh	toe
oo	boot
or	for
ow	out
oy	noise
th	thin
th (underlined)	this
u	put
uh	cut, about, section
ur	urge, butter
y	pie
zh	vision

Glossary

When an important science word is introduced in a unit, the word is printed in **boldface** type and defined. This glossary lists, in alphabetical order, all these important science words. Words used to define them that are also in the glossary appear in SMALL CAPITAL LETTERS. The chapter and section number where the word is found in the book appear in parentheses.

absolute zero: temperature (−273°C) at which the particles in a substance have their lowest possible energy. (18-1)
abyssal plain: flat floor of the OCEAN covering 60% of the ocean bottom. (14-6)
acceleration: change in VELOCITY divided by the period of time it takes for the change to happen. (16-1)
acid: COMPOUND that produces hydrogen IONS in SOLUTIONS. (9-5)
active metals: METALS that have a greater tendency than hydrogen to release ELECTRONS. (9-4)
adaptation: any TRAIT that enables an ORGANISM to live in a particular ENVIRONMENT. (5-1)
air mass: huge parcel of air that has the characteristics of the area over which it originated. (13-6)
alkali metals: group of ELEMENTS with ATOMIC NUMBERS always one more than those of the NOBLE GASES. (8-4)
alpha radiation: helium NUCLEI given off by RADIOACTIVE NUCLEI. (10-1)
alpine tundra: mountain BIOME that resembles Arctic TUNDRA but does not have PERMAFROST or a long period of darkness. (5-6)
alveoli: thin-walled air sacs in the lungs surrounded by tiny CAPILLARIES. (3-3)
amperes: units of ELECTRIC CURRENT. (19-4)
Animal Kingdom: multi-celled ORGANISMS that get their food by eating other organisms. (1-7)
aquifers: layers of rock in which large VOLUMES of ground water are stored. (11-5)
arteries: blood vessels that carry BLOOD away from the heart. (3-2)
asexual reproduction: type of REPRODUCTION in which new ORGANISMS are produced by a single parent. (2-3)
asteroids: minor planets. (15-3)
astronomy: study of everything in the UNIVERSE that can be observed from Earth or space. (15-1)
atmosphere: blanket of gases, about 900 km thick, that surrounds Earth. (13-1)
atmospheric pressure: FORCE caused by the ATMOSPHERE pressing down on Earth. (13-1)
atom: smallest particle of an individual ELEMENT that has all the PROPERTIES of that element. (7-1)
atomic mass: number of PROTONS plus the number of NEUTRONS in an ATOM. (7-4)
atomic number: number of PROTONS in the NUCLEUS of an ATOM. (7-4)
balanced equation: CHEMICAL EQUATION in which the REACTANTS and PRODUCTS contain the same number of each type of ATOM. (7-3)
barometer: device that measures ATMOSPHERIC PRESSURE. (13-1)
base: COMPOUND that produces hydroxide IONS in water. (9-5)
battery: two or more ELECTROCHEMICAL CELLS connected in SERIES, with the negative pole of one connected to the positive pole of the next. (19-5)
beta radiation: ELECTRONS moving near the speed of light, given off by RADIOACTIVE NUCLEI. (10-1)
binding energy: ENERGY given off when a NUCLEUS is formed from PROTONS and NEUTRONS. (10-5)
biomes: large areas of land, scattered over the globe, that contain similar COMMUNITIES. (5-2)

black hole: star completely collapsed into itself with a gravitational pull so great that light cannot escape from it. (15-5)
blood: LIQUID that carries substances to and from body CELLS. (3-2)
boiling point: TEMPERATURE at which a LIQUID changes to a GAS. (6-2)
capillaries: narrow, thin-walled blood vessels that connect the ARTERIES and VEINS. (3-2)
catalyst: substance that changes the rate of a CHEMICAL REACTION without being used up itself. (9-6)
cell: smallest unit able to carry out life activities of an ORGANISM. (1-2)
cell membrane: thin layer that surrounds the parts of a CELL. (1-2)
cell wall: structure, providing protection and support, found only in plant CELLS. (1-2)
centripetal force: inward-directed FORCE that keeps an object moving in a circle. (16-8)
cerebellum: part of the brain just below the CEREBRUM. (3-5)
cerebrum: largest part of the brain. (3-5)
chain reaction: series of reactions that occur when the PRODUCTS of one reaction start new reactions. (10-6)
chemical changes: changes in MATTER in which new substances, with different PROPERTIES, are formed. (6-5)
chemical equation: set of symbols representing the changes that take place in a CHEMICAL REACTION. (7-3)
chemical formula: series of CHEMICAL SYMBOLS and numbers representing the number and type of ATOMS in a MOLECULE. (7-2)
chemical reaction: process in which CHEMICAL CHANGES take place. (7-3)
chemical symbol: letter or pair of letters standing for the name of an ELEMENT. (7-2)
chlorophyll: green substance that enables plants to produce food. (1-2)
chloroplasts: CHLOROPHYLL-containing structures in plants. (1-2)
chromosomes: long, thin fibers containing instructions that control the activities of a CELL. (1-2)
cilia: tiny beating hairlike structures that propel paramecia. (1-3)
cirrus clouds: high, thin, wispy clouds. (13-4)
classification: system for separating objects into groups by similarities. (1-6)
climate: average weather of an area over a long period. (5-2, 13-7)
climax community: COMMUNITY that does not change except through natural disasters, sudden large CLIMATE changes, or human intervention. (4-5)
cold front: front formed when a cold AIR MASS moves into a warm air mass. (13-6)
comets: small objects near the edge of the SOLAR SYSTEM made up of fine dust and frozen water. (15-3)
community: interacting ORGANISMS in an ENVIRONMENT. (4-1)
complex machine: device made up of two or more SIMPLE MACHINES. (17-3)
compounds: pure substances that are combinations of two or more ELEMENTS and have different PROPERTIES from these elements. (6-5)
concave lens: LENS that is thicker at the edges than in the middle. (20-3)
concave mirror: mirror with a bowl-shaped surface that focuses or brings together parallel light rays. (20-2)
condensation: change in state from GAS to LIQUID. (6-2)
condensation nuclei: dust or salt particles on which water in the ATMOSPHERE condenses. (13-4)
conduction: transfer of HEAT from particle to particle. (18-4)
conductor: material through which HEAT passes quickly; material along which ELECTRONS move easily. (18-4, 19-3)
cone: pile of material that collects around the opening of a VOLCANO. (12-7)
constellations: patterns of stars visible only at certain times of the year. (15-1)
consumers: ORGANISMS that do not produce their own food. (4-2)
continental rise: gently sloping collection of sediments from land at the bottom of the CONTINENTAL SLOPE. (14-6)
continental shelf: extension of a continent that is covered by water. (14-6)
continental slope: steeply sloped area of the ocean floor beyond the CONTINENTAL SHELF. (14-6)
control group: group not subjected to the EXPERIMENTAL FACTOR in a CONTROLLED EXPERIMENT. (P-2)
controlled experiment: EXPERIMENT, consisting of an EXPERIMENTAL GROUP and a CONTROL GROUP, that measures the effect of only the factor being tested. (P-2)
convection: transfer of HEAT through movement of GASES or LIQUIDS due to differences of DENSITY. (18-4)
convex lens: LENS that is thicker in the middle than at the edges. (20-3)
convex mirror: dome-shaped mirror curving outward like part of the surface of a ball. (20-2)
Coriolis effect: change in the path of air currents caused by the ROTATION of Earth. (13-3)
cosmic rays: extremely high-ENERGY particles that shower Earth from outer space. (10-3)
covalent bonds: chemical bonds formed by sharing valence ELECTRONS. (8-6)
crater: opening at the top of a VOLCANO. (12-7)
crest: highest point in a wave. (14-2)
critical mass: smallest mass of fissionable

material that will support a CHAIN REACTION. (10-6)

crust: Earth's outermost SOLID layer. (11-1)

crystal: regular repeated arrangement of the particles of which it is made; regularly simple solid. (8-1, 11-1)

cumulus clouds: billowy clouds that resemble heaps of cotton. (13-4)

cyclones: winds circulating around a low pressure center. (13-5)

cytoplasm: matter that makes up most of the body of a CELL. (1-2)

data: measurements, changes observed, or other bits of information acquired during an EXPERIMENT. (P-2)

deciduous: shedding leaves annually in response to shortening periods of daylight. (5-3)

decomposers: ORGANISMS that break down dead organisms and animal waste and use them as food. (4-2)

deltas: DEPOSITS formed at the mouths of rivers. (12-4)

density: ratio of the MASS of a sample of MATTER to its VOLUME. (6-1)

deposits: materials that have been moved through EROSION and left in another place. (12-4)

desalination: removal of SALT from seawater. (14-7)

dew point: TEMPERATURE at which water vapor in air CONDENSES. (13-4)

diaphragm: strong, flat muscle, located at the base of the ribs below the lungs. (3-3)

diffusion: movement of a substance from an area where there is more of it to an area where there is less. (1-2)

digestion: breaking down of food into smaller and simpler substances. (3-4)

DNA: deoxyribonucleic acid, substance in CHROMOSOMES that contains information that controls all CELL activities. (2-7)

dominant gene: GENE that is always expressed when present in an ORGANISM. (2-6)

earthquake: sudden shaking of Earth's CRUST at a particular place. (12-6)

ecosystem: COMMUNITY and the nonliving parts of its ENVIRONMENT. (4-1)

efficiency: ratio of WORK output to work input. (17-4)

electric circuit: path from a region that has an excess of ELECTRONS to one where there are too few. (19-4)

electric current: stream of flowing ELECTRONS. (19-4)

electrochemical cell: device that produces electricity through OXIDATION-REDUCTION reactions. (9-4)

electrode: device by which ELECTRONS can flow into or out of an ELECTROCHEMICAL CELL. (9-4)

electromagnet: wire coil wound around an iron core which behaves like a MAGNET when electricity flows through the coil. (19-7)

electromagnetic spectrum: broad band of electromagnetic RADIATION. (20-5)

electron: NEGATIVELY CHARGED particle present in all MATTER. (7-4)

element: substance that cannot be separated into simpler substances during PHYSICAL and CHEMICAL CHANGES. (6-5)

endangered species: species that are close to being EXTINCT. (4-7)

endocrine glands: structures that produce HORMONES. (3-6)

endothermic: gaining ENERGY, describing a CHEMICAL REACTION that requires energy in order to take place. (9-2)

energy: ability or capacity to do WORK. (17-5)

energy levels: ENERGIES that an ELECTRON can have in an ATOM. (7-5)

environment: an ORGANISM'S surroundings, consisting of both living and nonliving things. (4-1)

epicenter: point on Earth's surface directly above the FOCUS of an EARTHQUAKE. (12-6)

erosion: process by which Earth materials are moved from one place to another, usually ending at lower levels of elevation. (12-3)

esophagus: long tube that connects the throat and the stomach. (3-4)

evaporation: change in state from LIQUID to GAS taking place below the BOILING POINT. (6-2)

evergreen forest: forest made up mostly of trees that stay green all year long. (5-3)

exothermic: releasing ENERGY, describing a CHEMICAL REACTION in which energy is lost. (9-2)

experiment: way of testing that makes use of measurements or OBSERVATIONS in the natural world. (P-2)

experimental factor: factor being tested in a CONTROLLED EXPERIMENT. (P-2)

experimental group: group subjected to the EXPERIMENTAL FACTOR in a CONTROLLED EXPERIMENT. (P-2)

extinct: no longer existing. (4-7)

fault: zone of weakness in Earth's CRUST along which movement of rock takes place. (12-6)

fertilization: process by which a sperm CELL and an egg cell fuse together. (2-4)

focal length: distance between a mirror or LENS and its PRINCIPAL FOCUS. (20-2)

focus: area in Earth's CRUST where ENERGY is released in an EARTHQUAKE. (21-6)

food chain: set of eating interactions within an ECOSYSTEM. (4-2)

food pyramid: illustration of the way ENERGY is passed along in an ECOSYSTEM. (4-3)

food web: FOOD CHAINS that overlap. (4-2)

force: push or pull. (16-2)

force of gravity: FORCE of attraction between objects. (16-5)

fossils: evidence of formerly living things. (11-3)

frequency: number of WAVELENGTHS or waves that pass a given point in one second. (20-4)
friction: FORCE that acts against the motion of one surface over another. (16-2)
fulcrum: point about which a lever turns or pivots. (17-2)
Fungi Kingdom: plant-like ORGANISMS, such as mushrooms and molds, that do not have CHLOROPHYLL and so must obtain food from other organisms. (1-7)
galaxies: groups of billions of stars. (15-4)
gametes: sex CELLS. (2-4)
gamma radiation: rays similar to x rays but with greater ENERGY and penetrating power, given off by RADIOACTIVE NUCLEI. (10-1)
gas: one of the STATES OF MATTER; a substance without definite shape and VOLUME. (6-2)
gene: segment of a CHROMOSOME. (2-6)
geologic map: map showing the various kinds of rock found at the surface or under the soil in a particular area. (11-6)
geothermal energy: ENERGY stored inside Earth. (18-7)
glacier: large, slowly moving mass of ice. (12-3)
greenhouse effect: warming of the ATMOSPHERE due to absorption by atmospheric carbon dioxide of long-wave ENERGY radiated from Earth. (13-2)
grounding: process by which a charge flows into Earth. (19-3)
group: vertical column in the PERIODIC TABLE. (8-5)
habitat: place where an ORGANISM lives in an ECOSYSTEM. (4-1)
half-life: time needed for half the ATOMS of a radioactive ISOTOPE to decay. (10-3)
halide: compound of a HALOGEN and another ELEMENT. (8-3)
halogens: group of ELEMENTS whose ATOMS have one less ELECTRON in their outermost shells than those of the NOBLE GASES. (8-3)
heat: ENERGY transferred to an object when it warms or from an object when it cools. (18-1)
heat engine: MACHINE that changes HEAT into mechanical ENERGY. (18-5)
heat of condensation: ENERGY given up when a GAS condenses to a LIQUID at its BOILING POINT. (9-1)
heat of vaporization: ENERGY required to change one gram of a LIQUID to a GAS at its BOILING POINT. (9-1)
hibernation: sleeplike state in which an animal's body TEMPERATURE, heartbeat and breathing rate decrease. (5-3)
hormones: chemicals released directly into the BLOOD by the ENDOCRINE GLANDS and responsible for changes in specific areas of the body. (3-6)
hybrid trait: TRAIT expressed when one of the GENES controlling it is DOMINANT and the other is RECESSIVE. (2-6)
hydrocarbon: COMPOUND that contains only hydrogen and carbon. (8-7)
hydropower: use of falling water to do WORK. (18-7)
hypothesis: tentative explanation for a scientific problem. (P-2)
igneous rocks: rocks formed from extremely hot melted rock material deep within Earth. (11-2)
impulse: "message" that travels through a NEURON or nerve CELL. (3-5)
indicator: COMPOUND that changes color in the presence of certain substances. (9-5)
induction: charging of an object by its placement near a charged object. (19-3)
inertia: resistance to change in motion. (16-2)
infrared radiation: invisible RADIATION contained in sunlight, with somewhat lower FREQUENCIES than those of red light. (20-5)
inland seas: seas almost completely surrounded by land. (14-1)
inner core: Earth's innermost layer, believed to be SOLID and composed mostly of nickel and iron. (12-5)
inorganic sediments: sediments on the ocean floor that come from nonliving sources. (14-5)
instincts: behaviors an ORGANISM is born with. (5-1)
insulator: material through which HEAT passes very slowly; material that keeps ELECTRONS from moving. (18-4, 19-3)
intensity: strength of an EARTHQUAKE at a particular place. (12-6)
involuntary muscle: muscles that are not under conscious control. (3-1)
ion: ATOM or group of atoms that has gained or lost one or more ELECTRONS. (8-3)
ionic bond: bond between oppositely charged IONS. (8-6)
ionization energy: ENERGY required to remove an ELECTRON from an ATOM. (8-5)
ionosphere: layer of ionized (charged) GASES in the upper ATMOSPHERE. (13-1)
isobars: lines on a weather map connecting areas of equal ATMOSPHERIC PRESSURE. (13-6)
isotherms: lines on a weather map connecting areas of equal TEMPERATURE. (13-6)
isotopes: ATOMS of the same ELEMENT that have different ATOMIC MASSES. (10-2)
jet streams: upper level winds found in middle latitudes about 10 km above Earth's surface. (13-3)
joint: point at which two bones meet. (3-1)
joule: WORK done when a FORCE of one newton acts over a distance of one meter. (17-1)
kidneys: bean-shaped ORGANS that filter waste and water from the BLOOD. (3-4)
kilogram: standard unit of MASS equal to 1000 grams. (6-1)
kinetic energy: ENERGY of a moving object due to its motion. (17-5)

lava: MAGMA that comes to Earth's surface when a VOLCANO erupts. (11-2)
law of conservation of energy: law stating that ENERGY can be transformed but that the total amount of energy remains the same. (18-5)
law of conservation of mass: law stating that MASS is neither created nor destroyed in a CHEMICAL REACTION. (7-3)
law of definite proportions: law stating that the ELEMENTS in a COMPOUND are combined in definite percentages by MASS. (6-5)
leeward: facing away from the wind. (13-7)
lens: TRANSPARENT object with at least one curved surface. (20-3)
ligaments: bands of tough, stretchy TISSUE that hold bones together. (3-1)
light-year: distance that light travels in one year. (15-5)
limiting factor: any factor that keeps a POPULATION from growing indefinitely. (4-4)
liquid: one of the STATES OF MATTER; something whose shape changes but whose VOLUME does not. (6-2)
liter: unit of VOLUME equal to 100 cm^3. (6-1)
longitudinal waves: waves that travel in the same direction as the vibrations of the medium through which they move. (20-8)
lunar eclipse: blocking out of the moon when Earth passes directly between it and the sun, casting Earth's shadow on the moon. (15-4)
machine: device that changes the direction or strength of a FORCE. (17-2)
magma: extremely hot melted rock material deep within Earth. (11-2)
magnet: object that has attraction for anything containing iron. (19-6)
magnetic field: region in which a MAGNET can affect an object. (19-6)
magnitude: amount of ENERGY released in an EARTHQUAKE; observed brightness of a star. (12-6, 15-5)
mantle: thick, mostly SOLID layer between Earth's OUTER CORE and CRUST. (12-5)
mariculture: farming of the sea. (14-7)
mass: measure of the amount of material in an object; ratio of an applied FORCE to the resulting ACCELERATION of an object. (6-1, 16-3)
matter: anything that takes up space and has MASS. (6-1)
meanders: looping curves in a river. (12-4)
mechanical advantage: ratio of resistance FORCE to effort force in a MACHINE. (17-4)
medulla: part of the brain, located at its base, which controls involuntary activities. (3-5)
meiosis: type of CELL division resulting in cells with only half the usual number of CHROMOSOMES. (2-4)
melting point: TEMPERATURE at which a SOLID changes to a LIQUID. (6-2)
metals: one of two basic groups of ELEMENTS having PROPERTIES in common that are different from those of NONMETALS. (8-1)
metamorphic rocks: rocks formed by changes in SEDIMENTARY or IGNEOUS ROCKS caused by large amounts of HEAT and pressure. (11-2)
meteor: fiery streak produced when a METEOROID burns up on entering Earth's ATMOSPHERE. (15-3)
meteorite: METEOROID that does not burn up in the ATMOSPHERE as it hits Earth. (15-3)
meteoroid: small piece of rock or mineral matter traveling through space. (15-3)
mid-ocean ridges: underwater mountain chains rising above the OCEAN floor. (12-5, 14-6)
minerals: naturally occurring solid ELEMENTS and COMPOUNDS. (11-1)
mitosis: process by which CHROMOSOMES duplicate themselves, then separate into two identical sets. (2-2)
mixture: physical combination of two or more substances. (6-3)
molecule: particle of MATTER made up of two or more chemically joined ATOMS. (7-1)
Moneran Kingdom: smallest one-celled ORGANISMS. (1-7)
moraines: deposits of glacial drift dropped in front of a GLACIER or along its sides. (12-4)
mutation: change in a GENE or CHROMOSOME. (2-8)
Nansen bottle: device used to collect a water sample at a specific depth and record its temperature. (14-1)
neap tides: lowest high tides each month. (14-3)
nebulas: huge clouds of GAS and dust in which stars form. (15-5)
negative charge: electric charge on an object that results from an excess of ELECTRONS. (19-2)
neurons: nerve CELLS. (3-5)
neutralization: CHEMICAL REACTION in which an ACID and BASE are mixed together forming a SALT and water. (9-5)
neutron: uncharged particle in the NUCLEUS of an ATOM. (7-4)
neutron star: very small, very dense star within the NEBULA produced by the explosion of a SUPERNOVA. (15-5)
Newton's first law of motion: objects remain in motion or at rest unless acted on by outside FORCES. (16-2)
Newton's second law of motion: when a force is applied to an object, the object accelerates in the direction of an applied FORCE; the ACCELERATION is greater when the force is greater and less when the MASS of the object is greater. (16-3)
Newton's third law of motion: for every action there is an equal but opposite reaction. (16-4)

niche: way of living of a species within a HABITAT. (4-1)

noble gases: helium, neon, argon, krypton, xenon, and radon, "inert" GASES with low MELTING and BOILING POINTS. (8-2)

nodules: solid particles formed when chemicals in seawater collect in lumps. (14-5)

nonmetals: one of two basic groups of ELEMENTS having PROPERTIES in common that are different from those of METALS. (8-1)

nuclear equation: set of symbols representing the process of RADIOACTIVE DECAY. (10-3)

nuclear fission: splitting of a NUCLEUS into two smaller nuclei. (10-6)

nuclear fusion: nuclear reaction in which lighter NUCLEI combine to form heavier nuclei; source of the sun's ENERGY. (10-8, 15-4)

nuclear reactor: device in which a controlled nuclear reaction is carried out. (10-7)

nucleus: "control center" of a CELL; center of an ATOM containing all its POSITIVE CHARGE and most of its MASS. (1-2, 7-4)

observations: measurements, changes observed, or other bits of information acquired during an EXPERIMENT. (P-2)

occluded front: front formed when a warm AIR MASS is lifted upward as two cold air masses meet. (13-6)

ocean currents: river-like movements of water running through the OCEANS. (14-4)

oceans: large areas of water covering almost 75% of Earth's surface. (14-1)

ohms: units of electrical RESISTANCE. (19-4)

opaque: having the capacity to absorb light so that it cannot pass through a substance. (20-1)

organ: two or more different types of TISSUE functioning together. (1-4)

organic sediments: sediments on the ocean floor that come from the remains of ORGANISMS that lived in the sea. (14-5)

organism: any living thing. (1-1)

outer core: hot LIQUID surrounding Earth's INNER CORE. (12-5)

oxidation: loss of ELECTRONS by an ATOM or ION. (9-3)

ozone layer: that part of the STRATOSPHERE where absorption of ULTRAVIOLET rays occurs. (13-1)

P waves: primary waves, waves of ENERGY released by an EARTHQUAKE that cause particles in a substance to move back and forth in the direction of wave motion. (12-6)

parallel circuit: ELECTRIC CIRCUIT in which ELECTRONS have more than one path to follow. (19-5)

period: horizontal row in the PERIODIC TABLE. (8-5)

periodic table: arrangement of ELEMENTS in which those with similar PROPERTIES are grouped together. (8-5)

permafrost: permanently frozen SOIL below the surface in the TUNDRA BIOME. (5-2)

***p*H scale:** scale used to measure ACID or BASE strength of a SOLUTION. (9-5)

phases: apparent shapes of the moon as it completes one REVOLUTION around Earth. (15-2)

photosynthesis: process by which plants and certain other ORGANISMS make their own food. (1-5)

physical changes: changes in MATTER in which no new substances are formed. (6-5)

pioneer community: first COMMUNITY to develop in a bare area. (4-5)

plane mirror: ordinary flat mirror. (20-2)

Plant Kingdom: multi-celled ORGANISMS that make their own food by PHOTOSYNTHESIS and have CELL WALLS. (1-7)

plasma: clear LIQUID, making up about half of the BLOOD, and consisting mostly of water. (3-2)

plate tectonics: theory that Earth's CRUST and upper MANTLE are divided into large, moving segments called plates. (12-5)

polar molecule: MOLECULE in which the ELECTRONS are not distributed evenly. (8-6)

pollination: process by which a pollen grain is transported from the stamen to the pistil of a flower. (2-4)

pollution: accumulation of harmful substances in an ENVIRONMENT. (4-6)

population: number of ORGANISMS of the same kind living within a particular area at the same time. (4-4)

positive charge: electric charge on an object that results from too few ELECTRONS to balance the PROTONS in an ATOM. (19-2)

potential energy: stored ENERGY of an object due to its position. (17-5)

power: amount of WORK done in a certain length of time. (17-7)

predators: ORGANISMS that catch and eat other organisms. (4-4)

pressure: ratio of a FORCE to the area over which it acts. (18-2)

prey: ORGANISMS that are caught and eaten by PREDATORS. (4-4)

principal focus: point at which rays from a CONCAVE MIRROR or LENS meet. (20-2)

producers: ORGANISMS that manufacture food. (4-2)

property: any quality that can help in identifying something. (6-1)

Protist Kingdom: one-celled ORGANISMS that have nuclear membranes. (1-7)

proton: POSITIVELY CHARGED particle in the NUCLEUS of an ATOM. (7-4)

pure trait: TRAIT expressed when both GENES of a pair are identical; for example, the genes may be either both DOMINANT or both RECESSIVE. (2-6)

radiation: transfer of ENERGY that does not require MATTER to transmit it. (18-4)
radioactive decay: change of a radioactive ISOTOPE into a different ELEMENT by the giving off of an alpha or a beta particle. (10-3)
radioactive nuclei: NUCLEI that give off RADIATION. (10-1)
radio telescopes: telescopes that gather radio waves from space. (15-6)
real image: image focused on a screen by a LENS or a CONCAVE MIRROR. (20-2)
recessive gene: GENE expressed in an ORGANISM only when the DOMINANT form of the gene is absent. (2-6)
red giant: stage in the life of a star caused by the slowing of NUCLEAR FUSION in which the star grows cooler and much larger. (15-5)
reduction: gain of ELECTRONS by an ATOM or ION. (9-3)
reflection: bouncing of light off an object. (20-1)
reflex: automatic response to a STIMULUS, controlled by the spinal cord, not the brain. (3-5)
refraction: bending of the path of light as it passes from one TRANSPARENT material into another. (20-3)
relative humidity: amount of water vapor in the air compared to the maximum amount the air could hold at that TEMPERATURE. (13-6)
renewable energy source: ENERGY source that is produced about as fast as it is used. (18-6)
replicate: to copy, to make a duplicate, as CHROMOSOMES do of themselves just before MITOSIS. (2-2)
reproduction: process by which ORGANISMS produce new organisms. (2-3)
resistance: limitation on the amount of ELECTRIC CURRENT that can flow through a CONDUCTOR. (19-4)
respiration: the process by which ENERGY is made available from food. (1-2)
revolution: movement of Earth around the sun. (15-1)
rock cycle: never-ending process of change from one rock type to another. (11-2)
rotation: spinning or turning about a straight line or axis. (15-1)
S waves: secondary waves, waves of ENERGY released by an EARTHQUAKE that cause rock particles to move from side to side at right angles to the direction of wave motion. (12-6)
salinity: amount of salt in seawater. (14-1)
salt: ionic COMPOUND formed when an ACID and BASE react. (9-5)
satellite: object that revolves around a planet. (15-2)
saturated zone: that part of the ground that contains ground water. (11-5)
savannas: grasslands that contain scattered trees, found in South America and Africa. (5-5)
scientific law: precise statement of a relationship in nature that can be expressed mathematically. (P-3)
scientific method: planned, logical way to develop and test explanations for scientific problems. (P-1)
scientific model: explanation of something unfamiliar or invisible in terms of something familiar or visible. (7-1)
seas: smaller areas of OCEAN, sometimes partly surrounded by land. (14-1)
sedimentary rocks: rocks make of MINERAL grains that become cemented together under pressure. (11-2)
seismograph: recording device that detects waves of ENERGY released by an EARTHQUAKE. (12-6)
series circuit: ELECTRIC CIRCUIT in which there is only one path that ELECTRONS can follow from the power supply and back to it. (19-5)
sex chromosomes: pair of CHROMOSOMES that contain GENES determining the sex of the ORGANISM. (2-6)
sex-linked trait: TRAIT controlled by GENES found on the SEX CHROMOSOMES. (2-6)
sexual reproduction: creation of a new ORGANISM by joining one male sex CELL with one female sex cell. (2-4)
simple machine: any of six devices, for example, the lever, from which all other MACHINES are made. (17-3)
soil: mixture of WEATHERED rock material and decayed, formerly living material which provides nutrients for plants. (12-2)
solar eclipse: blocking out of the sun's light when the moon passes directly between Earth and the sun. (15-4)
solar energy: ENERGY from the sun. (18-6)
solar system: sun and all objects in orbit around it. (15-3)
solid: one of the STATES OF MATTER; something whose VOLUME and shape do not change. (6-2)
solubility: amount of a substance in a given amount of solvent at a given TEMPERATURE. (6-3)
solution: MIXTURE in which the smallest particles of two or more substances are evenly distributed. (6-3)
sonar: device that maps the ocean floor by recording the time it takes for sound waves to be reflected from it. (14-6)
specific heat: amount of HEAT needed to raise the TEMPERATURE of 1 g of a substance 1°C. (18-3)
spectroscope: instrument used to produce a spectrum, or a series of colored bands or lines, and measure the WAVELENGTHS of light in the spectrum. (15-6)

speed: distance traveled in a period of time. (16-1)
spring tides: highest high tides each month. (14-3)
states of matter: the three states in which a substance can exist: GAS, LIQUID, and SOLID. (6-2)
static electricity: charges that stay on an object. (19-3)
stimulus: any "message" that produces a response by the body. (3-5)
stratosphere: layer of the ATMOSPHERE above the TROPOSPHERE. (13-1)
stratus clouds: layered clouds that usually cover the whole sky. (13-4)
structural formula: diagram showing the arrangement of ATOMS in a MOLECULE. (8-7)
succession: process by which an ECOSYSTEM changes over time. (4-5)
sunspots: dark spots on the surface of the sun. (15-4)
supernova: star that explodes, producing a NEBULA. (15-5)
synapse: gap between the end of one NEURON, or nerve CELL, and the beginning of another. (3-5)
system: ORGANS functioning together, such as the digestive system, circulatory system, etc. (1-4)
temperature: measure of the average KINETIC ENERGY of the particles of a substance. (18-1)
tendons: bands of TISSUE that attach muscles to bones. (3-1)
theory: HYPOTHESIS supported by experimental results. (P-3)
tissue: group of CELLS that are alike in appearance and function. (1-4)
topographic map: map showing the shape of the land surface by a system of contour lines. (11-6)
tracers: radioactive ISOTOPES used to trace a course of events. (10-4)
trachea: windpipe, tube that carries air from the nose and mouth to the lungs. (3-3)
traits: specific features of an ORGANISM. (2-1)
translucent: having the capacity to scatter light so that things cannot be seen clearly through a substance. (20-1)
transmutation: process by which ATOMS of an ELEMENT gain or lose PROTONS, forming a new element. (10-1)
transparent: having the capacity to allow most light to pass through so that objects can be seen clearly through a substance; clear glass and plastic are transparent. (20-1)
troposphere: layer of the ATMOSPHERE closest to Earth. (13-1)
trough: lowest point in a wave. (14-2)
tundra: BIOME in the polar CLIMATE zone with low TEMPERATURES, little rainfall, and long periods with no sunlight. (5-2)
ultraviolet radiation: invisible RADIATION contained in sunlight, with somewhat higher FREQUENCIES than those of violet light. (20-5)
universe: everything that exists. (15-1)
urine: LIQUID produced by the KIDNEYS from waste products and water brought to them by the BLOOD. (3-4)
vascular plants: plants that have a system of transport tubes. (1-7)
veins: blood vessels that carry BLOOD toward the heart. (3-2)
velocity: SPEED of an object in a particular direction. (16-1)
villi: fingerlike projections lining the walls of the small intestine through which digested food MOLECULES diffuse into the BLOOD. (3-4)
virtual image: images seen in a mirror but not able to be seen on a screen. (20-2)
visible spectrum: that part of the ELECTROMAGNETIC SPECTRUM consisting of the band of colors produced when white light passes through a prism. (20-5)
visual telescopes: telescopes that gather visual light. (15-6)
volcano: opening in Earth's CRUST through which MAGMA flows out as LAVA. (12-7)
volts: units of ENERGY of the electrons in an ELECTRIC CIRCUIT. (19-4)
volume: space taken up by an object. (6-1)
voluntary muscles: muscles that function when you want them to. (3-1)
warm front: front formed when a warm AIR MASS moves into a cold air mass. (13-6)
water cycle: movement of water from the land to the rivers and OCEANS, to the ATMOSPHERE, and back to the land again. (11-4)
water table: irregular upper boundary of the SATURATED ZONE. (11-5)
watt: unit of POWER equal to one JOULE per second. (17-7)
wavelength: distance from the CREST of one wave to the CREST of the next. (20-4)
weathering: process by which rocks break down. (12-1)
weight: gravitational FORCE that Earth exerts on an object. (16-6)
white dwarf: stage in the life of a star following the RED GIANT stage. (15-5)
windward: facing the wind. (13-7)
work: FORCE acting on an object multiplied by the distance the object moves. (17-1)
zygote: fertilized egg, formed by the fusing of a sperm CELL and an egg cell. (2-4)

Index

A

B

C

D

Illustration Credits

Arthur Ciccone: 1-4, 1-12, 33-1; **Marsha Dohrman:** 3-6, 3-11, 3-14, 3-15, 3-16, 3-17, 3-18, 3-19, 3-20, 3-24, 3-26; **Frank Fretz:** 1-14, 2-12, 2-13, 4-6, 4-17, 11-12, 13-6, 13-7, 14-5, 14-16, 14-17, 14-18; **Bob Jackson:** 2-7, 11-13, 11-14, 16-3, 16-5, 16-13, 17-4, 17-6, 17-17, 18-8, 18-10; **Omnigraphics:** 1-3, 1-7, 1-13, 2-2, 2-3, 2-8, 2-14, 2-15, 2-16, 2-18, 2-19, 4-7, 4-9, 4-10, 4-11, 4-18, 5-7, 5-8, 6-2, 6-3, 6-6, 6-8, 6-12, 6-13, 7-2, 7-3, 7-6, 7-13, 7-15, 8-8, 8-9, 8-10, 8-11, 8-13, 8-15, 8-16, 8-17, 8-18, 8-20, Feature 8-1, Feature 8-2, 9-1, 9-2, 9-12, Feature 9-1, Feature 9-2, 12-3, 12-4, 12-9, 12-13, 12-14, 12-15, 12-18, 12-19, 12-20, 12-21, 12-22, 13-2, 13-3, 13-8, 13-9, 13-10, 13-14, 13-18, 13-19, 13-21, 13-22, 14-3, 15-3, 15-4, 15-8, 15-9, 15-10, 15-15, 15-17, 15-18, 16-1, 16-7, 16-11, 16-12, 16-14, 17-7, 17-9, 17-11, 17-12, 17-15, 17-16, 17-18, 17-19, 17-22, 17-23, 17-24, 18-2, 18-3, 18-21, 20-6, 20-7, 20-8, 20-9, 20-14, 20-15, 20-16, 20-18, 20-19, 20-21, 20-22, 20-27, 20-28, 20-29, 20-31, Feature 20-1, Appendix Figures B-2, B-3, B-4, D-1, D-2, D-3, E-1, E-2, E-3; **Roget Ink:** 4-15, 7-4, 7-5, 7-9, 7-10, 7-14, 10-2, 10-3, 10-5, 10-6, 10-11, 10-13, 11-9, 12-16, 13-4, 13-5, 13-17, 14-4, 14-7, 14-8, 14-10, 14-11, 14-15, 16-18, 18-5, 18-11, 18-15, 18-20, 19-4, 19-5, 19-8, 19-9, 19-10, 19-11, 19-12, 19-14, 19-16, 19-17, 19-18, Activity Section Figures 5, 6, 7, 9-1, 9-2, 9-3, 10, 13-1, 13-2, 14-1, 14-2, 16, 18, 19, 22-2, 25, Appendix Figure B-1; **James Teason:** 1-16, 1-21, 2-9, 2-10, 2-11, 4-1, 4-2, 4-3, 4-4, 4-12, 5-1, 5-2, 5-9, 5-10, 5-11, 5-12, 5-13, 5-14, 5-15, 5-16, 5-18, 5-19, 5-20, Activity Section Figure 8; **Gary Torrisi:** 6-18, 17-13, 17-20, Activity Figures 1-1, 1-2, 2, 3, 4, 9-4, 11, 12, 13-3, 13-4, 15-1, 15-2, 15-3, 17, 20, 21-1, 21-2, 22-1, 23-1, 23-2, 24, 26, 27, Appendix Figures A-1, B-5, B-6; **Lydia Vickers:** 1-2, 3-1, 3-2, 3-3, 3-4, 3-12, 3-13, 3-21, 3-22, 3-23, 3-25, 19-19, 19-20, 19-21; **Paul Weiner:** 15-20

Photo Credits

Animals, Animals: *Zig Leszczynski* 6-10, C.W. *Perkins* 1-20c, *Carl Roessler* 1-19a, *Leonard Lee Rue* III 1-20e, 2-21; **Peter Arnold, Inc.:** *Fred Bavendam* 1-19f, *Dieter Blum* 14-19, *Bonnie Geller* 14-1, *Phillip* A. *Harrington* 1-4a, W.H. *Hodge* 1-18h, *Manfred Kage* 1-1a, 1-9, 1-10c, 1-11a, 1-18a, Feature 1-1, Feature 1-2, 3-10, 7 Opener, *Yoram Kahama* Unit 3 Careers b, *Malcom* S. *Kirk* 10-7, *Stephen* J. *Krasemann* 12-6, *John MacGregory* 1-20d, C. *Allan Morgan* 4 Feature; **D. Balkwill, D. Maratea:** Feature 19-1; **Lester V. Bergman:** 3-8; **Bettmann Archives:** 1-5, 1-10; **Brown Brothers:** 7-1, 19-3, 19-6; **Dr. James J. Childress, University of California, Santa Barbara:** Feature 5-1, Feature 5-2; **Bruce Coleman:** *Don Brewster* 8-6, *Vic Beunza* 5 Opener, *Bob and Clara Calhoun* 5-4b, E.R. *Degginger* 3-5b, 3-5c, 13-15, *Phil Degginger* 3-5a, *Nicholas de Vore* III 14 Opener, *Robert Donne* 1-11b, 1-15a, 1-15b, 1-18d, *Jessica Ehlers* 14-9a, 14-9b, *Michael Fogden* 5-4a, *Barry Parker* 13-16, *Mark Sherman* 2-22, *Norman Owen Tomalin* 13-13; **David Conklin:** 8 Opener, 16-4, 16-16, 17-5, 19-2a, 19-2b, 19-24, 20-1, 20-30; **Paul Conklin:** 1-17a, 1-18b, 1-18e, 1-20f, 2-1, 2-23b, 3-7a, 3-7b, Feature 3-2, 4-14, 4-16, Unit 1 Careers a, b, c, Feature 6, 6-7a, 6-7b, 6-14a, 6-14c, 6-15a–f, 6-16a, 6-16c, 7-7a, 7-7b, 7-7c, 7-8, Feature 7-1, Feature 7-2, Feature 7-3, 8-1, 8-2, 8-3, 8-4, 8-7a, 8-7b, 8-7c, 8-14, 8-19, 9 Opener, 9-4a, 9-4b, 9-5, 9-6a, 9-6b, 9-7, 9-8a–d, 9-10a, 9-10b, 9-11a, 9-11b, 9-13, 9-14, 10-1a, 10-1b, 10-14, Feature 10, Unit 2 Careers a, b, c, d, 11 Opener, 11-1a, 11-1b, 11-1c, 11-2, 11-3, 11-4a, 11-4b, 11-5a–d, 11-6a, 11-6b, 11-6c, 11-7a, 11-7b, 11-7c, Feature 11, 11-16, 12-5, 12-7, 13 Opener, 13-2a, 13-11a, 13-20a, 13-20b, Feature 13, 15-12, Unit 3 Careers d, 16-9, 16-10, 17-2, 17-8a, 17-8b, 17-21, 17-25b, 18-4a, 18-4b, 18-16, 18-17, 19-1, 19-13, 19-22, 19-23, 20 Opener, 20-4, 20-13, 20-25, 20-30, Unit 4 Careers b, c, d; **Deep Ocean Engineering, Inc.:** Feature 14-2; **E.R. Degginger:** 1-11c, 1-19d, 2-6, 2-17, 2-23a, 4-5, 5-5, 5-6b, 6-4, 13-15, 16-2, 18-7; **Phil Degginger:** 4-13a, 4-13b, 12-23a, 12-23b; **Earth Scenes:** *Adrienne* T. *Gibson* 1-18g; **Courtesy, Dr. Edgerton/M.I.T.:** 16-18; **William E. Ferguson:** 4 Opener, 11-8; **Hanson Planetarium:** 15-13; **Courtesy, D.C. Heath and Co.:** 20-23, 20-24; **Grant Heilman:** *Barry* L. *Runk* 1-18f, 1-22b, 3-9, 6-14d, 11-11, 14-13, *Runk/Schoenberger* 1-1b, 1-6, 1-10a, 1-10b, 2-5, 3-27a, 3-27b; **Randall Hyman:** 12-24; **The Image Bank:** *Michael Melford* Unit 1 Careers d; **Index Stone:** Unit 3 Opener; **Lick Observatory, University of California, Santa Cruz:** 10-16, 15-7a–f, 15-11, 15-16; **Life Magazine:** 10-4, *Huntington Potter and David Dressler* Feature 2-2; **Los Alamos National Laboratory:** 10-12; **Michael Philip Manheim:** 19 Opener; **David Muench:** 6 Opener; **Josef Muench:** 11-15; **NASA:** 12-10, 15-1, Feature 15, 16-15, 16-21, Feature 16; **Lennart Nilsson:** 20-17; **Photo Researchers:** *Robert* C. *Herms* 1-8, 1-19e, R. *Knauft* 14-21, Roger Tory Peterson 2 Opener; **The Picture Cube:** *Jeff Albertson* 12-2, *Barkam* 20-20, L. *Lorisso* 1-17b; **Rainbow:** *Dan McCoy* Feature 2-1, 10-17, 15-19, *Pierce* 10-8; **J.H. Robinson:** 5-3, 5-4c, 5-5a; **Jeffrey Rotman:** 1 Opener, 1-19b, 1-19c, 1-19f, 1-19g, 1-20a, 1-20b, 1-20g; **Dave Schaefer:** Unit 2 Opener, 6-1, 6-7c, 6-16b, 6-17, 7-11a, 7-11b, 7-11c, 7-12a, 7-12b, 8-5, 8-12a, 8-12b, 9-9, 9-15a, 9-15b, 16-6, 16-8, 17-3, 17-10, 17-14, 18-1, 18-12, Feature 18, 19-15, 20-2, 20-3, 20-5, 20-12, Unit 4 Careers a; **Spectra-Action:** *Lewis Portnoy* 3 Opener, Feature 17-1, Feature 17-2; **Dr. Fred M. Spiess, Scripps Institute of Oceanography, University of California, San Diego:** 6-9; **Tom Stack and Associates:** *Don Benson* 15-14, *John Gerlack* Feature 19, *Steve Martin* 12-17, *Gary Milburn* 19-7, *Spencer Swanger* 15-2; **Stock Boston:** *Michael Collier* 10 Opener, *Ira Kirschenbaum* 12-11, *Peter Menzel* Feature 12, 15 Opener, 18 Opener, 18-19, *Stacy Pick* 17 Opener, *Cary Wolinsky* Unit 1 Opener, Feature 4; **Michael Sullivan and Associates:** *George* O. *Miller* 12-12, *Patricia Smith-Fox* 12-1; **Sygma:** *Chuck Baker* 18-14, *Tom Zimberhoff* 18-6b; **Taurus Photos Inc.:** *Charles Marden Fitch* Unit 3 Careers b, *Eric Kroll* 15-5, 15-6, Unit 3 Careers a, Feature 20-3, T.R. *Leibel* 13-12; **Alfred Owczarzak** 1-1c, 2-4, *Alfred Pasieka* Unit 4 Opener, *Martin* M. *Rotker* 1-16a, Feature 3-1, *James Somers* 1-16b, 1-22a, 1-22c; **Uniphoto:** *Leah Bellas* 18-6a, R.H. *Confer* 6-5, *Dallas Heaton* 18-13, *Jerry Irwin* 16-17, *Richard Mata* 12 Opener, *Voscar* 12-8; **United States Steel Corporation:** 10-9; **USDA Soil Conservation Service:** 12-3b; **U.S. Department of Energy:** 10-15, 18-9, 18-18; **Valan Photos:** *Albert Kunigk* 1-18c, Brian Milne 5-17; **Mark Warner:** 13-11b; **Water Thermal and Chemical Technology Center, University of California, Berkeley:** 6-11; **Woods Hole Oceanographic Institute:** 14-2, 14-12, 14-14, Feature 14-1